# Progress in Mathematics

**SADLIER-OXFORD**

Rose Anita McDonnell

Catherine D. LeTourneau

Anne Veronica Burrows

Judith Ann Geschke

Francis H. Murphy

M. Winifred Kelly

*with*
Dr. Elinor R. Ford

## Series Consultants

Tim Mason
Math Specialist
Palm Beach County School District
West Palm Beach, FL

Margaret Mary Bell, S.H.C.J., Ph.D.
Director, Teacher Certification
Rosemont College
Rosemont, PA

Dennis W. Nelson, Ed.D.
Director of Basic Skills
Mesa Public Schools
Mesa, AZ

**Sadlier-Oxford**
A Division of William H. Sadlier, Inc.

The publisher wishes to thank the following teachers and administrators, who read portions of the program prior to publication, for their comments and suggestions.

Mrs. Maria Bono
Whitestone, NY

Sr. Lynn Roebert
Covina, CA

Mrs. Ana M. Rodriguez
Miami, FL

Mrs. Madonna Atwood
Creve Coeur, MO

Mrs. Jennifer Fife
Yardley, PA

Ms. Anna Cano-Amato
Brooklyn, NY

Sr. Ruthanne Gypalo
East Rockaway, NY

Mrs. Marlene Kitrosser
Bronx, NY

Ms. Donna Violi
Melbourne, FL

Mr. Galen Chappelle
Los Angeles, CA

Sr. Anita O'Dwyer
North Arlington, NJ

## Acknowledgments

Every good faith effort has been made to locate the owners of copyrighted material to arrange permission to reprint selections. In several cases this has proved impossible. The publisher will be pleased to consider necessary adjustments in future printings.

Thanks to the following for permission to reprint the copyrighted materials listed below.

"Arithmetic" (text only), Anonymous.

"Exit x" (text only) by David McCord from ONE AT A TIME by David McCord. Copyright © 1986 by David McCord. By permission of Little, Brown and Company.

"Grandmother's Almond Cookies" (text only) by Janet S. Wong. Reprinted with the permission of Margaret K. McElderry Books, an imprint of Simon & Schuster Children's Publishing Division from A SUITCASE OF SEAWEED AND OTHER POEMS by Janet S. Wong. Copyright © 1996 Janet S. Wong.

From *The Greedy Triangle* (text only) by Marilyn Burns. Copyright © 1994 by Marilyn Burns Education Associates. Reprinted by permission of Scholastic Inc.

"Leaves" (text only) by Soseki from AN INTRODUCTION TO HAIKU by Harold G. Henderson. Copyright © 1958 by Harold G. Henderson. Used by permission of Doubleday, a division of Bantam Doubleday Dell Publishing Group, Inc.

"Lincoln Monument: Washington" (text only) by Langston Hughes from COLLECTED POEMS by Langston Hughes. Copyright © 1994 by the Estate of Langston Hughes. Reprinted by permission of Alfred A. Knopf Inc.

Excerpts from MATH CURSE (text only) by Jon Scieszka. Copyright © 1995 by Jon Scieszka. Used by permission of Viking Penguin, a division of Penguin Putnam Inc.

"A Microscopic Topic" (text only) by Jack Prelutsky from THE NEW KID ON THE BLOCK by Jack Prelutsky. Copyright © 1984 by Jack Prelutsky. By permission of Greenwillow Books, a division of William Morrow & Company, Inc.

From *Nine O'Clock Lullaby* (text only) by Marilyn Singer. Copyright © 1991 by Marilyn Singer. Used by permission of HarperCollins Publishers.

"The Runner" (text only) by Faustin Charles. © 1994 by Faustin Charles from the book A CARIBBEAN DOZEN edited by John Agard & Grace Nichols. Reproduced by permission of the publisher Walker Books Ltd., London. Published in the U.S. by Candlewick Press, Cambridge Press, Cambridge, MA.

"Sand Dollar" (text only) by Barbara Juster Esbensen. © 1992 by Barbara Juster Esbensen. Used by permission of HarperCollins Publishers.

"Smart" (text only) by Shel Silverstein. Copyright © 1974 by Evil Eye Music, Inc. Used by permission of HarperCollins Publishers.

"Speed" (text only) by Monica Kulling. Copyright © 1996 by Monica Kulling. Reprinted by permission of Marian Reiner for the author.

Anastasia Suen, Literature Consultant

All manipulative products generously provided by ETA, Vernon Hills, IL.

## Photo Credits

*Myrleen Cate:* 10 right, 10 top, 10 center, 46, 122, 230, 311, 373, 426.
*Neal Farris:* 89, 302, 430.
*Leo de Wys:* 143.
*FPG/*Ken Reid: 76; Michael Simpson: 77; Telegraph Colour Library: 118; Arthur Tilley: 156-157; Steve Hix: 332; Dennis Galante: 444-445.

*Richard Hutchings:* 30-31, 108-109.
*Image Bank/* Garry Gay: 124.
*Imageworks/* Bob Daemmrich: 336.
*Index Stock:* 286.
*International Stock/* Bill Tucker: 304.
*Greg Lord:* 188, 226.
*Clay Patrick McBride:* xii, 29, 65, 95, 133, 163, 197, 211, 225, 237,

243, 250, 267, 301, 325, 330, 351, 352, 362-363, 381, 396, 397, 404, 415, 439.
*Barbara Macmillian:* 307.
*Stock Market:* 184; Chris Hamilton: 39; Jose L. Pelaez: 59; Michal Heron: 100; Myron J. Dorf: 125; Lew Long: 339.

*Tony Stone Images:* 461; Philipp Engelhorn: 10 left; Mark Lewis: 37; Cameron Hervet: 231; Wayne Eastep: 82; Bob Krist: 103; Peter Dean: 104; Chris Baker: 112; G. Brad Lewis: 334; Steven Peters: 466 right
*Uniphoto:* 171.

## Illustrators

Diane Ali
Batelman Illustration
Bob Berry
Don Bishop
Robert Burger

Ken Coffelt
Adam Gordon
Dave Jonason
Robin Kachantones
Sommer Keller

Bea Leute
Blaine Martin
Kathy O'Connell
John Quinn
Fernando Rangel

Sintora Vanderhorst
Dirk Wunderlich

# Contents

# CHAPTER 1
# Place Value, Addition, and Subtraction

# CHAPTER 2
# Multiplication

＊Algebraic Reasoning

# CHAPTER 3
## Division

# CHAPTER 4
## Number Theory and Fractions

\* **Algebraic Reasoning**

# CHAPTER 5
## Fractions: Addition and Subtraction

∗ **Algebraic Reasoning**

# CHAPTER 6
## Fractions: Multiplication and Division

* Algebraic Reasoning

# CHAPTER 9
# Measurement Topics

# CHAPTER 10
# Decimals: Addition and Subtraction

＊ **Algebraic Reasoning**

# CHAPTER 11
# Decimals: Multiplication and Division

# CHAPTER 12
# Metric Measurement, Area, and Volume

＊Algebraic Reasoning

\* **Algebraic Reasoning**

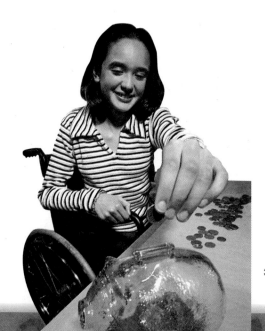

# Welcome to

# Progress in Mathematics

**W**hether you realize it or not, you *see* and use mathematics every day!

The lessons and activities in this textbook will help you enjoy mathematics *and* become a better mathematician.

This year you will build on the mathematical skills you already know, as you explore *new* ideas. Working in groups, you will solve problems using many different strategies. You will also have opportunities to make up your own problems and to keep a log of what you discover about math in your own personal Journal. You will learn more about fractions, decimals, geometry, measurement, probability, statistics, percents, and proportions—and even explore the world of algebra.

You will become a *Technowiz* by completing the computer and calculator lessons and activities. These not only will teach you valuable skills, but also expose you to how technology is used in different situations by many people.

And you can use the Skills Update section at the beginning of this book throughout the year to sharpen and review any skills you need to brush up on.

We hope that as you work through this program you will become aware of how mathematics really is a *big* part of your life.

# An Introduction to Skills Update

*Progress in Mathematics* includes a "handbook" of essential skills, Skills Update, at the beginning of the text. These one-page lessons review skills you learned in previous years. It is important for you to know this content so that you can succeed in math this year.

If you need to review a concept in Skills Update, your teacher can work with you, using ideas from the Teacher Edition. You can practice the skill using manipulatives, which will help you understand the concept better.

Your class may choose to do these one-page lessons at the beginning of the year so that you and your teacher can assess your understanding of these previously learned skills. Or you may choose to use Skills Update as a handbook throughout the year. Many lessons in your textbook refer to a particular page in the Skills Update. This means you can use that Skills Update lesson at the beginning of your math class as a warm-up activity. You may even want to practice those skills at home.

If you need more practice than what is provided on the Skills Update page, you can use exercises in the *Skills Update Practice Book*. It has an abundance of exercises for each lesson.

# Place Value to Thousands

▶ You can show 158,706 in a **place-value chart**. The value of each digit in a number depends on its **place** in the number. In 158,706 the value of:

Each period has three digits.

| Thousands Period | Ones Period |
|---|---|
| hundreds tens ones | hundreds tens ones |

**1 5 8, 7 0 6**

1 is 1 hundred thousand or 100,000.

5 is 5 ten thousands or 50,000.

8 is 8 thousands or 8000.

7 is 7 hundreds or 700.

0 is 0 tens or 0.

6 is 6 ones or 6.

▶ **Standard Form:** 158,706

Four-digit numbers may be written with or without a comma. In numbers *larger* than 9999, use a comma to separate the periods.

**Word Name:**

one hundred fifty-eight thousand,

seven hundred six

Write the place of the underlined digit. Then write its value.

**1.** 22<u>4</u>2
**2.** 6<u>3</u>,666
**3.** <u>1</u>99,999
**4.** 88<u>0</u>,888

Place a comma where needed in each. Then write the period name for the underlined digit.

**5.** 3 4 2 5 <u>9</u>
**6.** <u>1</u> 6 4 3 2
**7.** 2 0 <u>0</u> 0 6 0
**8.** <u>8</u> 0 5 0 2 7

Write the number in standard form.

**9.** forty-five thousand, seven hundred sixty-two
**10.** five thousand, six
**11.** nine hundred thousand, seven
**12.** ten thousand, nineteen

Write the word name for each number.

**13.** 7046
**14.** 37,008
**15.** 231,075
**16.** 923,780

# Numeration II

## Compare and Order Whole Numbers

Compare 363,420 and 381,787.

| | | Remember: |
|---|---|---|
| | | < means "is less than." |
| | | > means "is greater than." |

▶ **To compare whole numbers:**       363,420
  • Align the digits by place value.    381,787

  • Start at the left and find the first    363,420    3 = 3
    place where the digits are different.  381,787

  • Compare the value of these digits    363,420    8 > 6
    to find which number is greater.     381,787

So 381,787 > 363,420.          You could also say 363,420 < 381,787.

Order from greatest to least:  69,520; 19,478; 160,434; 63,215

▶ **To order whole numbers:**
  • Align the digits by place value.

  • Compare the digits in each place, starting with the greatest place.

| 69,520 | 69,520 | 69,520 |
|---|---|---|
| 19,478 | 19,478 | 19,478 |
| 160,434 | 160,434 | 160,434 |
| 63,215 | 63,215 | 63,215 |

There are no hundred thousands in the other numbers. 160,434 is the greatest.

6 = 6 and 1 < 6
19,478 is the least.

3 < 9
63,215 < 69,520

In order from greatest to least the numbers are:
160,434; 69,520; 63,215; 19,478

The order from least to greatest: 19,478; 63,215; 69,520; 160,434

Compare. Write < or >.

  **1.** 1563 __?__ 1519        **2.** 67,234 __?__ 67,243        **3.** 479,059 __?__ 479,056

Write in order from least to greatest.

  **4.** 9458; 9124; 948; 972              **5.** 3951; 3068; 369; 3547

  **6.** 99,407; 91,568; 90,999; 93,697        **7.** 216,418; 215,783; 213,614; 221,986

# Rounding Whole Numbers

**To round a number to a given place:**

- Find the place you are rounding to.

- Look at the digit to its right.
  If the digit is *less than 5*, round **down**.
  If the digit is *5 or more*, round **up**.

▶ Round 13,528 to the nearest *ten*.

13,528
↓
13,530

| 8 > 5 |
| Round **up** |
| to 13,530. |

▶ Round 13,528 to the nearest *hundred*.

13,528
↓
13,500

| 2 < 5 |
| Round **down** |
| to 13,500. |

▶ Round 13,528 to the nearest *thousand*.

13,528
↓
14,000

| 5 = 5 |
| Round **up** |
| to 14,000. |

Round to the nearest ten.

| **1.** 27 | **2.** 25 | **3.** 51 | **4.** 86 | **5.** 174 | **6.** 397 |
| **7.** 469 | **8.** 875 | **9.** 2587 | **10.** 4351 | **11.** 9289 | **12.** 3542 |

Round to the nearest hundred.

| **13.** 158 | **14.** 426 | **15.** 375 | **16.** 896 | **17.** 719 | **18.** 950 |
| **19.** 1047 | **20.** 3888 | **21.** 5942 | **22.** 6891 | **23.** 3098 | **24.** 8762 |
| **25.** 37,405 | **26.** 62,345 | **27.** 88,088 | **28.** 65,097 | **29.** 58,706 | **30.** 66,636 |

Round to the nearest thousand.

| **31.** 9155 | **32.** 7983 | **33.** 4550 | **34.** 6237 | **35.** 8396 |
| **36.** 33,888 | **37.** 15,942 | **38.** 93,192 | **39.** 87,983 | **40.** 46,237 |
| **41.** 326,150 | **42.** 145,706 | **43.** 357,029 | **44.** 563,498 | **45.** 807,476 |
| **46.** 821,593 | **47.** 450,513 | **48.** 435,127 | **49.** 205,120 | **50.** 761,604 |

# Whole Number Operations I

## Add and Subtract Whole Numbers

**To add or subtract whole numbers:**

• Estimate.

• Align the numbers. Add or subtract, starting with the ones. Regroup when necessary.

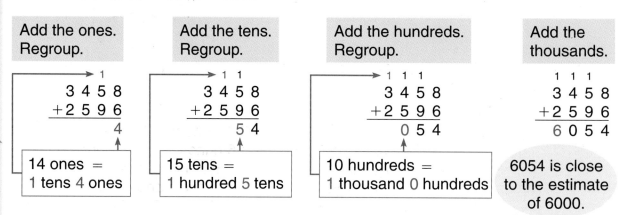

▶ Add: 3458 + 2596 = __?__

Estimate: 3000 + 3000 = 6000

| Add the ones. Regroup. | Add the tens. Regroup. | Add the hundreds. Regroup. | Add the thousands. |
|---|---|---|---|
| <br>      1<br>  3 4 5 8<br>+2 5 9 6<br>      4 | <br>    1 1<br>  3 4 5 8<br>+2 5 9 6<br>    5 4 | <br>  1 1 1<br>  3 4 5 8<br>+2 5 9 6<br>  0 5 4 | <br> 1 1 1<br>  3 4 5 8<br>+2 5 9 6<br> 6 0 5 4 |
| 14 ones = 1 tens 4 ones | 15 tens = 1 hundred 5 tens | 10 hundreds = 1 thousand 0 hundreds | 6054 is close to the estimate of 6000. |

▶ Subtract: 2842 − 1645 = __?__

Estimate: 3000 − 2000 = 1000

| More ones needed. Regroup. Subtract. | More tens needed. Regroup. Subtract. | Subtract. |
|---|---|---|
| <br>     3 12<br>  2 8 4̶ 2̶<br>−1 6 4 5<br>      7 | <br>     13<br>  7 3̶ 12<br>  2 8̶ 4̶ 2̶<br>−1 6 4 5<br>    9 7 | <br>     13<br>  7 3̶ 12<br>  2 8̶ 4̶ 2̶<br>−1 6 4 5<br>  1 1 9 7 |
| 4 tens 2 ones = 3 tens 12 ones | 8 hundreds 3 tens = 7 hundreds 13 tens | 1197 is close to the estimate of 1000. |

Estimate. Then add or subtract. (Watch for + or −.)

**1.** 215 + 687      **2.** 4306 + 3849      **3.** 6287 + 318

**4.** 659 − 286      **5.** 7583 − 2948      **6.** 3717 − 839

# Whole Number Operations II

## Multiplying One Digit

Multiply: $7 \times 816 = $ ?

First, estimate: $7 \times 816$
$$7 \times 800 = 5600$$

Then multiply.

| Multiply the ones. Regroup. | Multiply the tens. Add the regrouped tens. Regroup again. | Multiply the hundreds. Add the regrouped hundreds. |
|---|---|---|
| $\begin{array}{r} {}^{4}\phantom{0} \\ 8\,1\,6 \\ \times\ \ \ 7 \\ \hline 2 \end{array}$ | $\begin{array}{r} {}^{1\ 4} \\ 8\,1\,6 \\ \times\ \ \ 7 \\ \hline 1\,2 \end{array}$ | $\begin{array}{r} {}^{1\ 4} \\ 8\,1\,6 \\ \times\ \ \ 7 \\ \hline 5\,7\,1\,2 \end{array}$ |
| $7 \times 6$ ones $= 42$ ones<br>42 ones =<br>4 tens 2 ones | $7 \times 1$ ten $= 7$ tens<br>7 tens + 4 tens =<br>11 tens =<br>1 hundred 1 ten | $7 \times 8$ hundreds $= 56$ hundreds<br>56 hundreds + 1 hundred =<br>57 hundreds =<br>5 thousands 7 hundreds |

5712 is close to the estimate of 5600.

Estimate. Then multiply.

| | | | | |
|---|---|---|---|---|
| **1.** $\begin{array}{r} 25 \\ \times\ 3 \end{array}$ | **2.** $\begin{array}{r} 62 \\ \times\ 4 \end{array}$ | **3.** $\begin{array}{r} 58 \\ \times\ 5 \end{array}$ | **4.** $\begin{array}{r} 42 \\ \times\ 6 \end{array}$ | **5.** $\begin{array}{r} 19 \\ \times\ 7 \end{array}$ |
| **6.** $\begin{array}{r} 956 \\ \times\ \ \ 5 \end{array}$ | **7.** $\begin{array}{r} 619 \\ \times\ \ \ 8 \end{array}$ | **8.** $\begin{array}{r} 534 \\ \times\ \ \ 4 \end{array}$ | **9.** $\begin{array}{r} 519 \\ \times\ \ \ 5 \end{array}$ | **10.** $\begin{array}{r} 348 \\ \times\ \ \ 9 \end{array}$ |

Find the product.

| | | | | |
|---|---|---|---|---|
| **11.** $\begin{array}{r} 87 \\ \times\ 6 \end{array}$ | **12.** $\begin{array}{r} 93 \\ \times\ 7 \end{array}$ | **13.** $\begin{array}{r} 79 \\ \times\ 8 \end{array}$ | **14.** $\begin{array}{r} 41 \\ \times\ 5 \end{array}$ | **15.** $\begin{array}{r} 32 \\ \times\ 4 \end{array}$ |
| **16.** $\begin{array}{r} 759 \\ \times\ \ \ 3 \end{array}$ | **17.** $\begin{array}{r} 825 \\ \times\ \ \ 4 \end{array}$ | **18.** $\begin{array}{r} 329 \\ \times\ \ \ 6 \end{array}$ | **19.** $\begin{array}{r} 478 \\ \times\ \ \ 8 \end{array}$ | **20.** $\begin{array}{r} 976 \\ \times\ \ \ 9 \end{array}$ |

**21.** $9 \times 49$  **22.** $8 \times 93$  **23.** $7 \times 358$  **24.** $5 \times 953$

5

# Whole Number Operations III

## One-Digit Quotients

Divide: $73 \div 9 =$ ___?___

| Decide where to begin the quotient. | Divisor → $9\overline{)73}$ ← Dividend $9\overline{)73}$ | $9 > 7$  **Not enough tens** $9 < 73$  **Enough ones** |

The quotient begins in the ones place.

Estimate: About how many 9s are in 73?

$8 \times 9 = 72$
$9 \times 9 = 81$ ← 73 is between 72 and 81. Try 8.

| Divide the ones. | Multiply. | Subtract and compare. | Write the remainder. |
| --- | --- | --- | --- |
| $\dfrac{8}{9\overline{)7\ 3}}$ | $\begin{array}{r} \times 8 \\ 9\overline{)7\ 3} \\ \hookrightarrow 7\ 2 \end{array}$ | $\begin{array}{r} 8 \\ 9\overline{)7\ 3} \\ -7\ 2 \\ \hline 1 \end{array}$  $1 < 9$ | $\begin{array}{r} 8 \ \text{R1} \\ 9\overline{)7\ 3} \\ -7\ 2 \\ \hline 1 \end{array}$ ← Remainder |

Check by multiplying and adding.

$\begin{array}{r} 8 \ \text{← Quotient} \\ \times\ 9 \ \text{← Divisor} \\ \hline 7\ 2 \\ +\ 1 \ \text{← Remainder} \\ \hline 7\ 3 \ \text{← Dividend} \end{array}$

The remainder must be less than the divisor.

### Divide and check.

1. $5\overline{)47}$
2. $4\overline{)39}$
3. $3\overline{)25}$
4. $7\overline{)59}$
5. $8\overline{)76}$

6. $6\overline{)51}$
7. $9\overline{)87}$
8. $6\overline{)49}$
9. $7\overline{)60}$
10. $4\overline{)23}$

11. $4\overline{)31}$
12. $6\overline{)38}$
13. $5\overline{)33}$
14. $8\overline{)79}$
15. $7\overline{)68}$

### Find the quotient and the remainder.

16. $58 \div 6$
17. $65 \div 8$
18. $29 \div 4$
19. $62 \div 7$

20. $32 \div 7$
21. $49 \div 5$
22. $75 \div 8$
23. $89 \div 9$

24. $26 \div 3$
25. $51 \div 9$
26. $47 \div 6$
27. $53 \div 8$

# Whole Number Operations IV

## Two-Digit Quotients

Divide: $82 \div 3 = $ __?__

| Decide where to begin the quotient. | $3\overline{)82}$ | $3 < 8$ **Enough tens** |
|---|---|---|

The quotient begins in the tens place.

Estimate: About how many 3s are in 8?

$2 \times 3 = 6$ ← Try 2.
$3 \times 3 = 9$

| Divide the tens. | Multiply. | Subtract and compare. | Bring down the ones. |
|---|---|---|---|
| $\dfrac{2}{3\overline{)8\ 2}}$ | $\overset{\times 2}{3\overline{)8\ 2}}$ →6 | $\begin{array}{r} 2 \\ 3\overline{)8\ 2} \\ -6 \\ \hline 2 \end{array}$ ← $2 < 3$ | $\begin{array}{r} 2 \\ 3\overline{)8\ 2} \\ -6\downarrow \\ \hline 2\ 2 \end{array}$ |

**Repeat the steps to divide the ones.**

| Divide the ones. | Multiply. | Subtract and compare. | Check. |
|---|---|---|---|
| $\begin{array}{r} 2\ 7 \\ 3\overline{)8\ 2} \\ -6\downarrow \\ \hline 2\ 2 \end{array}$ | $\begin{array}{r} \times \\ 2\ 7 \\ 3\overline{)8\ 2} \\ -6\downarrow \\ \hline 2\ 2 \\ →2\ 1 \end{array}$ | $\begin{array}{r} 2\ 7\ R1 \\ 3\overline{)8\ 2} \\ -6\downarrow \\ \hline 2\ 2 \\ -2\ 1 \\ \hline 1 \end{array}$ ← $1 < 3$ | $\begin{array}{r} 27 \\ \times\ 3 \\ \hline 81 \\ +\ 1 \\ \hline 82 \end{array}$ |

Divide and check.

1. $2\overline{)58}$
2. $4\overline{)84}$
3. $6\overline{)96}$
4. $3\overline{)79}$
5. $7\overline{)89}$

6. $7\overline{)85}$
7. $5\overline{)73}$
8. $4\overline{)69}$
9. $6\overline{)93}$
10. $8\overline{)97}$

Find the quotient and the remainder.

11. $47 \div 3$
12. $85 \div 2$
13. $77 \div 5$
14. $59 \div 4$

15. $83 \div 6$
16. $91 \div 8$
17. $81 \div 7$
18. $74 \div 6$

# Fractions I

## Fractions

A **fraction** is a number that names one or more *equal parts* of a whole or region, or of a set.

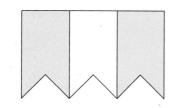

2 of the 3 equal parts of the banner are green.
$\frac{2}{3}$ of the banner is shaded.

2 of the 3 cars in this parking lot face right.
$\frac{2}{3}$ of the cars face right.

**numerator** ⟶ $\frac{2}{3}$ ⟵ number of equal parts being considered

**denominator** ⟶ ⟵ total number of equal parts in the whole or set

**Standard Form:**   $\frac{2}{3}$

**Word Name:**   two thirds

Write the fraction for the shaded part. Then write the fraction for the part that is *not* shaded.

**1.**

**2.**

**3.**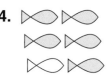

**4.**

Write the fraction in standard form.

**5.** three fifths

**6.** nine tenths

**7.** five twelfths

**8.** six elevenths

**9.** four twentieths

**10.** two twenty-fifths

**11.** The numerator is 6.
The denominator is 13.

**12.** The numerator is 8.
The denominator is 17.

Write the word name for each fraction.

**13.** $\frac{1}{2}$   **14.** $\frac{2}{7}$   **15.** $\frac{5}{9}$   **16.** $\frac{6}{11}$   **17.** $\frac{7}{8}$   **18.** $\frac{8}{13}$

# Equivalent Fractions

**Equivalent fractions** name the *same part* of a whole, a region, or a set.

One half ( $\frac{1}{2}$ ) of the whole is shaded blue.

Two fourths ( $\frac{2}{4}$ ) of the whole is shaded blue.

Four eighths ( $\frac{4}{8}$ ) of the whole is shaded blue.

$$\frac{1}{2} = \frac{2}{4} = \frac{4}{8}$$

$\frac{1}{2}$ , $\frac{2}{4}$ , and $\frac{4}{8}$ are equivalent fractions since they name the same part of the whole.

**Equivalent Fractions Chart**

| | | |
|---|---|---|
| 1 | | 1 whole |

| | | |
|---|---|---|
| $\frac{1}{2}$ | $\frac{1}{2}$ | 2 halves |
| $\frac{1}{3}$  $\frac{1}{3}$  $\frac{1}{3}$ | | 3 thirds |
| $\frac{1}{4}$  $\frac{1}{4}$  $\frac{1}{4}$  $\frac{1}{4}$ | | 4 fourths |
| $\frac{1}{5}$  $\frac{1}{5}$  $\frac{1}{5}$  $\frac{1}{5}$  $\frac{1}{5}$ | | 5 fifths |
| $\frac{1}{6}$  $\frac{1}{6}$  $\frac{1}{6}$  $\frac{1}{6}$  $\frac{1}{6}$  $\frac{1}{6}$ | | 6 sixths |
| $\frac{1}{8}$  $\frac{1}{8}$  $\frac{1}{8}$  $\frac{1}{8}$  $\frac{1}{8}$  $\frac{1}{8}$  $\frac{1}{8}$  $\frac{1}{8}$ | | 8 eighths |
| $\frac{1}{9}$  $\frac{1}{9}$  $\frac{1}{9}$  $\frac{1}{9}$  $\frac{1}{9}$  $\frac{1}{9}$  $\frac{1}{9}$  $\frac{1}{9}$  $\frac{1}{9}$ | | 9 ninths |
| $\frac{1}{10}$  $\frac{1}{10}$  $\frac{1}{10}$  $\frac{1}{10}$  $\frac{1}{10}$  $\frac{1}{10}$  $\frac{1}{10}$  $\frac{1}{10}$  $\frac{1}{10}$  $\frac{1}{10}$ | | 10 tenths |
| $\frac{1}{12}$  $\frac{1}{12}$  $\frac{1}{12}$  $\frac{1}{12}$  $\frac{1}{12}$  $\frac{1}{12}$  $\frac{1}{12}$  $\frac{1}{12}$  $\frac{1}{12}$  $\frac{1}{12}$  $\frac{1}{12}$  $\frac{1}{12}$ | | 12 twelfths |

$$1 = \frac{2}{2} = \frac{3}{3} = \frac{4}{4} = \frac{5}{5} = \frac{6}{6} = \frac{8}{8} = \frac{9}{9} = \frac{10}{10} = \frac{12}{12}$$

**Use the chart above to find equivalent fractions.**

**1.** $\frac{1}{2} = \frac{?}{6}$  **2.** $\frac{1}{3} = \frac{?}{6}$  **3.** $\frac{1}{4} = \frac{?}{8}$  **4.** $\frac{1}{5} = \frac{?}{10}$

**5.** $\frac{1}{3} = \frac{?}{9}$  **6.** $\frac{1}{4} = \frac{?}{12}$  **7.** $\frac{8}{10} = \frac{?}{5}$  **8.** $\frac{6}{9} = \frac{?}{12}$

**Use the chart above to compare. Write <, =, or >.**

**9.** $\frac{3}{4} \, ? \, \frac{6}{8}$  **10.** $\frac{1}{3} \, ? \, \frac{4}{9}$  **11.** $\frac{7}{10} \, ? \, \frac{4}{6}$  **12.** $\frac{6}{12} \, ? \, \frac{5}{10}$

**13.** $\frac{2}{8} \, ? \, \frac{1}{5}$  **14.** $\frac{3}{5} \, ? \, \frac{1}{2}$  **15.** $\frac{4}{6} \, ? \, \frac{8}{12}$  **16.** $\frac{3}{5} \, ? \, \frac{8}{10}$

**Write the missing number to complete the equivalent fraction.**

**17.** $\frac{1}{5} = \frac{?}{10}$  **18.** $\frac{3}{4} = \frac{6}{?}$  **19.** $\frac{2}{10} = \frac{?}{5}$  **20.** $\frac{3}{5} = \frac{?}{10}$  **21.** $\frac{4}{6} = \frac{?}{12}$

**22.** $\frac{3}{6} = \frac{6}{?}$  **23.** $\frac{3}{4} = \frac{?}{12}$  **24.** $\frac{4}{8} = \frac{?}{12}$  **25.** $\frac{2}{3} = \frac{6}{?}$  **26.** $\frac{6}{9} = \frac{8}{?}$

# Fractions III

## Adding and Subtracting Fractions: Like Denominators

Add: $\frac{2}{4} + \frac{1}{4} = $ ___?___

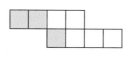

▶ To **add fractions** with *like* denominators:

- Add the numerators.

  $\frac{2}{4} + \frac{1}{4} = \quad^{3}$

- Write the sum over the common denominator.

  $\frac{2}{4} + \frac{1}{4} = \frac{3}{4}$

$2 + 1 = 3$

Subtract: $\frac{3}{5} - \frac{1}{5} = $ ___?___

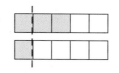

▶ To **subtract fractions** with *like* denominators:

- Subtract the numerators.

  $\frac{3}{5} - \frac{1}{5} = \quad^{2}$

- Write the difference over the common denominator.

  $\frac{3}{5} - \frac{1}{5} = \frac{2}{5}$

$3 - 1 = 2$

Study these examples.

$$\begin{array}{r} \frac{5}{9} \\ + \frac{2}{9} \\ \hline \frac{7}{9} \end{array}$$

$\dfrac{5+2}{9}$

$$\begin{array}{r} \frac{8}{9} \\ - \frac{2}{9} \\ \hline \frac{6}{9} \end{array}$$

$\dfrac{8-2}{9}$

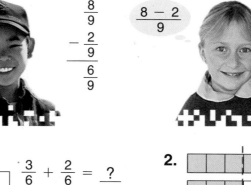

Add or subtract.

**1.** 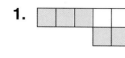 $\quad \frac{3}{6} + \frac{2}{6} = $ ___?___

**2.** 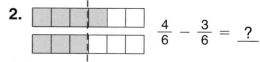 $\quad \frac{4}{6} - \frac{3}{6} = $ ___?___

**3.** $\frac{5}{9} + \frac{3}{9}$ 　　**4.** $\frac{5}{8} + \frac{2}{8}$ 　　**5.** $\frac{8}{10} - \frac{5}{10}$ 　　**6.** $\frac{4}{5} - \frac{2}{5}$

**7.** $\begin{array}{r} \frac{7}{10} \\ + \frac{2}{10} \\ \hline \end{array}$ 　　**8.** $\begin{array}{r} \frac{1}{5} \\ + \frac{3}{5} \\ \hline \end{array}$ 　　**9.** $\begin{array}{r} \frac{4}{9} \\ + \frac{4}{9} \\ \hline \end{array}$ 　　**10.** $\begin{array}{r} \frac{7}{8} \\ - \frac{3}{8} \\ \hline \end{array}$ 　　**11.** $\begin{array}{r} \frac{10}{12} \\ - \frac{8}{12} \\ \hline \end{array}$

10

# Tenths and Hundredths

A number less than one can be written either as a fraction or as a **decimal**.

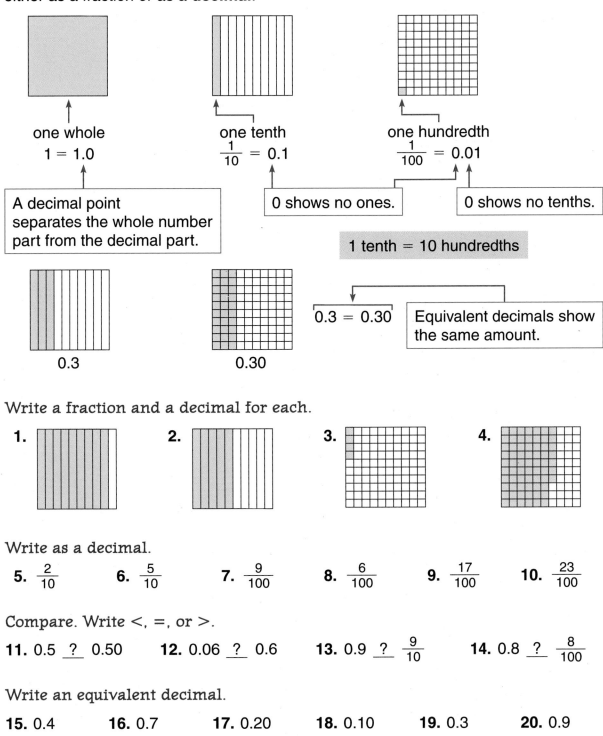

one whole
$1 = 1.0$

one tenth
$\frac{1}{10} = 0.1$

one hundredth
$\frac{1}{100} = 0.01$

A decimal point separates the whole number part from the decimal part.

0 shows no ones.

0 shows no tenths.

1 tenth = 10 hundredths

$0.3 = 0.30$   Equivalent decimals show the same amount.

0.3

0.30

Write a fraction and a decimal for each.

**1.**   **2.**   **3.**   **4.**

Write as a decimal.

**5.** $\frac{2}{10}$   **6.** $\frac{5}{10}$   **7.** $\frac{9}{100}$   **8.** $\frac{6}{100}$   **9.** $\frac{17}{100}$   **10.** $\frac{23}{100}$

Compare. Write $<$, $=$, or $>$.

**11.** 0.5 $\underline{\ ?\ }$ 0.50   **12.** 0.06 $\underline{\ ?\ }$ 0.6   **13.** 0.9 $\underline{\ ?\ }$ $\frac{9}{10}$   **14.** 0.8 $\underline{\ ?\ }$ $\frac{8}{100}$

Write an equivalent decimal.

**15.** 0.4   **16.** 0.7   **17.** 0.20   **18.** 0.10   **19.** 0.3   **20.** 0.9

# Geometry 1

## Lines and Angles

| Description | Figure | Symbol | Read As |
|---|---|---|---|
| A **line** is a set of points that extends infinitely in opposite directions. | A      B | $\overrightarrow{AB}$ or $\overleftrightarrow{BA}$ | line *AB* |
| A **line segment** is part of a line with two endpoints. | C      D | $\overline{CD}$ or $\overline{DC}$ | line segment *CD* or *DC* |
| A **ray** is part of a line with one endpoint. | E      F | $\overrightarrow{EF}$ | ray *EF* |
| An **angle** is formed by two rays with a common endpoint called the *vertex*. | vertex  G  H  I | $\angle GHI$ or $\angle H$ or $\angle IHG$ | angle *GHI* or angle *H* or angle *IHG* |
| **Intersecting lines** are lines that meet at a common point. | A  D  C  P  B | $\overleftrightarrow{AB}$ and $\overleftrightarrow{CD}$ intersect at *P*. | Line *AB* and line *CD* intersect at point *P*. |
| **Parallel lines** are lines in the same plane that never meet. | E  F  G  H | $\overleftrightarrow{EF} \parallel \overleftrightarrow{GH}$ | Line *EF* is parallel to line *GH*. |

Identify each figure. Then name it using symbols.

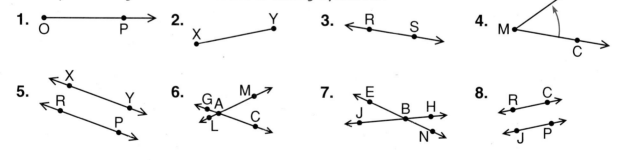

**1.** O ——● P

**2.** X ——— Y

**3.** R ——— S

**4.** M ——— A, C

**5.** X, R ——— Y, P

**6.** G, A, M, L, C

**7.** E, J, B, H, N

**8.** R, C, J, P

Draw and label each figure. You may use dot paper.

**9.** $\overline{DM}$     **10.** $\overleftrightarrow{XY}$     **11.** $\overrightarrow{FE}$     **12.** $\angle R$     **13.** $\angle SQR$

**14.** lines *EM* and *DR* intersecting at *X*          **15.** parallel lines *XR* and *YT*

# Identifying Polygons

▶ A **polygon** is a closed plane figure formed by line segments. The line segments are called *sides*. Pairs of sides meet at a point called a *vertex* (plural: vertices).

▶ Polygons are classified by the number of sides or vertices (or angles).

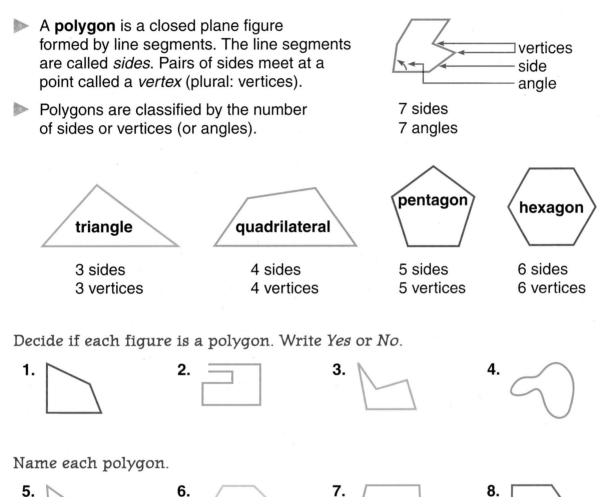

7 sides
7 angles

**triangle**
3 sides
3 vertices

**quadrilateral**
4 sides
4 vertices

**pentagon**
5 sides
5 vertices

**hexagon**
6 sides
6 vertices

Decide if each figure is a polygon. Write *Yes* or *No*.

**1.** **2.** **3.** **4.**

Name each polygon.

**5.** **6.** **7.** **8.**

Copy and complete the table.

| | Figure | Name | Number of sides | Number of vertices |
|---|---|---|---|---|
| **9.** | | ? | ? | ? |
| **10.** | ? | ? | ? | 5 |
| **11.** | ? | ? | 6 | ? |
| **12.** | | ? | ? | ? |

13

# Measurement I

## Customary Units of Length

The **inch (in.)**, **foot (ft)**, **yard (yd)**, and
**mile (mi)** are customary units of length.

| 12 inches (in.) = 1 foot (ft) |
| 36 inches = 1 yard (yd) |
| 3 feet = 1 yard |
| 5280 feet = 1 mile (mi) |
| 1760 yards = 1 mile |

about 1 in. long          about 1 ft long

about 1 yd wide          The distance a person can walk in 20 minutes is about 1 mile.

Before you can compare measurements in different units,
you need to **rename** units.

Compare: 4 ft  _?_  52 in.

You can make a table.

| ft | 1 | 2 | 3 | 4 | 5 |
|----|----|----|----|----|----|
| in. | 12 | 24 | 36 | 48 | 60 |

4 ft = 48 in.          48 < 52          So 4 ft < 52 in.

Which unit would you use to measure? Write *in.*, *ft*, *yd*, or *mi*.

1. length of an eraser
2. width of a board
3. distance between 2 cities
4. height of a desk
5. length of a soccer field
6. width of a quarter

Write the letter of the best estimate.

7. length of a pencil          **a.** 4 yd          **b.** 4 in.          **c.** 4 ft
8. height of a basketball player          **a.** 6 ft          **b.** 6 in.          **c.** 6 yd

Compare. Use <, =, or >.

9. 8 ft  _?_  96 in.          10. 6 yd  _?_  2 ft          11. 1 mi  _?_  3000 yd

# Customary Units of Capacity and Weight

The **cup (c)**, **pint (pt)**, **quart (qt)**, and **gallon (gal)** are customary units of liquid capacity.

2 cups = 1 pint (pt)
2 pints = 1 quart (qt)
2 quarts = 1 half gallon
4 quarts = 1 gallon (gal)

1 c

1 pt

1 qt

1 half gal

1 gal

The **ounce (oz)** and **pound (lb)** are customary units of weight.

16 ounces (oz) = 1 pound (lb)

about 1 oz

about 1 lb

Which unit would you use to measure? Write c, pt, qt, or gal.

1. juice in a pitcher
2. ice cream in a carton
3. paint in a can
4. water in a swimming pool
5. milk in a recipe
6. water in a bucket

Which unit would you use to measure the weight of each? Write oz or lb.

7. a toaster
8. a television
9. a dog
10. an envelope
11. a feather
12. a bag of oranges

Copy and complete each table.

13.

| pt | 1 | 2 | ? | 4 | 5 | 6 |
|---|---|---|---|---|---|---|
| c | 2 | ? | 6 | ? | ? | ? |

14.

| oz | 16 | 32 | ? | 64 | ? | 96 | ? | ? |
|---|---|---|---|---|---|---|---|---|
| lb | 1 | 2 | 3 | ? | 5 | ? | 7 | 8 |

# Measurement III

## Metric Units of Length

The **centimeter (cm)**, **decimeter (dm)**, **meter (m)**, and **kilometer (km)** are metric units of length.

| | | |
|---|---|---|
| 1 m | = | 100 cm |
| 1 m | = | 10 dm |
| 1 km | = | 1000 m |

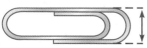

about 1 cm wide

about 1 dm long

about 1 m long

The Brooklyn Bridge in New York is about 1 km long.

Which metric unit of length is best to measure each?
Write *cm*, *m*, or *km*.

**1.** length of a car     **2.** depth of the ocean     **3.** height of a person

**4.** width of a tape     **5.** thickness of a sandwich

Write the letter of the best estimate.

**6.** length of an umbrella     **a.** 1 m     **b.** 1 dm     **c.** 1 km

**7.** width of a postage stamp     **a.** 0.22 cm     **b.** 2.2 cm     **c.** 22 cm

Copy and complete each table.

**8.**

| dm | 1 | 2 | 3 | ? | 5 | 6 |
|----|---|---|---|---|---|---|
| cm | 10 | ? | ? | 40 | ? | ? |

**9.**

| km | 1 | 2 | ? | 4 | 5 |
|----|---|---|---|---|---|
| m | 1000 | ? | 3000 | ? | ? |

Compare. Write <, =, or >.

**10.** 5 m _?_ 48 dm     **11.** 100 cm _?_ 2 m     **12.** 1000 m _?_ 1 km

# Metric Units of Capacity and Mass

▶ The **milliliter (mL)** and **liter (L)** are metric units of liquid capacity.

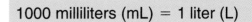

| 1000 milliliters (mL) = 1 liter (L) |

20 drops of water is about 1 mL.

about 1 L

▶ The **gram (g)**, and **kilogram (kg)** are metric units of mass.

| 1000 grams (g) = 1 kilogram (kg) |

A paper clip has a mass of about 1 g.

A hardcover dictionary has a mass of about 1 kg.

Which metric unit is best to measure the capacity of each? Write m*L* or *L*.

1. a bucket
2. a perfume bottle
3. a test tube
4. a bathtub
5. a can of juice
6. an eyedropper

Which metric unit is best to measure the mass of each? Write *g* or *kg*.

7. a computer
8. a peanut
9. an electric iron
10. a sugar cube
11. a comb
12. a bowling ball

Copy and complete each table.

13.

| L | 1 | 2 | ? | ? | ? | ? | ? | 8 |
|----|------|---|---|------|---|---|---|---|
| mL | 1000 | ? | ? | 4000 | ? | ? | ? | ? |

14.

| kg | 1 | ? | 3 | ? | ? | ? | ? | 8 |
|----|------|---|---|------|---|---|---|---|
| g | 1000 | ? | ? | 4000 | ? | ? | ? | ? |

# Statistics 1

## Making Pictographs

Make a **pictograph** to organize the data at the right.

▶ To make a pictograph:

- List each kind of book.

- If necessary, round the data to nearby numbers.

  298 ⟶ 300    54 ⟶ 50

- Choose a symbol or picture to represent the number of books for each kind to make the *key*.
  Let ⬜ = 100 books.

- Draw symbols to represent the number for each kind of book.

- Label the pictograph. Write the *title* and the *key*.

| Books in the Jackson Public Library | |
|---|---|
| **Kind** | **Number of Books** |
| Science | 298 |
| Medicine | 54 |
| Biography | 195 |
| Art | 147 |
| Fiction | 554 |
| History | 256 |

These are about 150 art books.

| Books in the Jackson Public Library | |
|---|---|
| Science | ⬜ ⬜ ⬜ |
| Medicine | ⬜ |
| Biography | ⬜ ⬜ |
| Art | ⬜ ⬜ |
| Fiction | ⬜ ⬜ ⬜ ⬜ ⬜ |
| History | ⬜ ⬜ ⬜ |
| Key: ⬜ = 100 books   ⬜ = 50 books | |

Make a pictograph for each set of data.

**1.**

| Students Taking Part in After-School Activities | |
|---|---|
| **Activities** | **Number of Students** |
| Clubs | 50 |
| Sports | 63 |
| Chorus | 38 |
| School Paper | 14 |
| Student Council | 7 |

**2.**

| Compact Disc Sales | |
|---|---|
| **Music** | **Compact Discs Sold** |
| Classical | 105 |
| Country | 886 |
| Jazz | 212 |
| Rap | 384 |
| Rock | 790 |
| R & B/Soul | 450 |

# Making Bar Graphs

Organize the data at the right in a **horizontal bar graph**.

▶ To make a horizontal bar graph:

- Use the data from the table to choose an appropriate scale.

- Draw and label the scale on the horizontal axis. Start at 0.

- Draw and label the vertical axis. List the name of each item.

- Draw horizontal bars to represent each number.

- Write the title of the bar graph.

▶ Make a **vertical bar graph** by placing the scale along the vertical axis and the items along the horizontal axis.

| Heights of Some U.S. Waterfalls | |
|---|---|
| **Name** | **Height in Feet** |
| Akaka | 442 |
| Bridalveil | 620 |
| Lower Yellowstone | 310 |
| Niagara | 182 |
| Silver Strand | 1170 |

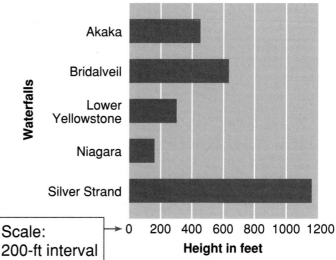

**Heights of Some U.S. Waterfalls**

Scale: 200-ft interval

Make a horizontal bar graph for the data listed below.

**1.**

| Results of Canned Food Drive | |
|---|---|
| Class | Number of Cans |
| 3A | 125 |
| 3B | 102 |
| 4A | 96 |
| 4B | 85 |
| 5A | 141 |
| 5B | 115 |

Make a vertical bar graph for the data listed below.

**2.**

| Favorite Sports Activity | |
|---|---|
| Sport | Number of Students |
| Baseball/Softball | 25 |
| Basketball | 18 |
| Gymnastics | 14 |
| Soccer | 28 |
| Tennis | 12 |

# Probability

## Equally/Not Equally Likely Outcomes

For each of the spinners *A* and *B* there are three different possible results or **outcomes**: red, blue, green.

► Spinner *A* is divided into 3 equal sections, and each section is a different color. Since there is 1 equal section of each color, each color has the same chance of occurring. The outcomes are **equally likely**.

Since there is 1 red section out of a total of 3 sections, the probability of landing on red is 1 out of 3.

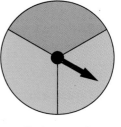

Spinner A

► Spinner *B* is divided into 6 equal sections. Since there is *not* an equal number of sections for each color, each color does not have the same chance of occurring. The outcomes are **not equally likely**.

Since there are 3 red sections, the spinner is more likely to land on red than on green or blue.

Since there are 3 red sections out of a total of 6 sections, the probability of landing on red is 3 out of 6.

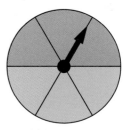

Spinner B

List the different outcomes. Then write whether the outcomes are *equally likely* or *not equally likely*.

1.    2.    3.    4.

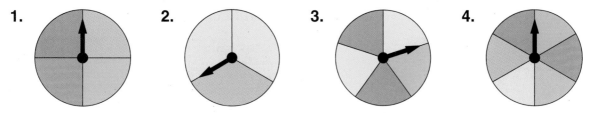

Use the spinner on the right to find the probability of landing on:

5. red      6. blue

7. green      8. yellow

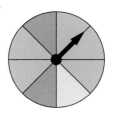

# Calculating Money

You can use a **calculator** to perform operations with money.

Use the decimal point key, ·, to separate dollars and cents.

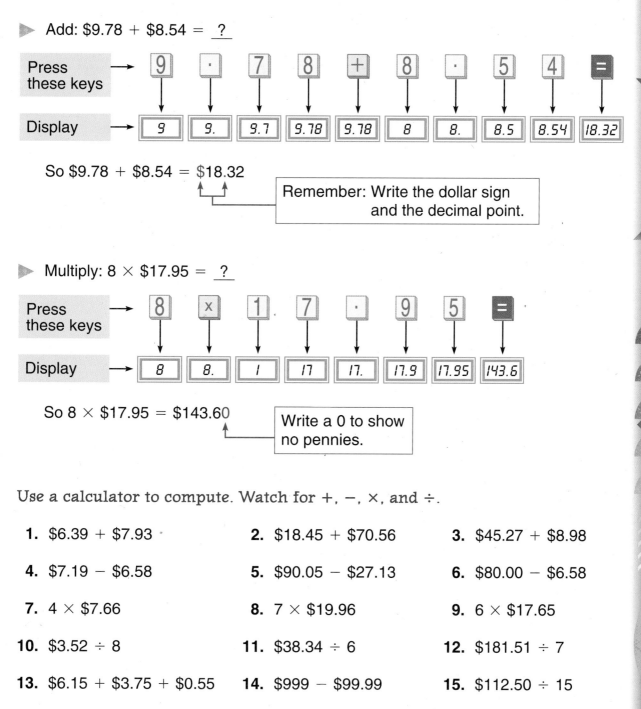

▶ Add: $9.78 + $8.54 = ?

So $9.78 + $8.54 = $18.32

Remember: Write the dollar sign and the decimal point.

▶ Multiply: 8 × $17.95 = ?

So 8 × $17.95 = $143.60

Write a 0 to show no pennies.

Use a calculator to compute. Watch for +, −, ×, and ÷.

1. $6.39 + $7.93
2. $18.45 + $70.56
3. $45.27 + $8.98
4. $7.19 − $6.58
5. $90.05 − $27.13
6. $80.00 − $6.58
7. 4 × $7.66
8. 7 × $19.96
9. 6 × $17.65
10. $3.52 ÷ 8
11. $38.34 ÷ 6
12. $181.51 ÷ 7
13. $6.15 + $3.75 + $0.55
14. $999 − $99.99
15. $112.50 ÷ 15

# Introduction to Problem Solving

## Dear Student,

Problem solvers are super sleuths. We invite you to become a super sleuth by using these *five* steps when solving problems.

**1 ▸ IMAGINE**
Create a mental picture.

**2 ▸ NAME**
List the facts and the questions.

**3 ▸ THINK**
Choose and outline a plan.

**4 ▸ COMPUTE**
Work the plan.

**5 ▸ CHECK**
Test that the solution is reasonable.

Sleuths use clues to find a solution to a problem. When working together to solve a problem, you may choose to use one or more of these *strategies* as clues:

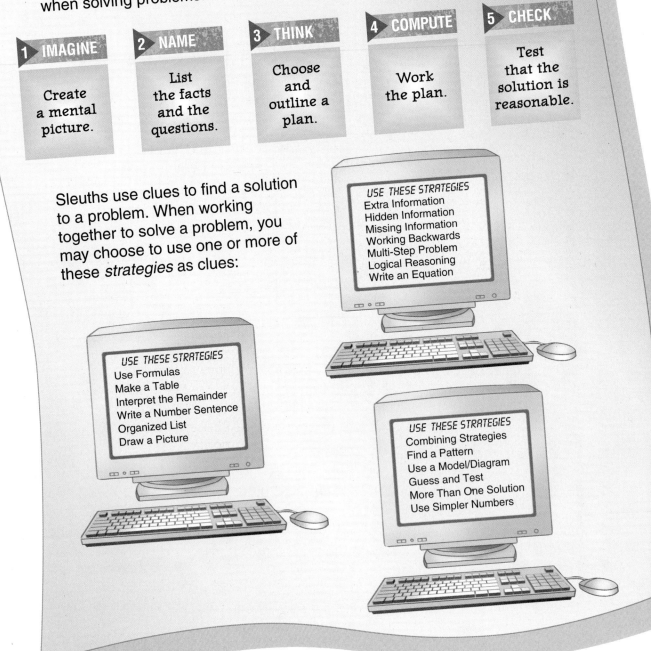

USE THESE STRATEGIES
Extra Information
Hidden Information
Missing Information
Working Backwards
Multi-Step Problem
Logical Reasoning
Write an Equation

USE THESE STRATEGIES
Use Formulas
Make a Table
Interpret the Remainder
Write a Number Sentence
Organized List
Draw a Picture

USE THESE STRATEGIES
Combining Strategies
Find a Pattern
Use a Model/Diagram
Guess and Test
More Than One Solution
Use Simpler Numbers

**1 ▶ IMAGINE**

Create
a mental
picture.

As you read a problem, create a picture in your mind. Make believe you are there in the problem. This will help you think about:

- what facts you will need;
- what the problem is asking;
- how you will solve the problem.

After reading the problem, it might be helpful to sketch the picture you imagined so that you can refer to it.

**2 ▶ NAME**

List
the facts
and the
questions.

Name or list all the facts given in the problem. Be aware of *extra* information not needed to solve the problem. Look for *hidden* information to help solve the problem. Name the question or questions the problem asks.

**3 ▶ THINK**

Choose
and
outline a
plan.

Think about how to solve the problem by:

- looking at the picture you drew;
- thinking about what you did when you solved similar problems;
- choosing a strategy or strategies for solving the problem.

**4 ▶ COMPUTE**

Work
the plan.

Work with the listed facts and the strategy to find the solution. Sometimes a problem will require you to add, subtract, multiply, or divide. Two-step problems require more than one choice of operation or strategy. It is good to *estimate* the answer before you compute.

**5 ▶ CHECK**

Test
that the
solution is
reasonable.

Ask yourself:

- "Have I answered the question?"
- "Is the answer reasonable?"

Check the answer by comparing it to the estimate. If the answer is not reasonable, check your computation. You may use a calculator.

# Problem Solving

## Strategy: Logical Reasoning

**Problem:** Tom, Roger, and Sue each had a different fruit for lunch today. One had a banana, one had an apple, and one had an orange. Tom and the boy who had a banana are cousins. Sue did *not* have an apple. What did each person have for lunch?

**1 IMAGINE** Create a mental picture.

**2 NAME**

*Facts:* Each had a different fruit.
Tom and the boy who had a banana are cousins. Sue did not have an apple.

*Question:* What did each person have?

**3 THINK** To solve the problem, make a table and use logical reasoning to eliminate the false conclusions.

|  | Banana | Apple | Orange |
|---|---|---|---|
| Tom |  |  |  |
| Roger |  |  |  |

When you write **yes** in a box, write **no** in the corresponding boxes in both that row and that column.

**4 COMPUTE** Since Tom and the boy who had a banana are cousins, write **yes** under "Banana" across from "Roger" and write the **no**s in the corresponding boxes. Since Sue did *not* have an apple, write **no** across from Sue under that column.

|  | Banana | Apple | Orange |
|---|---|---|---|
| Tom | no | yes | no |
| Roger | yes | no | no |
| Sue | no | no | yes |

So Sue did have an orange.
(Write the **yes** and **no** in the "Orange" column)

So Tom did have an apple.
(Write **yes** in the remaining box.)

**5 CHECK** Compare the completed chart to the facts given in the problem.

# Strategy: Interpret the Remainder

**Problem:** Ms. Cooper needs 115 decorations for cakes. Decorations come 9 to a box. How many boxes should she order? When the cakes are decorated, how many decorations from the last box will *not* be used?

**1 IMAGINE** Picture yourself in the problem.

**2 NAME**

*Facts:* 115 decorations needed
9 decorations per box

*Questions:* How many boxes should be ordered? How many decorations will *not* be used?

**3 THINK** To find how many boxes should be ordered, divide:

$$115 \div 9 = \underline{?}$$
decorations    per box    boxes

To find how many decorations will *not* be used, subtract the remainder from 9.

**4 COMPUTE**

```
      1 2  R 7
  9)1 1 5
   − 9 ↓
      2 5
    −1 8
       7
```

Twelve boxes will not be enough.
So she will need to order 13 boxes of decorations.

$$9 - 7 = 2$$

Two decorations will *not* be used from the last box.

**5 CHECK** Check division by using multiplication and addition.
$$12 \times 9 + 7 = 115$$

25

# Problem Solving

## Strategy: Missing Information

**Problem:**  Marvin read 128 pages of his book. He skipped the 19 pages of maps. How many more pages does he have left to read to finish the book?

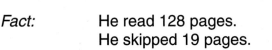

read
128 pages
skipped
19 pages

? pages left to read

**1 IMAGINE**  Picture yourself in the problem.

**2 NAME**

*Fact:*  He read 128 pages.
He skipped 19 pages.

*Question:*  How many more pages does he have left to read?

**3 THINK**  To find the number of pages left, subtract. Substitute the given information.

As stated, this problem can *not* be solved. There is missing information.

Do you have enough information to solve the problem?

No, the *number of total book pages* is not given.

To find a possible solution, choose a 3-digit number, like 341, for the total number of pages.

First add the pages Marvin read and skipped.

Then subtract that sum from the total number of pages.

**4 COMPUTE**

```
    1
  1 2 8  pages read
+   1 9  pages skipped
  -----
  1 4 7
```

```
  2 13 11
  3̸ 4̸ 1̸  total number of pages
- 1 4 7
  -----
  1 9 4
```

If there are 341 pages in the book, Marvin has 194 pages left to read.

**5 CHECK**  Use a calculator to check your computation.

26

## Strategy: More Than One Solution

**Problem:** Rory multiplied a 2-digit number by a 1-digit number greater than 1. The product was between 40 and 45. What were the numbers?

$\times$ ▢

product

**1 IMAGINE**

Picture yourself multiplying two factors.

**2 NAME**

*Facts:* The factors were a 2-digit number and a 1-digit number greater than 1. The product was between 40 and 45.

*Question:* What were the numbers?

**3 THINK**

To find the numbers (factors), list:
- 1-digit numbers greater than 1. (2, 3, 4, . . .)
- 2-digit numbers. (10, 11, 12, . . .)

Multiply the factors to find the products that equal 41, 42, 43, or 44.

> Remember: Not all problems have just one solution.

**4 COMPUTE**

It may help to record the factors and products in a table.

| Factors | 21 × 2 | 22 × 2 | 23 × 2 | 13 × 3 | 14 × 3 | 15 × 3 | 11 × 4 |
|---------|------|------|------|------|------|------|------|
| Product | 42 Yes | 44 Yes | 46 No | 39 No | 42 Yes | 45 No | 44 Yes |

Since $5 \times 8 = 40$ and $5 \times 9 = 45$, the 1-digit number is less than 5.

So there is *more than one solution*.
The factors are $2 \times 21$, $2 \times 22$, $3 \times 14$, and $4 \times 11$.

**5 CHECK**

Reread the problem. Are all the solutions reasonable? Yes. Check your computation with a calculator.

27

# Problem Solving

## Applications

Choose a strategy from the list or use another strategy you know to solve each problem.

1. The Stellar Circus built a new tent. It used 110 more yards of green canvas than yellow canvas. How many yards of green canvas were used?

2. Mr. Posio has 123 circus stickers to pass out to his class. He gives 4 stickers to each student. How many students are there in Mr. Posio's class? How many stickers are left over?

3. The circus attendance in April was less than the attendance in May but greater than the attendance in June. The circus attendance in July was between the attendances in April and in May. Write these months in increasing order of attendance.

4. The circus has tigers named Leo, Clem and Fred. Gary, Mary, and Barry are the trainers. Mary does *not* train Clem. She watches Gary train Leo before her act. Match the trainers with their tigers.

5. Patrick was paid $85 for each 5-day work week at the circus. How much was his total pay while working at the circus?

6. Ms. Gretchen needs 69 fruit bars for her students at the circus. There are 8 fruit bars in a box. How many boxes of fruit bars should Ms. Gretchen order?

7. There can be between 20 and 40 clowns in the act. The number of clowns in the act is divisible by 4, but *not* by 6. How many clowns can be in the act?

USE THESE STRATEGIES

Logical Reasoning
Interpret the Remainder
Missing Information
More Than One Solution

# Place Value, Addition, and Subtraction

**1**

## Lincoln Monument: Washington

Let's go see old Abe
Sitting in the marble and the moonlight,
Sitting lonely in the marble and the moonlight,
Quiet for ten thousand centuries, old Abe.
Quiet for a million, million years.

Quiet—

And yet a voice forever
Against the
Timeless walls
Of time—
Old Abe.

*Langston Hughes*

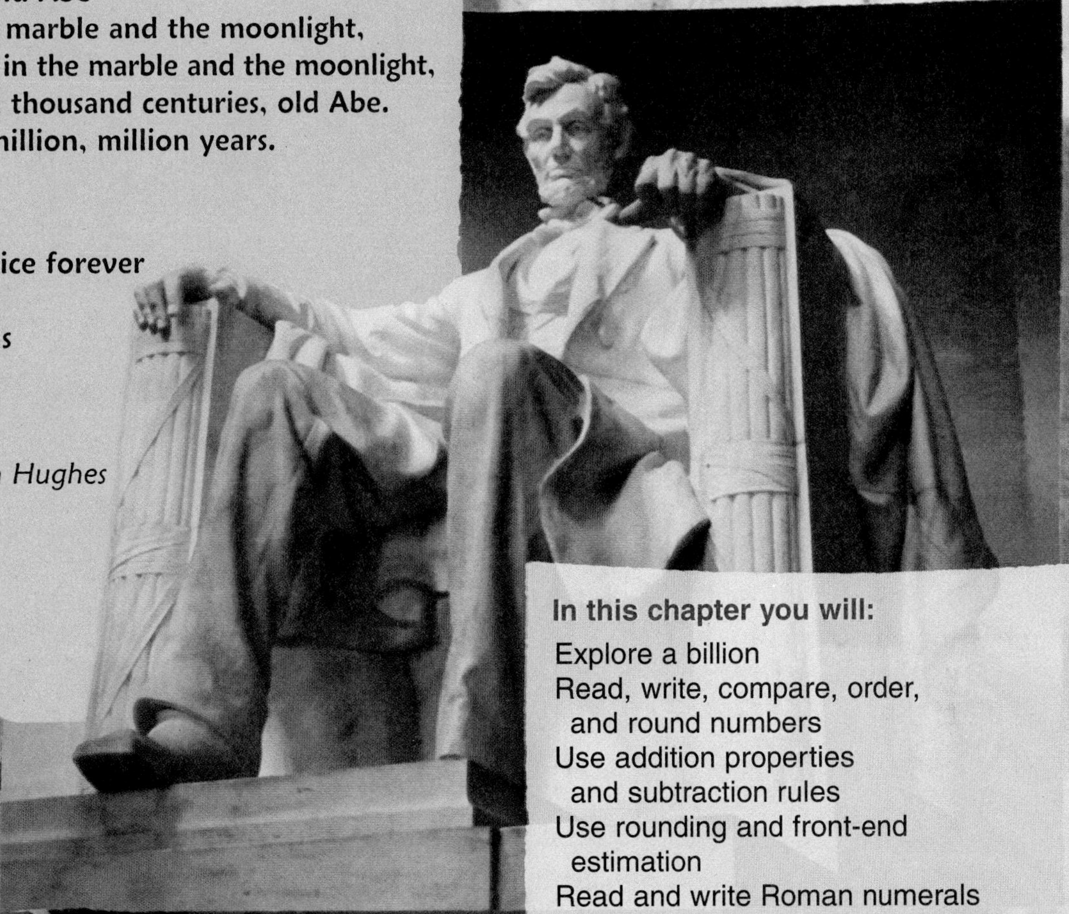

### In this chapter you will:

Explore a billion
Read, write, compare, order, and round numbers
Use addition properties and subtraction rules
Use rounding and front-end estimation
Read and write Roman numerals
Learn about flowcharts
Solve by the Guess-and-Test strategy

### Critical Thinking/Finding Together

In 1863 Abraham Lincoln began a speech, "Four score and seven years ago...." In 1922 the Lincoln Memorial in Washington, DC, was built. If *score* means 20, use *score* to describe the number of years between the year Lincoln was referring to when he gave the speech and 1922.

## 1-1  What Is a Billion?

### Discover Together

**Materials Needed:** calculator, paper, pencil, base ten cube stamp, construction paper, almanac, newspapers, magazines

Use the calculator to compute exercise 1.
Record each number sentence and the answer.
Look for a pattern.

1. $10 \times 1 = \underline{\ ?\ }$
   $10 \times 10 = \underline{\ ?\ }$
   $10 \times 100 = \underline{\ ?\ }$
   $10 \times 1000 = \underline{\ ?\ }$
   $10 \times 10{,}000 = \underline{\ ?\ }$
   $10 \times 100{,}000 = \underline{\ ?\ }$
   $10 \times 1{,}000{,}000 = \underline{\ ?\ }$
   Predict the product of $10 \times 10{,}000{,}000$;
   $10 \times 100{,}000{,}000$.

2. Describe the pattern in the products when
   10 is multiplied by a multiple of 10.

The number that is $10 \times 100{,}000{,}000$ is *one billion*, or
1,000,000,000. One billion is the next counting number
after 999,999,999.

3. How is 1,000,000,000 like 1,000,000; 10,000,000;
   and 100,000,000? How is it different?

4. If $1{,}000{,}000{,}000 = 10$ hundred millions, then
   $1{,}000{,}000{,}000 = 100$ ten millions.
   How many millions is one billion equal to?
   how many thousands?

Use the base ten cube as a thousand model. Stamp 10 base
ten cubes on a sheet of construction paper.

5. How many sheets of paper each with 10 base ten cubes
   pictured would be needed for 10 thousand? 100 thousand?
   1 million? 10 million? 100 million? 1 billion?

Use the calculator to find the answers to questions 6–8.

If you could travel 1 mile per second, you could get to places very quickly. At 1 mile per second:

6. About how many minutes would it take you to travel 1000 miles? 1,000,000 miles? 1,000,000,000 miles?

7. About how many hours would it take you to travel 1000 miles? 1,000,000 miles? 1,000,000,000 miles?

8. About how many days would it take you to travel 1000 miles? 1,000,000 miles? 1,000,000,000 miles?

## Communicate

Discuss ✓

9. How did you discover how many minutes it would take you to travel 1000 miles; 1,000,000 miles; and 1,000,000,000 miles at 1 mile per second?

10. How did you discover how many hours it would take you to travel 1000 miles; 1,000,000 miles; and 1,000,000,000 miles at 1 mile per second?

11. How did you discover how many days it would take you to travel 1000 miles; 1,000,000 miles; and 1,000,000,000 miles at 1 mile per second?

## Project

Use the almanac, newspapers, and magazines to find numbers in the billions.

12. Write a short description of the kinds of activities that involve references to billions.

# 1-2 Place Value to Billions

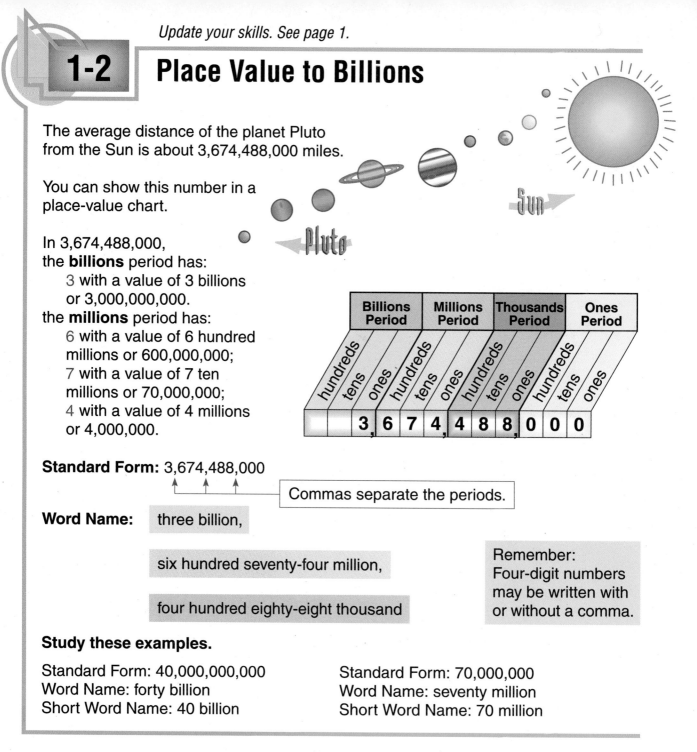

The average distance of the planet Pluto from the Sun is about 3,674,488,000 miles.

You can show this number in a place-value chart.

In 3,674,488,000,
the **billions** period has:
  3 with a value of 3 billions or 3,000,000,000.
the **millions** period has:
  6 with a value of 6 hundred millions or 600,000,000;
  7 with a value of 7 ten millions or 70,000,000;
  4 with a value of 4 millions or 4,000,000.

| Billions Period | | | Millions Period | | | Thousands Period | | | Ones Period | | |
|---|---|---|---|---|---|---|---|---|---|---|---|
| hundreds | tens | ones | hundreds | tens | ones | hundreds | tens | ones | hundreds | tens | ones |
| | | 3 | 6 | 7 | 4 | 4 | 8 | 8 | 0 | 0 | 0 |

**Standard Form:** 3,674,488,000

Commas separate the periods.

**Word Name:** three billion,

six hundred seventy-four million,

four hundred eighty-eight thousand

Remember:
Four-digit numbers may be written with or without a comma.

**Study these examples.**

Standard Form: 40,000,000,000
Word Name: forty billion
Short Word Name: 40 billion

Standard Form: 70,000,000
Word Name: seventy million
Short Word Name: 70 million

## Write the place of the underlined digit. Then write its value.

1. <u>5</u>,476,807,139
2. 3,9<u>6</u>0,135,741
3. 7,7<u>0</u>8,304,016
4. 9,4<u>2</u>8,001,230
5. 16,350<u>,</u>846,760
6. 39,714,062,0<u>3</u>0
7. 2<u>4</u>,398,407,268
8. 9<u>0</u>,165,270,000
9. 365,123,145,00<u>0</u>
10. 190,477,<u>6</u>53,002
11. <u>4</u>01,743,000,295
12. 839,200,43<u>0</u>,000

**Write the number in standard form.**

**13.** three million, five hundred forty thousand, thirty-seven

**14.** forty million, one hundred thousand, two hundred five

**15.** two hundred twenty million, five thousand, eight

**16.** three billion, six hundred six million, seventy-seven thousand, four hundred three

**17.** seventy-nine billion, one

**18.** eighty-one million

**19.** nine hundred forty billion

**20.** thirteen million, two

**21.** 800 million

**22.** 40 billion

**23.** 500 billion

**Write the word name for each number.**

**24.** 1,042,003,051

**25.** 4,725,000,000

**26.** 72,200,000,020

**27.** 12,025,617,809

**28.** 500,476,807,139

**29.** 23,539,417,148

**Write the short word name for each number.**

**30.** 6,000,000

**31.** 100,000,000

**32.** 20,000,000

**33.** 30,000,000,000

**34.** 6,000,000,000

**35.** 500,000,000,000

**PROBLEM SOLVING**

**36.** The average distance from Earth to the planet Saturn is about 773,119,750 miles. Write the word name of this number.

**37.** At times, the planet Pluto is about five billion miles from Earth. Write this number in standard form.

**Critical Thinking**

**Rearrange the digits in the given statement to make new true statements.**

**38.** 7234 < 7243

42 _?_ _?_ < _?_ _?_ 24

_?_ 7 _?_ _?_ > _?_ _?_ 4 _?_

**39.** 62,249 < 63,975

69, _?_ _?_ 2 > 69, _?_ 7 _?_

_?_ _?_, _?_ 42 < 9 _?_, _?_ _?_ 3

# Expanded Form

The value of each digit of a number can be shown by writing the number in **expanded form**.

| Billions Period | | | Millions Period | | | Thousands Period | | | Ones Period | | |
|---|---|---|---|---|---|---|---|---|---|---|---|
| hundreds | tens | ones | hundreds | tens | ones | hundreds | tens | ones | hundreds | tens | ones |
| | | | | | | | 8, | 6 | 3 | 0 | | |
| | | | | | | 8 | 6, | 3 | 0 | 2 | |
| | | | | | 8 | 6 | 3, | 0 | 2 | 0 | |
| | | | | 8, | 6 | 3 | 0, | 2 | 0 | 1 | |
| | | 8, | 6 | 3 | 0, | 2 | 0 | 1, | 0 | 0 | 0 |

Place that holds a zero may be omitted in expanded form.

$(8 \times 1000) + (6 \times 100) + (3 \times 10) + (0 \times 1)$

$(8 \times 10{,}000) + (6 \times 1000) + (3 \times 100) + (2 \times 1)$

$(8 \times 100{,}000) + (6 \times 10{,}000) + (3 \times 1000) + (2 \times 10)$

$(8 \times 1{,}000{,}000) + (6 \times 100{,}000) + (3 \times 10{,}000) + (2 \times 100) + (1 \times 1)$

$(8 \times 1{,}000{,}000{,}000) + (6 \times 100{,}000{,}000) + (3 \times 10{,}000{,}000) + (2 \times 100{,}000) + (1 \times 1000)$

| Standard Form | Expanded Form |
|---|---|
| 8630 | 8000 + 600 + 30 |
| 86,302 | 80,000 + 6000 + 300 + 2 |
| 863,020 | 800,000 + 60,000 + 3000 + 20 |
| 8,630,201 | 8,000,000 + 600,000 + 30,000 + 200 + 1 |
| 8,630,201,000 | 8,000,000,000 + 600,000,000 + 30,000,000 + 200,000 + 1000 |

**Copy and complete.**

1. $1487 = (\underline{\ ?\ } \times 1000) + (\underline{\ ?\ } \times 100) + (\underline{\ ?\ } \times 10) + (\underline{\ ?\ } \times 1)$

2. $87{,}020 = (\underline{\ ?\ } \times 10{,}000) + (\underline{\ ?\ } \times 1000) + (\underline{\ ?\ } \times 10)$

3. $180{,}764 = (1 \times \underline{\ ?\ }) + (8 \times \underline{\ ?\ }) + (7 \times \underline{\ ?\ }) + (6 \times \underline{\ ?\ }) + (4 \times \underline{\ ?\ })$

4. $32{,}530{,}008 = (3 \times \underline{\ ?\ }) + (2 \times \underline{\ ?\ }) + (5 \times \underline{\ ?\ }) + (3 \times \underline{\ ?\ }) + (8 \times \underline{\ ?\ })$

5. $4{,}700{,}930{,}002 = (4 \times \underline{\ ?\ }) + (7 \times \underline{\ ?\ }) + (9 \times \underline{\ ?\ })$
   $+ (\underline{\ ?\ } \times 10{,}000) + (\underline{\ ?\ } \times 1)$

**Write each in standard form.**

**6.** 4000 + 500 + 60 + 9

**7.** 20,000 + 2000 + 900 + 80 + 7

**8.** 400,000+ 300 + 50

**9.** 3,000,000 + 9000 + 40 + 8

**10.** 60,000,000 + 3,000,000 + 400,000 + 5000 + 7

**11.** 1,000,000,000 + 200,000,000 + 50,000,000 + 300 + 9

**Write in expanded form.**

**12.** 8998

**13.** 6745

**14.** 15,243

**15.** 37,418

**16.** 672,115

**17.** 350,001

**18.** 700,946

**19.** 2,200,002

**20.** 13,004,205

**21.** 604,003,020

**22.** 2,005,940,000

**Choose the correct answer.**

**23.** In the number 62,725, the 6 means:
**a.** $6 \times 1000$    **b.** $6 \times 100$    **c.** $6 \times 100,000$    **d.** $6 \times 10,000$

**24.** In the number 2,784,349, the 2 means:
**a.** $2 \times 1000$    **b.** $2 \times 10,000$    **c.** $2 \times 1,000,000$    **d.** $2 \times 100,000,000$

**25.** In the number 34,056,971,000, the 3 means:
**a.** $3 \times 1000$    **b.** $3 \times 10,000$    **c.** $3 \times 10,000,000$    **d.** $3 \times 10,000,000,000$

**PROBLEM SOLVING**

**26.** The distance from the center of Earth to the center of the Sun is 92,955,807 miles. Write this number in expanded form.

**Mental Math**

**Use the number 14,567,903,104. What number is:**

**27.** 10,000 greater?

**28.** 1,000,000 less?

**29.** 10,000,000,000 greater?

**30.** 100,000,000 less?

## 1-4 Thousandths

Each one of the ten parts of 0.01 is 0.001.

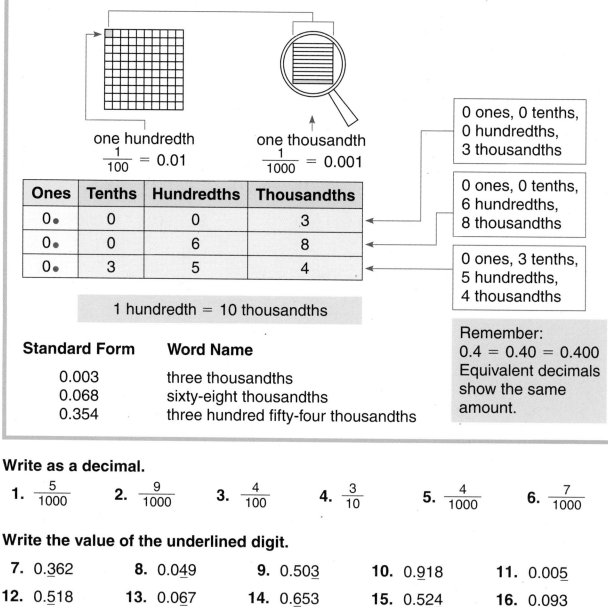

one hundredth
$\frac{1}{100}$ = 0.01

one thousandth
$\frac{1}{1000}$ = 0.001

| Ones | Tenths | Hundredths | Thousandths |
|------|--------|------------|-------------|
| 0● | 0 | 0 | 3 |
| 0● | 0 | 6 | 8 |
| 0● | 3 | 5 | 4 |

0 ones, 0 tenths,
0 hundredths,
3 thousandths

0 ones, 0 tenths,
6 hundredths,
8 thousandths

0 ones, 3 tenths,
5 hundredths,
4 thousandths

1 hundredth = 10 thousandths

**Remember:**
0.4 = 0.40 = 0.400
Equivalent decimals
show the same
amount.

| Standard Form | Word Name |
|---------------|-----------|
| 0.003 | three thousandths |
| 0.068 | sixty-eight thousandths |
| 0.354 | three hundred fifty-four thousandths |

**Write as a decimal.**

1. $\frac{5}{1000}$
2. $\frac{9}{1000}$
3. $\frac{4}{100}$
4. $\frac{3}{10}$
5. $\frac{4}{1000}$
6. $\frac{7}{1000}$

**Write the value of the underlined digit.**

7. 0.3<u>6</u>2
8. 0.04<u>9</u>
9. 0.50<u>3</u>
10. 0.9<u>1</u>8
11. 0.00<u>5</u>
12. 0.<u>5</u>18
13. 0.06<u>7</u>
14. 0.6<u>5</u>3
15. 0.52<u>4</u>
16. 0.09<u>3</u>

**Write the decimal in standard form.**

17. seven thousandths
18. nine hundred four thousandths
19. fifty-six thousandths
20. sixty-three thousandths
21. one hundred three thousandths
22. three hundred two thousandths

36

**Write the word name for each decimal.**

**23.** 0.461     **24.** 0.159     **25.** 0.009     **26.** 0.112     **27.** 0.258

**28.** 0.053     **29.** 0.158     **30.** 0.002     **31.** 0.273     **32.** 0.419

**Write an equivalent decimal.**

**33.** 0.9     **34.** 0.09     **35.** 0.23     **36.** 0.25     **37.** 0.72

**38.** 0.80     **39.** 0.50     **40.** 0.650     **41.** 0.300     **42.** 0.010

**Write the letter of the correct answer.**

**43.** Three hundred three thousandths is  _?_
    **a.** 303,000     **b.** 0.303     **c.** 303     **d.** 0.33

**44.** One hundred thirteen thousandths is  _?_
    **a.** 0.113     **b.** 0.013     **c.** 113,000     **d.** 113

**45.** Four hundred fifty-seven thousandths is  _?_
    **a.** 0.407     **b.** 457     **c.** 0.457     **d.** 457,000

## PROBLEM SOLVING

**46.** Minerva walked a distance of forty-five thousandths of a kilometer to the museum. Write this distance in standard form.

**47.** Mike rides 0.8 km on his bicycle. Write this distance as thousandths of a kilometer.

**48.** A car travels at a speed of 0.917 mile per minute. Write the word name of this speed.

### Calculator Activity

**Try this on your calculator.**

ON/AC   +   ·   0   0   1   =   =   =

**49.** What number did you get?

**50.** How many times would you have to press =
    **a.** 0.008     **b.** 0.01     **c.** 0.015     **d.** 0.02     **e.** 0.029

# Decimals Greater Than One

You can write a number greater than one as a decimal.

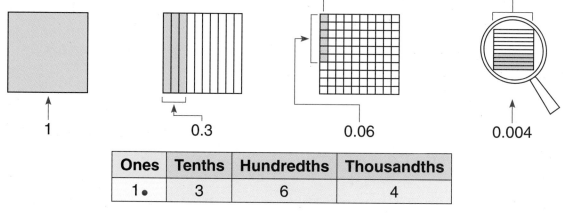

1        0.3                0.06              0.004

| Ones | Tenths | Hundredths | Thousandths |
|------|--------|------------|-------------|
| 1●   | 3      | 6          | 4           |

▶ A place-value chart can help you read decimals.
  • If there is a whole number, read the whole number first. Then read the decimal point as *and*.
  • Read the decimal as a whole number before reading the place value of the last digit.

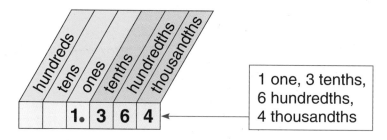

1 one, 3 tenths, 6 hundredths, 4 thousandths

**Standard Form**          **Word Name**
    1.364                  one and three hundred sixty-four thousandths

**Read the number.**

1. 0.392      2. 2.307      3. 19.3      4. 1.002      5. 17.017

6. 53.147     7. 103.551    8. 317.03    9. 37.730     10. 932.73

**Write the place of the underlined digit. Then write its value.**

11. 7.6̲78     12. 75.196̲    13. 8̲0.103    14. 35.64̲3    15. 13̲8.2

16. 4̲25.13    17. 90̲.121    18. 6.2̲31     19. 9.47̲8     20. 1.411̲

38

**Write the number in standard form.**

**21.** seven and fourteen hundredths

**22.** one and two thousandths

**23.** sixty-three and two tenths

**24.** three and five hundredths

**25.** three and four thousandths

**26.** forty-five and six tenths

**27.** one hundred forty-five and two thousandths

**28.** sixty-one and three hundred eighteen thousandths

**29.** one hundred thirty-eight and five hundred forty-one thousandths

**Write the word name for each number.**

**30.** 10.392    **31.** 2.307    **32.** 19.3    **33.** 1.002    **34.** 8.017

**35.** 3.147    **36.** 12.551    **37.** 37.03    **38.** 5.730    **39.** 319.723

**Use the number 958.826. What number is:**

**40.** one tenth greater?    **41.** one hundredth less?    **42.** one thousandth greater?

**43.** three and one tenth less?    **44.** twenty and two thousandths greater?

## PROBLEM SOLVING

**45.** Marla's time for the bicycle race was fifty-nine and one hundred twenty-two thousandths seconds. Write this time in standard form.

**46.** Steve's time for the bicycle race was 48.235 seconds. Write the word name for his time.

**Challenge**

**Complete the patterns.**

Algebra ✓

**47.** 0.3, 0.4, 0.5, _?_ , _?_

**48.** 0.6, 0.5, 0.4, _?_ , _?_

**49.** 1.9, 2, 2.1, _?_ , _?_

**50.** 0.09, 0.08, 0.07, _?_ , _?_

**51.** 0.005, 0.006, 0.007, _?_ , _?_

**52.** 3.26, 3.25, 3.24, _?_ , _?_

## 1-6 Compare and Order Numbers

Compare 8,532,314,516 and 8,539,417,148.
Which is greater?

> **Remember:**
> < means "is less than."
> > means "is greater than."

▶ You can compare whole numbers by comparing the digits in each place-value position. Start at the left and check each place until the digits are different.

8,532,314,516
8,539,417,148

| 8 = 8 |
| 5 = 5 |
| 3 = 3 |

9 > 2

8,539,417,148 > 8,532,314,516   or   8,532,314,516 < 8,539,417,148

Order from least to greatest:
1,353,678,945; 1,359,712,148; 358,643,208; 1,353,432,816

▶ You can order whole numbers by comparing them in the same way.

| 1,353,678,945 | 1,353,678,945 | 1,353,678,945 |
| 1,359,712,148 | 1,359,712,148 | 1,359,712,148 |
| → 358,643,208 | 358,643,208 | 358,643,208 |
| 1,353,432,816 | 1,353,432,816 | 1,353,432,816 |

No billions.
358,643,208
is least.

3 = 3 and 9 > 3
1,359,712,148 is
greatest.

6 > 4
1,353,678,945 > 1,353,432,816

The order from least to greatest:
358,643,208; 1,353,432,816; 1,353,678,945; 1,359,712,148

The order from greatest to least:
1,359,712,148; 1,353,678,945; 1,353,432,816; 358,643,208

**Compare. Write < or >.**

1. 479,059 ? 479,056     2. 2,873,303 ? 2,808,323     3. 2,124,371 ? 256,721

4. 2,356,100,910 ? 2,561,009,102     5. 7,495,851,787 ? 7,489,987,565

6. 3,410,999,246 ? 3,410,989,243     7. 6,355,601,501 ? 999,031,276

**Write in order from least to greatest.**

8. 4,767,831; 4,984,321; 4,113,121; 4,801,125

9. 9,238,456,348; 9,760,816; 989,507,555; 9,238,940,067

**Write in order from greatest to least.**

10. 162,550,743; 99,927,483; 159,294,604; 162,475,988

11. 2,458,599,763; 2,196,536,401; 2,423,038,972; 2,314,043,179

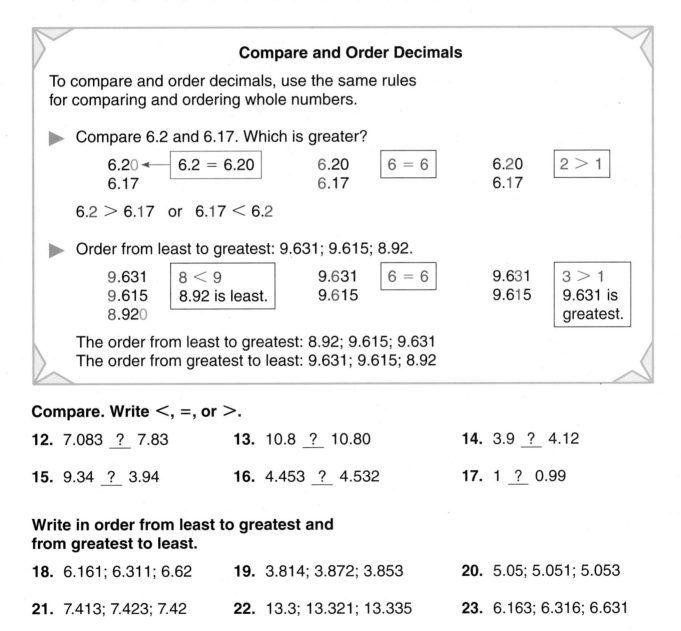

## Compare and Order Decimals

To compare and order decimals, use the same rules for comparing and ordering whole numbers.

▶ Compare 6.2 and 6.17. Which is greater?

$$6.20 \leftarrow \boxed{6.2 = 6.20}$$
6.17

$$\begin{matrix} 6.20 \\ 6.17 \end{matrix} \quad \boxed{6 = 6}$$

$$\begin{matrix} 6.20 \\ 6.17 \end{matrix} \quad \boxed{2 > 1}$$

6.2 > 6.17   or   6.17 < 6.2

▶ Order from least to greatest: 9.631; 9.615; 8.92.

$$\begin{matrix} 9.631 \\ 9.615 \\ 8.920 \end{matrix} \quad \boxed{\begin{matrix} 8 < 9 \\ \text{8.92 is least.} \end{matrix}}$$

$$\begin{matrix} 9.631 \\ 9.615 \end{matrix} \quad \boxed{6 = 6}$$

$$\begin{matrix} 9.631 \\ 9.615 \end{matrix} \quad \boxed{\begin{matrix} 3 > 1 \\ \text{9.631 is} \\ \text{greatest.} \end{matrix}}$$

The order from least to greatest: 8.92; 9.615; 9.631
The order from greatest to least: 9.631; 9.615; 8.92

**Compare. Write <, =, or >.**

12. 7.083 ? 7.83

13. 10.8 ? 10.80

14. 3.9 ? 4.12

15. 9.34 ? 3.94

16. 4.453 ? 4.532

17. 1 ? 0.99

**Write in order from least to greatest and from greatest to least.**

18. 6.161; 6.311; 6.62

19. 3.814; 3.872; 3.853

20. 5.05; 5.051; 5.053

21. 7.413; 7.423; 7.42

22. 13.3; 13.321; 13.335

23. 6.163; 6.316; 6.631

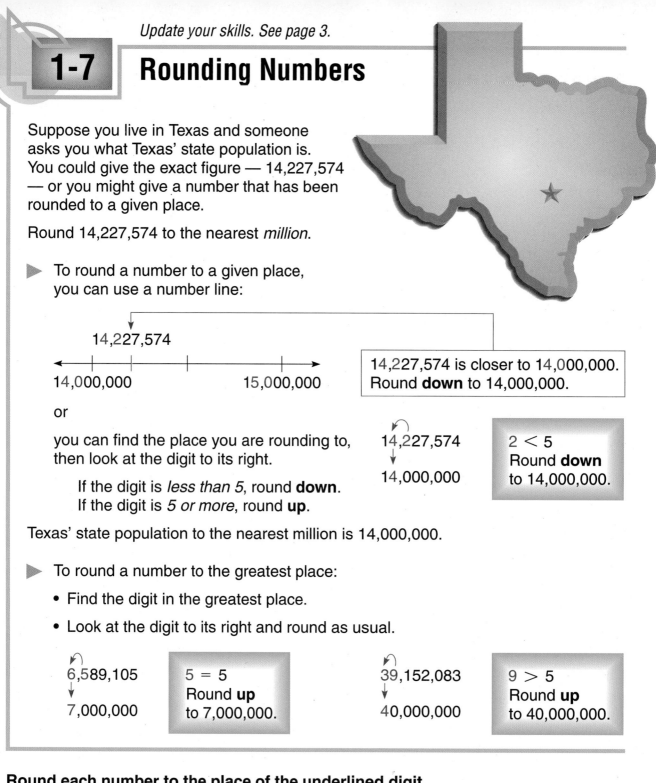

*Update your skills. See page 3.*

## 1-7 Rounding Numbers

Suppose you live in Texas and someone asks you what Texas' state population is. You could give the exact figure — 14,227,574 — or you might give a number that has been rounded to a given place.

Round 14,227,574 to the nearest *million*.

▶ To round a number to a given place, you can use a number line:

14,227,574

14,000,000          15,000,000

14,227,574 is closer to 14,000,000. Round **down** to 14,000,000.

or

you can find the place you are rounding to, then look at the digit to its right.

If the digit is *less than 5*, round **down**.
If the digit is *5 or more*, round **up**.

14,227,574
↓
14,000,000

2 < 5
Round **down**
to 14,000,000.

Texas' state population to the nearest million is 14,000,000.

▶ To round a number to the greatest place:

• Find the digit in the greatest place.

• Look at the digit to its right and round as usual.

6,589,105
↓
7,000,000

5 = 5
Round **up**
to 7,000,000.

39,152,083
↓
40,000,000

9 > 5
Round **up**
to 40,000,000.

**Round each number to the place of the underlined digit.**
You may use a number line to help you.

1. 1<u>6</u>3,128

2. <u>9</u>25,684

3. <u>1</u>,675,213

4. 6,<u>5</u>89,105

5. <u>3</u>6,813,431

6. 12,4<u>3</u>5,129

7. 2<u>3</u>5,198,051

8. 84,19<u>3</u>,103

**Round to the greatest place.**

**9.** 53,678      **10.** 99,407      **11.** 783,229      **12.** 359,048

**13.** 114,726      **14.** 5,748,111      **15.** 1,098,093      **16.** 7,523,670

**17.** 20,248,973      **18.** 37,561,444      **19.** 86,124,826      **20.** 15,543,901

---

### Rounding Decimals and Money

To round decimals and money amounts, use
the same rules for rounding whole numbers.

▶ Round 36.375 to the nearest:

| **Whole Number** | **Tenth** | | **Hundredth** |
|---|---|---|---|
| 36.375 | 36.375 | | 36.375 |
| ↓ | ↓ | | ↓ |
| 36 | 36.4 | Do not write zeros to the right. | 36.38 |

▶ Round $473.28 to the nearest:

| **Ten Cents** | **Dollar** | **Ten Dollars** | **Hundred Dollars** |
|---|---|---|---|
| $ 473.28 | $473.28 | $473.28 | $473.28 |
| ↓ | ↓ | ↓ | ↓ |
| $ 473.30 | $473 | $470 | $500 |

---

**Round each to the nearest *whole number*, *tenth*, and *hundredth*.**

**21.** 6.148      **22.** 1.792      **23.** 3.732      **24.** 24.873      **25.** 39.925

**26.** 73.159      **27.** 29.866      **28.** 548.501      **29.** 112.549      **30.** 332.532

**Round each to the nearest *ten cents*, *dollar*, *ten dollars*,
and *hundred dollars*.**

**31.** $427.89      **32.** $642.87      **33.** $792.46      **34.** $225.98      **35.** $146.72

**36.** $119.28      **37.** $542.76      **38.** $125.58      **39.** $918.92      **40.** $699.45

### PROBLEM SOLVING

**41.** The world's largest rock crystal ball weighs 106.75 pounds. Round this weight to the nearest tenth.

**42.** Julie bought two books for $14.98 and $19.45. Find the total cost of the books to the nearest dollar.

# Addition Properties/Subtraction Rules

▶ The properties of addition can help you add quickly and correctly.

addend + addend = sum

- Changing the *order* of the addends does not change the sum. (*Commutative property of addition*)

Think: "order."

$$6 + 9 = 15 \qquad \begin{array}{r} 6 \\ +9 \\ \hline 15 \end{array} \quad \begin{array}{r} 9 \\ +6 \\ \hline 15 \end{array}$$

$$9 + 6 = 15$$

- The sum of *zero* and a number is the same as that number. (*Identity property of addition*)

Think: "same."

$$9 + 0 = 9 \qquad \begin{array}{r} 9 \\ +0 \\ \hline 9 \end{array} \quad \begin{array}{r} 0 \\ +9 \\ \hline 9 \end{array}$$

$$0 + 9 = 9$$

- Changing the *grouping* of the addends does not change the sum. (*Associative property of addition*)

Think: "grouping."

$$(2 + 3) + 6 = 2 + (3 + 6)$$
$$5 + 6 = 2 + 9$$
$$11 = 11$$

▶ Use the properties to find shortcuts when adding more than two numbers.

**Change the order.**

Add down.

$$\begin{array}{r} 3 \\ 0 \quad 3 \\ 4 \quad 7 \\ 7 \quad 14 \\ +6 \quad 20 \\ \hline 20 \end{array}$$

Add up.

$$\begin{array}{r} 3 \quad 20 \\ 0 \quad 17 \\ 4 \quad 17 \\ 7 \quad 13 \\ +6 \\ \hline 20 \end{array}$$

**Change the order and the grouping.**

$$\begin{array}{r} 3 \\ 4 \\ 6 \\ +7 \\ \hline 20 \end{array} \quad 10 \quad 10$$

$$(3 + 7) + (4 + 6) = 20$$
$$10 \quad + \quad 10 \quad = 20$$

**Find the missing number. Name the property of addition that is used.**

*Algebra* ✓

1. $8 + 7 = \square + 8$

2. $8 = 0 + \square$

3. $(6 + 1) + 9 = 6 + (1 + \square)$

4. $\square + 4 = 4$

5. $5 + \square = 6 + 5$

6. $3 + (5 + 6) = (3 + \square) + 6$

**Add.** Use the properties of addition to find shortcuts.

| **7.** | 9 | **8.** | 4 | **9.** | 5 | **10.** | 4 | **11.** | 1 | **12.** | 9 |
|---|---|---|---|---|---|---|---|---|---|---|---|
| | 3 | | 2 | | 4 | | 7 | | 2 | | 4 |
| | 7 | | 6 | | 5 | | 6 | | 6 | | 1 |
| | +1 | | +8 | | +3 | | +2 | | +8 | | +5 |

**13.** 2 + 7 + 0 + 5 + 3     **14.** 1 + 6 + 5 + 0 + 4     **15.** 2 + 0 + 4 + 8 + 1

---

### Subtraction Rules

Subtraction is the *inverse* of addition. It "undoes" addition.

$$7 + 4 = 11$$
$$11 - 4 = 7$$

| 7 | 11 |
|---|---|
| +4 | − 4 |
| 11 | 7 |

The **rules of subtraction** can help you subtract quickly and correctly.

- The *subtrahend* is subtracted from the *minuend* to find the difference:

$$5 \leftarrow \textbf{minuend}$$
$$-3 \leftarrow \textbf{subtrahend}$$
$$2 \leftarrow \textbf{difference}$$

- When the minuend is equal to the subtrahend, the difference is always *zero*.

$$9 - 9 = 0$$

| 9 |
|---|
| −9 |
| 0 |

- When zero is the subtrahend, the difference is equal to the *minuend*.

$$9 - 0 = 9$$

| 9 |
|---|
| −0 |
| 9 |

---

**Find the missing addend.**

*Algebra*

**16.** 7 + ☐ = 11  Think: 11 − 7 = 4
   So 7 + 4 = 11

**17.** 6 + ☐ = 15   **18.** ☐ + 9 = 18

**19.** 8 + ☐ = 14   **20.** ☐ + 4 = 12   **21.** 7 + ☐ = 7   **22.** 8 + ☐ = 13

### PROBLEM SOLVING

**23.** There are 16 books on a shelf. Hannah takes 7 books from the shelf. How many books are left on the shelf?

**24.** Ramon puts 14 books in a box. Eight of the books are textbooks. How many books are *not* textbooks?

**25.** Bianca reads 12 pages in her science book. Alison reads 8 pages. How many more pages does Bianca read than Alison?

**26.** Luis has 9 points in his reading class. He needs 12 points to win a star. How many more points does he need to win a star?

## 1-9 Estimating Sums and Differences

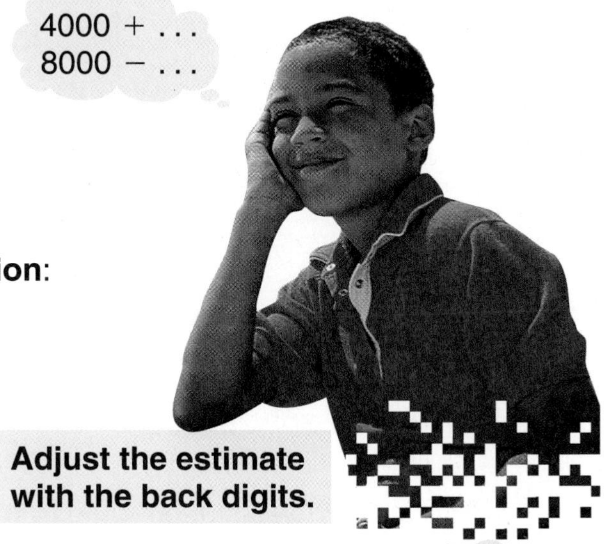

Mr. Blackwell asked his class to estimate
the sum: 4164 + 987 + 4213
the difference: 8365 − 3821

4000 + ...
8000 − ...

You can use front-end estimation to estimate
sums and differences.

▶ To *estimate sums* using **front-end estimation**:
  • Add the front digits. Then write zeros
    for the other digits.
  • Adjust the estimate with the back digits.

| Add the front digits. Write zeros for the other digits. | Adjust the estimate with the back digits. |
|---|---|
| 4164<br>987<br>3895<br>+ 4213<br>about 11,000 | 4164 ⎫ about 1000<br>987 ⎭<br>3895 ⎫ about 1000<br>+ 4213 ⎭     11,000<br>1,000<br>+ 1,000<br>13,000 |

Rough estimate: 11,000

Adjusted estimate:
11,000 + 1000 + 1000 = 13,000

The estimated sum is 13,000.

▶ To *estimate differences* using **front-end estimation**:
  • Subtract the front digits.
  • Write zeros for the other digits.

8365
− 3821
about   5000

The estimated difference is 5000.

**Study these examples.**

$324.54 ⟶ $324.54 ⎫ about $100
  276.37     276.37 ⎭
+  436.93   +  436.93
about  $900.00

9561
− 742
about   9000

$943.86
−137.13
about   $800.00

Rough estimate: $900
Adjusted estimate:
  $900 + $100 = $1000

**Estimate the sum or difference.** Use front-end estimation.

| 1. | 2. | 3. | 4. | 5. |
|---|---|---|---|---|
| 4987 | 6325 | 232 | $115.27 | $947.60 |
| 2526 | 3691 | 7625 | 372.62 | 25.89 |
| + 2844 | + 2236 | + 3475 | +236.91 | + 550.09 |

| 6. | 7. | 8. | 9. | 10. |
|---|---|---|---|---|
| 6626 | 7242 | 8934 | $887.56 | $932.55 |
| − 4813 | − 5759 | − 812 | − 259.60 | − 47.28 |

**11.** 6325 + 3632 + 8422 + 1362 **12.** 7459 + 1359 + 813 + 5231

---

### Estimation by Rounding

▶ **Rounding** is another estimation strategy.
To estimate by rounding:

- Round each number to the greatest place of the least number.
- Add or subtract the rounded numbers.

| | | | |
|---|---|---|---|
| | 1 1 | | |
| 6917 ⟶ 6920 | $5.78 ⟶ $5.80 | 5931 ⟶ 5900 | |
| 78 ⟶ 80 | 3.26 ⟶ 3.30 | − 723 ⟶ − 700 | |
| + 434 ⟶ + 430 | + 0.83 ⟶ + 0.80 | about 5200 | |
| about 7430 | about $9.90 | | |

▶ When an estimated
difference is **zero**, round
to the next greatest place.

| | |
|---|---|
| $39.48 ⟶ $40.00 | $39.48 ⟶ $39.00 |
| − 35.62 ⟶ − 40.00 | − 35.62 ⟶ − 36.00 |
| about $ 0 | about $3.00 |

---

**Estimate the sum or difference.** Use rounding.

| 13. | 14. | 15. | 16. | 17. |
|---|---|---|---|---|
| 2732 | 3257 | 4239 | $ 4.67 | $41.07 |
| 6146 | 612 | 624 | 15.08 | 92.53 |
| + 7378 | + 5701 | + 38 | + 41.13 | + 3.12 |

| 18. | 19. | 20. | 21. | 22. |
|---|---|---|---|---|
| 7893 | 8934 | 9434 | $83.72 | $932.55 |
| − 5421 | − 819 | − 9251 | − 8.44 | − 47.48 |

**23.** 2357 + 4612 + 5318 + 675 **24.** 6531 + 7735 + 943 + 39

## Critical Thinking

Solve and explain the
method you used.

*Communicate* ✓

**25.** Find the greatest number and the least number that round to
400,000 when rounded to the nearest hundred thousand.

# 1-10  Addition: Three or More Addends

How many pairs of sneakers did Allan Sporting Goods store sell during the three-month period?

First estimate the sum.

100 + 200 + 100 = 400

To find how many pairs of sneakers the store sold, add: 119 + 206 + 94 = __?__

| Month | Pairs of Sneakers Sold |
|-------|------------------------|
| April | 119 |
| May | 206 |
| June | 94 |

### Add the ones. Regroup.

```
    1
  119
  206
+  94
────
    9
```

19 ones = 1 ten 9 ones

### Add the tens. Regroup.

```
   11
  119
  206
+  94
────
   19
```

11 tens = 1 hundred 1 ten

### Add the hundreds.

```
   11
  119
  206
+  94
────
  419
```

419 is close to the estimate of 400.

Allan Sporting Goods store sold 419 pairs of sneakers.

### Study these examples.

```
  111
  1715
  4673
+ 2586
──────
  8974
```

```
  111
  2358
   793
  4312
+ 6135
──────
 13,598
```

```
   1 1
  $3.59
   1.43
+  0.85
──────
  $5.87
```

```
  11 2
 $13.59
  24.38
  47.15
+ 32.23
───────
 $117.35
```

## Estimate. Then add.

**1.**
```
  54
  32
+ 23
```

**2.**
```
  43
  25
+ 31
```

**3.**
```
 183
 214
+302
```

**4.**
```
 516
 242
+321
```

**5.**
```
 624
 143
+232
```

**6.**
```
 501
 243
+ 76
```

**7.**
```
 251
  39
+490
```

**8.**
```
 3429
 5182
+2404
```

**9.**
```
 3297
 4356
+1579
```

**10.**
```
 6783
 3452
+ 594
```

**Estimate. Then find the sum.**

| 11. | $26.34 | 12. | $19.57 | 13. | $52.09 | 14. | $23.21 | 15. | $56.25 |
|---|---|---|---|---|---|---|---|---|---|
| | 14.72 | | 70.46 | | 43.17 | | 17.64 | | 9.18 |
| | + 37.18 | | + 13.12 | | + 17.45 | | + 1.92 | | + 13.46 |

| 16. | $16.83 | 17. | $29.54 | 18. | $95.12 | 19. | $45.73 | 20. | $ 8.75 |
|---|---|---|---|---|---|---|---|---|---|
| | 23.19 | | 47.21 | | 3.81 | | 18.92 | | 19.16 |
| | 41.62 | | 25.38 | | 19.09 | | 21.45 | | 27.32 |
| | + 19.18 | | + 31.09 | | + 21.35 | | + 3.28 | | + 3.26 |

**Align and add.**

**21.** 2386 + 1396 + 2176 + 7266

**22.** 5449 + 2176 + 2347 + 3248

**23.** 3829 + 1760 + 1857 + 704

**24.** 8176 + 45 + 589 + 1259

**25.** 1105 + 1075 + 589 + 2863

**26.** 2749 + 3890 + 917 + 44

## PROBLEM SOLVING

**27.** Three rivers form a river system and have lengths of 513 miles, 247 miles, and 397 miles. Through how many miles do these rivers run?

**28.** Linda has 107 stamps from North America, 319 stamps from Africa, 43 stamps from Asia, and 168 stamps from Europe. How many stamps does Linda have in all?

## Choose a Computation Method

Look carefully at the numbers in a problem. The size and type of numbers will help you decide which computation method to use when an exact answer is needed.

**Computation Methods**
- Mental Math
- Paper and Pencil
- Calculator

**Add. Use mental math, paper and pencil, or calculator. Explain the method you used.**

**29.** 274 + 289 + 87 + 300

**30.** 7000 + 100 + 600 + 17

**31.** 117 + 117 + 147 + 1570

**32.** 5389 + 126 + 3427 + 8653

**33.** 6000 + 500 + 40 + 3

**34.** 5734 + 3268 + 521 + 1614

# 1-11 Subtraction with Zeros

Julia collected 4000 pennies for the charity drive. Raymond collected 3135 pennies. How many more pennies did Julia collect than Raymond?

First estimate the difference.
4000 − 3000 = 1000

▶ To **subtract** when the minuend has zeros:

- Regroup as many times as necessary before starting to subtract.

- Subtract.

| More hundreds, tens, and ones are needed. Regroup all. | Subtract. | Check. |
|---|---|---|
| 9 9<br>3 1̶0̶ 1̶0̶ 10<br>4̶ 0̶ 0̶ 0̶<br>− 3 1 3 5 | 9 9<br>3 1̶0̶ 1̶0̶ 10<br>4̶ 0̶ 0̶ 0̶<br>− 3 1 3 5<br>8 6 5 | 1 1 1<br>865<br>+ 3135<br>4000 |

4 thousands =
3 thousands 10 hundreds   0 tens   0 ones =
3 thousands   9 hundreds 10 tens   0 ones =
3 thousands   9 hundreds   9 tens 10 ones

865 is close to the estimate of 1000.

Julia collected 865 more pennies than Raymond.

## Study these examples.

| 9 9<br>6 1̶0̶ 1̶0̶ 12<br>7̶ 0̶ 0̶ 2̶<br>− 3 2 5 8<br>3 7 4 4 | 9 15<br>8 1̶0̶ 1̶6̶ 13<br>9̶ 0̶ 6̶ 3̶<br>− 4 3 7 6<br>4 6 8 7 | 9 9<br>4 1̶0̶ 1̶0̶ 10<br>5̶ 0̶ 0̶ 0̶<br>−   6 9 8<br>4 3 0 2 | 9 9<br>8 1̶0̶ 1̶0̶ 10<br>$9̶ 0̶.0̶ 0̶<br>−   7 2.5 6<br>$1 7.4 4 |

## Estimate. Then subtract.

| 1. | 2. | 3. | 4. | 5. |
|---|---|---|---|---|
| 40<br>− 26 | 80<br>− 29 | 90<br>− 35 | 70<br>− 18 | 60<br>− 42 |

**Estimate. Then find the difference.**

| 6. | 800<br>− 526 | 7. | 700<br>− 439 | 8. | 300<br>− 124 | 9. | 902<br>− 514 | 10. | 600<br>− 78 |
|---|---|---|---|---|---|---|---|---|---|

| 11. | 9000<br>− 4572 | 12. | 8000<br>− 2333 | 13. | 6006<br>− 1737 | 14. | 8060<br>− 5274 | 15. | 3000<br>− 543 |
|---|---|---|---|---|---|---|---|---|---|

| 16. | $7.00<br>− 5.21 | 17. | $6.00<br>− 3.92 | 18. | $8.00<br>− 2.97 | 19. | $5.09<br>− 1.35 | 20. | $4.00<br>− 0.83 |
|---|---|---|---|---|---|---|---|---|---|

| 21. | $87.00<br>− 64.27 | 22. | $93.00<br>− 78.42 | 23. | $60.03<br>− 14.59 | 24. | $48.00<br>− 7.03 | 25. | $30.20<br>− 4.53 |
|---|---|---|---|---|---|---|---|---|---|

**Align and subtract.**

26. 4000 − 784          27. 9000 − 8762          28. 5003 − 1784

29. 7020 − 4721          30. 7200 − 6548          31. 5081 − 329

32. 8700 − 421          33. 9300 − 7842          34. 4800 − 703

**Find the missing minuend.**

Algebra

| 35. | ?<br>− 764<br>136 | 36. | ?<br>− 459<br>241 | 37. | ?<br>− 623<br>278 | 38. | ?<br>− 596<br>257 | 39. | ?<br>− 861<br>263 |
|---|---|---|---|---|---|---|---|---|---|

| 40. | ?<br>− 5278<br>2722 | 41. | ?<br>− 4927<br>1073 | 42. | ?<br>− 3452<br>3548 | 43. | ?<br>− 1777<br>1226 | 44. | ?<br>− 2182<br>1848 |
|---|---|---|---|---|---|---|---|---|---|

**PROBLEM SOLVING**

45. Bobby has 2000 international coins. One hundred twenty-three coins are from Asia. How many coins are *not* from Asia?

46. Carla had $30.00. She bought a book for $7.95. How much money did she have left?

## Finding Together

Discuss

**Use the digits 0, 1, 2, 3.**

47. Write as many 4-digit numbers as you can without repeating digits. Then subtract the least from the greatest of these numbers.

**Larger Sums and Differences**

Study these examples. First estimate.
Then add or subtract as usual.

Add: 115,463 + 97,912 + 122,877 = __?__

TODAY'S NEWS
115,463
Watch Game!

| Estimate. | Add. Regroup where necessary. |
|---|---|
| | 1 1 2  1 1 |
| 100,000 | 115,463 |
| 100,000 | 97,912 |
| + 100,000 | + 122,877 |
| about    300,000 | 336,252 |

336,252 is close to
the estimate of 300,000.

Subtract: 820,410 − 647,635 = __?__

| Estimate. | Subtract. Regroup. |
|---|---|
| | 11  9 13 10 |
| | 7 12 10 14 11 10 |
| 800,000 | 8 2 0,4 1 0 |
| − 600,000 | − 6 4 7,6 3 5 |
| about    200,000 | 1 7 2,7 7 5 |

172,775 is close to
the estimate of 200,000.

▶ You can use a calculator to add or subtract larger numbers.

• Enter:   115,463 ⊞ 97,912 ⊞ 122,877 ▬

   Display:   ⎧ *336252.* ⎫ ◀─────

• Enter:   820,410 ⊟ 647,635 ▬

   Display:   ⎧ *172775.* ⎫ ◀─────

The decimal point
may not appear
in the display.

**Estimate. Then add or subtract.** (Watch for + or −.) You may use a calculator.

| | | | |
|---|---|---|---|
| **1.**    115,609 | **2.**    356,789 | **3.**    471,009 | **4.**    365,786 |
|    205,399 |    141,217 |    180,007 |    274,982 |
| + 411,111 | + 222,888 | + 277,777 | + 186,214 |

| | | | |
|---|---|---|---|
| **5.**    672,244 | **6.**    681,337 | **7.**    524,700 | **8.**    938,400 |
| − 456,688 | − 278,456 | − 316,672 | − 619,711 |

**Estimate. Then find the sum or difference.** (Watch for + or −.)
You may use a calculator.

| 9. | $247.00<br>+ 166.72 | 10. | $621.21<br>− 354.25 | 11. | $516.83<br>+ 378.35 | 12. | $700.01<br>− 549.34 |
|----|----|----|----|----|----|----|----|

**Align. Then add or subtract.** (Watch for + or −.)

**13.** 33,624 + 6109 + 34,200

**14.** 117,618 + 6004 + 27,906

**15.** 338,400 − 19,711

**16.** 146,502 − 78,781

**17.** 45,162 + 215 + 3614 + 7

**18.** 204,106 + 403 + 7000 + 10,691

**19.** 746,500 − 28,781

**20.** 978,432 − 739,853

**Write each group of numbers in order from greatest to least. Then add and subtract the two greatest numbers.** You may use a calculator.

**21.** 38,745; 39,547; 37,845; 39,845

**22.** 77,178; 71,718; 77,781; 71,871

**23.** 40,060; 40,600; 40,006; 46,000

**24.** 54,980; 54,908; 54,809; 54,890

## PROBLEM SOLVING
Use the tables. You may use a calculator.

**25.** What is the combined seating capacity of Veterans Stadium and Royals Stadium?

**26.** What is the combined seating capacity of Municipal Stadium and Anaheim Stadium?

**27.** How many more U. S. troops are there in East Asia and the Pacific than in the Western Hemisphere?

**28.** If 109,888 U.S. troops are in Germany, how many U.S. troops in Europe are *not* in Germany?

| Arena | Seating Capacity |
|----|----|
| Veterans Stadium, Philadelphia | 62,382 |
| Municipal Stadium, Cleveland | 74,208 |
| Royals Stadium, Kansas City | 40,625 |
| Anaheim Stadium, California | 64,573 |

| Place | Number of U.S. Troops |
|----|----|
| Europe | 171,904 |
| East Asia and the Pacific | 110,054 |
| Western Hemisphere | 16,204 |

 **Share Your Thinking**

**29.** Write a letter to a classmate giving some examples that would show that sometimes it is better *not* to use a calculator to do computations.

*Communicate* ✓

# 1-13 Roman Numerals

The ancient Romans used letters to write numbers.
Study this table of **Roman numerals** and their values.

| I | II | III | IV | V | VI | VII | VIII | IX | X |
|---|----|-----|----|---|----|-----|------|----|---|
| 1 | 2 | 3 | 4 | 5 | 6 | 7 | 8 | 9 | 10 |
| V | X | XV | XX | XXV | XXX | XXXV | XL | XLV | L |
| 5 | 10 | 15 | 20 | 25 | 30 | 35 | 40 | 45 | 50 |
| X | XX | XXX | XL | L | LX | LXX | LXXX | XC | C |
| 10 | 20 | 30 | 40 | 50 | 60 | 70 | 80 | 90 | 100 |
| C | CC | CCC | CD | D | DC | DCC | DCCC | CM | M |
| 100 | 200 | 300 | 400 | 500 | 600 | 700 | 800 | 900 | 1000 |

▶ To find the value of a Roman numeral,
add:
  • if the letter is repeated.
      XX = 10 + 10 = 20
   CCC = 100 + 100 + 100 = 300

  • if a letter with a smaller value comes
    *after* a letter with a larger value.
      XV = 10 + 5 = 15
   DCX = 500 + 100 + 10 = 610

subtract:
  • if a letter with a smaller value comes
    *before* a letter with a larger value.
      XL = 50 − 10 = 40
   CM = 1000 − 100 = 900

> A letter is never repeated
> more than three times.

Sometimes you must both add and subtract.

$$CDLXIV = (500 - 100) + (50 + 10) + (5 - 1)$$
$$400 \quad + \quad 60 \quad + \quad 4 \quad = \quad 464$$

**Copy and complete.**

1. CCLXIII = 100 + _?_ + 50 + _?_ + _?_ + _?_ + _?_ = _?_

2. CMXCIV = (1000 − _?_ ) + ( _?_ − 10) + ( _?_ − _?_ ) = _?_

**Write the Roman numeral in standard form.**

3. XXXIV 4. MVII 5. LV 6. DXXI

7. CCLXX 8. DCCXC 9. XCIX 10. MDIII

11. XLVII 12. MCCLVI 13. CXLV 14. MDCCXCI

15. MMCLI 16. MMDCCCIII 17. MDCCLXXXV 18. MDCCCXLV

**Write each as a Roman numeral.**

19. 18 20. 24 21. 31 22. 52 23. 14 24. 73

25. 180 26. 193 27. 387 28. 504 29. 919 30. 623

31. 731 32. 876 33. 415 34. 327 35. 613 36. 287

37. 1321 38. 1449 39. 2001 40. 3555 41. 2765 42. 3046

**Write the date of the admittance of each state into the Union as a standard numeral.**

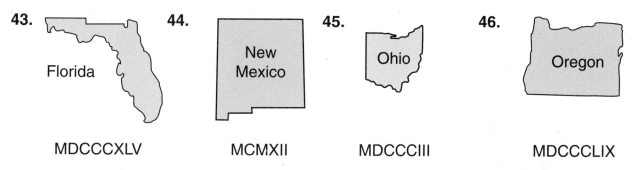

43. Florida    MDCCCXLV

44. New Mexico    MCMXII

45. Ohio    MDCCCIII

46. Oregon    MDCCCLIX

## PROBLEM SOLVING

47. The Statue of Liberty was dedicated in 1886. Write this date as a Roman numeral.

48. Dr. Evans saw the date MDIX on a building in Rome. Write this number as a standard numeral.

**Make Up Your Own**

*Communicate* ✓

49. Use some of the digits 1, 3, 5, 7, 9 only once to write 5 numbers less than 2000 and then express each number as a Roman numeral. Share your work with a classmate.

## Flowcharts

A **flowchart** displays the steps needed to complete a task.
Flowcharts show how a computer follows the instruction
of a program. Symbols are used to show the different
steps of a program. Each symbol shows one step.

Below are the standard flowchart symbols and their meanings.

| oval | parallelogram | rectangle | diamond |
|------|---------------|-----------|---------|
| Starts or Stops Instruction | Inputs or Outputs Information | Processes Information | Asks for a Yes or No decision |

**Study this flowchart and its output.** Arrows are
used to show what step to do next.

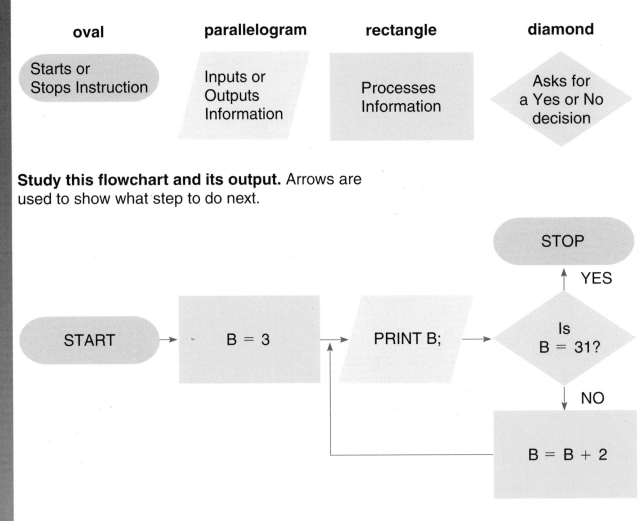

**Output:**  3  5  7  9  11  13  15  17  19  21  23  25  27  29  31

**Write in your Math Journal the answer to each question.**

1. Name the symbol used to input or print information.

2. Which symbol is used to perform a computation?

3. How is the diamond symbol used in a flowchart?

4. How many steps of a program can be in one symbol?

**Use the flowchart on page 56 to answer questions 5–8.**

5. When will the program stop?

6. How many "Yes" decisions will be made?

7. When B = 31, which step or steps would you skip?

8. How would you change the flowchart so that the output is the even numbers from 2 to 30?

**Copy and complete the flowchart using the program to the right.**

9.

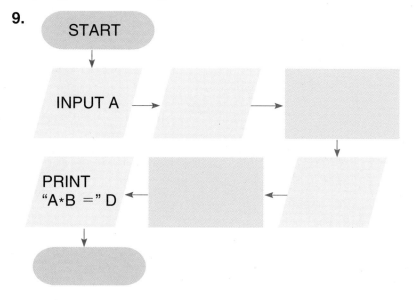

```
10  INPUT A
20  INPUT B
30  LET C = A + B
40  PRINT "A + B =" C
50  LET D = A*B
60  PRINT "A*B =" D
70  END
```

10. Create a flowchart that would give an output of the numbers from 1 to 100. Include all the symbols at the top of page 56.

# 1-15 Problem Solving: Guess and Test

**Problem:** Ed needs to take his cat, bird, and snake to the veterinarian. His car can hold only 2—1 pet and himself. If left alone together, the cat (*C*) will eat the bird (*B*), and the snake (*S*) will eat the bird (*B*). How many trips will Ed (*E*) need to make?

**1 IMAGINE** Put yourself in the problem.

**2 NAME** *Facts:* Ed and 3 pets to the veterinarian.
*C* and *B* or *B* and *S* cannot be left alone together.
Only 1 pet and Ed fit into the car.

*Question:* How many trips does he need to make?

**3 THINK** Make a guess. Draw a picture or act it out to test each guess.

**4 COMPUTE**

| | | **Home** | | **Veterinarian** |
|---|---|---|---|---|
| **1st** | Ed takes the bird, because the cat will not eat the snake. | *C, S* | $\xrightarrow{E, B}$ 1st | |
| **2nd** | Ed returns, leaving the bird. | *C, S* | $\xleftarrow{E}$ 2nd | *B* |
| **3rd** | Ed takes the cat and leaves it at the veterinarian. | *S* | $\xrightarrow{E, C}$ 3rd | *B* |
| **4th** | Ed returns with the bird. | *S* | $\xleftarrow{E, B}$ 4th | *C* |
| **5th** | Ed takes the snake and leaves the bird home. | *B* | $\xrightarrow{E, S}$ 5th | *C* |
| **6th** | Ed returns after leaving the snake with the cat. | *B* | $\xleftarrow{E}$ 6th | *C, S* |
| **7th** | Ed takes the bird. Now the 3 pets are at the veterinarian. | | $\xrightarrow{E, B}$ 7th | *C, S* |

So Ed needs to make 7 trips.

**5 CHECK** Did more than two go in the car? No.
Was the cat ever left alone with the bird? No.
Was the snake ever left alone with the bird? No.

**Use Guess and Test to solve each problem.**

1. Pat's dad is 2 ft 1 in. taller than Pat. The sum of their heights is 10 ft 5 in. How tall is Pat?

| **IMAGINE** | Create a mental picture. |

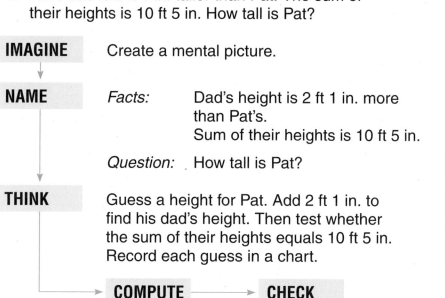

**NAME**    *Facts:*    Dad's height is 2 ft 1 in. more than Pat's.
Sum of their heights is 10 ft 5 in.

          *Question:*  How tall is Pat?

**THINK**    Guess a height for Pat. Add 2 ft 1 in. to find his dad's height. Then test whether the sum of their heights equals 10 ft 5 in. Record each guess in a chart.

| Pat | 4 ft |
|---|---|
| Dad | 6 ft 1 in. |
| Sum | 10 ft 1 in. |

**COMPUTE** ⟶ **CHECK**

2. Drew wrote a 4-digit number less than 2000. The sum of its digits is 20. Only the digits in the ones place and hundreds place are even. The digit in the ones place is double the digit in the thousands place. What number did Drew write?

3. Grace has a cat, a bird, and a package of birdseed. She wants to get all three home safely, but her bicycle basket will hold only *one* at a time. The cat will eat the bird if the two are left alone together. The bird will eat the birdseed if they are left alone. How many trips does Grace need to make to get everything home safely?

4. Five coins fell out of Doug's pocket. He lost 27¢. What coins did Doug lose?

5. In the subtraction example at the right, each letter stands for a different digit. Find the value of X, Y, and Z.

$$\begin{array}{r} X\,Y\,X \\ -\ \ Z\,X \\ \hline X\,Y \end{array}$$

 **Make Up Your Own**

 Communicate

6. Write a problem that requires you to use the Guess and Test strategy. Then solve it. Share your work with a classmate.

# 1-16 Problem-Solving Applications

## Connections: Social Studies

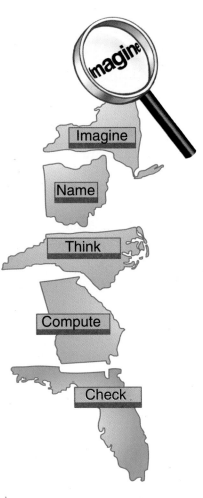

Imagine

Name

Think

Compute

Check

**Solve each problem and explain the method you used.**

1. The first U.S. census was taken in 1790. At that time, the population was recorded as 3,929,000. How many times greater is the 9 in the hundred thousands place than the 9 in the thousands place?

2. By the 1800 census the population had reached 5,308,000. Is this an increase of more or less than 2 million over the 1790 population? Explain.

3. By 1810, the population had increased to 7,240,000. What is the increase over the 1800 census?

4. The center of population in 1980 was 0.25 miles west of De Soto, Missouri. Write 0.25 as a fraction. Write its word name.

5. In 1990, the center of population moved southwest by $\frac{5}{10}$ of a mile more than 39 miles. Write this distance as a decimal.

6. Between 1790 and 1990, the center of population for the United States shifted 818.6 miles. What is 818.6 rounded to the nearest one?

7. Write the year 1790, when the first U.S. census was taken, in Roman numerals.

8. This chart shows the census population of the ten most populated states in 1990. Write the states in order from greatest to least population.

9. Which states have populations of about 11 million?

10. Which states have populations of between 9 million and 12 million?

11. Which state has about double the population of Georgia?

### 1990 U.S. Census

| State | Population |
| --- | --- |
| California | 29,760,021 |
| Florida | 12,937,926 |
| Georgia | 6,478,216 |
| Illinois | 11,430,602 |
| Michigan | 9,295,297 |
| New Jersey | 7,730,188 |
| New York | 17,990,455 |
| North Carolina | 6,628,637 |
| Ohio | 10,847,115 |
| Texas | 16,986,510 |

**Choose a strategy from the list or use another strategy you know to solve each problem.**

USE THESE STRATEGIES
More Than One Solution
Guess and Test
Logical Reasoning
Use a Graph
Missing Information

**12.** The fourth census took place in a year that can be written as a Roman numeral using these letters: *X, C, D, C, M, X, C.* What is the standard numeral for the year of the fourth census?

**13.** A rural village's population is between 800 and 1000. The sum of the digits in its population is 21, and the digits in the ones and the hundreds places are the same. What might be the population of the village?

**14.** In 1990, Alaska's population was less than Virginia's but greater than Wyoming's. Hawaii's population was between Alaska's and Virginia's. Write these states in increasing order of population.

**15.** Between 1790 and 1990, the U.S. population increased by 244,780,873. The population was almost 250,000,000 in 1990. If the population increases in the next 200 years, will the population in 2190 be more than 1 billion? Explain.

**Use the circle graph for problems 16–18.**

**16.** Which age group represented more than half the U.S. population in 1990? Explain.

**17.** What percent of the U.S. population was under the age of 18 in 1990?

**18.** Which age group represented between 10% and 25% of the population?

**U.S. Population
Age Distribution 1990**
(percent)

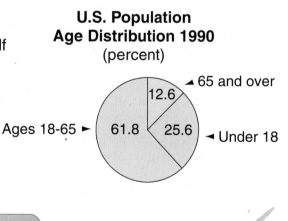

65 and over

12.6

Ages 18-65 ► 61.8  25.6  ◄ Under 18

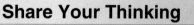
**Share Your Thinking**

Math Journal

**19.** Write in your Math Journal which problems you solved using the same strategy and explain why. Then write a problem modeled on these problems and have a classmate solve it.

# Chapter Review and Practice

**In the number 308,610,547,823, write the digit in the:**    *(See pp. 30–33.)*

**1.** ten-billions place　　　　**2.** millions place　　　　**3.** hundred-thousands place

**Write the number in standard form.**    *(See pp. 30–39, 54–55.)*

**4.** three hundred four billion, six hundred thousand　　　**5.** CCLXI

**6.** 1,000,000,000 + 40,000 + 80 + 3　　　**7.** eight and twelve thousandths

**Write the word name for each number.**

**8.** 360,071　　　**9.** 1,009,124,008　　　**10.** 6.71　　　**11.** 0.531　　　**12.** CMLXI

**Compare. Write $<$, $=$, or $>$.**    *(See pp. 40–41.)*

**13.** 185,035,013 _?_ 185,503,013　　　**14.** 10.09 _?_ 10.1　　　**15.** 9.63 _?_ 9.630

**Write in order from least to greatest.**

**16.** 6,135,936;　6,315,396;　6,531,639;　6,153,693　　　**17.** 3.12;　31.2;　0.312

**Round each number to the place of the underlined digit.**    *(See pp. 42–43.)*

**18.** 474,198,575　　**19.** 313,983,156　　**20.** 145.728　　**21.** $766.13

**Find the missing addend.**    *(See pp. 44–45.)*

**22.** 8 + ☐ = 15　　**23.** ☐ + 9 = 17　　**24.** 14 = ☐ + 7　　**25.** 11 = 6 + ☐

**Estimate. Then add or subtract.**    *(See pp. 46–53.)*

| **26.** | **27.** | **28.** | **29.** | **30.** |
|---|---|---|---|---|
| 25,736 | 503,149 | $235.17 | 600,000 | $907.15 |
| 12,548 | 180,590 | 137.23 | − 421,351 | − 35.43 |
| + 36,985 | + 248,762 | + 427.45 | | |

## PROBLEM SOLVING    *(See pp. 58–60.)*

**31.** The sum of two numbers is 34. Their difference is 18. What are the two numbers?

(See *Still More Practice*, p. 477.)

## VENN DIAGRAMS

**Venn diagrams** are drawings, usually circles, that show relationships.

This Venn diagram shows that:

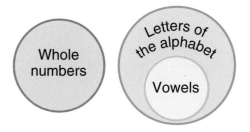

All vowels are letters of the alphabet.

Some letters of the alphabet are vowels.

No whole numbers are letters of the alphabet.

▶ Venn diagrams may be used in solving problems.

Of 35 students, 23 like math, 21 like science, and 9 like both subjects. How many students like math only? science only?

- Draw and label two overlapping circles, Math and Science.

- Write 9, the number of students that like both subjects, in the overlapping portions of the circles.

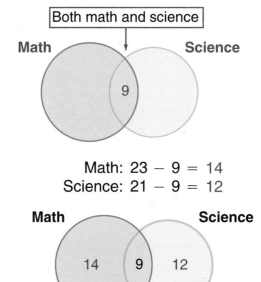

- Subtract 9 from the total number of students that like each subject.

Math: 23 − 9 = 14
Science: 21 − 9 = 12

- Write each difference in the remaining portion of the corresponding circle.

There are 14 students who like math only and 12 students who like science only.

14 + 9 + 12 = 35

## Draw a Venn diagram to illustrate each statement.

1. All roses are flowers.

2. No triangles are squares.

**PROBLEM SOLVING**  Use a Venn diagram.

3. Of 120 people, 80 enjoy classical music, 74 enjoy the theater, and 34 enjoy both classical music and the theater. How many people enjoy classical music only? the theater only?

# Check Your Mastery

**Use front-end estimation and rounding to estimate the answers.**
Tell which estimation strategy produces an estimate closer
to the actual answer and explain why.

**1.** $90,043 + 53,621 + 1,285 = $ _?_

**2.** $300.06 - $181.09 = $ _?_

**In the number 21,825,493,076, write the digit in the:**

**3.** hundred-thousands place

**4.** billions place

**5.** ten-millions place

**Write each number in standard form.**

**6.** three billion, two million, forty-five thousand, eighty-three

**7.** nine and twenty-one thousandths

**8.** $8,000,000 + 4000 + 60 + 2$

**Write the word name for each number.**

**9.** 1,000,935,009

**10.** 10.08

**11.** 9.036

**Compare. Write** $<$, $=$, **or** $>$.

**12.** 800,905,174 _?_ 800,905,147

**13.** 3.215 _?_ 3.125

**14.** 9.07 _?_ 9.070

**Write in order from greatest to least.**

**15.** 1,745,236;  1,475,236;  1,745,632;  1,475,263

**16.** 9.47;  9.56;  9.37;  9.68

**Round each number to the place of the underlined digit.**

**17.** 14,6̲73,584

**18.** 35.96̲8

**19.** $4̲0.35

**Find the missing number. Name the property
of addition that is used.**

**20.** $9 + 5 = \square + 9$

**21.** $7 = 0 + \square$

**22.** $(5 + 2) + 3 = 5 + (2 + \square)$

**Write each as a Roman numeral.**

**23.** 999

**24.** 1750

**PROBLEM SOLVING**   *Use a strategy you have learned.*

**25.** The area of Oregon is 97,073 square miles and the area of California
is 158,706 square miles. What is the total area of the two states?

# The Runner

Run, run, runner man,
As fast as you can,
Faster than the speed of light,
Smoother than a bird in flight.
Run, run, runner man,
No one can catch the runner man,
Swifter than an arrow,
Outrunning his own shadow.
Run, run, runner man,
Faster than tomorrow.
Run, run, runner man,
Quicker than a rocker!
Into deep space spinning a comet!
Run, run, runner man,
Lighting the heavens of the night,
Run, run, runner man,
Out of sight,
Run, run, runner man, run!

*Faustin Charles*

# Multiplication

# 2

**In this chapter you will:**

Use properties, special factors, and patterns
Estimate and multiply up to 3-digit numbers and money
Learn about IF-THEN statements
Solve problems with hidden information

**Critical Thinking/Finding Together**

You are training for a marathon. Each week you need
to run five miles more than the previous week. If you
need to run a total of 130 miles, how many miles will
you run during each of the next four weeks?

## 2-1 Meaning of Multiplication

There are 5 packs. Each pack contains 6 cans of juice. How many cans of juice are there in all?

To find how many cans in all, you can add:

6 + 6 + 6 + 6 + 6 = 30

$$\begin{array}{r} 6 \\ 6 \\ 6 \\ 6 \\ +6 \\ \hline 30 \end{array}$$

or

you can multiply since there are equal sets.

|     |          |     |   |     |
|-----|----------|-----|---|-----|
| 5 sixes |      |     | = | 30  |
| 5   | ×        | 6   | = | 30  |

number of sets    number in each set    number in all

Multiplication is repeated addition.

$$\begin{array}{r} 6 \longleftarrow \textbf{in each set} \\ \times 5 \longleftarrow \textbf{sets} \\ \hline 30 \longleftarrow \textbf{in all} \end{array}$$

There are 30 cans of juice in all.

### Study this example.

8 + 8 + 8 + 8 = 32
4 eights = 32
4 × 8 = 32

factor   factor   product

$$\begin{array}{r} 8 \longleftarrow \textbf{factor} \\ \times 4 \longleftarrow \textbf{factor} \\ \hline 32 \longleftarrow \textbf{product} \end{array}$$

4 × 8 = 32 is a multiplication number sentence.

## Write the multiplication fact.

**1.** 9 + 9 + 9

**2.** 4 + 4 + 4 + 4 + 4

**3.** 3 + 3 + 3 + 3 + 3 + 3

**4.** 2 + 2 + 2 + 2 + 2 + 2 + 2 + 2

**5.** 5 + 5 + 5 + 5

**6.** 8 + 8 + 8 + 8 + 8 + 8 + 8 + 8 + 8

**7.** 7 + 7 + 7 + 7 + 7 + 7 + 7

**8.** 6 + 6

**Find the product.**

| 9. | 10. | 11. | 12. | 13. | 14. |
|---|---|---|---|---|---|
| 8 | 7 | 5 | 0 | 1 | 9 |
| ×3 | ×4 | ×5 | ×6 | ×7 | ×9 |

| 15. | 16. | 17. | 18. | 19. | 20. |
|---|---|---|---|---|---|
| 2 | 7 | 3 | 6 | 5 | 6 |
| ×6 | ×7 | ×9 | ×8 | ×3 | ×6 |

**Find the missing factor.**

| 21. | Think: | 22. | 23. | 24. | 25. |
|---|---|---|---|---|---|
| 8 | 6 × 8 = 48 | 7 | ? | ? | 4 |
| ×? | | ×? | ×9 | ×6 | ×? |
| 48 | | 42 | 54 | 18 | 0 |

| 26. | 27. | 28. | 29. | 30. | 31. |
|---|---|---|---|---|---|
| ? | ? | ? | ? | ? | 9 |
| ×5 | ×4 | ×3 | ×6 | ×7 | ×? |
| 35 | 36 | 24 | 0 | 63 | 81 |

**Compare. Write <, =, or >.**

**32.** 6 × 3 _?_ 3 × 7

**33.** 9 × 0 _?_ 8 × 0

**34.** 5 × (2 × 3) _?_ 5 × (3 × 3)

**35.** 9 × 7 _?_ 8 × 8

**36.** (3 × 2) × 6 _?_ 3 × (2 × 4)

**37.** (2 × 3) × 6 _?_ 2 × (3 × 2)

**PROBLEM SOLVING**

**38.** Each pack holds 4 videotapes. How many videotapes are in 9 packs?

**39.** Nine large books will fit on one shelf. How many large books will fit on 8 shelves?

**40.** When you multiply 7 by itself, what is the product?

**41.** Two factors are 8 and 9. What is the product?

**42.** The product is 81. One factor is 9. What is the other factor?

**43.** The product is 36. One factor is 6. What is the other factor?

**Mental Math**

**Compute. Work from left to right.**

**44.** 6 × 6 + 4 − 2

**45.** 9 × 8 + 6 − 10

**46.** 6 × 7 − 8 − 6

**47.** 2 × 3 × 5 − 8

**48.** 7 × 1 + 6 − 1

**49.** 6 × 5 + 7 + 3

**Properties of Multiplication**

The properties of multiplication can help you multiply quickly and correctly.

- Changing the **order** of the factors does not change the product. *(Commutative property of multiplication)*

Think: "order."

$$9 \times 6 = 54$$
$$6 \times 9 = 54$$

$$\begin{array}{cc} 6 & 9 \\ \times 9 & \times 6 \\ \hline 54 & 54 \end{array}$$

- Changing the **grouping** of the factors does not change the product. *(Associative property of multiplication)*

Think: "grouping."

$$(2 \times 3) \times 3 = 2 \times (3 \times 3)$$
$$6 \times 3 = 2 \times 9$$
$$18 = 18$$

- The product of **one** and a number is the same as that number. *(Identity property of multiplication)*

Think: "same."

$$1 \times 7 = 7$$
$$7 \times 1 = 7$$

$$\begin{array}{cc} 7 & 1 \\ \times 1 & \times 7 \\ \hline 7 & 7 \end{array}$$

- The product of **zero** and a number is zero. *(Zero property of multiplication)*

Think: "0 product."

$$0 \times 4 = 0$$
$$4 \times 0 = 0$$

$$\begin{array}{cc} 4 & 0 \\ \times 0 & \times 4 \\ \hline 0 & 0 \end{array}$$

**Name the property of multiplication used.**

**1.** $5 \times 2 = 2 \times 5$

**2.** $9 \times 0 = 0$

**3.** $3 \times (2 \times 4) = (3 \times 2) \times 4$

**4.** $1 \times 8 = 8$

**5.** $0 \times 6 = 0$

**6.** $(2 \times 2) \times 4 = 2 \times (2 \times 4)$

**7.** $4 \times 1 = 4$

**8.** $9 \times 8 = 8 \times 9$

**9.** $0 \times 0 = 0$

**10.** $1 \times 1 = 1$

**Find the missing number.** Use the properties of multiplication.

**11.** $\underline{\ ?\ } \times 4 = 4 \times 6$

**12.** $9 \times \underline{\ ?\ } = 9$

**13.** $2 \times \underline{\ ?\ } = 0$

**14.** $1 \times 7 = \underline{\ ?\ }$

**15.** $6 \times 8 = 8 \times \underline{\ ?\ }$

**16.** $0 \times 6 = \underline{\ ?\ }$

**17.** $3 \times (2 \times 4) = (3 \times \underline{\ ?\ }) \times 4$

**18.** $(4 \times 2) \times 4 = \underline{\ ?\ } \times (2 \times 4)$

---

### Distributive Property

When the same factor is distributed across two addends, the product does not change. *(Distributive property of multiplication over addition)*

Think: "same factor across addends."

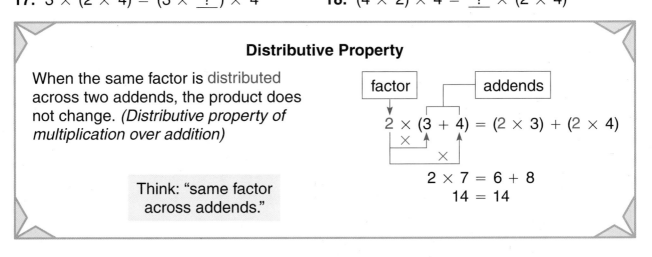

$$2 \times (3 + 4) = (2 \times 3) + (2 \times 4)$$
$$2 \times 7 = 6 + 8$$
$$14 = 14$$

---

**Copy and complete.**

**19.** $3 \times (5 + 2) = (3 \times 5) + (\underline{\ ?\ } \times 2)$

**20.** $\underline{\ ?\ } \times (4 + 2) = (6 \times 4) + (6 \times 2)$

**21.** $2 \times (3 + 6) = (\underline{\ ?\ } \times 3) + (\underline{\ ?\ } \times 6)$

**22.** $5 \times (\underline{\ ?\ } + \underline{\ ?\ }) = (5 \times 2) + (5 \times 3)$

**23.** $4 \times (2 + 3) = (4 \times \underline{\ ?\ }) + (4 \times \underline{\ ?\ })$

**24.** $6 \times (5 + 2) = (\underline{\ ?\ } \times \underline{\ ?\ }) + (\underline{\ ?\ } \times \underline{\ ?\ })$

### PROBLEM SOLVING

**25.** The product is 0 and one factor is 3. What is the other factor?

**26.** The product is 8 and one factor is 8. What is the other factor?

**27.** If $6 \times 13 = 78$, what is the product of $13 \times 6$?

**28.** If $4 \times 12 = 48$, what is $4 \times (3 + 9)$ equal to?

 **Share Your Thinking**

**29.** Tell your teacher:
- How does the commutative property of multiplication differ from the associative property of multiplication?
- How can the associative property or the distributive property be helpful to you in mental math computation?

# Mental Math: Special Factors

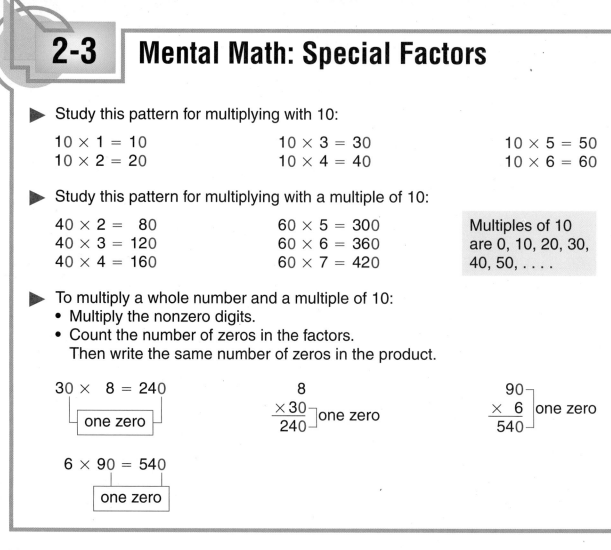

▶ Study this pattern for multiplying with 10:

| | | |
|---|---|---|
| 10 × 1 = 10 | 10 × 3 = 30 | 10 × 5 = 50 |
| 10 × 2 = 20 | 10 × 4 = 40 | 10 × 6 = 60 |

▶ Study this pattern for multiplying with a multiple of 10:

40 × 2 =  80          60 × 5 = 300
40 × 3 = 120          60 × 6 = 360
40 × 4 = 160          60 × 7 = 420

Multiples of 10 are 0, 10, 20, 30, 40, 50, . . . .

▶ To multiply a whole number and a multiple of 10:
 • Multiply the nonzero digits.
 • Count the number of zeros in the factors.
  Then write the same number of zeros in the product.

30 × 8 = 240
one zero

8
×30  one zero
240

90
× 6  one zero
540

6 × 90 = 540
one zero

## Find the products.

**1.** 10 × 7
10 × 8
10 × 9

**2.** 10 × 2
20 × 2
30 × 2

**3.** 20 × 6
30 × 6
40 × 6

**4.** 30 × 8
40 × 8
50 × 8

**5.** 4 × 10
5 × 10
6 × 10

**6.** 6 × 10
6 × 20
6 × 30

**7.** 7 × 20
7 × 30
7 × 40

**8.** 9 × 30
9 × 40
9 × 50

## Multiply.

**9.**    20
    × 4

**10.**    50
    × 7

**11.**    40
    × 8

**12.**    60
    × 9

**13.**    70
    × 6

**14.** 6 × 50

**15.** 8 × 60

**16.** 9 × 20

**17.** 3 × 40

**18.** 2 × 70

**19.** 7 × 80

**20.** 5 × 30

**21.** 4 × 60

**Find the product.**

| | | | | |
|---|---|---|---|---|
| **22.** $\begin{array}{r} 2 \\ \times\,30 \end{array}$ | **23.** $\begin{array}{r} 7 \\ \times\,50 \end{array}$ | **24.** $\begin{array}{r} 8 \\ \times\,40 \end{array}$ | **25.** $\begin{array}{r} 9 \\ \times\,60 \end{array}$ | **26.** $\begin{array}{r} 8 \\ \times\,70 \end{array}$ |
| **27.** $\begin{array}{r} 6 \\ \times\,90 \end{array}$ | **28.** $\begin{array}{r} 8 \\ \times\,80 \end{array}$ | **29.** $\begin{array}{r} 5 \\ \times\,50 \end{array}$ | **30.** $\begin{array}{r} 4 \\ \times\,30 \end{array}$ | **31.** $\begin{array}{r} 9 \\ \times\,40 \end{array}$ |

**32.** $8 \times 50$   **33.** $7 \times 30$   **34.** $8 \times 20$   **35.** $6 \times 40$

**36.** $70 \times 2$   **37.** $30 \times 5$   **38.** $80 \times 7$   **39.** $5 \times 60$

**40.** $60 \times 4$   **41.** $90 \times 6$   **42.** $70 \times 4$   **43.** $40 \times 2$

**44.** $3 \times 70$   **45.** $7 \times 50$   **46.** $9 \times 90$   **47.** $8 \times 80$

## PROBLEM SOLVING

**48.** The movie theater in Milmont Shopping Center has 40 rows of seats with 9 seats in each row. How many people in all can the theater seat?

**49.** The theater sold 6 cartons of popcorn at the Saturday matinee. If there were 30 bags in each carton, how many bags of popcorn in all did it sell?

**50.** The theater sold 40 orange drinks at each of 2 shows each night for 5 nights. How many orange drinks in all did it sell?

**51.** The theater sold 30 sandwiches at each of 3 shows each day for 5 days. How many sandwiches in all did it sell?

## Critical Thinking

**52.** Name two factors of 10 whose sum is 7.

**53.** Name two factors of 30 whose difference is 7.

**54.** Name two factors of 20 whose sum is 12.

**55.** Name two factors of 12 whose difference is 4.

**Patterns in Multiplication**

Study these patterns for multiplying with 100, 1000, or their multiples:

| | | |
|---|---|---|
| $1 \times 7 = 7$ | $2 \times 8 = 16$ | $4 \times 5 = 20$ |
| $10 \times 7 = 70$ | $20 \times 8 = 160$ | $40 \times 5 = 200$ |
| $100 \times 7 = 700$ | $200 \times 8 = 1600$ | $400 \times 5 = 2000$ |
| $1000 \times 7 = 7000$ | $2000 \times 8 = 16,000$ | $4000 \times 5 = 20,000$ |

| | | |
|---|---|---|
| $10 \times 70 = 700$ | $20 \times 80 = 1600$ | $40 \times 50 = 2000$ |
| $100 \times 70 = 7000$ | $200 \times 80 = 16,000$ | $400 \times 50 = 20,000$ |
| $1000 \times 70 = 70,000$ | $2000 \times 80 = 160,000$ | $4000 \times 50 = 200,000$ |

▶ To multiply a whole number and 100, 1000, or their multiples:
- Multiply the nonzero digits.
- Count the number of zeros in the factors.
  Then write the same number of zeros in the product.

$600 \times 3 = 1800$    **2 zeros**

$$\begin{array}{r} 3 \\ \times\, 600 \\ \hline 1800 \end{array}\text{2 zeros}$$

$8000 \times 40 = 320,000$    **4 zeros**

$$\begin{array}{r} 40 \\ \times\, 8000 \\ \hline 320,000 \end{array}\text{4 zeros}$$

**Find the products.**

**1.** $10 \times 6$
$100 \times 6$
$1000 \times 6$

**2.** $10 \times 8$
$100 \times 8$
$1000 \times 8$

**3.** $20 \times 3$
$200 \times 3$
$2000 \times 3$

**4.** $60 \times 5$
$600 \times 5$
$6000 \times 5$

**5.** $10 \times 4$
$100 \times 4$
$1000 \times 4$

**6.** $10 \times 9$
$100 \times 9$
$1000 \times 9$

**7.** $30 \times 7$
$300 \times 7$
$3000 \times 7$

**8.** $50 \times 8$
$500 \times 8$
$5000 \times 8$

**9.** $10 \times 40$
$100 \times 40$
$1000 \times 40$

**10.** $30 \times 70$
$300 \times 70$
$3000 \times 70$

**11.** $20 \times 50$
$200 \times 50$
$2000 \times 50$

**12.** $90 \times 40$
$900 \times 40$
$9000 \times 40$

**Multiply.**

| 13. | 14. | 15. | 16. | 17. |
|---|---|---|---|---|
| 7<br>× 400 | 9<br>× 300 | 8<br>× 4000 | 6<br>× 7000 | 3<br>× 8000 |

| 18. | 19. | 20. | 21. | 22. |
|---|---|---|---|---|
| 10<br>× 900 | 30<br>× 600 | 20<br>× 5000 | 80<br>× 3000 | 90<br>× 2000 |

**23.** 8 × 600    **24.** 6 × 400    **25.** 5 × 3000    **26.** 9 × 6000

**27.** 700 × 6    **28.** 200 × 9    **29.** 6000 × 8    **30.** 7000 × 5

**31.** 4 × 300    **32.** 6 × 500    **33.** 8 × 30,000    **34.** 6 × 60,000

**35.** 20 × 3000    **36.** 30 × 2000    **37.** 10 × 40,000    **38.** 20 × 20,000

## PROBLEM SOLVING

Use the pictograph for problems 39–45.

How many books of each type
were sold?

**39.** romance        **40.** biography

**41.** mystery        **42.** classics

**43.** How many books in all
were sold?

**Books Sold at a Bookstore**

| Romance | 🔲 🔲 🔲 🔲 🔲 🔲 |
|---|---|
| Biography | 🔲 🔲 🔲 🔲 🔲 |
| Mystery | 🔲 🔲 🔲 🔲 🔲 🔲 |
| Classics | 🔲 🔲 🔲 |
| Key: | 🔲 = 100 books |

**44.** How many more romance books
were sold than biography books?

**45.** How many books were sold
that were *not* classics?

**46.** There are 50 parcels of flyers.
Each parcel contains 100 flyers.
How many flyers are there in all?

**47.** There are 60 reams of paper.
Each ream contains 500 sheets.
How many sheets are there in all?

## Challenge

**Find each product.**

**48.** 10 × 20 × 30    **49.** 20 × 40 × 50    **50.** 20 × 30 × 40

**51.** 80 × 10 × 700    **52.** 60 × 50 × 200    **53.** 30 × 50 × 100

**54.** 40 × 50 × 8000    **55.** 20 × 30 × 6000    **56.** 20 × 40 × 9000

# Estimating Products

About how many pounds will 487 boxes of tools weigh if a box of tools weighs 213 pounds?

To find about how many pounds, estimate: 487 × 213

▶ To estimate the product of two numbers:
- Round each factor to its greatest place.
- Multiply.

$$
\begin{array}{r}
213 \longrightarrow 200 \\
\times\,487 \longrightarrow \times\,500 \\
\hline
\text{about} \quad 100{,}000
\end{array}
$$

$$
487 \times 213
$$
$$
500 \times 200 = \text{about } 100{,}000
$$

The boxes of tools weigh about 100,000 pounds.

## Study these examples.

$$
\begin{array}{r}
657 \longrightarrow 700 \\
\times\;91 \longrightarrow \times\;90 \\
\hline
\text{about} \quad 63{,}000
\end{array}
$$

$$
\begin{array}{r}
\$48.36 \longrightarrow \$50.00 \\
\times\quad 674 \longrightarrow \times\quad 700 \\
\hline
\text{about} \quad \$35{,}000.00
\end{array}
$$

Write $ and . in the product.

## Estimate each product.

| | | | | |
|---|---|---|---|---|
| **1.** 72 ×16 | **2.** 87 ×11 | **3.** 61 ×27 | **4.** 56 ×19 | **5.** 29 ×38 |
| **6.** 383 ×162 | **7.** 627 ×215 | **8.** 783 ×457 | **9.** 919 ×189 | **10.** 502 ×305 |
| **11.** 114 × 25 | **12.** 162 × 33 | **13.** 139 × 21 | **14.** 124 × 15 | **15.** 219 × 38 |
| **16.** $8.75 × 7 | **17.** $7.61 × 47 | **18.** $2.17 × 23 | **19.** $29.93 × 174 | **20.** $36.45 × 238 |
| **21.** $7.17 × 23 | **22.** $9.61 × 57 | **23.** $59.37 × 245 | **24.** $78.12 × 343 | **25.** $98.23 × 478 |

**Choose the best estimate.**

**26.** 2463 × 79     **a.** 100,000     **b.** 21,000     **c.** 160,000     **d.** 31,000

**27.** 78 × $24.32     **a.** $1600     **b.** $1400     **c.** $2400     **d.** $2100

---

### Estimation by Clustering

When a number of addends "cluster" around a certain number, an estimate for the sum may be obtained by multiplying that number by the number of addends.

*Think: Addends "cluster" around 700.*

Estimate: 692 + 703 + 711 + 691 + 708

        700 + 700 + 700 + 700 + 700

          → 5 × 700 = 3500 ← **estimated sum**

Estimate: $18.92 + $21.37 + $23.46 + $19.31

      $20 + $20 + $20 + $20

        → 4 × $20 = $80 ← **estimated sum**

---

**Estimate the sum.** Use clustering.

**28.** 23 + 19 + 24 + 17     **29.** 102 + 96 + 98 + 103     **30.** 823 + 790 + 799

**31.** $10.12 + $9.99 + $10.45       **32.** $32.54 + $29.43 + $30.21

## Choose a Computation Method

**Solve and explain the method you used. Write whether you estimated or found an exact answer.**

**33.** One carton of apples weighs 32 pounds. How many pounds will 200 cartons of apples weigh?

**34.** One box of oranges weighs 48 pounds. Will 550 boxes of oranges weigh less than 25,000 pounds?

**35.** Ms. Chan bought 18 baskets of fruit at $10.85 a basket. Did she spend more than $200?

**36.** A pound of potatoes costs $1.19. About how much will 54 pounds of potatoes cost?

## 2-6 Zeros in the Multiplicand

Each of three classes uses 2708 mL of distilled water in a science experiment. How much distilled water is used altogether by the three classes?

First, estimate: 3 × 2708
             ↓      ↓
          3 × 3000 = 9000

To find how much distilled water is used, multiply: 3 × 2708 = __?__

**Multiply the ones.
Regroup.**

```
        2
  2 7 0 8
×       3
        4
```

| 3 × 8 ones = 24 ones |
| = 2 tens 4 ones |

**Multiply the tens.
Then add the regrouped tens.**

```
        2
  2 7 0 8
×       3
      2 4
```

| 3 × 0 tens = 0 tens |
| 0 tens + 2 tens = 2 tens |

**Multiply the hundreds.
Regroup.**

```
    2   2
  2 7 0 8
×       3
    1 2 4
```

| 3 × 7 hundreds |
| = 21 hundreds |
| = 2 thousands 1 hundred |

**Multiply the thousands.
Then add the regrouped thousands.**

```
    2   2
  2 7 0 8  ←——— multiplicand
×       3  ←——— multiplier
  8 1 2 4  ←——— product
```

| 3 × 2 thousands = 6 thousands |
| 6 thousands + 2 thousands |
| = 8 thousands |

The three classes use 8124 mL of distilled water.

8124 is close to the estimate of 9000.

### Study these examples.

```
      7
  6 0 8 0
×       9
5 4,7 2 0
```

Use the distributive property:
6 × 90,500 = 6 × (90,000 + 500) = (6 × 90,000) + (6 × 500)
             = 540,000 + 3000 = 543,000

**Estimate. Then multiply.**

| | | | | |
|---|---|---|---|---|
| **1.** 1109 <br> × 3 | **2.** 6043 <br> × 4 | **3.** 5180 <br> × 7 | **4.** 9205 <br> × 5 | **5.** 6089 <br> × 8 |
| **6.** 4009 <br> × 5 | **7.** 8400 <br> × 8 | **8.** 3090 <br> × 6 | **9.** 7008 <br> × 9 | **10.** 9060 <br> × 4 |
| **11.** 23,016 <br> × 5 | **12.** 68,509 <br> × 8 | **13.** 40,243 <br> × 7 | **14.** 52,050 <br> × 4 | **15.** 80,403 <br> × 6 |
| **16.** 83,600 <br> × 3 | **17.** 90,053 <br> × 5 | **18.** 40,070 <br> × 8 | **19.** 80,003 <br> × 7 | **20.** 89,000 <br> × 9 |

**Find the product.** You may use the distributive property.

**21.** 6 × 9081   **22.** 9 × 3014   **23.** 7 × 4209   **24.** 5 × 4870

**25.** 4 × 20,859   **26.** 8 × 68,806   **27.** 5 × 70,042   **28.** 3 × 68,006

**29.** 8 × 25,070   **30.** 9 × 90,506   **31.** 6 × 76,080   **32.** 7 × 58,004

**33.** 9 × 91,006   **34.** 4 × 78,500   **35.** 5 × 90,003   **36.** 8 × 79,000

**37.** 3 × 70,008   **38.** 7 × 90,098   **39.** 4 × 170,009   **40.** 6 × 703,007

## PROBLEM SOLVING

**41.** A train travels an average of 9075 miles per day. How many miles does it travel in 6 days?

**42.** A factory can make 6500 boxes in an hour. How many boxes can it make in 5 hours?

**43.** How many days are there in 3600 weeks?

**44.** How many feet are there in 8003 yards?

## Connections: Science

**45.** Due to Earth's rotation, a point on the equator travels about 1700 km every hour. How far does a point on the equator travel in 9 hours?

**46.** If a satellite travels 6500 km in one hour, how far does it travel in 7 hours?

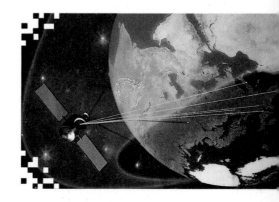

## 2-7 Multiplying Two Digits

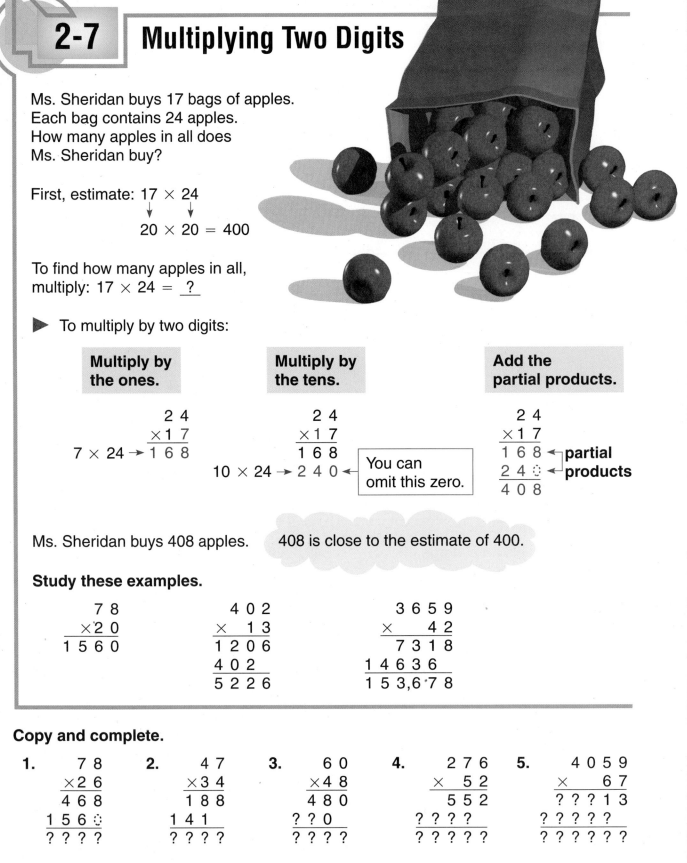

Ms. Sheridan buys 17 bags of apples.
Each bag contains 24 apples.
How many apples in all does
Ms. Sheridan buy?

First, estimate: 17 × 24
             ↓      ↓
         20 × 20 = 400

To find how many apples in all,
multiply: 17 × 24 = ?

▶ To multiply by two digits:

| Multiply by the ones. | Multiply by the tens. | Add the partial products. |
|---|---|---|

Multiply by the ones.

```
              2 4
            × 1 7
7 × 24 →    1 6 8
```

Multiply by the tens.

```
             2 4
           × 1 7
             1 6 8
10 × 24 →  2 4 0
```
You can omit this zero.

Add the partial products.

```
     2 4
   × 1 7
     1 6 8  ← partial
     2 4 0  ← products
     4 0 8
```

Ms. Sheridan buys 408 apples.     408 is close to the estimate of 400.

### Study these examples.

```
      7 8            4 0 2            3 6 5 9
    × 2 0          ×   1 3          ×     4 2
    1 5 6 0          1 2 0 6          7 3 1 8
                     4 0 2          1 4 6 3 6
                     5 2 2 6        1 5 3,6 7 8
```

### Copy and complete.

1.
```
      7 8
    × 2 6
      4 6 8
    1 5 6 ⊙
    ? ? ? ?
```

2.
```
      4 7
    × 3 4
      1 8 8
    1 4 1
    ? ? ? ?
```

3.
```
      6 0
    × 4 8
      4 8 0
    ? ? 0
    ? ? ? ?
```

4.
```
      2 7 6
    ×   5 2
      5 5 2
    ? ? ? ?
    ? ? ? ? ?
```

5.
```
      4 0 5 9
    ×     6 7
    ? ? ? 1 3
    ? ? ? ? ?
    ? ? ? ? ? ?
```

78

**Estimate. Then multiply.**

6.  62
    × 18

7.  54
    × 26

8.  46
    × 37

9.  70
    × 52

10. 83
    × 64

11. 413
    ×  48

12. 572
    ×  63

13. 620
    ×  44

14. 206
    ×  37

15. 639
    ×  58

16. 2741
    ×   35

17. 1052
    ×   29

18. 8506
    ×   74

19. 7009
    ×   86

20. 6927
    ×   67

**Find the product.**

21. 27 × 429

22. 30 × 625

23. 47 × 804

24. 92 × 520

25. 50 × 3693

26. 74 × 6240

27. 23 × 4127

28. 48 × 3219

29. 90 × 4120

30. 83 × 7059

31. 76 × 9008

32. 39 × 7853

**PROBLEM SOLVING** Use the bar graph.

33. How many cherries are there in 32 cartons?

34. How many plums are there in 48 cartons?

35. How many kiwis are there in 56 cartons?

36. How many strawberries are there in 67 cartons?

37. Which contain more fruit: 40 cartons of strawberries or 50 cartons of cherries?

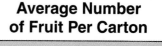

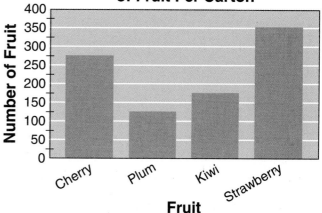

## Skills to Remember

**Align and add.**

38. 1425 + 5700 + 28,500

39. 2428 + 6070 + 121,400

40. 2912 + 8320 + 124,800

41. 2125 + 29,750 + 127,500

42. 2616 + 6540 + 130,800

43. 8532 + 56,880 + 663,600

# Multiplying Three Digits

Norma's father has a vegetable farm of 126 rows of tomato plants. Each row has 178 plants. How many tomato plants are on the farm?

First, estimate: $126 \times 178$

$$100 \times 200 = 20,000$$

To find how many tomato plants are on the farm, multiply: $126 \times 178 = \underline{\ ?\ }$

▶ To multiply by three digits:

| Multiply by the ones. | Multiply by the tens. | Multiply by the hundreds. Add the partial products. |
|---|---|---|
| 1 7 8 <br> × 1 2 6 <br> 1 0 6 8 ← 6 × 178 | 1 7 8 <br> × 1 2 6 <br> 1 0 6 8 <br> 3 5 6 ○ ← 20 × 178 | 1 7 8 <br> × 1 2 6 <br> 1 0 6 8 <br> 3 5 6 ○ <br> 1 7 8 ○ ○ ← 100 × 178 <br> 2 2,4 2 8 |

There are 22,428 tomato plants on the farm.

22,428 is close to the estimate of 20,000.

▶ You can use a calculator to multiply large numbers.

Multiply:
$528 \times 6350 = \underline{\ ?\ }$

Enter: 528 ×ₓ 6350 =

Display: | 3352800. |

$528 \times 6350 = 3,352,800$

## Copy and complete.

1.
```
    4 2 7
  × 3 2 4
  1 7 0 8
  8 5 4 ○
? ? ? ? ○ ○
? ? ? ? ? 8
```

2.
```
    6 0 7
  × 2 1 4
  2 4 2 8
  6 0 7
? ? ? 4
? ? ? ? ? ?
```

3.
```
    3 7 0
  × 8 6 3
  1 1 1 0
  2 2 2 0
? ? ? ?
? ? ? ? ? ?
```

4.
```
    5 1 9 2
  ×   2 7 4
  2 0 7 6 8
  3 6 3 4 4
? ? ? ? ?
? ? ? ? ? ? ?
```

**Estimate. Then multiply.**

| 5. | 541<br>× 122 | 6. | 345<br>× 211 | 7. | 217<br>× 115 | 8. | 431<br>× 134 | 9. | 501<br>× 272 |
|---|---|---|---|---|---|---|---|---|---|
| 10. | 244<br>× 152 | 11. | 420<br>× 135 | 12. | 305<br>× 271 | 13. | 360<br>× 417 | 14. | 742<br>× 343 |
| 15. | 328<br>× 274 | 16. | 523<br>× 249 | 17. | 362<br>× 275 | 18. | 853<br>× 418 | 19. | 672<br>× 415 |

**Find the product.** You may use a calculator.

20. $354 \times 120$

21. $417 \times 131$

22. $252 \times 204$

23. $475 \times 218$

24. $624 \times 382$

25. $728 \times 618$

26. $236 \times 1143$

27. $962 \times 4085$

28. $819 \times 2709$

29. $567 \times 6009$

30. $415 \times 5186$

31. $725 \times 6390$

## PROBLEM SOLVING

32. There are 245 rows of corn plants. Each row has 125 plants. How many corn plants are there in all?

33. There are 135 baskets of potatoes. Each basket holds 115 potatoes. How many potatoes are there in all?

34. Dennis picks an average of 465 baskets of apples during the season. If each basket holds 378 apples, how many apples does Dennis pick during the season?

35. A supermarket receives 625 cases of oranges. Each case holds 135 oranges. How many oranges in all does the supermarket receive?

36. In your Math Journal explain why:
    - there are 3 partial products in exercises 1–4;
    - the zeros are written in the partial products in exercise 3.

Math Journal

**Project**

37. Look at newspapers, magazines, books, or store signs to find instances in daily life that would involve multiplication of whole numbers. Make a poster illustrating these situations.

## 2-9 Zeros in the Multiplier

A theater sold out all 405 seats
for each play performance.
If there were 698 performances,
how many seats were sold?

First, estimate: 405 × 698

400 × 700 = 280,000

To find how many seats were sold,
multiply: 405 × 698 = __?__

### Long Way

```
      6 9 8
    × 4 0 5
    3 4 9 0  ←——— 5 × 698
    0 0 0 0  ←——— 0 × 698
  2 7 9 2 0 0 ←——— 400 × 698
  2 8 2,6 9 0
```

The theater sold 282,690 seats.

### Short Way

```
      6 9 8
    × 4 0 5
    3 4 9 0  ←——— 5 × 698
  2 7 9 2 0 0 ←——— 400 × 698
  2 8 2,6 9 0
```

There are 0 tens in 405,
so omit the second
partial product.

282,690 is close to the estimate of 280,000.

### Study these examples.

```
    3 0 0 2
  ×     7 0 0
  2,1 0 1,4 0 0
```

700 has 0 ones
and 0 tens, so
omit the zeros.

Remember to write
this digit directly
under the multiplier place.

```
      3 2 5 6
    ×   3 5 0
    1 6 2 8 0 0  ←——— 50 × 3256
    9 7 6 8 0 0  ←——— 300 × 3256
  1,1 3 9,6 0 0
```

350 has 0 ones,
so omit the zeros.

**Copy and complete.** Use the short way.

**1.**
```
      7 1 4
    × 6 0 0
  ? ? ? ? 0 0
```

**2.**
```
      4 0 2
    × 3 0 7
    2 8 1 4
  ? ? ? 6 0 0
  ? ? ? ? ? ?
```

**3.**
```
      9 5 6
    × 5 8 0
    7 6 4 8 0
  ? ? 8 0
  ? ? ? ? ? ?
```

**4.**
```
      3 5 8 0
    ×   7 0 6
    2 1 4 8 0
  ? ? ? 6 0
  ? ? ? ? ? ?
```

**Estimate. Then multiply.**

| 5. | 219<br>× 304 | 6. | 391<br>× 104 | 7. | 604<br>× 206 | 8. | 508<br>× 709 | 9. | 760<br>× 306 |
| --- | --- | --- | --- | --- | --- | --- | --- | --- | --- |
| 10. | 360<br>× 703 | 11. | 362<br>× 202 | 12. | 937<br>× 209 | 13. | 846<br>× 407 | 14. | 928<br>× 607 |
| 15. | 457<br>× 320 | 16. | 936<br>× 430 | 17. | 869<br>× 650 | 18. | 947<br>× 730 | 19. | 898<br>× 860 |

**Find the product.** You may use a calculator.

20. 600 × 739    21. 900 × 846    22. 700 × 4004    23. 500 × 8009

24. 720 × 365    25. 740 × 438    26. 860 × 549    27. 930 × 714

28. 507 × 367    29. 604 × 863    30. 708 × 905    31. 403 × 870

32. 230 × 1258    33. 470 × 2479    34. 605 × 4059    35. 209 × 7086

36. 601 × 3583    37. 807 × 7859    38. 920 × 7003    39. 640 × 8705

## PROBLEM SOLVING

40. The art guild had its exhibit for 105 days. It sold 436 tickets for each day. How many tickets did it sell for its exhibit?

41. The average family uses 370 gallons of water a day. How many gallons of water does the average family use in 120 days?

42. A bar of iron weighs 500 pounds. How many pounds will 738 bars of iron weigh?

43. A machine produces 420 chips in one minute. How many chips does it produce in 150 minutes?

## Finding Together

*Discuss* ✓

**Write the multiplication sign in the right place to get the given product.** You may use a calculator.

44. 1 2 3 4 5 6 = 56,088

45. 3 3 3 3 3 3 = 109,989

46. 1 3 5 7 9 0 = 122,130

47. 1 0 2 4 6 8 = 81,968

48. 2 2 4 4 6 6 = 98,252

49. 9 8 7 6 5 4 = 533,304

## 2-10 Multiplication with Money

Marion bought 8 boxes of greeting cards at $6.95 a box. How much did Marion pay for the greeting cards?

First, estimate: 8 × $6.95

8 × $7.00 = $56.00

To find how much Marion paid, multiply: 8 × $6.95 = __?__

▶ To multiply an amount of money:
- Multiply as usual.
- Write a decimal point in the product two places from the right.
- Write the dollar sign in the product.

$$\begin{array}{r} \$6.9\ 5 \\ \times \qquad 8 \\ \hline \$5\ 5.6\ 0 \end{array}$$

2 places from the right

Write the dollar sign.

Marion paid $55.60.    $55.60 is close to the estimate of $56.00.

### Study these examples.

$$\begin{array}{r} \$0.8\ 9 \\ \times \quad 4\ 6 \\ \hline 5\ 3\ 4 \\ 3\ 5\ 6 \\ \hline \$4\ 0.9\ 4 \end{array}$$

$$\begin{array}{r} \$2.1\ 5 \\ \times \quad 9\ 7 \\ \hline 1\ 5\ 0\ 5 \\ 1\ 9\ 3\ 5 \\ \hline \$2\ 0\ 8.5\ 5 \end{array}$$

$$\begin{array}{r} \$2.0\ 4 \\ \times \quad 3\ 2\ 0 \\ \hline 4\ 0\ 8\ 0 \\ 6\ 1\ 2\ 0 \\ \hline \$6\ 5\ 2.8\ 0 \end{array}$$

Write the dollar sign and the decimal point.

Use a calculator to find the product:
375 × $65.86 = __?__

Enter:    375 ☒ 65.86 ▣

There is no $ key on the calculator.

Display:    24697.5

The display does *not* show final zeros to the right of the decimal point.

375 × $65.86 = $24,697.50

**Copy and complete.** Write the dollar sign and decimal point.

1.  $0.8 3
    ×      9
    —————
    ? ? 7

2.  $3.5 4
    ×      9
    —————
    ? ? 8 6

3.  $0.6 5
    ×      4
    —————
    ? ? ?

4.  $7.3 8
    ×      6
    —————
    ? ? ? ?

5.  $8.6 9
    ×      8
    —————
    ? ? ? ?

**Estimate. Then find the product.**

6.  $9.03
    ×     5

7.  $7.80
    ×     6

8.  $0.57
    ×    38

9.  $2.90
    ×    70

10. $9.80
    ×    55

11. $4.50
    ×   605

12. $2.18
    ×   340

13. $9.06
    ×   214

14. $7.24
    ×   416

15. $6.18
    ×   524

**Multiply.** You may use a calculator.

16. 540 × $4.09

17. 215 × $6.07

18. 432 × $7.80

19. 279 × $84.27

20. 514 × $34.65

21. 483 × $65.19

**PROBLEM SOLVING** You may use a calculator.

22. Pancho earns $6.75 an hour as a laboratory assistant. How much does he earn in 32 hours?

23. Mr. Montes buys 29 copies of books for his class. Each book costs $9.75. How much do the books cost?

24. A class of 38 students goes on a field trip. Each student pays $8.65 for the trip. How much does the class pay for the trip?

25. Tina bought 15 pounds of cherries at $0.84 per pound. Roy bought 14 pounds of cherries at $0.90 per pound. Who paid more?

 **Calculator Activity**

**Write a multiplication example for each.**

26. Multiply a 2-digit number by a 2-digit number so that the product is a:   **a.** 3-digit number
    **b.** 4-digit number   [1] [2] [×] [3] [0] [=] [ *360.* ]

27. Multiply a 3-digit number by a 3-digit number so that the product is a:   **a.** 5-digit number
    **b.** 6-digit number

## IF-THEN Statements

Here is a list of important commands used in BASIC.

**READ**-takes entries one by one from a DATA statement and assigns each to a variable.

**GOTO**-branches the flow of the program to a specified line number.

**FOR . . . NEXT**-tells the computer to repeat lines between the FOR and NEXT statements a given number of times.

**DATA**-stores a list of items used by READ statements.

**INPUT**-waits for a response from the user.

**LET**-assigns a value to a variable.

**IF-THEN**-makes a decision that affects the flow of the program based on the result of an expression.

An IF-THEN statement is similar to the diamond symbol in a flowchart.

**Study this flowchart and its correlating program.**

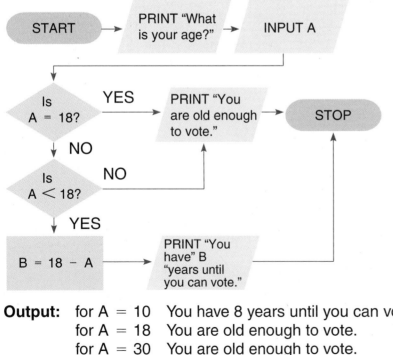

```
10  PRINT "What is your age?"
20  INPUT A
30  IF A = 18 THEN GOTO 50
40  IF A < 18 THEN GOTO 70
50  PRINT "You are old enough
            to vote."
60  GOTO 90
70  LET B = 18 − A
80  PRINT "You have" B "years
            until you can vote."
90  END
```

**Output:**  for A = 10  You have 8 years until you can vote.
  for A = 18  You are old enough to vote.
  for A = 30  You are old enough to vote.

## PROBLEM SOLVING

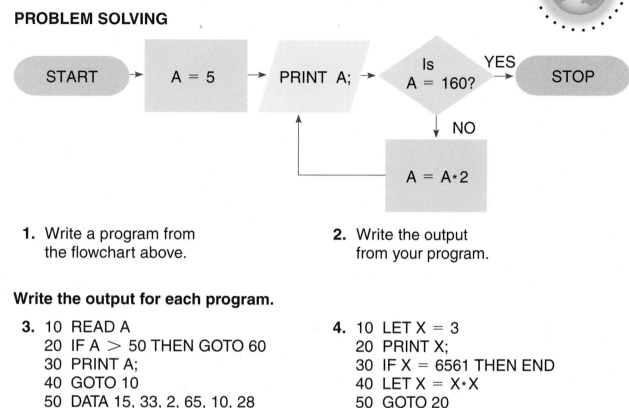

1. Write a program from the flowchart above.

2. Write the output from your program.

### Write the output for each program.

3. 10  READ A
   20  IF A > 50 THEN GOTO 60
   30  PRINT A;
   40  GOTO 10
   50  DATA 15, 33, 2, 65, 10, 28
   60  END

4. 10  LET X = 3
   20  PRINT X;
   30  IF X = 6561 THEN END
   40  LET X = X*X
   50  GOTO 20

5. 10  FOR X = 1 to 5
   20  READ A, B
   30  IF A*B = 495 THEN PRINT A "and" B "are factors of 495."
   40  NEXT X
   50  DATA 12, 33, 15, 33, 15, 15, 99, 5, 22, 15, 33, 15
   60  END

6. 10  LET A = 50
   20  READ B
   30  IF B = 0 THEN GOTO 60
   40  PRINT A "divided by" B "=" A/B
   50  GOTO 20
   60  PRINT "Cannot divide" A "by" B
   70  GOTO 20
   80  DATA 10, 2, 25, 0, 5, 50

7. 10  LET A = 4
   20  LET B = 5
   30  LET C = A + 5*B
   40  PRINT C;
   50  IF C > 100 THEN END
   60  LET A = B
   70  LET B = B + 1
   80  GOTO 30

8. Write a program that will print every other even number from 2 through 200.

9. Write a program that will print the multiples of 3 from 3 to 99.

# 2-12 Problem Solving: Hidden Information

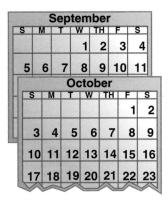

**Problem:** Kenny swims 23 laps every day. How many laps did he swim in the months of September and October?

**1 IMAGINE** You are recording Kenny's laps on a calendar.

**2 NAME** *Facts:* Swims 23 laps every day
Swims months of September and October

*Question:* How many laps did he swim in September and October?

**3 THINK** Is all the information you need listed in the problem? No.
Is there hidden information in the problem? Yes.

> There are 30 days in September and 31 days in October.

First add to find the total number of days in the two months: $30 + 31 = 61$ days

Then multiply to find the number of laps Kenny swam: $61 \times 23 = \underline{\ ?\ }$

**4 COMPUTE** First estimate. $60 \times 20 = 1200$
about 1200 laps

Then multiply.
$$
\begin{array}{r}
23 \\
\times\,61 \\
\hline
23 \\
138\phantom{0} \\
\hline
1403
\end{array}
$$

The product 1403 is close to the estimate of 1200.

Kenny swam 1403 laps.

**5 CHECK** Did you answer the question asked? Yes.

Check your computation by changing the order of the factors.
$$
\begin{array}{r}
61 \\
\times\,23 \\
\hline
183 \\
122\phantom{0} \\
\hline
1403
\end{array}
$$
The answer checks.

**Find the hidden information to solve each problem.**

1. Jan earns $12.50 every week baby-sitting.
   How much money will Jan make in a year?

**IMAGINE**    Put yourself in the problem.

**NAME**

*Facts:*     $12.50 each week
baby-sitting

*Question:*     How much will Jan
earn in a year?

**THINK**    Is there information not stated in
the problem? Yes.

There are 52 weeks in a year.

To find the amount of money Jan
will earn, multiply: 52 × $12.50 = __?__

**COMPUTE** ⟶ **CHECK**

2. A farmer takes 30 minutes to harvest 1 row of tomatoes. At this
   rate, how many hours will it take the farmer to harvest 18 rows?

3. Mr. Hudson uses three cups of flour in every loaf of bread. He bakes
   67 loaves of bread a day. How many loaves of bread does he bake
   in a week? How many cups of flour does he use in a week?

4. There are 73 shelves of books of fiction in the library.
   About $1\frac{1}{2}$ dozen books fit on each shelf. About how
   many books of fiction can fit on the shelves?

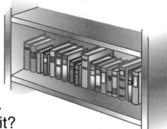

5. Hector can fit two dozen coins on a page of his coin album.
   If his album has 125 pages, how many coins can he put in it?

6. Frank types 125 words per minute. If it took him $1\frac{1}{4}$ hours
   to type a report, about how many words are in the report?

**Make Up Your Own**

*Communicate* ✓

7. Write and solve a problem that has hidden
   information. Have someone solve it.

## 2-13 | Problem-Solving Applications

**Solve each problem and explain the method you use.**

1. KidCo's first product is beaded bracelets. Each bracelet uses 9 in. of bead wire. Will 1500 in. of wire be enough for 150 bracelets?

2. Each bracelet uses 30 beads. How many beads are needed to make this first batch?

3. The next KidCo product is matching necklaces. Each uses 120 beads. How many beads will be needed to produce 75 necklaces?

4. Each necklace uses 72 in. of wire. Will a 5000-in. roll of wire be enough to make 75 necklaces? If not, how much more wire will be needed?

5. The total cost of materials is $3 for each bracelet. KidCo plans to sell the bracelets for $5 each. How much profit will it make if it sells all 150 bracelets?

6. Each necklace costs $11.25 to make. How much will it cost to make 75 necklaces?

7. KidCo rented a booth at Town Hall Market. It sold 18 pairs of earrings at $4.50 each and 8 belts at $8.05 each. How much money did KidCo collect from the sales?

8. KidCo owners had flyers printed. Each word costs 12 cents to set. About how much did it cost to set this flyer?

9. Bulk mail costs 16¢ a piece. KidCo mailed 750 flyers. How much did the owners pay for this service?

10. On Saturday morning there were 205 people at the Town Hall Market. There were double that number in the afternoon. How many people came to the Town Hall Market on Saturday?

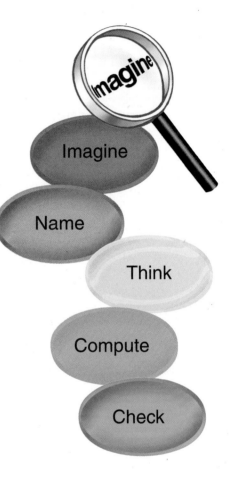

Announcing:

**KidCo Products**

Town Hall Market

*Every Saturday*

• Beaded Belts •
Bracelets • Earrings
• Baked Goods •
and More!

**Choose a strategy from the list or use another strategy you know to solve each problem.**

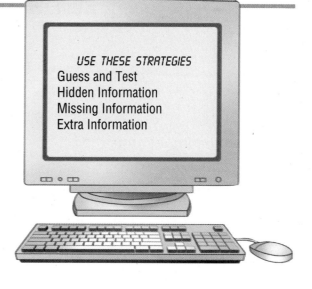

USE THESE STRATEGIES
Guess and Test
Hidden Information
Missing Information
Extra Information

11. KidCo owners disagreed on how much to charge for necklaces. Some wanted to charge $15, and others wanted to charge $16.50. How much more will they collect on 75 necklaces if they charge the higher price?

12. Shawn took in $69.15 for 2 hours work on Saturday, selling belts for $8.05 and earrings for $4.50. How many of each did Shawn sell?

13. KidCo belts are made of braided cords. Each belt uses 96 in. of cording. Will a 120-ft roll of cording be enough to make a dozen belts?

14. KidCo sold 20 belts last week. Each belt sells for $8.05. How much did KidCo make?

15. Renting a booth at the market costs $15.75 per Saturday or $53 for four Saturdays. If a booth is rented every Saturday from May 2 to June 20 at the lower rate, what will the savings be?

16. Each person works about 75 minutes a day on KidCo projects. About how long does each person work in a week? in a month?

**New KidCo Products**

| Product | Teddy Bears | Corn Muffins |
|---|---|---|
| Cost of Materials | $7 | $0.89 per doz |
| Product Price | $11 | $1.80 per doz |
| Time Required | 85 min each | 200 min for 25 doz |
| Number Made | 80 | 25 doz |

**Use the chart for problems 17–18.**

17. How much does it cost to make all the teddy bears?

18. All the corn muffins were sold. How much was earned?

**Share Your Thinking**

Math Journal

19. Write in your Math Journal which problem you solved using two strategies and explain why. Then write a problem modeled on this problem and have a classmate solve it.

# Chapter Review and Practice

**Write the corresponding multiplication fact.**  *(See pp. 66–67.)*

**1.** $7 + 7 + 7 + 7 + 7$

**2.** $6 + 6 + 6 + 6 + 6 + 6 + 6$

**Find the missing factor.**

**3.** $4 \times \underline{\ ?\ } = 20$

**4.** $\underline{\ ?\ } \times 7 = 42$

**5.** $9 \times \underline{\ ?\ } = 54$

**Name the property of multiplication used.**  *(See pp. 68–69.)*

**6.** $2 \times 6 = 6 \times 2$

**7.** $0 \times 8 = 0$

**8.** $4 \times (3 \times 2) = (4 \times 3) \times 2$

**9.** $1 \times 4 = 4$

**10.** $3 \times (2 + 4) = (3 \times 2) + (3 \times 4)$

**Find the product.**  *(See pp. 70–73, 76–85.)*

**11.** $8 \times 30$

**12.** $6 \times 20$

**13.** $40 \times 9$

**14.** $50 \times 7$

**15.** $15 \times 67$

**16.** $32 \times 83$

**17.** $215 \times 356$

**18.** $605 \times 4582$

**19.** $372 \times \$1.59$

**20.** $625 \times \$4.37$

**21.** $394 \times \$7.85$

**Estimate. Then multiply.**  *(See pp. 74–85.)*

**22.**  $\begin{array}{r} 86 \\ \times\, 24 \\ \hline \end{array}$

**23.**  $\begin{array}{r} 246 \\ \times\, 26 \\ \hline \end{array}$

**24.**  $\begin{array}{r} 607 \\ \times\, 47 \\ \hline \end{array}$

**25.**  $\begin{array}{r} 318 \\ \times\, 64 \\ \hline \end{array}$

**26.**  $\begin{array}{r} 215 \\ \times\, 31 \\ \hline \end{array}$

**27.**  $\begin{array}{r} 416 \\ \times\, 258 \\ \hline \end{array}$

**28.**  $\begin{array}{r} 346 \\ \times\, 517 \\ \hline \end{array}$

**29.**  $\begin{array}{r} 237 \\ \times\, 608 \\ \hline \end{array}$

**30.**  $\begin{array}{r} 6289 \\ \times\, 413 \\ \hline \end{array}$

**31.**  $\begin{array}{r} 7385 \\ \times\, 329 \\ \hline \end{array}$

**32.**  $\begin{array}{r} \$4.29 \\ \times\, 32 \\ \hline \end{array}$

**33.**  $\begin{array}{r} \$7.48 \\ \times\, 62 \\ \hline \end{array}$

**34.**  $\begin{array}{r} \$26.42 \\ \times\, 104 \\ \hline \end{array}$

**35.**  $\begin{array}{r} \$72.48 \\ \times\, 320 \\ \hline \end{array}$

**36.**  $\begin{array}{r} \$6.75 \\ \times\, 342 \\ \hline \end{array}$

## PROBLEM SOLVING  *(See pp. 84–85, 88–91.)*

**37.** The drama club sold 364 tickets. The tickets cost $2.75 each. How much money did the club make on the ticket sales?

**38.** Pencils are packed 12 dozen per box. How many pencils are there in 30 boxes?

(See *Still More Practice*, p. 478.)

## EXPONENTS

When a number is used as a factor several times, it can be written with an **exponent**. The exponent tells how many times the number, called the *base*, is used as a factor.

$$2 \times 2 \times 2 \times 2 \times 2 \times 2 = 2^6 \leftarrow \text{exponent}$$

2 used as a factor 6 times | base

The example shows that 2 is used as a factor six times and the product is 64.

$$2^6 = 2 \times 2 \times 2 \times 2 \times 2 \times 2 = 64 \leftarrow \text{product}$$

Read: "two to the sixth power"

### Study these examples.

$$7^1 = 7$$

Read: "seven to the first power"

$$7^2 = 7 \times 7 = 49$$

Read: "seven squared"

$$7^3 = 7 \times 7 \times 7 = 343$$

Read: "seven cubed"

### Write each product using an exponent.

**1.** $2 \times 2 \times 2$  **2.** $4 \times 4 \times 4 \times 4$  **3.** $8 \times 8$

**4.** $9 \times 9 \times 9 \times 9 \times 9$  **5.** $6 \times 6 \times 6$  **6.** $10 \times 10 \times 10 \times 10$

**7.** $1 \times 1 \times 1 \times 1 \times 1 \times 1 \times 1 \times 1$  **8.** $2 \times 2 \times 2 \times 2 \times 2 \times 2 \times 2 \times 2 \times 2$

**9.** $7 \times 7 \times 7 \times 7 \times 7 \times 7 \times 7$  **10.** $5 \times 5 \times 5 \times 5 \times 5 \times 5 \times 5 \times 5 \times 5 \times 5$

### Find the product.

**11.** $2^2$  **12.** $3^4$  **13.** $4^3$  **14.** $5^3$  **15.** $1^{10}$  **16.** $10^4$

**17.** $9^1$  **18.** $6^3$  **19.** $8^3$  **20.** $2^7$  **21.** $4^5$  **22.** $10^9$

**23.** $4^6$  **24.** $2^{10}$  **25.** $5^4$  **26.** $3^5$  **27.** $6^2$  **28.** $10^6$

# Check Your Mastery

## Performance Assessment

**Choose the computation method.**
Use the numbers in the box to write multiplication
sentences that you can solve using:

| 101 | 33 |
|-----|-----|
| 2853 | 50 |

**1.** mental math    **2.** calculator    **3.** paper and pencil

**Find the missing factor.**

**4.** $6 \times \underline{\ ?\ } = 42$

**5.** $64 = 8 \times \underline{\ ?\ }$

**6.** $15 = \underline{\ ?\ } \times 3$

**Name the property of multiplication used.**

**7.** $9 \times 1 = 9$

**8.** $3 \times 4 = 4 \times 3$

**9.** $(5 \times 7) \times 2 = 5 \times (7 \times 2)$

**10.** $12 \times 0 = 0$

**11.** $4 \times (5 + 3) = (4 \times 5) + (4 \times 3)$

**Find the product.**

**12.** $4 \times 4$
$4 \times 40$
$4 \times 400$

**13.** $9 \times 6$
$9 \times 60$
$9 \times 600$

**14.** $3 \times 7$
$3 \times 70$
$3 \times 700$

**15.** $8 \times 5$
$8 \times 50$
$8 \times 500$

**16.**
$\begin{array}{r} 164 \\ \times\ 56 \\ \hline \end{array}$

**17.**
$\begin{array}{r} 279 \\ \times\ 34 \\ \hline \end{array}$

**18.**
$\begin{array}{r} 312 \\ \times 284 \\ \hline \end{array}$

**19.**
$\begin{array}{r} 673 \\ \times 406 \\ \hline \end{array}$

**20.**
$\begin{array}{r} 5120 \\ \times\ 700 \\ \hline \end{array}$

**Estimate. Then multiply.**

**21.**
$\begin{array}{r} 4076 \\ \times\ \ \ 4 \\ \hline \end{array}$

**22.**
$\begin{array}{r} 428 \\ \times\ 47 \\ \hline \end{array}$

**23.**
$\begin{array}{r} 5085 \\ \times\ \ 68 \\ \hline \end{array}$

**24.**
$\begin{array}{r} 547 \\ \times 305 \\ \hline \end{array}$

**25.**
$\begin{array}{r} 7457 \\ \times\ 263 \\ \hline \end{array}$

**26.**
$\begin{array}{r} \$35.24 \\ \times\ \ \ \ 6 \\ \hline \end{array}$

**27.**
$\begin{array}{r} \$2.15 \\ \times\ \ 52 \\ \hline \end{array}$

**28.**
$\begin{array}{r} \$0.25 \\ \times\ \ 43 \\ \hline \end{array}$

**29.**
$\begin{array}{r} \$11.42 \\ \times\ \ 579 \\ \hline \end{array}$

**30.**
$\begin{array}{r} \$26.75 \\ \times\ \ 489 \\ \hline \end{array}$

**PROBLEM SOLVING**    *Use a strategy you have learned.*

**31.** Corn sells for $2.50 a bushel.
How much will Farmer Zeke
make if he sells 1250 bushels?

**32.** Ms. Daly bought 8 books for her
class. If each book cost $5.95,
how much did she spend?

# Division 3

## A Microscopic Topic

I am a paramecium
that cannot do a simple sum,
and it's a rather well-known fact
I'm quite unable to subtract.

If I'd an eye, I'd surely cry
about the way I multiply,
for though I've often tried and tried,
I do it backward...and divide.

*Jack Prelutsky*

### In this chapter you will:

Use the meanings of division and patterns
Explore divisibility rules and short division
Estimate using compatible numbers
  Learn about the order of operations
  Make a table and find a pattern to
    solve problems

### Critical Thinking/Finding Together

The first minute you look at a
  slide under a microscope
  you see 5 bacteria. The
  number of bacteria doubles
every minute. If you look at the
slide every minute, how many
bacteria will you see in the
tenth minute?

95

# Understanding Division

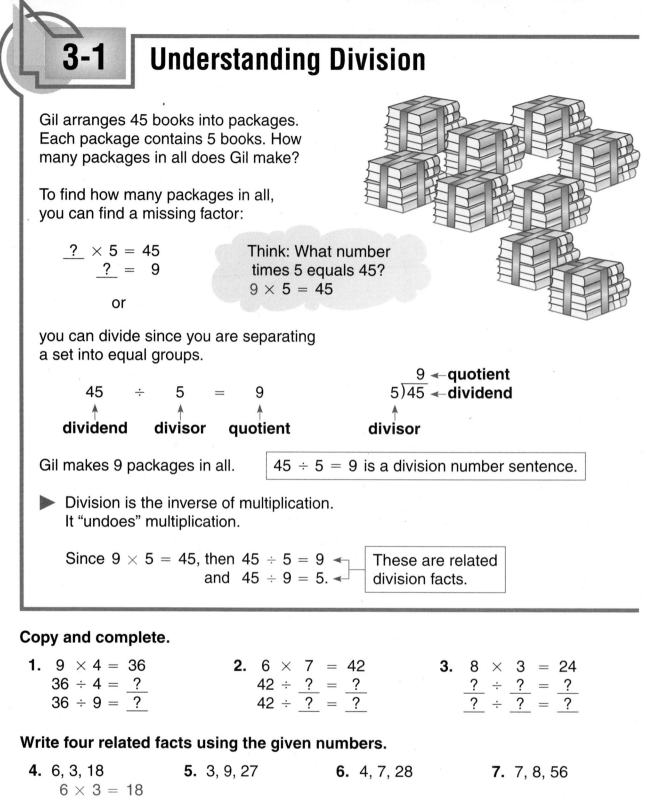

Gil arranges 45 books into packages. Each package contains 5 books. How many packages in all does Gil make?

To find how many packages in all, you can find a missing factor:

$$\underline{\phantom{?}?\phantom{?}} \times 5 = 45$$
$$\underline{\phantom{?}?\phantom{?}} = 9$$

Think: What number times 5 equals 45?
$9 \times 5 = 45$

or

you can divide since you are separating a set into equal groups.

$$45 \div 5 = 9$$

**dividend**   **divisor**   **quotient**

$$\begin{array}{r} 9 \leftarrow \textbf{quotient} \\ 5)\overline{45} \leftarrow \textbf{dividend} \end{array}$$

**divisor**

Gil makes 9 packages in all.    $45 \div 5 = 9$ is a division number sentence.

▶ Division is the inverse of multiplication. It "undoes" multiplication.

Since $9 \times 5 = 45$, then $45 \div 5 = 9$
and $45 \div 9 = 5$.

These are related division facts.

## Copy and complete.

**1.** $9 \times 4 = 36$
$36 \div 4 = \underline{\phantom{?}?\phantom{?}}$
$36 \div 9 = \underline{\phantom{?}?\phantom{?}}$

**2.** $6 \times 7 = 42$
$42 \div \underline{\phantom{?}?\phantom{?}} = \underline{\phantom{?}?\phantom{?}}$
$42 \div \underline{\phantom{?}?\phantom{?}} = \underline{\phantom{?}?\phantom{?}}$

**3.** $8 \times 3 = 24$
$\underline{\phantom{?}?\phantom{?}} \div \underline{\phantom{?}?\phantom{?}} = \underline{\phantom{?}?\phantom{?}}$
$\underline{\phantom{?}?\phantom{?}} \div \underline{\phantom{?}?\phantom{?}} = \underline{\phantom{?}?\phantom{?}}$

## Write four related facts using the given numbers.

**4.** 6, 3, 18
$6 \times 3 = 18$
$3 \times 6 = 18$
$18 \div 3 = 6$
$18 \div 6 = 3$

**5.** 3, 9, 27

**6.** 4, 7, 28

**7.** 7, 8, 56

**8.** 8, 9, 72

**9.** 5, 8, 40

**10.** 6, 8, 48

**11.** 2, 8, 16

**12.** 7, 6, 42

**13.** 5, 4, 20

**14.** In your Math Journal write what you notice about the four related facts in exercises 4–13. What does this tell you about multiplication and division?

## Rules of Division

Here are some rules of division that can help you divide quickly and correctly.

- When the divisor is *one,* the quotient is the same as the dividend.

$8 \div 1 = 8$

$\dfrac{8}{1\overline{)8}}$

- When the divisor and the dividend are the *same* number, the quotient is always one.

$7 \div 7 = 1$

$\dfrac{1}{7\overline{)7}}$

- When the dividend is *zero,* the quotient is zero.

$0 \div 3 = 0$

$\dfrac{0}{3\overline{)0}}$

- The divisor can *never* be zero.

**Divide.**

**15.** $43\overline{)43}$  **16.** $37\overline{)0}$  **17.** $1\overline{)97}$  **18.** $91\overline{)0}$  **19.** $58\overline{)58}$  **20.** $35\overline{)35}$

**21.** $39\overline{)0}$  **22.** $51\overline{)51}$  **23.** $85\overline{)85}$  **24.** $1\overline{)65}$  **25.** $98\overline{)0}$  **26.** $49\overline{)49}$

**27.** $561 \div 561$  **28.** $0 \div 483$  **29.** $612 \div 612$  **30.** $0 \div 165$

## PROBLEM SOLVING

**31.** The quotient is 1. The divisor is 60. What is the dividend?

**32.** The dividend is 49. The quotient is 7. What is the divisor?

**33.** The dividend is 40. The divisor is 8. What is the quotient?

**34.** The divisor is 16. The quotient is 0. What is the dividend?

**35.** One side of a record plays for 45 minutes. Each song is 5 minutes long. How many songs are on one side of a record?

**36.** Ada has a record case that holds 32 records. She divides the case into 4 equal sections. How many records does each section hold?

## 3-2 Division Patterns

Use division facts and patterns with zero
to divide with multiples of 10, 100, or 1000.

Study these division patterns:

| | |
|---|---|
| Fact: 8 ÷ 2 = 4 | Fact: 30 ÷ 6 = 5 |
| 80 ÷ 2 = 40 | 300 ÷ 6 = 50 |
| 800 ÷ 2 = 400 | 3000 ÷ 6 = 500 |
| 8000 ÷ 2 = 4000 | 30,000 ÷ 6 = 5000 |
| 80,000 ÷ 2 = 40,000 | 300,000 ÷ 6 = 50,000 |

**Remember:**
Look for a basic division fact when dividing with multiples of 10, 100, or 1000.

| | | |
|---|---|---|
| Fact: 24 ÷ 8 = 3 | Fact: 7 ÷ 1 = 7 | Fact: 18 ÷ 2 = 9 |
| 240 ÷ 80 = 3 | 70 ÷ 10 = 7 | 180 ÷ 2 = 90 |
| 2400 ÷ 80 = 30 | 700 ÷ 100 = 7 | 1800 ÷ 20 = 90 |
| 24,000 ÷ 80 = 300 | 7000 ÷ 1000 = 7 | 18,000 ÷ 200 = 90 |
| 240,000 ÷ 80 = 3000 | 70,000 ÷ 10,000 = 7 | 180,000 ÷ 2000 = 90 |

**Find the quotients.** Look for a pattern.

| | | |
|---|---|---|
| **1.**   9 ÷ 3 | **2.**   48 ÷ 6 | **3.**   30 ÷ 5 |
| 90 ÷ 3 | 480 ÷ 6 | 300 ÷ 50 |
| 900 ÷ 3 | 4800 ÷ 6 | 3000 ÷ 50 |
| 9000 ÷ 3 | 48,000 ÷ 6 | 30,000 ÷ 50 |
| 90,000 ÷ 3 | 480,000 ÷ 6 | 300,000 ÷ 50 |

| | | |
|---|---|---|
| **4.**   12 ÷ 4 | **5.**   45 ÷ 9 | **6.**   56 ÷ 7 |
| 120 ÷ 40 | 450 ÷ 90 | 560 ÷ 70 |
| 1200 ÷ 40 | 4500 ÷ 900 | 5600 ÷ 700 |
| 12,000 ÷ 40 | 45,000 ÷ 9000 | 56,000 ÷ 7000 |
| 120,000 ÷ 40 | 450,000 ÷ 90,000 | 560,000 ÷ 70,000 |

**Divide.**

**7.** 7⟌350       **8.** 9⟌720       **9.** 3⟌1800       **10.** 8⟌6400

**11.** 80⟌240       **12.** 60⟌420       **13.** 50⟌300       **14.** 30⟌120

**15.** 50⟌2000       **16.** 40⟌2800       **17.** 60⟌3600       **18.** 20⟌18,000

**19.** 40⟌16,000       **20.** 700⟌49,000       **21.** 800⟌480,000       **22.** 300⟌270,000

**Find the quotient.**

**23.** 720 ÷ 9          **24.** 60 ÷ 3          **25.** 800 ÷ 2          **26.** 2400 ÷ 6

**27.** 1200 ÷ 40      **28.** 3500 ÷ 70      **29.** 2800 ÷ 40      **30.** 6300 ÷ 90

**31.** 30,000 ÷ 50    **32.** 64,000 ÷ 80    **33.** 45,000 ÷ 90    **34.** 54,000 ÷ 60

**Compare. Write <, =, or >.**

**35.** 3600 ÷ 6 ? 4000 ÷ 8          **36.** 4200 ÷ 70 ? 4800 ÷ 80

**37.** 3000 ÷ 5 ? 2000 ÷ 4          **38.** 1800 ÷ 30 ? 4200 ÷ 60

**39.** 70,000 ÷ 7 ? 80,000 ÷ 2      **40.** 45,000 ÷ 90 ? 25,000 ÷ 5

**PROBLEM SOLVING** Use the bar graph.

The graph shows the different distances traveled by 4 families in 5 days. If each family traveled the same distance each day, how many miles did each family travel per day?

**41.** Ayala          **42.** Tan

**43.** Ford           **44.** Smith

**Distance Traveled in 5 Days**

**Skills to Remember**

**Write the digit in the tens place.**

**45.** 39          **46.** 247          **47.** 6531          **48.** 78,093          **49.** 189,704

**Write the digit in the hundreds place.**

**50.** 563          **51.** 849          **52.** 7442          **53.** 65,104          **54.** 282,312

**Write the place of the red digit.**

**55.** 9472          **56.** 8435          **57.** 67,892          **58.** 60,948          **59.** 349,925

**60.** 17,539          **61.** 417,058          **62.** 502,931          **63.** 896,127          **64.** 642,573

## 3-3  Three-Digit Quotients

Manuel has 866 baseball cards in all.
He divides them equally among his
7 friends. How many cards does each
friend get? How many cards are left over?

To find how many cards each friend gets,
divide: $866 \div 7 = $  ?

▶ Use the division steps.

• Decide where to begin
  the quotient.

$7\overline{)866}$

$7 < 8$
**Enough hundreds**

Divide the
hundreds first.

• Divide the hundreds.
  Estimate:   ? $\times 7 = 8$
       $1 \times 7 = 7$
       $2 \times 7 = 14$
       Try 1.

$$\begin{array}{r} 1\phantom{66} \\ 7\overline{)8\ 6\ 6} \\ -7\downarrow\phantom{6} \\ \hline 1\ 6 \end{array}$$

**Division Steps**
• Decide where to begin
  the quotient.
• Estimate.
• Divide.
• Multiply.
• Subtract and compare.
• Bring down.
• Repeat the steps
  as necessary.
• Check.

• Divide the tens.
  Estimate:   ? $\times 7 = 16$
       $2 \times 7 = 14$
       $3 \times 7 = 21$
       Try 2.

$$\begin{array}{r} 1\ 2\phantom{6} \\ 7\overline{)8\ 6\ 6} \\ -7\downarrow\phantom{6} \\ \hline 1\ 6\phantom{6} \\ -1\ 4\downarrow \\ \hline 2\ 6 \end{array}$$

• Divide the ones.
  Estimate:   ? $\times 7 = 26$
       $3 \times 7 = 21$
       $4 \times 7 = 28$
       Try 3.

$$\begin{array}{r} 1\ 2\ 3 \ \text{R5} \\ 7\overline{)8\ 6\ 6} \\ -7\downarrow\phantom{6} \\ \hline 1\ 6\phantom{6} \\ -1\ 4\downarrow \\ \hline 2\ 6 \\ -2\ 1 \\ \hline 5 \end{array}$$

Remember:
Write the remainder
in the quotient.

• Check: $123 \times 7 + 5 = 866$

Each friend gets 123 baseball cards.
There are 5 baseball cards left over.

**Divide and check.**

1. $3\overline{)372}$  2. $4\overline{)568}$  3. $2\overline{)295}$  4. $6\overline{)999}$

5. $4\overline{)872}$  6. $6\overline{)712}$  7. $7\overline{)917}$  8. $8\overline{)904}$

9. $3\overline{)2184}$  10. $5\overline{)4455}$  11. $2\overline{)1168}$  12. $5\overline{)4366}$

13. $7\overline{)2436}$  14. $6\overline{)4559}$  15. $8\overline{)7462}$  16. $9\overline{)1098}$

**Find the quotient and the remainder. Then check.**

17. $568 \div 3$  18. $907 \div 8$  19. $817 \div 7$  20. $694 \div 4$

21. $857 \div 2$  22. $762 \div 5$  23. $805 \div 6$  24. $877 \div 3$

25. $3739 \div 6$  26. $1841 \div 5$  27. $4039 \div 9$  28 $3964 \div 5$

29. $1379 \div 4$  30. $2167 \div 8$  31. $2586 \div 6$  32. $3048 \div 7$

**PROBLEM SOLVING**

33. There are 3150 canceled stamps in 9 boxes. If each box contains the same number of canceled stamps, how many canceled stamps are in each box?

34. The Art Guild has 1438 flyers to give out. If 5 members of the Guild share the job equally, how many flyers will each give out? How many flyers will be left over?

35. Ms. Fox needs to put 1032 books on shelves. If a shelf holds 8 books, what is the least number of shelves Ms. Fox needs?

36. Seven ticket agents sold 4662 tickets. Each agent sold the same number of tickets. How many tickets were sold by each agent?

Communicate

 **Share Your Thinking**

37. Tell your parent(s) or another adult:
    • What was easy for you in this lesson on three-digit quotients; what was difficult for you. Explain why.
    • What you needed to know before you studied today's math lesson on three-digit quotients.

# Larger Quotients

To divide large dividends, keep repeating the division steps until the division is completed.

- Divide: 44,776 ÷ 6 = ?

```
        7 4 6 2  R4
  6)4 4,7 7 6
   − 4 2 ↓
      2 7
    − 2 4 ↓
        3 7
      − 3 6 ↓
          1 6
        − 1 2
            4
```

Check:
6 × 7462 + 4 = 44,776

- Divide: 480,897 ÷ 9 = ?

```
        5 3,4 3 3
  9)4 8 0,8 9 7
   − 4 5 ↓
      3 0
    − 2 7 ↓
        3 8
      − 3 6 ↓
          2 9
        − 2 7 ↓
            2 7
          − 2 7
              0
```

Check:
9 × 53,433 = 480,897

▶ You can use a calculator to help you divide large numbers.

Divide: 76,438 ÷ 8 = ?

- For calculators with an [INT÷] key,

    Enter: 76,438 [INT÷] 8 [=]

    Display: | $_{Q}$9554 | | R 6 |

    Quotient   Remainder

- For other calculators,

    **Step 1**

    Enter: 76,438 [÷] 8 [=]

    Display: | 9554.75 |   The remainder is a decimal number.

    **Step 2**

    Enter: 76,438 [−] 9554 [×] 8 [=]

    Display: | 6. |   The remainder is expressed as a whole number.

$$76,438 ÷ 8 = 9554 \text{ R6} \quad \text{or} \quad 8)\overline{76,438} \quad 9\,554 \text{ R6}$$

**Divide and check.**

1. $5\overline{)34{,}061}$
2. $6\overline{)38{,}558}$
3. $7\overline{)43{,}511}$
4. $8\overline{)50{,}519}$

5. $9\overline{)19{,}014}$
6. $8\overline{)35{,}356}$
7. $5\overline{)42{,}736}$
8. $7\overline{)25{,}361}$

9. $6\overline{)211{,}994}$
10. $8\overline{)670{,}197}$
11. $5\overline{)349{,}782}$
12. $9\overline{)767{,}893}$

13. $7\overline{)596{,}081}$
14. $9\overline{)850{,}609}$
15. $3\overline{)295{,}058}$
16. $4\overline{)230{,}178}$

**Use a calculator to find the quotient.**

17. $8\overline{)65{,}714}$
18. $5\overline{)39{,}719}$
19. $6\overline{)52{,}736}$
20. $7\overline{)93{,}712}$

21. $7\overline{)43{,}296}$
22. $9\overline{)48{,}732}$
23. $6\overline{)73{,}501}$
24. $8\overline{)36{,}098}$

25. $5\overline{)102{,}315}$
26. $4\overline{)362{,}003}$
27. $3\overline{)271{,}514}$
28. $6\overline{)483{,}015}$

29. $4\overline{)675{,}153}$
30. $9\overline{)869{,}563}$
31. $5\overline{)686{,}347}$
32. $7\overline{)532{,}456}$

## PROBLEM SOLVING

33. There were 12,744 people who attended the 6 performances of a play presented by a theater guild. If an equal number of people attended each of the performances, how many people attended each performance?

34. In 1990, the British Library published its *General Catalogue of Printed Books to 1975* on a set of three CD-ROMs. A typical reader would need 6 months to scan 198,000 catalog pages in 360 volumes. If a typical reader can scan an equal number of catalog pages each month, how many catalog pages can he scan in one month?

**Share Your Thinking**

Communicate ✓

35. Tell your teacher:
    • When it is helpful to use a calculator to divide whole numbers; when it is better to use another method.

36. The ⬚6 key on your calculator does not work and you are to divide 125,802 by 6 using a calculator. Describe a way you can divide on your broken calculator.

## 3-5 Zeros in the Quotient

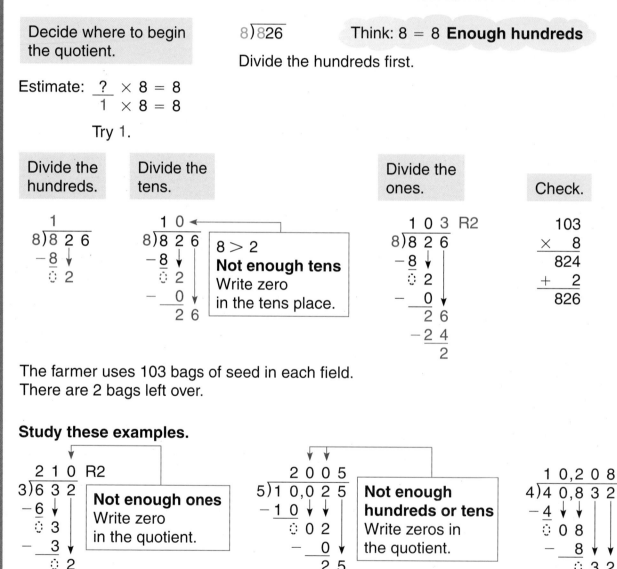

A farmer has 826 bags of seed to plant in 8 fields. He uses the same number of bags of seed in each field. How many bags of seed does he use in each field? How many bags of seed are left over?

To find how many bags of seed are used in each field, divide: 826 ÷ 8 = ?

| Decide where to begin the quotient. | 8)826 | Think: 8 = 8 **Enough hundreds** |
|---|---|---|

Divide the hundreds first.

Estimate: _?_ × 8 = 8
　　　　　 1 × 8 = 8

　　　　Try 1.

| Divide the hundreds. | Divide the tens. | Divide the ones. | Check. |
|---|---|---|---|

Divide the hundreds.

```
      1
  8)8 2 6
  − 8 ↓
    ⋮ 2
```

Divide the tens.

```
      1 0 ◄
  8)8 2 6
  − 8 ↓ ↓
    ⋮ 2 ↓
  −   0 ↓
      2 6
```
8 > 2
**Not enough tens**
Write zero
in the tens place.

Divide the ones.

```
      1 0 3  R2
  8)8 2 6
  − 8 ↓
    ⋮ 2 ↓
  −   0 ↓
      2 6
    − 2 4
        2
```

Check.

```
    103
  ×   8
    824
  +   2
    826
```

The farmer uses 103 bags of seed in each field. There are 2 bags left over.

**Study these examples.**

```
    2 1 0  R2
  3)6 3 2
  − 6 ↓
    ⋮ 3 ↓
  −   3 ↓
    ⋮   2
```
**Not enough ones**
Write zero
in the quotient.

```
      2 0 0 5
  5)1 0,0 2 5
  − 1 0 ↓ ↓ ↓
    ⋮ 0 2 ↓
  −     0 ↓
        2 5
      − 2 5
          0
```
**Not enough
hundreds or tens**
Write zeros in
the quotient.

```
      1 0,2 0 8
  4)4 0,8 3 2
  − 4 ↓ ↓ ↓
    ⋮ 0 8 ↓ ↓
  −     8 ↓ ↓
        ⋮ 3 2
      − 3 2
          0
```

**Divide and check.**

1. 4)830
2. 6)652
3. 5)604
4. 3)722

5. 6)662
6. 5)537
7. 8)828
8. 9)927

9. 6)6120
10. 8)2565
11. 5)1545
12. 7)7063

13. 8)1200
14. 6)1248
15. 3)18,162
16. 4)20,172

17. 6)36,570
18. 5)25,065
19. 7)21,030
20. 9)40,582

## PROBLEM SOLVING

21. A farmer plants 2745 tomato plants in 9 rows. Each row has the same number of tomato plants. How many tomato plants are in each row?

22. Mr. Rivera plants 2800 corn plants in 8 rows. Each row has an equal number of corn plants. How many corn plants are in each row?

23. Julia stores 3535 cans of juice on 7 shelves in a stockroom. Each shelf has the same number of cans of juice stored on it. How many cans of juice are stored on each shelf?

24. A vendor packs 784 plums in 6 cases. Each case holds the same number of plums. How many plums are in each case? How many plums are left over?

25. Mr. O'Brien needs 250 fruit bars for all the children in the camp. There are 8 fruit bars in a box. How many boxes of fruit bars should Mr. O'Brien order?

26. Ms. Murphy buys juice for the 308 students in the camp. There are 6 cans of juice in a pack. How many packs of juice should Ms. Murphy order?

27. In your Math Journal write:

    • When a zero must be placed in the quotient.

    • What a zero in the quotient indicates.

Math Journal

**Critical Thinking**

Algebra

28. • Find the missing factor:

    $\underline{\phantom{?}} \times 0 = 9$    $\underline{\phantom{?}} \times 0 = 0$

    • How do these multiplication sentences show that you can *never* divide by 0?

# Short Division

A travel agent sold an equal number of tickets for each of 7 destinations. If the agent sold a total of 2996 tickets, how many tickets did he sell for each destination?

To find how many tickets were sold for each destination, divide: $2996 \div 7 = \underline{\ ?\ }$

▶ To divide using short division:

- Divide to find the first digit of the quotient.

$$7\overline{)2\ 9\ ^19\ 6}$$ → 4

| $4 \times 7 = 28$ |
| $29 - 28 = 1$ |

- Multiply and subtract mentally.

$$7\overline{)2\ 9\ ^19\ ^56}$$ → 4 2

| $2 \times 7 = 14$ |
| $19 - 14 = 5$ |

- Write each remainder in front of the next digit in the dividend.

$$7\overline{)2\ 9\ ^19\ ^56}$$ → 4 2 8

| $8 \times 7 = 56$ |
| $56 - 56 = 0$ |

- Repeat the steps until the division is completed.

- Check.  $7 \times 428 = 2996$

The travel agent sold 428 tickets for each destination.

**Study these examples.**

$$8\overline{)6\ 4,4\ ^44\ ^48}$$  8 0 5 6

**Not enough hundreds** Write zero.

$$7\overline{)7\ 5,^52\ ^37\ ^24\ 3}$$  1 0, 7 5 3 R3

Write the remainder.

**Use short division to divide. Then check.**

1. $2\overline{)723}$  2. $4\overline{)965}$  3. $3\overline{)756}$  4. $5\overline{)125}$

5. $4\overline{)7364}$  6. $5\overline{)8740}$  7. $3\overline{)6147}$  8. $7\overline{)6566}$

9. $9\overline{)47,376}$  10. $8\overline{)56,365}$  11. $6\overline{)85,742}$  12. $4\overline{)104,232}$

**Use short division to find the quotient.**

13. $2\overline{)806}$    14. $4\overline{)408}$    15. $2\overline{)614}$    16. $7\overline{)749}$

17. $4\overline{)8360}$    18. $3\overline{)3015}$    19. $6\overline{)6246}$    20. $7\overline{)7630}$

21. $9\overline{)93,609}$    22. $8\overline{)80,416}$    23. $6\overline{)72,186}$    24. $8\overline{)84,008}$

25. $5\overline{)78,025}$    26. $4\overline{)28,084}$    27. $9\overline{)96,228}$    28. $7\overline{)14,357}$

29. $3\overline{)120,066}$    30. $5\overline{)303,450}$    31. $7\overline{)284,914}$    32. $9\overline{)162,459}$

## PROBLEM SOLVING

33. An airplane travels 3920 miles in 7 hours. How many miles does it travel in one hour?

34. There are 3488 greeting cards in packs. Each pack holds 8 cards. How many packs of cards are there in all?

35. A loaf of bread uses 9 ounces of flour. How many loaves of bread can 3501 ounces of flour make?

36. A machine produces 3360 clips in 8 minutes. How many clips does it produce in one minute?

37. There are 378 people going on a field trip. Nine buses are hired for the trip. If the same number of people ride in each bus, how many people ride in each bus?

38. There were 10,050 tickets sold for a 3-game series. If the same number of tickets were sold for each game, how many tickets were sold for each game?

39. A bicyclist is planning a 1500-mile trip. His average speed is 8 miles per hour. Will the trip take more or less than 200 hours? Explain your answer.

### Mental Math

**Find the quotient.**

40. $9000 \div 30$    41. $1600 \div 40$    42. $3600 \div 60$

43. $15,000 \div 500$    44. $81,000 \div 900$    45. $63,000 \div 700$

46. $240,000 \div 800$    47. $180,000 \div 200$    48. $420,000 \div 600$

49. $480,000 \div 8000$    50. $540,000 \div 6000$    51. $630,000 \div 9000$

# Exploring Divisibility

### Discover Together

| 316 | 520 |
|-----|-----|
| 8634 | 1722 |
| 95,628 | 68,616 |

**Materials Needed:** calculator, paper, pencil

A number is divisible by another number when you divide and the remainder is zero.

The list above shows numbers that are divisible by 2. Examine the numbers.

1. What do the ones digits of all the numbers have in common?

2. What rule can you write to show that a number is divisible by 2?

3. Use your rule to write some numbers that are divisible by 2. Check by using a calculator to divide.

Suppose you want to test the numbers in the list for divisibility by 4.

4. Use a calculator to find which numbers are divisible by 4.

5. Find the number formed by the tens and ones digits of each number divisible by 4. What do you notice about each of these numbers?

6. What rule can you write to show that a number is divisible by 4?

7. Use your rule to write some numbers that are divisible by 4. Check by using a calculator to divide.

Now test the numbers in the list for divisibility by 3.

8. Use a calculator to find which numbers are divisible by 3.

**9.** Find the sum of the digits of each number divisible by 3. What do you notice about each of these sums?

**10.** What rule can you write to show that a number is divisible by 3?

**11.** Use the rule for divisibility by 3 as a model to write a rule for divisibility by 9.

Pick any five numbers different from those in the list. Multiply each number by 5. Examine the numbers.

**12.** Are the products divisible by 5? Why?

**13.** What are the ones digits of the products?

**14.** What rule can you write to show that a number is divisible by 5?

**15.** Use the rule for divisibility by 5 as a model to write a rule for divisibility by 10.

## Communicate
*Discuss* ✓

**16.** If a number is divisible by 4, is it always divisible by 2? Explain your answer.

**17.** If a number is divisible by 10, is it always divisible by 5? Explain your answer.

**18.** If a number is divisible by 9, is it always divisible by 3? Explain your answer.

**19.** If a number is divisible by 3 and 2, by what number is it also divisible? How do you know?

**Project**

**20.** Create boxes such as these. Then challenge your classmates to find what number does *not* belong and explain why.

| 21 | 72 |
|----|----|
| 105 | 202 |

| 75 | 20 |
|----|----|
| 4 | 45 |

| 18 | 603 |
|----|----|
| 906 | 27 |

| 22 | 72 |
|----|----|
| 52 | 36 |

# 3-8 Divisibility and Mental Math

**Divisibility rules** can help you decide if one number is divisible by another number.

The chart below shows the divisibility rules for 2, 5, 10, 4, 3, 9, and 6.

| Rule<br>A number is divisible | Example | |
|---|---|---|
| **by 2** if its ones digit is divisible by 2. | 20, 42, 84, 936, 1048 are divisible by 2. | All even numbers are divisible by 2. |
| **by 5** if its ones digit is 0 or 5. | 60, 135, 4890, 53,965 are divisible by 5. | |
| **by 10** if its ones digit is 0. | 70, 860, 4050, 96,780 are divisible by 10. | |
| **by 4** if its tens and ones digits form a number that is divisible by 4. | 6128 → 28 ÷ 4 = 7.<br>31,816 → 16 ÷ 4 = 4.<br>6128 and 31,816 are divisible by 4. | |
| **by 3** if the sum of its digits is divisible by 3. | 27 → 2 + 7 = 9 and 9 ÷ 3 = 3.<br>3591 → 3 + 5 + 9 + 1 = 18 and 18 ÷ 3 = 6.<br>27 and 3591 are divisible by 3. | |
| **by 9** if the sum of its digits is divisible by 9. | 216 → 2 + 1 + 6 = 9 and 9 ÷ 9 = 1.<br>5058 → 5 + 0 + 5 + 8 = 18 and 18 ÷ 9 = 2.<br>216 and 5058 are divisible by 9. | |
| **by 6** if it is divisible by both 2 and 3. | 516 is divisible by both 2 and 3.<br>516 is divisible by 6. | |

**Tell which numbers are divisible by 2.**

**1.** 24    **2.** 47    **3.** 98    **4.** 436    **5.** 569    **6.** 760

**7.** 6135    **8.** 9842    **9.** 7764    **10.** 57,961    **11.** 79,778    **12.** 490,893

**Tell which numbers are divisible by 5. Tell which are divisible by 10.**

**13.** 65    **14.** 90    **15.** 873    **16.** 745    **17.** 4000    **18.** 9154

**19.** 35,960    **20.** 45,782    **21.** 73,590    **22.** 94,615    **23.** 870,520    **24.** 791,621

**Tell which numbers are divisible by 4.**

**25.** 96    **26.** 82    **27.** 324    **28.** 422    **29.** 3820    **30.** 9416

**31.** 79,131    **32.** 83,536    **33.** 20,904    **34.** 72,072    **35.** 131,616    **36.** 806,300

**Tell which numbers are divisible by 3. Tell which numbers are divisible by 9.**

**37.** 69    **38.** 87    **39.** 135    **40.** 159    **41.** 4320    **42.** 3519

**43.** 71,415    **44.** 83,721    **45.** 95,580    **46.** 81,693    **47.** 100,512    **48.** 560,373

**Tell which numbers are divisible by 6.**

**49** 84    **50.** 93    **51.** 204    **52.** 396    **53.** 1029    **54.** 5415

**55.** 11,712    **56.** 30,609    **57.** 28,514    **58.** 72,144    **59.** 503,640    **60.** 712,820

**Write whether each number is divisible by 2, 3, 4, 5, 6, 9, and/or 10.**

**61.** 1425    **62.** 2360    **63.** 4390    **64.** 6570    **65.** 8735    **66.** 9822

**67.** 12,360    **68.** 19,585    **69.** 23,130    **70.** 335,412    **71.** 240,120    **72.** 350,262

**PROBLEM SOLVING** Use the chart.

A number of students at Kennedy School are to be divided into equal groups for activities during the school's field day.

Can the number of students be divided into groups of 2? groups of 3? groups of 4? groups of 5? groups of 6? groups of 9? groups of 10?

| | Number of students | Number of Students in Each Group | | | | | | |
|---|---|---|---|---|---|---|---|---|
| | | 2 | 3 | 4 | 5 | 6 | 9 | 10 |
| **73.** | 48 | ? | ? | ? | ? | ? | ? | ? |
| **74.** | 180 | ? | ? | ? | ? | ? | ? | ? |
| **75.** | 315 | ? | ? | ? | ? | ? | ? | ? |
| **76.** | 1080 | ? | ? | ? | ? | ? | ? | ? |

**Finding Together**

*Discuss* ✓

**77.** How many numbers between 200 and 225 are divisible by 10? by 5? by 2? by 3? by 9? by 4? by 6?

**78.** How many numbers between 150 and 200 are divisible by both 3 and 5? by both 4 and 10? by both 6 and 9?

# Estimation: Compatible Numbers

Seven Siberian tigers at the city zoo eat 2075 pounds of meat each week. If the tigers eat equal amounts, about how many pounds of meat does each tiger eat each week?

To find about how many pounds, estimate: $2075 \div 7$

▶ To estimate quotients using compatible numbers:

- Think of nearby numbers that are compatible.

- Divide.

$$2075 \div 7$$
$$\downarrow \qquad \downarrow$$
Think: $2100 \div 7 = 300$

**Compatible numbers** are numbers that are easy to compute mentally.

Each tiger eats about 300 pounds of meat each week.

▶ Compatible-number estimation may use different sets of numbers to estimate a quotient.

Estimate: $17,652 \div 4$

Think: $17,652 \div 4$ $\begin{array}{l} \rightarrow 16,000 \div 4 = 4000 \\ \rightarrow 20,000 \div 4 = 5000 \end{array}$

So $17,652 \div 4$ is about 4000
or $17,652 \div 4$ is about 5000.

**Study these examples.**

Estimate: $8325 \div 41$
$$\qquad \downarrow \qquad \downarrow$$
Think: $8000 \div 40 = 200$

So $8325 \div 41$ is about 200.

Estimate: $63,356 \div 56$
$$\qquad \downarrow \qquad \downarrow$$
Think: $60,000 \div 60 = 1000$

So $63,356 \div 56$ is about 1000.

**Write each division using compatible numbers.**

**1.** 1758 ÷ 4　　　**2.** 3951 ÷ 5　　　**3.** 7453 ÷ 8　　　**4.** 8326 ÷ 9

**5.** 538 ÷ 52　　　**6.** 439 ÷ 36　　　**7.** 5650 ÷ 64　　　**8.** 7749 ÷ 81

**9.** 9875 ÷ 23　　**10.** 4282 ÷ 34　　**11.** 5565 ÷ 42　　**12.** 5875 ÷ 51

**13.** 13,579 ÷ 12　**14.** 42,572 ÷ 44　**15.** 63,792 ÷ 59　**16.** 84,796 ÷ 78

**Estimate the quotient.**

**17.** 1957 ÷ 4　　**18.** 4893 ÷ 5　　**19.** 6397 ÷ 8　　**20.** 3319 ÷ 9

**21.** 2679 ÷ 83　**22.** 8529 ÷ 92　**23.** 4813 ÷ 68　**24.** 7945 ÷ 94

**25.** 83,592 ÷ 94　**26.** 39,125 ÷ 58　**27.** 61,958 ÷ 75　**28.** 38,958 ÷ 49

**Estimate to compare. Write <, =, or >.**

**29.** 27,903 ÷ 7 _?_ 35,903 ÷ 9　　　**30.** 5798 ÷ 3 _?_ 11,938 ÷ 6

**31.** 2829 ÷ 23 _?_ 4173 ÷ 13　　　**32.** 12,636 ÷ 24 _?_ 15,296 ÷ 32

**33.** 46,879 ÷ 18 _?_ 49,362 ÷ 19　　**34.** 69,135 ÷ 27 _?_ 56,238 ÷ 16

**PROBLEM SOLVING**

**35.** Jane earned $557 for a 5-day job. About how much did she earn each day?

**36.** Mr. Duffy earns $38,796 a year. About how much does he earn in one month?

**37.** Twenty-two agents distributed 38,525 pencils to schools. If each agent distributed the same number of pencils, about how many pencils did each agent distribute?

 **Connections: Art**

**38.** In an auction of antiques on June 24, 1981, at Christie's, London, the highest price paid for spoons was $238,000 for a set of 13 Henry VIII Apostle spoons owned by Lord Astor of Hever. About how much did one of the spoons cost?

# Teens as Divisors

You may have to change your estimate more than once when the divisor is a number from 11 through 19.

Divide: 11,378 ÷ 13 = __?__

| Decide where to begin the quotient. | $13\overline{)11{,}378}$ | Think: 13 > 11 **Not enough thousands** |
|---|---|---|
| | $13\overline{)11{,}378}$ | 13 < 113 **Enough hundreds** |

The quotient begins in the hundreds place.

Estimate: $13\overline{)11{,}378}$

Try 9.

Think: $10\overline{)100}$ → 10

A quotient digit *cannot* be greater than 9.

**Divide the hundreds.**

```
        9
13)1 1,3 7 8
→ 1 1 7
   Too large.
   Try 8.
```

```
           → 8
13)1 1,3 7 8
 − 1 0 4 ↓
       9 7
```

**Divide the tens.**

```
        8 7
13)1 1,3 7 8
 − 1 0 4 ↓
       9 7
     − 9 1 ↓
         6 8
```

**Divide the ones.**

```
        8 7 5 R3
13)1 1,3 7 8
 − 1 0 4 ↓
       9 7
     − 9 1 ↓
         6 8
       − 6 5
           3
```

**Check.**

```
        8 7 5
×          1 3
      2 6 2 5
        8 7 5
      1 1 3 7 5
+            3
      1 1,3 7 8
```

**Copy and complete.**

**1.**
```
         9
15)1 3 5
 − ? ? ?
       ?
```

**2.**
```
         8 R  ?
17)1 4 9
 − ? ? ?
       ?
```

**3.**
```
            9
17)1 5 2 1
   1 5 3
 Try 8.
```
```
          → 8 ? R  ?
17)1 5 2 1
 − ? ? ?
     ? ? ?
   − ? ? ?
         ?
```

**Divide and check.**

**4.** $14\overline{)129}$  **5.** $18\overline{)144}$  **6.** $13\overline{)403}$  **7.** $15\overline{)780}$

**8.** $19\overline{)950}$  **9.** $15\overline{)498}$  **10.** $14\overline{)747}$  **11.** $17\overline{)884}$

**Find the quotient and the remainder. Then check.**

**12.** 19)1578     **13.** 17)1462     **14.** 18)1693     **15.** 15)1159

**16.** 18)3427     **17.** 17)2869     **18.** 14)3609     **19.** 13)3921

**20.** 12)10,512     **21.** 18)16,038     **22.** 17)13,243     **23.** 19)18,981

**24.** 14)73,501     **25.** 12)13,732     **26.** 13)13,296     **27.** 15)16,438

**28.** 11)115,932     **29.** 13)148,732     **30.** 14)193,475     **31.** 16)167,652

## PROBLEM SOLVING

**32.** A tank containing 336 gallons of fuel can be emptied in 12 minutes. How many gallons of fuel can be emptied in one minute?

**33.** A plane uses 570 gallons of gasoline in a 15-hour trip. How many gallons of gasoline does it use in one hour?

**34.** Melissa puts 420 photos in an album. Each page of the album holds 14 photos. How many pages does she fill?

**35.** Albert traveled 4000 miles in 16 days. If he traveled the same number of miles each day, how many miles did he travel each day?

**36.** There are 540 children enrolled in Valley School. If there are 18 classrooms in the school, what is the average number of students in each classroom?

**Critical Thinking**

**Find the errors in the division process.
Then make the corrections.**

**37.**
```
        9  R3
12)1 0 8 3
 -1 0 8
         3
```

**38.**
```
       1 3
17)1 7 5 1
 -1 7
       5 1
     -5 1
        0
```

**39.**
```
       1 5  1
13)1 9,5 1 3
 -1 3
     6 5
   -6 5
         1 3
       -1 3
          0
```

**40.**
```
       8 6 3
18)1 5 5,3 4 0
 -1 4 4
     1 1 3
   -1 0 8
         5 4
       -5 4
          0
```

115

## 3-11 Two-Digit Divisors

Mr. Jansen has 1825 tickets to distribute to his 23 salespersons. If he distributes the tickets equally among his salespersons, how many tickets does each one receive? How many tickets are left over?

To find how many tickets each salesperson receives, divide: $1825 \div 23 = \underline{\ ?\ }$

| Decide where to begin the quotient. | $23\overline{)1825}$ | Think: $23 > 18$ **Not enough hundreds** |
|---|---|---|
| | $23\overline{)1825}$ | $23 < 182$ **Enough tens** |

The quotient begins in the tens place.

Estimate: $23\overline{)1825}$
Try 9.

Think: $20\overline{)180}$ → 9

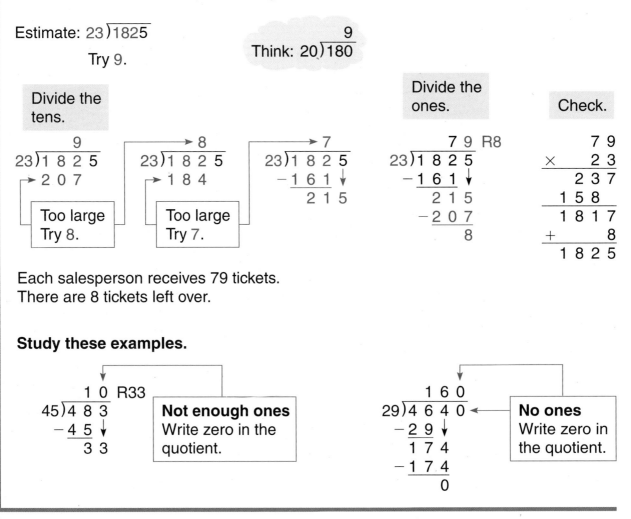

**Divide the tens.**

$$\begin{array}{r} 9 \\ 23\overline{)1825} \\ 207 \end{array}$$
Too large
Try 8.

$$\begin{array}{r} 8 \\ 23\overline{)1825} \\ 184 \end{array}$$
Too large
Try 7.

$$\begin{array}{r} 7 \\ 23\overline{)1825} \\ -161 \\ \hline 215 \end{array}$$

**Divide the ones.**

$$\begin{array}{r} 79 \text{ R8} \\ 23\overline{)1825} \\ -161 \\ \hline 215 \\ -207 \\ \hline 8 \end{array}$$

**Check.**

$$\begin{array}{r} 79 \\ \times\ 23 \\ \hline 237 \\ 158 \\ \hline 1817 \\ +\ 8 \\ \hline 1825 \end{array}$$

Each salesperson receives 79 tickets.
There are 8 tickets left over.

**Study these examples.**

$$\begin{array}{r} 10 \text{ R33} \\ 45\overline{)483} \\ -45 \\ \hline 33 \end{array}$$
**Not enough ones**
Write zero in the quotient.

$$\begin{array}{r} 160 \\ 29\overline{)4640} \\ -29 \\ \hline 174 \\ -174 \\ \hline 0 \end{array}$$
**No ones**
Write zero in the quotient.

116

**Copy and complete.**

1.
```
      2 R ?
41)8 6
 - ? ?
     ?
```

2.
```
      2 ?
32)6 7 2
 - ? ?
    ? ?
  - ? ?
      ?
```

3.
```
        9                    ? ? R ?
47)4 2 1 6          47)4 2 1 6
   4 2 3               - ? ? ?
                          ? ? ?
  Try  ?                 - ? ? ?
                            ? ?
```

**Divide and check.**

4. 32)96

5. 22)88

6. 41)205

7. 17)153

8. 61)854

9. 43)688

10. 34)680

11. 27)621

12. 51)358

13. 65)201

14. 82)331

15. 46)283

16. 35)1019

17. 76)3733

18. 44)1456

19. 63)3792

20. 59)1193

21. 36)2884

22. 43)3886

23. 72)4332

24. 45)9542

25. 62)6905

26. 81)9729

27. 76)9884

## PROBLEM SOLVING

28. Roy feeds the birds in the zoo 6500 ounces of birdseed in 52 weeks. How many ounces of birdseed does he feed the birds each week?

29. If 6036 people visit the zoo in 12 days, what is the average number of people who visit the zoo each day?

30. A club collected $5500 in annual membership fees. The annual membership fee is $25. How many club members paid their fees?

Bird Seed

**Calculator Activity**

**Find the number.**

31. A number between 130 and 140 when divided by 12 has a quotient that contains the same two digits and no remainder.

32. A number between 2700 and 2800 when divided by 25 has a quotient that contains three odd digits and has no remainder.

117

## 3-12 Dividing Larger Numbers

Buses transported 162,448 fans to games for a season. If 52 fans went on each bus trip, how many trips did the buses make?

To find how many trips,
divide: 162,448 ÷ 52 = __?__

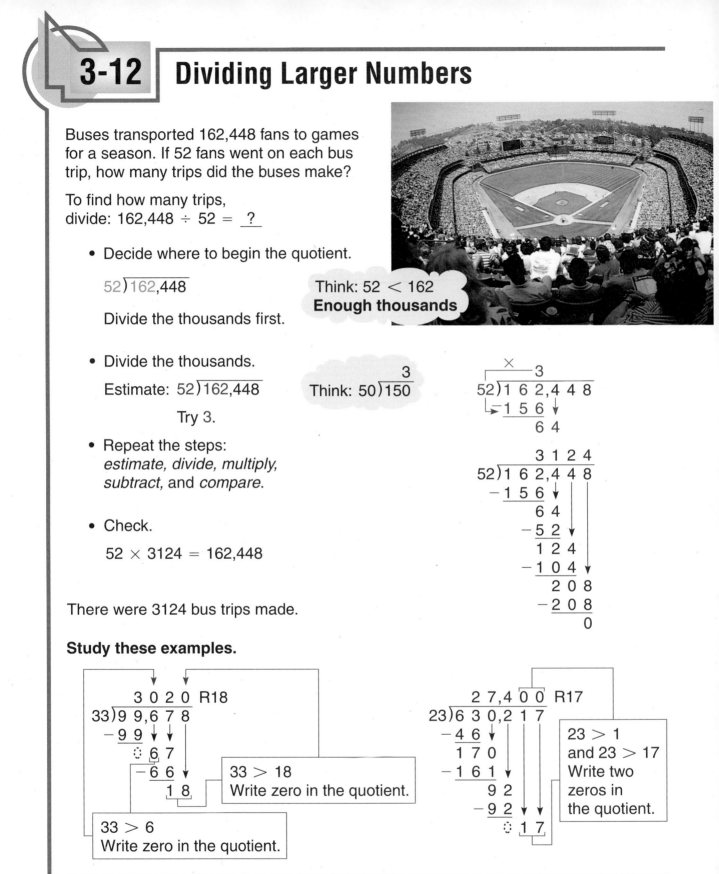

- Decide where to begin the quotient.

$$52\overline{)162{,}448}$$

Divide the thousands first.

**Think:** 52 < 162
**Enough thousands**

- Divide the thousands.

Estimate: $52\overline{)162{,}448}$

**Think:** $50\overline{)150}$

Try 3.

$$\begin{array}{r} \times\quad 3 \\ 52\overline{)162{,}448} \\ -156\phantom{000} \\ \hline 64\phantom{00} \end{array}$$

- Repeat the steps: *estimate, divide, multiply, subtract,* and *compare.*

$$\begin{array}{r} 3124 \\ 52\overline{)162{,}448} \\ -156\phantom{000} \\ \hline 64\phantom{00} \\ -52\phantom{00} \\ \hline 124\phantom{0} \\ -104\phantom{0} \\ \hline 208 \\ -208 \\ \hline 0 \end{array}$$

- Check.

52 × 3124 = 162,448

There were 3124 bus trips made.

### Study these examples.

$$\begin{array}{r} 3020 \text{ R18} \\ 33\overline{)99{,}678} \\ -99\phantom{000} \\ \hline 67\phantom{0} \\ -66\phantom{0} \\ \hline 18 \end{array}$$

33 > 18
Write zero in the quotient.

33 > 6
Write zero in the quotient.

$$\begin{array}{r} 27{,}400 \text{ R17} \\ 23\overline{)630{,}217} \\ -46\phantom{0000} \\ \hline 170\phantom{00} \\ -161\phantom{00} \\ \hline 92\phantom{0} \\ -92\phantom{0} \\ \hline 17 \end{array}$$

23 > 1
and 23 > 17
Write two zeros in the quotient.

**Divide and check.**

1. $63\overline{)31,550}$
2. $34\overline{)32,200}$
3. $57\overline{)22,850}$
4. $72\overline{)56,890}$

5. $62\overline{)29,145}$
6. $43\overline{)42,145}$
7. $54\overline{)37,841}$
8. $92\overline{)82,890}$

9. $27\overline{)553,529}$
10. $16\overline{)521,613}$
11. $29\overline{)884,560}$
12. $21\overline{)430,629}$

**Estimate. Then use a calculator to divide.**

13. $42\overline{)193,242}$
14. $32\overline{)876,821}$
15. $27\overline{)105,595}$
16. $26\overline{)174,590}$

17. $58\overline{)349,334}$
18. $64\overline{)493,444}$
19. $91\overline{)364,460}$
20. $82\overline{)582,692}$

21. $77\overline{)273,080}$
22. $47\overline{)991,985}$
23. $39\overline{)928,210}$
24. $53\overline{)483,651}$

**Divide. Use mental math, paper and pencil, or calculator. Explain the method you used.**

*Computation Method* ✓

25. $90\overline{)63,000}$
26. $39\overline{)45,164}$
27. $80\overline{)32,320}$
28. $41\overline{)12,500}$

29. $56\overline{)420,810}$
30. $45\overline{)180,000}$
31. $27\overline{)101,520}$
32. $17\overline{)170,006}$

**PROBLEM SOLVING**

33. There are 43,560 apples to be shipped to stores. If 72 apples are packed in each box, how many boxes of apples are to be shipped?

34. There are 38,912 pears to be boxed. If each box contains 64 pears, how many boxes are needed for the pears?

35. A stadium has 98,400 seats in all. How many rows of seats does the stadium have if each row has 96 seats?

36. If there are 32 nails in a box, how many boxes are needed to pack 65,852 nails?

37. If a bus seats 52 passengers, how many buses will be needed to transport 162,478 fans to games for the entire season?

**Challenge**

*Communicate* ✓

38. What is the greatest number of digits you can have in a quotient if you divide a 6-digit number by a 2-digit number? What is the least number? Explain how you found your answers.

## 3-13 Dividing Money

Ms. Taylor paid $133.65 for 27 identical boxes of greeting cards. How much did she pay for each box of greeting cards?

To find the cost of a box of greeting cards, divide: $133.65 ÷ 27 = __?__

▶ To divide money:

- Place the dollar sign and the decimal point in the quotient.

$$\begin{array}{r} \$ \qquad . \quad \\ 27\overline{)\$1\ 3\ 3.6\ 5} \end{array}$$

- Divide as usual.

$$\begin{array}{r} \$ \quad 4.9\ 5 \\ 27\overline{)\$1\ 3\ 3.6\ 5} \\ -1\ 0\ 8\phantom{\downarrow} \\ \hline 2\ 5\ 6 \\ -2\ 4\ 3\phantom{\downarrow} \\ \hline 1\ 3\ 5 \\ -1\ 3\ 5 \\ \hline 0 \end{array}$$

- Check: 27 × $4.95 = $133.65

Ms. Taylor paid $4.95 for each box.

**Study these examples.**

$$\begin{array}{r} \$ \quad 3.0\ 7 \\ 42\overline{)\$1\ 2\ 8.9\ 4} \\ -1\ 2\ 6\phantom{\downarrow\downarrow} \\ \hline 2\ 9\ 4 \\ -2\ 9\ 4 \\ \hline 0 \end{array}$$

29 < 42
Write zero in the quotient.

$$\begin{array}{r} \$ \quad 2.3\ 0 \\ 53\overline{)\$1\ 2\ 1.9\ 0} \\ -1\ 0\ 6\phantom{\downarrow} \\ \hline 1\ 5\ 9 \\ -1\ 5\ 9 \\ \hline 0 \end{array}$$

There are no pennies. Write zero in the quotient.

**Copy and complete.**

1.
$$\begin{array}{r} \$ \quad 2.?\ ? \\ 6\overline{)\$1\ 6.1\ 4} \\ -1\ 2\phantom{} \\ \hline 4\ 1 \\ -?\ ? \\ \hline ?\ ? \\ -?\ ? \\ \hline ? \end{array}$$

2.
$$\begin{array}{r} \$ \quad 7.?\ ? \\ 13\overline{)\$9\ 7.7\ 6} \\ -?\ ?\phantom{} \\ \hline ?\ ? \\ -?\ ? \\ \hline ?\ ? \\ -?\ ? \\ \hline ? \end{array}$$

3.
$$\begin{array}{r} \$ \quad ?.?\ ? \\ 29\overline{)\$2\ 0\ 0.6\ 8} \\ -1\ 7\ 4\phantom{} \\ \hline ?\ ?\ ? \\ -?\ ?\ ? \\ \hline ?\ ? \\ -?\ ? \\ \hline ? \end{array}$$

120

**Divide and check.**

4. $4 \overline{)\$15.12}$  5. $3 \overline{)\$6.27}$  6. $8 \overline{)\$159.60}$  7. $7 \overline{)\$107.10}$

8. $54 \overline{)\$14.04}$  9. $47 \overline{)\$39.95}$  10. $67 \overline{)\$62.31}$  11. $24 \overline{)\$13.68}$

12. $19 \overline{)\$114.00}$  13. $26 \overline{)\$208.00}$  14. $15 \overline{)\$139.50}$  15. $42 \overline{)\$153.30}$

16. $28 \overline{)\$157.92}$  17. $31 \overline{)\$186.62}$  18. $53 \overline{)\$365.70}$  19. $85 \overline{)\$177.65}$

20. $34 \overline{)\$173.06}$  21. $47 \overline{)\$325.24}$  22. $32 \overline{)\$322.56}$  23. $11 \overline{)\$250.25}$

24. $17 \overline{)\$402.05}$  25. $23 \overline{)\$530.15}$  26. $19 \overline{)\$1179.71}$  27. $21 \overline{)\$1997.10}$

## PROBLEM SOLVING
**Use the table for problems 28–31.**

How much does each box of each kind of card cost?

28. Thank you cards

29. Get well cards

30. Birthday cards

31. Anniversary cards

| Quantity | Item | Total Cost |
|----------|------|------------|
| 25 boxes | Thank you cards | $ 86.25 |
| 32 boxes | Get well cards | $155.20 |
| 46 boxes | Birthday cards | $273.70 |
| 18 boxes | Anniversary cards | $125.10 |

**Solve and explain the method you used. Write whether you estimated or found an exact answer.**

Computation Method

32. Mark earned $536.10 in 6 days. If he earned the same amount of money each day, how much did he earn each day?

33. Fifteen part-time workers earned $424.80. About how much did each worker receive if the money was divided equally?

34. An art supply kit costs $67.75 per student per year. Is $2300 enough to supply an art class of 28?

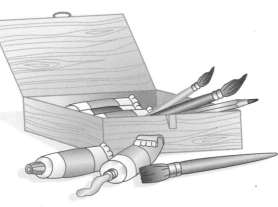

# Order of Operations

Compute: $3 + 7 \times 1 - 4 \div 2 = \underline{\ ?\ }$

- *First* multiply or divide.
  Work from left to right.

- *Then* add or subtract.
  Work from left to right.

$$3 + \underset{\downarrow}{\underline{7 \times 1}} - \underset{\downarrow}{\underline{4 \div 2}}$$

$$\underset{\downarrow}{\underline{3 + 7}} \quad - \quad 2$$

$$10 \quad - \quad 2 \quad = 8$$

▶ When there are parentheses, do the operations within the parentheses *first*.

Compute: $(8 \times 3) \div (4 + 2) = \underline{\ ?\ }$

$$\underset{\downarrow}{\underline{(8 \times 3)}} \div \underset{\downarrow}{\underline{(4 + 2)}}$$

$$24 \quad \div \quad 6 \quad = 4$$

## Study these examples.

| | |
|---|---|
| $(56 \div 8) - 2 + (5 + 6) \times 3 = \underline{\ ?\ }$ | $19 + 21 \div 7 \times 8 - 13 = \underline{\ ?\ }$ |
| $7 - 2 + 11 \times 3$ | $19 + 3 \times 8 - 13$ |
| $7 - 2 + 33$ | $19 + 24 - 13$ |
| $5 + 33 = 38$ | $43 - 13 = 30$ |

## Compute.

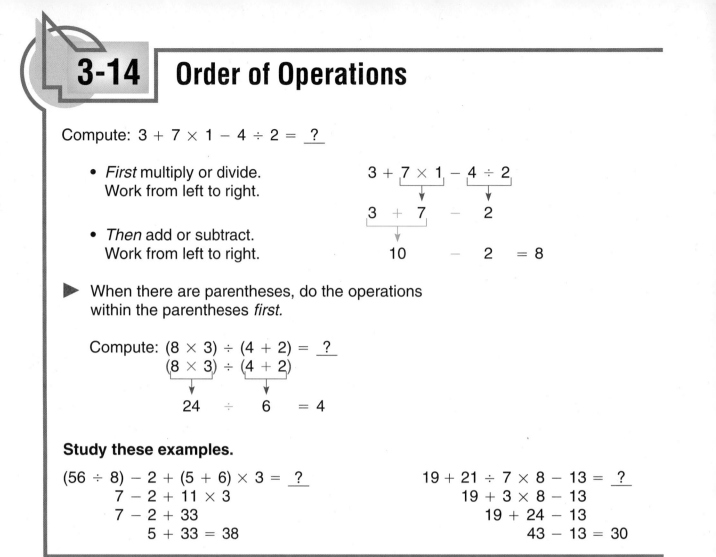

1. $4 - 3 + 2$
2. $10 + 8 - 5$

3. $16 \div 2 \times 3$
4. $8 \times 2 \div 4$

5. $4 \times 6 + 3$
6. $2 \times 7 - 4$

7. $81 \div 9 - 3$
8. $64 \div 8 + 5$

9. $8 + 3 \times 4 - 5$
10. $9 + 45 \div 5 - 3$

11. $9 \times 4 \div 6 + 7$
12. $48 \div 6 \times 3 - 5$

13. $27 - 16 \div 4 + 2$
14. $18 - 3 \times 2 + 9$

15. $81 \div 9 - 2 \times 3$
16. $6 \times 5 + 27 \div 3$

**Use the order of operations to compute.**

**17.** $4 - 9 \div 3 - 1$  **18.** $16 \div 4 + 2 \times 6$

**19.** $(3 \times 7) + (64 \div 8)$  **20.** $(18 - 9) \div (1 + 2)$

**21.** $20 + 6 \div 3 - 7$  **22.** $24 - 8 \div 4 \times 3$

**23.** $18 \times (11 - 6)$  **24.** $7 + (19 - 2) \times 3$

**25.** $2 + (3 \times 6) + 10$  **26.** $(12 + 72) \div 6$

**27.** $(28 + 32) \times 4$  **28.** $(9 \times 8) - (3 \times 6)$

**29.** $8 \times 2 \div 2 + 24$  **30.** $6 + 2 - 3 \times 6 \div 9$

**31.** $3 + 5 \times 10 \div 2 + 8$  **32.** $17 + 63 \div 3 \times 6 - 9$

**33.** $59 - 45 \div 5 \times 3 + 41$  **34.** $134 - 8 \div 4 \times 2$

**35.** $10 \times 4 + (49 \div 7) \times 2$  **36.** $(35 \div 5) \times 2 + 3 \times 6$

**37.** $18 - 3 \div 3 + (63 \div 3) - 6$  **38.** $19 - 4 \times 2 + (19 - 3) \div 4$

**39.** $(28 \div 7) + 5 - 3 + (7 \times 2)$  **40.** $4 + (29 - 2) \div 9 + (16 + 2)$

**41.** $(4 \times 8) - 5 + (0 \div 6)$  **42.** $(24 \div 6) - 3 + (2 \times 4)$

**Rewrite each number sentence using parentheses to make it true.**

**43.** $25 - 5 \times 10 \div 2 = 0$  **44.** $19 - 4 + 3 \div 7 = 18$

**45.** $3 + 6 \times 5 + 5 = 50$  **46.** $9 - 5 \times 2 + 6 = 14$

**47.** $8 + 24 \div 14 - 8 = 12$  **48.** $27 - 5 + 4 \div 3 = 24$

**49.** $9 + 5 \div 2 - 4 = 3$  **50.** $4 \times 3 + 5 - 2 = 30$

### Make Up Your Own

**Use two different operations to write an exercise for each number.**

**51.** $3 = $ _?_      **52.** $5 = $ _?_      **53.** $8 = $ _?_      **54.** $11 = $ _?_

123

# 3-15 | Problem Solving: Make a Table/Find a Pattern

**Problem:** A shop rents bicycles and 3-wheeled buggies. Every day Larry checks the 60 wheels on the 25 vehicles for safety. How many of each type of vehicle does he have?

**1 IMAGINE**  Picture the worker checking the wheels on each of the vehicles.

**2 NAME**

*Facts:* Shop rents bicycles and 3-wheeled buggies.
There is a total of 25 vehicles.
There is a total of 60 wheels.

*Question:* How many of each vehicle does the shop rent?

**3 THINK**  Make a table to find the different combinations of bicycles and buggies. Look for a pattern to find the combination that has exactly 60 wheels.

| Bicycles | 10 | 11 | 12 | 13 |
|---|---|---|---|---|
| Buggies | 15 | 14 | 13 | 12 |
| Wheels | 20 + 45 = 65 | 22 + 42 = 64 | 24 + 39 = 63 | 26 + 36 = 62 |

Notice the pattern in the table. As the number of buggies decreases by 1, so does the total number of wheels. So to get from 65 wheels to 60 wheels, subtract: $65 - 60 = 5$.

**4 COMPUTE**  To find the number of buggies, subtract: $15 - 5 = 10$.
There are 10 buggies $(15 - 5)$
and 15 bicycles $(10 + 5)$.

**5 CHECK**

15 bicycles:  $2 \times 15 = 30$ wheels
10 buggies:  $3 \times 10 = 30$ wheels
$25 \stackrel{?}{=} 10 + 15$  Yes.  $60 \stackrel{?}{=} 30 + 30$  Yes.

**Make a table and find a pattern to solve each problem.**

1. Cassie's grandparents gave her $1 for her first birthday. Each year after, they gave her $1 more than the year before. How much money will they have given her by her 20th birthday?

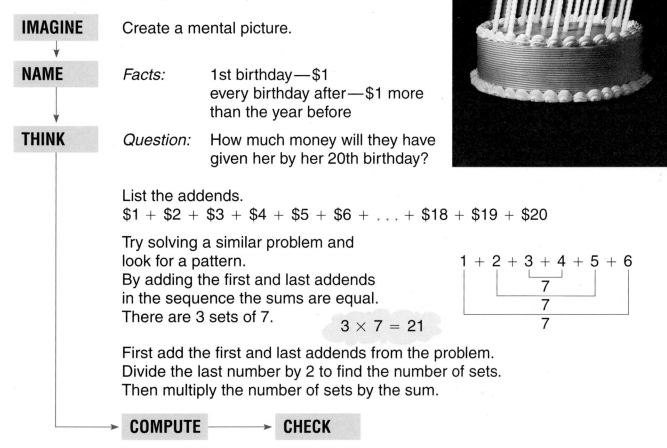

**IMAGINE**     Create a mental picture.

**NAME**     *Facts:*     1st birthday—$1
                          every birthday after—$1 more
                          than the year before

**THINK**     *Question:*     How much money will they have given her by her 20th birthday?

List the addends.
$1 + $2 + $3 + $4 + $5 + $6 + ... + $18 + $19 + $20

Try solving a similar problem and look for a pattern.
By adding the first and last addends in the sequence the sums are equal.
There are 3 sets of 7.

$$1 + 2 + 3 + 4 + 5 + 6$$
$$7$$
$$7$$
$$7$$

$$3 \times 7 = 21$$

First add the first and last addends from the problem.
Divide the last number by 2 to find the number of sets.
Then multiply the number of sets by the sum.

**COMPUTE** ⟶ **CHECK**

2. Nancy bought a bag of red, white, and blue balloons for the party. There were 49 balloons in the bag. If there are 2 times as many red as blue and half as many white as blue, how many of each color balloon are in the bag?

3. The temperature at 10:00 P.M. was 37°F. If it dropped 2°F every hour until 4:00 A.M. and then rose 4°F each hour after that, what was the temperature at noon the next day?

**Make Up Your Own**

*Communicate* ✓

4. Write a problem using the Make a Table/Find a Pattern strategy. Have someone solve it.

# 3-16 | Problem-Solving Applications

**Solve each problem. Explain the method you used.**

1. The Stampton Post Office sold 3768 stamps yesterday. The office was open for 8 hours, and business was steady all day. About how many stamps were sold each hour?

2. Mae came to the post office and bought a sheet of 40 stamps for $7.60. What is the cost of each stamp?

3. Allen bought a sheet of 50 stamps for $26.00. How much did each stamp cost?

4. In a busy hour, 3 clerks can each serve about the same number of customers. There are 90 customers. About how many customers can each clerk serve in an hour?

5. The office has 444 post office boxes arranged in rows. There are 37 equal rows of boxes. How many boxes are in each row?

6. There are 12 mail carriers in Stampton. Monday, they delivered 24,780 letters. Each carrier delivered the same number of letters. How many letters did each carrier deliver?

7. A new commemorative Earth stamp is produced on sheets of 40 stamps. One clerk has 840 of the stamps at her station. A customer wants to buy 22 sheets of stamps. Does the clerk have enough?

8. Mr. Jared delivered 8456 letters. He delivered the same number of letters each hour during a 7-hour period. About how many letters did he deliver each hour?

9. Mr. Jared's mail truck logged 51 mi, 47 mi, 63 mi, 54 mi, 44 mi, and 65 mi. What is the average number of miles the truck traveled each day in one workweek?

Imagine

Name

Think

Compute

Check

**Choose a strategy from the list or use another strategy you know to solve each problem.**

USE THESE STRATEGIES
Hidden Information
Make a Table/Find a Pattern
Interpret the Remainder
Use a Graph
More Than One Solution
Extra Information

10. A new stamp booklet holds 20 stamps, each showing a different type of tree. Ten tellers at the Stampton office sold 13,640 of the stamps. How many booklets were sold?

11. A machine can cancel 120 letters per minute. How many letters can it cancel in one hour?

12. A postal clerk can work a 6-hour or an 8-hour shift. If she worked 44 hours one week, how many shifts of each length did she work?

13. It costs $168 to rent a post office box for one year. At that rate how much does it cost to rent a box for one month?

14. Each page of Cathy's stamp album holds 12 stamps. If she has 377 stamps to put in her album, how many more stamps does she need to fill a page?

15. Danielle sorts letters into bins. She sorts 302 letters into the first bin, 413 letters into the second bin, and 524 letters into the third. If the pattern continues, how many letters will she put into the sixth and seventh bins?

**Use the bar graph for problems 16 and 17.**

16. Letters are delivered 6 days a week. About how many letters were delivered each day during the first week of August? during the fourth week?

17. What is the average number of letters delivered each week in August?

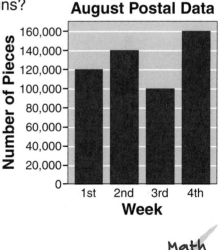

**August Postal Data**

*Number of Pieces* (y-axis: 0 to 160,000 in increments of 20,000)

Week (x-axis): 1st, 2nd, 3rd, 4th

**Share Your Thinking**

Math Journal

18. Zip codes help postal workers sort mail. How many 5-digit zip codes begin with the digits 100_ _? Write in your Math Journal about the strategies you use to solve this problem.

# Chapter Review and Practice

**Write four related facts using the given numbers.** *(See pp. 96–97.)*

**1.** 6, 7, 42      **2.** 5, 9, 45      **3.** 8, 9, 72      **4.** 3, 4, 12

**Find the quotients.** *(See pp. 98–99.)*

**5.**
63 ÷ 9
630 ÷ 9
6300 ÷ 9
63,000 ÷ 9

**6.**
54 ÷ 6
540 ÷ 60
5400 ÷ 600
54,000 ÷ 6000

**7.**
35 ÷ 7
350 ÷ 70
3500 ÷ 70
35,000 ÷ 70

**Divide and check.** *(See pp. 100–107, 114–121.)*

**8.** 4)542      **9.** 5)798      **10.** 3)893      **11.** 6)754

**12.** 8)5040      **13.** 7)6239      **14.** 6)6128      **15.** 5)3005

**16.** 9)3027      **17.** 8)5866      **18.** 24)49      **19.** 41)984

**20.** 31)1836      **21.** 15)945      **22.** 86)68,906      **23.** 73)78,146

**24.** 28)$56.56      **25.** 17)$35.02      **26.** 26)$286.26      **27.** 64)$204.80

**Write whether each number is divisible by 2, 3, 4, 5, 6, 9, and/or 10.** *(See pp. 108–111.)*

**28.** 90      **29.** 795      **30.** 4152      **31.** 6252      **32.** 70,320

**Estimate the quotient.** *(See pp. 112–113.)*

**33.** 845 ÷ 9      **34.** 1015 ÷ 29      **35.** 1836 ÷ 15

**Use the order of operations to compute.** *(See pp. 122–123.)*

**36.** 36 − 3 × 7 + 10 ÷ 5      **37.** (35 ÷ 7) + 2 × 3 − 4

## PROBLEM SOLVING

*(See pp. 114–115, 124–127.)*

**38.** Lois has 36 colored pencils. They are either green or red. For every green pencil, Lois has 3 red pencils. How many red pencils does Lois have?

**39.** Ralph put 1620 canceled stamps in 18 boxes. If he put the same number of stamps in each box, how many stamps were in one box?

(See *Still More Practice*, p. 479.)

Algebra ✓

## VARIABLES

A **variable** is a symbol, usually a letter, that is used to represent a number.

An *algebraic expression* uses one or more variables and the operation symbols $+, -, \times, \div$.

$$n + 3 \qquad b - 2 \qquad a \times b \text{ or } ab \qquad n \div m \text{ or } \frac{n}{m}$$

▶ A *word phrase* can be *translated* into an *algebraic expression* by using variables and the operation symbols.

| Word phrase | Algebraic expression |
|---|---|
| *a* more than 7 | $7 + a$ |
| *b* less than 5 | $5 - b$ |
| the product of $7m$ and *n* | $7m \times n$ or $7mn$ |
| the sum of *c* and *d*, divided by 9 | $(c + d) \div 9$ or $\dfrac{c + d}{9}$ |

**Translate the word phrase into an algebraic expression.**

1. *c* increased by 6

2. 5 decreased by *b*

3. twice the product of *m* and *n*

4. *x* divided by *y*

5. 4 less than *a*

6. *m* more than *n*

**Represent by an algebraic expression.**

7. a distance that is 10 meters shorter than *d* meters

8. a number that is 6 less than a number *n*

9. the cost of *x* suits if each suit costs $150

10. a weight that is 25 lb heavier than *m* lb

11. an amount of money that is twice *y* dollars

12. the width of a rectangle that is half of its length $\ell$

13. the total number of days in *w* weeks and *d* days

129

# Check Your Mastery

## Performance Assessment

Explain where you can place one set of parentheses in the expression at the right to result in an answer

$$30 - 3 \times 10 + 9 \div 3$$

1. greater than 100

2. between 10 and 30

**Complete.**

3. $4 \times 7 = \underline{\ ?\ }$
   $\underline{\ ?\ } \div 7 = 4$

4. $5 \times \underline{\ ?\ } = 30$
   $30 \div 5 = \underline{\ ?\ }$

5. $\underline{\ ?\ } \times 9 = 72$
   $72 \div \underline{\ ?\ } = 9$

**Divide.**

6. $30 \overline{)900}$

7. $8 \overline{)5600}$

8. $5875 \div 4$

9. $5050 \div 3$

**Find the quotient and check.**

10. $7 \overline{)935}$

11. $5 \overline{)6789}$

12. $3 \overline{)2718}$

13. $8 \overline{)5982}$

14. $17 \overline{)8891}$

15. $51 \overline{)1377}$

16. $68 \overline{)53,176}$

17. $57 \overline{)\$182.40}$

18. $23 \overline{)\$276.92}$

**Write whether each number is divisible by 2, 3, 4, 5, 6, 9, and/or 10.**

19. 360

20. 7155

21. 8472

22. 43,140

**Estimate to compare. Write $<$, $=$, or $>$.**

23. $298 \div 3 \ \underline{\ ?\ } \ 282 \div 4$

24. $1392 \div 7 \ \underline{\ ?\ } \ 1821 \div 6$

**PROBLEM SOLVING**  *Use a strategy you have learned.*

25. The scoutmaster ordered 14 buses for 952 people. If he assigned the same number of people to each bus, how many passengers were in each bus?

# Cumulative Review I

**Choose the best answer.**

1. In 10,234,567,890 which digit is in the ten-millions place?

   **a.** 0  **b.** 1
   **c.** 3  **d.** 9

2. Choose the standard form of the number.

   seven billion, three hundred six thousand

   **a.** 7,000,306,000
   **b.** 7,000,360,000
   **c.** 7,306,000,000
   **d.** 7,360,000,000

3. Which is ordered greatest to least?

   **a.** 8.524; 8.534; 8.53
   **b.** 8.534; 8.53; 8.524
   **c.** 8.53; 8.534; 8.524
   **d.** none of these

4. Which shows 15,695,823 rounded to its greatest place?

   **a.** 10,000,000
   **b.** 16,000,000
   **c.** 200,000,000
   **d.** 20,000,000

5. Estimate.

   563,682
   472,289
   + 186,451

   **a.** 130,000
   **b.** 930,000
   **c.** 1,100,000
   **d.** 1,300,000

6. Subtract.

   726,423
   − 318,619

   **a.** 231,516
   **b.** 407,804
   **c.** 914,722
   **d.** 417,804

7.  3046
   × 6

   **a.** 18,276
   **b.** 21,276
   **c.** 33,412
   **d.** 18,876

8. 217 × $25.81

   **a.** $326.98
   **b.** $1410.77
   **c.** $5600.77
   **d.** not given

9. Which are divisible by 3?

   **a.** 18,585; 325,714; 1823
   **b.** 69,132; 276,204; 2301
   **c.** 418,608; 45,806; 2002
   **d.** 115,321; 35,432; 2106

10. Which compatible numbers are used to estimate

    $19\overline{)3817}$ ?

    **a.** $20\overline{)4000}$
    **b.** $9\overline{)3600}$
    **c.** $40\overline{)2000}$
    **d.** not given

11. $44\overline{)112,928}$

    **a.** 810
    **b.** 2160 R1
    **c.** 2566 R24
    **d.** 2516 R14

12. $32\overline{)\$2398.40}$

    **a.** $36.81
    **b.** $112.14
    **c.** $174.95
    **d.** not given

13. Compute.
    Use the order of operations.

    $2 \times 6 + 36 \div 9 - 5$

    **a.** $\frac{1}{3}$  **b.** 11
    **c.** 16  **d.** 24

14. Which number is 1000 more than

    $4\overline{)81,608}$ ?

    **a.** 1242
    **b.** 3402
    **c.** 20,402
    **d.** 21,402

131

## For Your Portfolio

**Solve each problem. Explain the steps and the strategy or strategies you used for each. Then choose one from problems 1–4 for your Portfolio.**

1. Jamal had 918 cards. He gave an equal number to each of 17 classmates. At most, how many cards did Jamal give to each classmate?

2. The cafeteria charges $.75 for a slice of pizza. On Monday, 324 slices were sold. About how much money was received from the sale of pizza?

3. The fifth grade collected recyclables worth $36.18. The sixth grade collected recyclables worth $198. The seventh grade collected the same amount as the sixth grade. How much money was collected by the three grades?

4. Ada has 47 red buttons in her collection. She has three times as many blue buttons as she has green buttons. If there are 175 buttons of all three colors in the collection, how many blue buttons does Ada have?

### Tell about it.

5. **a.** Did you use multiplication to help you solve problem 3? problem 4? Explain.

   **b.** How is the hidden information in problem 3 different from the hidden information in problem 4?

Communicate

## For Rubric Scoring

**Listen for information on how your work will be scored.**

6. Solve the following riddle: I am a number. If you multiply me by 2, the result is 346 more than the result of multiplying me by 0. What number am I?

7. Solve the following riddle: I am a number. If you divide me by myself, the result is 252 less than what number?

8. Ari writes a number pattern in which the following are true:
   • The first number in the pattern is divisible by 2,
   • The second number is divisible by 3,
   • The third number is divisible by 9,
   • The pattern then repeats itself.
   Which of the numbers below could be the 12th number in Ari's pattern? Explain.

   240          250          260          270

# Number Theory and Fractions

**4**

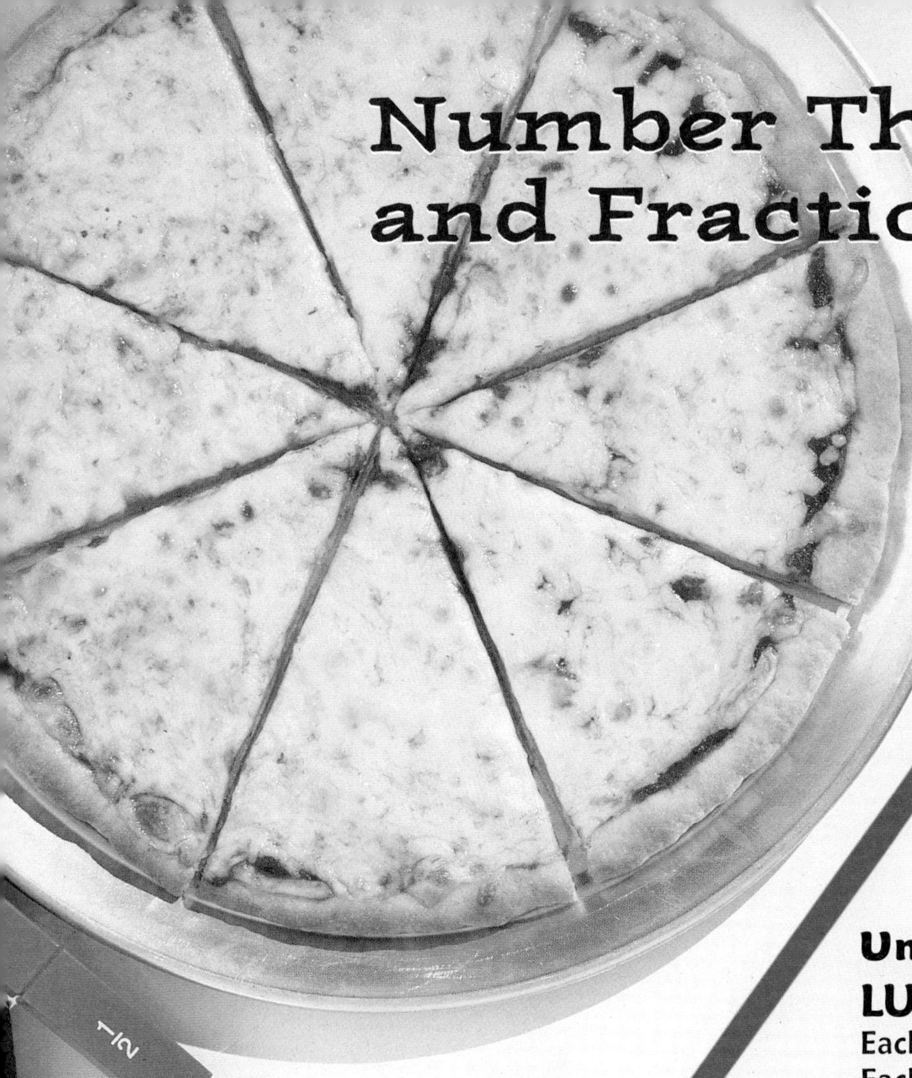

**Unfortunately for me,
LUNCH** is pizza and apple pie.
Each pizza is cut into 8 equal slices.
Each pie is cut into 6 equal slices.
And you know what that means:
### fractions

From *Math Curse* by Jon Scieszka

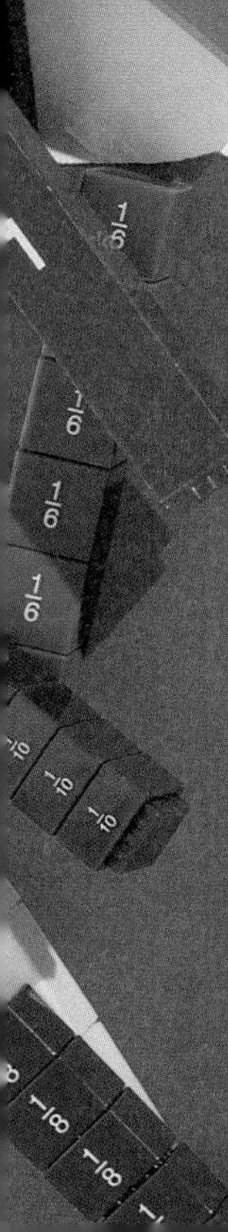

**In this chapter you will:**

Explore factors, primes,
  composites, and multiples

Rename equivalent fractions, improper
  fractions, and mixed numbers

Find whether a fraction is closer
  to 0, $\frac{1}{2}$, or 1

Compare and order fractions

Explore prime numbers and patterns
  on a calculator

Solve problems using organized lists

**Critical Thinking/Finding Together**

You ate $\frac{1}{4}$ of a pizza and your friend
ate $\frac{1}{6}$ of the remainder. What fraction
of the pizza was left?

133

# Exploring Prime and Composite Numbers

## Discover Together

**Materials Needed:** color tiles, paper, pencil

Any nonzero whole number can be represented by a set of arrays that form a rectangle. This is done by considering the nonzero whole number as a number of objects arranged in rows with an equal number of objects in each row.

1. Use color tiles to show all the rectangles in which 12 tiles can be arranged. (The figure above shows one rectangle.)

2. How many rectangles can be formed with 12 tiles?

Rectangles can be described by naming their length times their width. The rectangle above is a 4 × 3 rectangle.

3. Name all the rectangles formed with 12 tiles.

The length and width of each rectangle are factors of the number. Both 4 and 3 are factors of 12.

4. Name all the factors of 12.

5. How many factors does 12 have?

6. What do you notice about the number of rectangles formed with 12 tiles and the number of factors of 12?

Use color tiles to show all rectangles represented by each number. Then name the rectangles and factors for each number.

| | | | | |
|---|---|---|---|---|
| **7.** 5 | **8.** 9 | **9.** 3 | **10.** 8 | **11.** 10 |
| **12.** 4 | **13.** 15 | **14.** 7 | **15.** 13 | **16.** 6 |

Refer to exercises 7–16.

**17.** Which numbers have exactly 2 rectangles?
more than 2 rectangles?

**18.** Which numbers have exactly 2 factors?
more than 2 factors?

If a whole number is represented by exactly
2 rectangles, then the number is a *prime number*.

**19.** Which of the numbers in exercises 7–16 are
prime numbers?

**20.** How many factors does a prime number have?

If a whole number is represented by more than
2 rectangles, then the number is a *composite number*.

**21.** Which of the numbers in exercises 7–16 are
composite numbers?

**22.** How many factors does a composite number have?

## Communicate

Discuss ✓

**23.** What do you notice about the number of rectangles
and the number of factors of a whole number?

**24.** Use color tiles to show all rectangles represented
by 1. Is 1 a prime number or a composite number?
Explain why.

**25.** Is 2 a prime number or a composite number?
Explain your answer.

## Skills to Remember

### Find the missing factor.

 Algebra

**26.** $4 \times \underline{\ ?\ } = 32$

**27.** $7 \times \underline{\ ?\ } = 56$

**28.** $5 \times \underline{\ ?\ } = 40$

**29.** $\underline{\ ?\ } \times 6 = 48$

**30.** $\underline{\ ?\ } \times 9 = 81$

**31.** $\underline{\ ?\ } \times 10 = 90$

**32.** $6 \times \underline{\ ?\ } = 42$

**33.** $9 \times \underline{\ ?\ } = 45$

**34.** $3 \times \underline{\ ?\ } = 27$

# 4-2 Factors, Primes, and Composites

**Factors** are numbers that are multiplied to find a product.

$$5 \times 6 = 30 \qquad\qquad 5 \times 2 \times 3 = 30$$

factors · · · · · · · factors

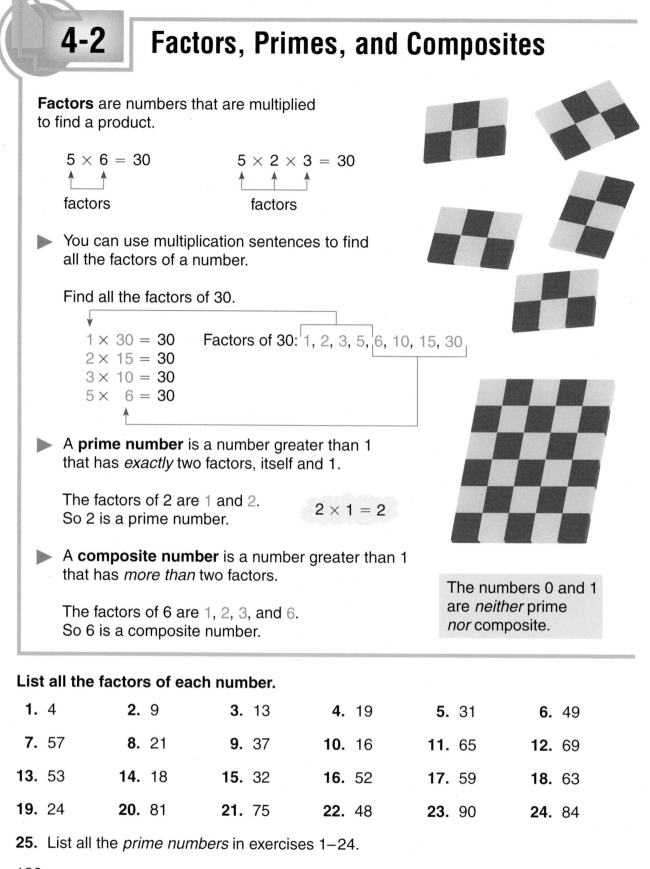

▶ You can use multiplication sentences to find all the factors of a number.

Find all the factors of 30.

$1 \times 30 = 30$ 
$2 \times 15 = 30$ 
$3 \times 10 = 30$ 
$5 \times\ \ 6 = 30$ 

Factors of 30: 1, 2, 3, 5, 6, 10, 15, 30

▶ A **prime number** is a number greater than 1 that has *exactly* two factors, itself and 1.

The factors of 2 are 1 and 2.
So 2 is a prime number.

$2 \times 1 = 2$

▶ A **composite number** is a number greater than 1 that has *more than* two factors.

The factors of 6 are 1, 2, 3, and 6.
So 6 is a composite number.

The numbers 0 and 1 are *neither* prime *nor* composite.

## List all the factors of each number.

1. 4 **2.** 9 **3.** 13 **4.** 19 **5.** 31 **6.** 49

**7.** 57 **8.** 21 **9.** 37 **10.** 16 **11.** 65 **12.** 69

**13.** 53 **14.** 18 **15.** 32 **16.** 52 **17.** 59 **18.** 63

**19.** 24 **20.** 81 **21.** 75 **22.** 48 **23.** 90 **24.** 84

**25.** List all the *prime numbers* in exercises 1–24.

# Prime Factorization

A composite number can be shown as the product of prime factors. This is called **prime factorization**.

▶ You can use a *factor tree* to find the prime factorization of a number.

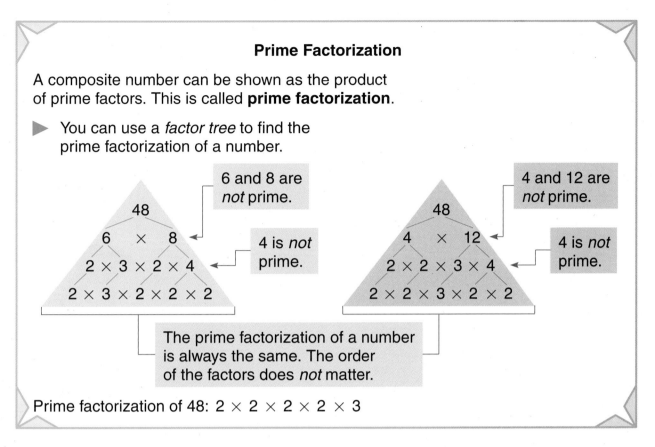

6 and 8 are *not* prime.

4 is *not* prime.

4 and 12 are *not* prime.

4 is *not* prime.

The prime factorization of a number is always the same. The order of the factors does *not* matter.

Prime factorization of 48: $2 \times 2 \times 2 \times 2 \times 3$

**Copy and complete.**

**26.**
```
      18
     /  \
    3 × 6
   / \ / \
  ? × ? × ?
```

**27.**
```
      45
     /  \
    5 × ?
   / \ / \
  ? × ? × ?
```

**28.**
```
          72
        /    \
       9  ×   8
      /|\    /|\
     3 × ? × 2 × ?
    /|\   /|\  |
   ? × ? × ? × ? × ?
```

**29.**
```
          80
        /    \
       8  ×   10
      /|\    /  \
     ? × ? × ? × ?
    /|\  |  |   |
   ? × ? × ? × ? × ?
```

**Use a factor tree to find the prime factorization of each.**

**30.** 12    **31.** 40    **32.** 54    **33.** 63    **34.** 84    **35.** 24

**36.** 78    **37.** 90    **38.** 28    **39.** 75    **40.** 50    **41.** 96

**42.** 120    **43.** 128    **44.** 108    **45.** 132    **46.** 138    **47.** 144

## Challenge

**48.** Write all the even prime numbers from 2 to 50.

**49.** Write all the odd composite numbers from 2 to 50.

# Greatest Common Factor

The **greatest common factor (GCF)** of two or more numbers is the greatest number that is a factor of these numbers.

▶ To find the greatest common factor (GCF):
- List the factors of each number.
- List the common factors of the numbers.
- Find which common factor is the greatest.

Find the greatest common factor (GCF) of 12 and 27.

$1 \times 12 = 12$      $1 \times 27 = 27$
$2 \times 6 = 12$      $3 \times 9 = 27$
$3 \times 4 = 12$

Factors of 12:      Factors of 27:
1, 2, 3, 4, 6, 12      1, 3, 9, 27

Common factors of 12 and 27: 1, 3
Greatest common factor (GCF) of 12 and 27: 3

## Study this example.

Find the greatest common factor (GCF) of 16, 28, and 32.

$1 \times 16 = 16$      $1 \times 28 = 28$      $1 \times 32 = 32$
$2 \times 8 = 16$      $2 \times 14 = 28$      $2 \times 16 = 32$
$4 \times 4 = 16$      $4 \times 7 = 28$      $4 \times 8 = 32$

Factors of 16:      Factors of 28:      Factors of 32:
1, 2, 4, 8, 16      1, 2, 4, 7, 14, 28      1, 2, 4, 8, 16, 32

Common factors of 16, 28, and 32: 1, 2, 4
Greatest common factor (GCF) of 16, 28, and 32: 4

## Copy and complete the table.

| | Number | Factors | Common Factors | GCF |
|---|---|---|---|---|
| **1.** | 6 | ? ? ? ? | ? ? | ? |
| | 10 | ? ? ? ? | | |
| **2.** | 18 | ? ? ? ? ? ? | ? ? ? ? | ? |
| | 24 | ? ? ? ? ? ? ? ? | | |

**List the factors of each number. Then underline the common factors of each pair of numbers.**

**3.** 6 and 9      **4.** 3 and 15      **5.** 4 and 11      **6.** 10 and 24

**7.** 8 and 20      **8.** 11 and 26      **9.** 8 and 12      **10.** 10 and 30

**List the common factors of each set of numbers. Then underline the GCF.**

**11.** 15 and 21      **12.** 24 and 32      **13.** 12 and 72      **14.** 27 and 36

**15.** 24 and 36      **16.** 16 and 20      **17.** 14 and 32      **18.** 18 and 36

**19.** 3, 9, and 15      **20.** 4, 8, and 12      **21.** 6, 9, and 12

**Find the GCF of each set of numbers.**

**22.** 45 and 60      **23.** 24 and 40      **24.** 18 and 21      **25.** 16 and 48

**26.** 30 and 45      **27.** 48 and 56      **28.** 36 and 63      **29.** 36 and 42

**30.** 12, 15, and 18      **31.** 7, 35, and 49      **32.** 16, 20, and 24

## PROBLEM SOLVING

**33.** Ms. Durkin wants to package 16 math books and 28 science books equally without mixing the books and with none left over. What is the greatest number of books she can put in each package?

**34.** Mr. Diaz wants to group the 18 girls and 24 boys at the summer camp separately into teams. To be able to match boys with girls during games, the team sizes have to be the same. What is the greatest team size the boys and girls can form?

## Critical Thinking

Communicate ✓

**Write *True* or *False* for each statement. Explain your answer.**

**35.** One is a common factor of every set of numbers.

**36.** Zero can be a common factor of a set of numbers.

**37.** Two numbers can have no common factors.

**38.** The greatest common factor of two prime numbers is 1.

# 4-4 Fraction Sense: Closer to 0, $\frac{1}{2}$, 1

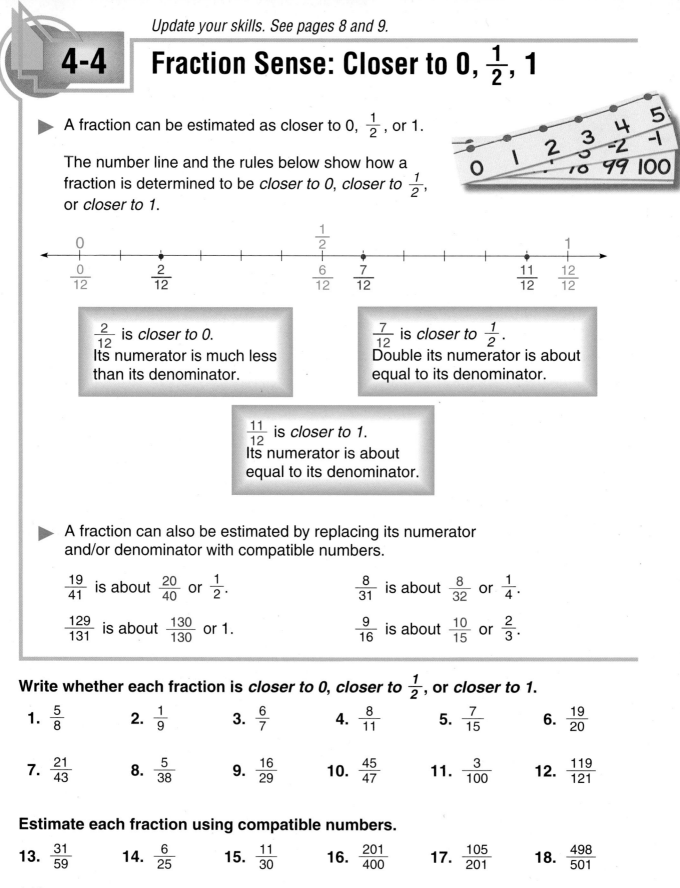

▶ A fraction can be estimated as closer to 0, $\frac{1}{2}$, or 1.

The number line and the rules below show how a fraction is determined to be *closer to 0, closer to $\frac{1}{2}$, or closer to 1.*

$\frac{2}{12}$ is *closer to 0.*
Its numerator is much less than its denominator.

$\frac{7}{12}$ is *closer to $\frac{1}{2}$.*
Double its numerator is about equal to its denominator.

$\frac{11}{12}$ is *closer to 1.*
Its numerator is about equal to its denominator.

▶ A fraction can also be estimated by replacing its numerator and/or denominator with compatible numbers.

$\frac{19}{41}$ is about $\frac{20}{40}$ or $\frac{1}{2}$.

$\frac{8}{31}$ is about $\frac{8}{32}$ or $\frac{1}{4}$.

$\frac{129}{131}$ is about $\frac{130}{130}$ or 1.

$\frac{9}{16}$ is about $\frac{10}{15}$ or $\frac{2}{3}$.

**Write whether each fraction is *closer to 0, closer to $\frac{1}{2}$, or closer to 1.***

1. $\frac{5}{8}$
2. $\frac{1}{9}$
3. $\frac{6}{7}$
4. $\frac{8}{11}$
5. $\frac{7}{15}$
6. $\frac{19}{20}$

7. $\frac{21}{43}$
8. $\frac{5}{38}$
9. $\frac{16}{29}$
10. $\frac{45}{47}$
11. $\frac{3}{100}$
12. $\frac{119}{121}$

**Estimate each fraction using compatible numbers.**

13. $\frac{31}{59}$
14. $\frac{6}{25}$
15. $\frac{11}{30}$
16. $\frac{201}{400}$
17. $\frac{105}{201}$
18. $\frac{498}{501}$

# Finding Equivalent Fractions

▶ You can *multiply* or *divide* the numerator and denominator by the *same nonzero number* to find equivalent fractions.

$$\frac{2 \times 2}{6 \times 2} = \frac{4}{12} \qquad \frac{2 \div 2}{6 \div 2} = \frac{1}{3} \qquad \frac{2}{6} = \frac{4}{12} = \frac{1}{3} \leftarrow$$ These are equivalent fractions.

▶ You can also multiply or divide the numerator and denominator by the same nonzero number to find a missing numerator or denominator in equivalent fractions.

$$\frac{5}{8} = \frac{?}{24}$$

Think:
$$8 \times \underline{?} = 24$$
$$8 \times 3 = 24$$

$$\frac{5 \times 3}{8 \times 3} = \frac{15}{24}$$

$$\frac{5}{8} = \frac{15}{24}$$

$$\frac{18}{27} = \frac{2}{?}$$

Think:
$$18 \div \underline{?} = 2$$
$$18 \div 9 = 2$$

$$\frac{18 \div 9}{27 \div 9} = \frac{2}{3}$$

$$\frac{18}{27} = \frac{2}{3}$$

**Write three equivalent fractions for each.**

19. $\frac{1}{9}$    20. $\frac{2}{5}$    21. $\frac{3}{7}$    22. $\frac{7}{9}$    23. $\frac{5}{6}$    24. $\frac{7}{8}$

**Write the missing number to complete the equivalent fraction.**

*Algebra*

25. $\frac{4}{5} = \frac{?}{25}$    26. $\frac{7}{8} = \frac{21}{?}$    27. $\frac{21}{49} = \frac{?}{7}$    28. $\frac{32}{40} = \frac{4}{?}$

29. $\frac{2}{3} = \frac{4}{?} = \frac{8}{?}$    30. $\frac{5}{8} = \frac{10}{?} = \frac{15}{?}$    31. $\frac{6}{7} = \frac{12}{?} = \frac{18}{?}$

32. Explain in your Math Journal why you can multiply or divide the numerator and denominator of a fraction by the same number without changing its value.

*Math Journal*

## PROBLEM SOLVING

33. Four ninths of the class watched the glee club concert. Is the class attendance at the concert less than or more than $\frac{1}{2}$ of the class?

34. The fifth grade's class banner is $\frac{7}{8}$ yd long. The sixth grade's class banner is $\frac{14}{16}$ yd long. Which banner is longer? Explain.

*Update your skills. See page 9.*

## 4-5 Fractions in Lowest Terms

Eighteen of the 24 stamps in Ben's collection are foreign. Write a fraction in lowest terms to show what part of the stamps in Ben's collection are foreign.

▶ A fraction is in **lowest terms**, or in **simplest form**, when its numerator and denominator have no common factor other than 1.

To **rename a fraction** as an equivalent fraction in lowest terms, or in simplest form:
- Find the greatest common factor (GCF) of the numerator and the denominator.
- Divide the numerator and the denominator by their greatest common factor (GCF).

Factors of 18: 1, 2, 3, 6, 9, 18
Factors of 24: 1, 2, 3, 4, 6, 8, 12, 24
GCF of 18 and 24: 6

$$\frac{18}{24} = \frac{18 \div 6}{24 \div 6} = \frac{3}{4} \longleftarrow \text{lowest terms}$$

In lowest terms, $\frac{3}{4}$ of the stamps in Ben's collection are foreign.

Remember: When 1 is the GCF of the numerator and denominator, the fraction is in lowest terms.

**Is each fraction in lowest terms? Write *Yes* or *No*.**

1. $\frac{3}{5}$    2. $\frac{2}{6}$    3. $\frac{2}{9}$    4. $\frac{2}{4}$    5. $\frac{6}{8}$    6. $\frac{4}{7}$

7. $\frac{5}{10}$    8. $\frac{2}{11}$    9. $\frac{2}{10}$    10. $\frac{4}{8}$    11. $\frac{7}{8}$    12. $\frac{3}{12}$

13. $\frac{6}{15}$    14. $\frac{12}{31}$    15. $\frac{10}{19}$    16. $\frac{7}{21}$    17. $\frac{10}{25}$    18. $\frac{23}{26}$

**Write the letter of the equivalent fraction in lowest terms.**

19. $\frac{6}{8}$        a. $\frac{2}{3}$        b. $\frac{1}{3}$        c. $\frac{2}{4}$        d. $\frac{3}{4}$

20. $\frac{9}{45}$        a. $\frac{1}{5}$        b. $\frac{2}{10}$        c. $\frac{2}{5}$        d. $\frac{4}{10}$

21. $\frac{18}{27}$        a. $\frac{1}{3}$        b. $\frac{2}{3}$        c. $\frac{3}{9}$        d. $\frac{4}{6}$

**Write each fraction in lowest terms.**

22. $\frac{3}{6}$    23. $\frac{6}{9}$    24. $\frac{4}{10}$    25. $\frac{3}{12}$    26. $\frac{5}{15}$    27. $\frac{8}{24}$

28. $\frac{6}{18}$    29. $\frac{9}{12}$    30. $\frac{8}{20}$    31. $\frac{6}{24}$    32. $\frac{4}{22}$    33. $\frac{8}{12}$

34. $\frac{5}{25}$    35. $\frac{4}{20}$    36. $\frac{7}{21}$    37. $\frac{4}{18}$    38. $\frac{6}{15}$    39. $\frac{9}{63}$

**Write each fraction in simplest form.**

40. $\frac{30}{40}$    41. $\frac{20}{80}$    42. $\frac{16}{24}$    43. $\frac{24}{48}$    44. $\frac{20}{28}$    45. $\frac{24}{36}$

46. $\frac{28}{35}$    47. $\frac{24}{32}$    48. $\frac{32}{44}$    49. $\frac{18}{63}$    50. $\frac{45}{72}$    51. $\frac{33}{66}$

52. $\frac{34}{51}$    53. $\frac{14}{42}$    54. $\frac{20}{32}$    55. $\frac{35}{40}$    56. $\frac{18}{45}$    57. $\frac{36}{72}$

**PROBLEM SOLVING** Write each answer in simplest form.

58. There were 8 stamp collections at the Hobby Fair. If there were 24 hobbies in all, what fractional part of the hobbies were stamp collections?

59. Three out of 30 visitors to the Hobby Fair are stamp collectors. What fractional part of the visitors are stamp collectors?

60. Seven out of 28 stamps in Kyle's collection are from Europe. What fractional part of Kyle's collection is *not* from Europe?

61. At a recent spelling bee, 15 out of 24 contestants were girls. What fractional part of the contestants were boys?

62. A scientist worked 36 hours on an experiment last week. She spent 15 hours doing research and 12 hours recording data. The rest of the time she spent writing her report. What fractional part of her time was spent writing her report?

 **Share Your Thinking**    Communicate ✓

63. Find the greatest common factor of the numerator and the denominator of fractions in lowest terms. Then explain to your teacher how you can identify when a fraction is in simplest form.

# 4-6 Fractions in Higher Terms

A fraction is in **higher terms** than its equivalent fraction when its numerator and denominator are *greater than* the numerator and denominator of its equivalent fraction.

$$\frac{1}{2} = \frac{3}{6} = \frac{6}{12}$$

$\frac{3}{6}$ and $\frac{6}{12}$ are higher terms than $\frac{1}{2}$.    $\frac{1}{2}$, $\frac{3}{6}$, and $\frac{6}{12}$ are equivalent.

▶ To **rename a fraction** as an equivalent fraction in higher terms, *multiply* the numerator and the denominator by the *same* number.

$$\frac{1}{2} = \frac{1 \times 3}{2 \times 3} = \frac{3}{6}$$ ◀ higher-terms fraction    $$\frac{1}{2} = \frac{1 \times 6}{2 \times 6} = \frac{6}{12}$$ ◀ higher-terms fraction

▶ To find a missing numerator or denominator in a higher-terms fraction:
  • Find what number the given numerator or denominator was multiplied by.
  • Multiply the other term of the given fraction by the same number.

$$\frac{2}{3} = \frac{14}{?}$$    Think: $2 \times \underline{\;?\;} = 14$
$2 \times \underline{\;7\;} = 14$

$$\frac{2 \times 7}{3 \times 7} = \frac{14}{21} \quad \text{So } \frac{2}{3} = \frac{14}{21}.$$

**Study these examples.**

$$\frac{6}{8} = \frac{?}{64}$$    Think: $8 \times \underline{\;?\;} = 64$
$8 \times \underline{\;8\;} = 64$

$$\frac{20}{25} = \frac{40}{?}$$    Think: $20 \times \underline{\;?\;} = 40$
$20 \times \underline{\;2\;} = 40$

$$\frac{6 \times 8}{8 \times 8} = \frac{48}{64} \quad \text{So } \frac{6}{8} = \frac{48}{64}.$$

$$\frac{20 \times 2}{25 \times 2} = \frac{40}{50} \quad \text{So } \frac{20}{25} = \frac{40}{50}.$$

**Write the letter of the equivalent fraction in higher terms.**

1. $\frac{1}{5}$    **a.** $\frac{3}{16}$    **b.** $\frac{4}{20}$    **c.** $\frac{3}{10}$    **d.** $\frac{5}{10}$

2. $\frac{3}{4}$    **a.** $\frac{10}{12}$    **b.** $\frac{9}{10}$    **c.** $\frac{5}{8}$    **d.** $\frac{12}{16}$

**Find the missing term.**

**3.** $\dfrac{6}{8} = \dfrac{?}{16}$   **4.** $\dfrac{2}{3} = \dfrac{?}{9}$   **5.** $\dfrac{4}{6} = \dfrac{12}{?}$   **6.** $\dfrac{7}{8} = \dfrac{21}{?}$

**7.** $\dfrac{4}{5} = \dfrac{?}{45}$   **8.** $\dfrac{3}{4} = \dfrac{15}{?}$   **9.** $\dfrac{3}{5} = \dfrac{15}{?}$   **10.** $\dfrac{7}{10} = \dfrac{?}{50}$   **11.** $\dfrac{6}{8} = \dfrac{?}{64}$

**12.** $\dfrac{7}{10} = \dfrac{?}{20}$   **13.** $\dfrac{2}{3} = \dfrac{24}{?}$   **14.** $\dfrac{4}{9} = \dfrac{20}{?}$   **15.** $\dfrac{7}{12} = \dfrac{49}{?}$   **16.** $\dfrac{10}{15} = \dfrac{20}{?}$

**17.** $\dfrac{8}{10} = \dfrac{?}{60}$   **18.** $\dfrac{2}{5} = \dfrac{16}{?}$   **19.** $\dfrac{3}{4} = \dfrac{36}{?}$   **20.** $\dfrac{8}{20} = \dfrac{?}{80}$   **21.** $\dfrac{6}{11} = \dfrac{?}{55}$

**22.** $\dfrac{5}{8} = \dfrac{?}{32}$   **23.** $\dfrac{5}{7} = \dfrac{40}{?}$   **24.** $\dfrac{8}{9} = \dfrac{72}{?}$   **25.** $\dfrac{3}{11} = \dfrac{9}{?}$   **26.** $\dfrac{7}{12} = \dfrac{28}{?}$

**Find equivalent fractions.**

**27.** $\dfrac{1}{3} = \dfrac{2}{6} = \dfrac{?}{9} = \dfrac{?}{12} = \dfrac{5}{?}$   **28.** $\dfrac{3}{4} = \dfrac{?}{8} = \dfrac{9}{?} = \dfrac{?}{16} = \dfrac{15}{?}$

**29.** $\dfrac{3}{5} = \dfrac{6}{?} = \dfrac{?}{15} = \dfrac{12}{?} = \dfrac{?}{25}$   **30.** $\dfrac{5}{6} = \dfrac{10}{?} = \dfrac{?}{18} = \dfrac{20}{?} = \dfrac{?}{30}$

**31.** $\dfrac{4}{7} = \dfrac{?}{14} = \dfrac{?}{28} = \dfrac{32}{?} = \dfrac{12}{?}$   **32.** $\dfrac{8}{9} = \dfrac{?}{18} = \dfrac{24}{?} = \dfrac{40}{?} = \dfrac{?}{72}$

**33.** $\dfrac{1}{2} = \dfrac{?}{8} = \dfrac{6}{?} = \dfrac{?}{16} = \dfrac{?}{20} = \dfrac{12}{?}$   **34.** $\dfrac{2}{3} = \dfrac{?}{6} = \dfrac{6}{?} = \dfrac{8}{?} = \dfrac{?}{15} = \dfrac{?}{18}$

**35.** $\dfrac{4}{5} = \dfrac{?}{10} = \dfrac{?}{15} = \dfrac{?}{20} = \dfrac{20}{?} = \dfrac{24}{?}$   **36.** $\dfrac{3}{7} = \dfrac{?}{14} = \dfrac{?}{21} = \dfrac{12}{?} = \dfrac{15}{?} = \dfrac{18}{?}$

## PROBLEM SOLVING

**37.** Eden had $\dfrac{1}{3}$ of a pie left. She cuts this into two pieces. Write a fraction that shows the two pieces as part of the whole pie.

**38.** Seven twelfths of the flowers in the box are red. Write a fraction in higher terms to show what part of the flowers in the box are red.

**Project**

**39.**  • Research how the following musical notes are represented: whole note, half note, quarter note, eighth note.

• Make a poster entitled "Musical Equivalences" that shows pictures of notes having equivalent values. Display this in your classroom.

# Multiples: LCM and LCD

▶ The **multiples** of a number are the products of that number and 0, 1, 2, 3, 4, . . .

| Multiples of 3 | $\begin{array}{r}3\\ \times0\\\hline 0\end{array}$ | $\begin{array}{r}3\\ \times1\\\hline 3\end{array}$ | $\begin{array}{r}3\\ \times2\\\hline 6\end{array}$ | $\begin{array}{r}3\\ \times3\\\hline 9\end{array}$ | $\begin{array}{r}3\\ \times4\\\hline 12\end{array}$ | $\begin{array}{r}3\\ \times5\\\hline 15\end{array}$ | and so on. |

| Multiples of 4 | $\begin{array}{r}4\\ \times0\\\hline 0\end{array}$ | $\begin{array}{r}4\\ \times1\\\hline 4\end{array}$ | $\begin{array}{r}4\\ \times2\\\hline 8\end{array}$ | $\begin{array}{r}4\\ \times3\\\hline 12\end{array}$ | $\begin{array}{r}4\\ \times4\\\hline 16\end{array}$ | $\begin{array}{r}4\\ \times5\\\hline 20\end{array}$ | and so on. |

▶ Nonzero multiples that are *the same* for two or more numbers are called **common multiples**.

Multiples of 3: 3, 6, 9, 12, 15, 18, 21, 24, . . .
Multiples of 4: 4, 8, 12, 16, 20, 24, 28, 32, . . .
Common multiples of 3 and 4: 12, 24, . . .

▶ The **least common multiple (LCM)** of two or more numbers is the *least number* that is a *multiple* of those numbers.

Least common multiple (LCM) of 3 and 4: 12

**Study this example.**

Multiples of 2: 2, 4, 6, 8, 10, 12, . . .
Multiples of 3: 3, 6, 9, 12, 15, . . .
Multiples of 6: 6, 12, 18, 24, . . .

Common multiples of 2, 3, and 6: 6, 12, . . .
Least common multiple (LCM) of 2, 3, and 6: 6

**List the first twelve nonzero multiples of each number.**

**1.** 5     **2.** 7     **3.** 8     **4.** 9     **5.** 1     **6.** 10

**List the first four common multiples of each set of numbers.**

**7.** 3, 5     **8.** 6, 9     **9.** 4, 8     **10.** 3, 9     **11.** 3, 4, 9

**Find the least common multiple (LCM) of each set of numbers.**

**12.** 2, 4      **13.** 6, 8      **14.** 9, 12      **15.** 3, 10      **16.** 10, 15

**17.** 3, 4, and 9      **18.** 5, 6, and 10      **19.** 2, 7, and 8      **20.** 12, 16, and 18

---

### Least Common Denominator (LCD)

The **least common denominator (LCD)** of two or more fractions is the least common multiple (LCM) of the denominators.

Find the least common denominator (LCD) of $\frac{3}{4}$, $\frac{2}{5}$, and $\frac{9}{10}$.

- Find the common multiples of the denominators.

  Multiples of 4: 4, 8, 12, 16, 20, . . .
  Multiples of 5: 5, 10, 15, 20, . . .
  Multiples of 10: 10, 20, 30, . . .

- Find the LCM of the denominators. This is the least common denominator (LCD).

  LCM of 4, 5, and 10: 20
  So LCD of $\frac{3}{4}$, $\frac{2}{5}$, and $\frac{9}{10}$: 20

---

**Find the least common denominator (LCD) of each set of fractions.**

**21.** $\frac{1}{2}$, $\frac{3}{4}$      **22.** $\frac{2}{3}$, $\frac{1}{9}$      **23.** $\frac{1}{3}$, $\frac{3}{5}$      **24.** $\frac{3}{4}$, $\frac{1}{6}$      **25.** $\frac{5}{6}$, $\frac{5}{8}$

**26.** $\frac{1}{3}$, $\frac{7}{10}$      **27.** $\frac{5}{8}$, $\frac{7}{12}$      **28.** $\frac{3}{10}$, $\frac{2}{15}$      **29.** $\frac{2}{3}$, $\frac{3}{11}$      **30.** $\frac{2}{9}$, $\frac{4}{15}$

**31.** $\frac{3}{4}$, $\frac{2}{5}$, and $\frac{9}{20}$      **32.** $\frac{1}{3}$, $\frac{5}{6}$, and $\frac{7}{12}$      **33.** $\frac{1}{12}$, $\frac{3}{16}$, and $\frac{5}{18}$

### PROBLEM SOLVING

**34.** Blue paper sells in multiples of 5 sheets, and green paper sells in multiples of 10 sheets. What is the least number of sheets of each color Ted can buy to have the same number of each color?

**35.** Trisha colors every third square in her art design yellow and every fourth square in her art design red. Of 36 squares in the design, how many will be colored both red and yellow?

### Share Your Thinking

Communicate ✓

**36.** Tell a classmate what method you use and how you solve the following problem:
The least common multiple of three consecutive numbers is 60. The sum of the numbers is 15. What are the numbers?

# Mixed Numbers

Rodney feeds his kittens two and three fourths cups of milk each day.

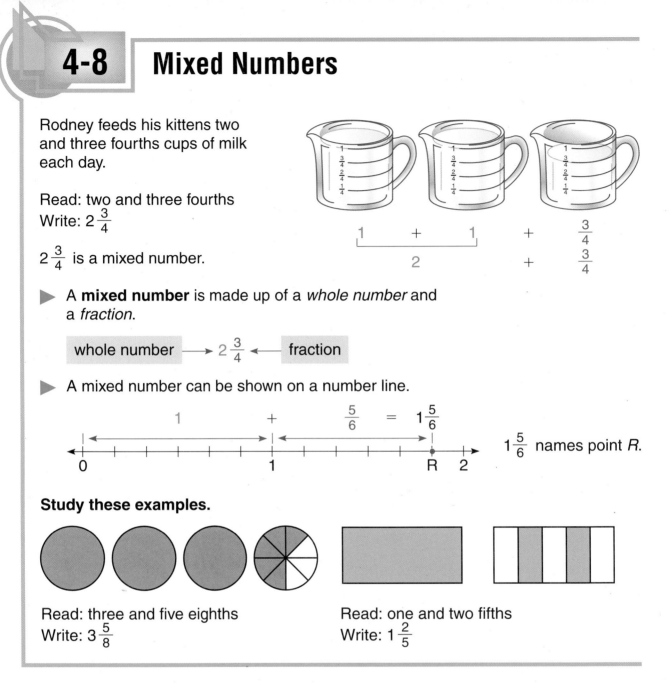

Read: two and three fourths
Write: $2\frac{3}{4}$

$2\frac{3}{4}$ is a mixed number.

$1 + 1$
$2$
$+ \frac{3}{4}$
$+ \frac{3}{4}$

▶ A **mixed number** is made up of a *whole number* and a *fraction*.

whole number ⟶ $2\frac{3}{4}$ ⟵ fraction

▶ A mixed number can be shown on a number line.

$1 + \frac{5}{6} = 1\frac{5}{6}$

$1\frac{5}{6}$ names point *R*.

## Study these examples.

Read: three and five eighths
Write: $3\frac{5}{8}$

Read: one and two fifths
Write: $1\frac{2}{5}$

## Write the mixed number that represents the shaded part.

**1.**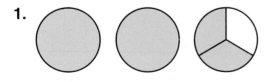

**2.**

## Write as a mixed number.

**3.** seven and one sixth

**4.** four and five eighths

**5.** eleven and four fifths

**6.** nine and six sevenths

**Write the mixed number for each point.**

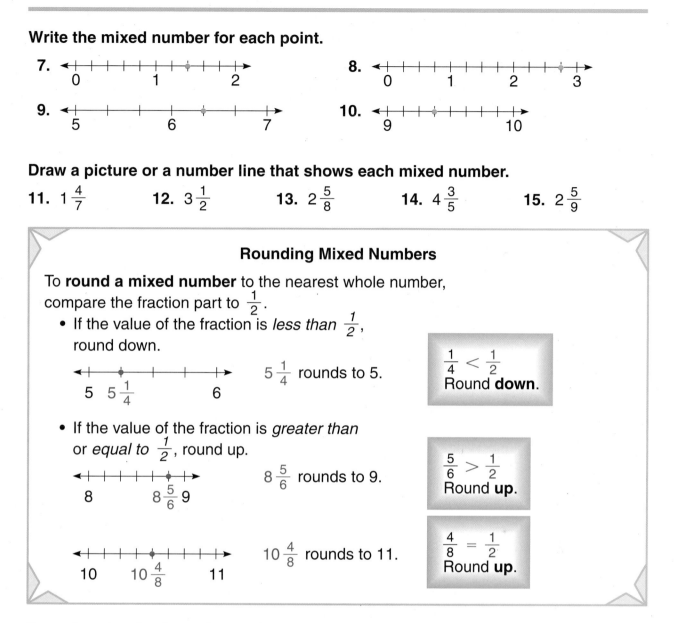

7.
0    1    2

8.
0    1    2    3

9.
5    6    7

10.
9    10

**Draw a picture or a number line that shows each mixed number.**

11. $1\frac{4}{7}$    12. $3\frac{1}{2}$    13. $2\frac{5}{8}$    14. $4\frac{3}{5}$    15. $2\frac{5}{9}$

---

**Rounding Mixed Numbers**

To **round a mixed number** to the nearest whole number, compare the fraction part to $\frac{1}{2}$.

- If the value of the fraction is *less than* $\frac{1}{2}$, round down.

  5    $5\frac{1}{4}$    6          $5\frac{1}{4}$ rounds to 5.

  $\frac{1}{4} < \frac{1}{2}$
  Round **down**.

- If the value of the fraction is *greater than* or *equal to* $\frac{1}{2}$, round up.

  8    $8\frac{5}{6}$ 9          $8\frac{5}{6}$ rounds to 9.

  $\frac{5}{6} > \frac{1}{2}$
  Round **up**.

  10    $10\frac{4}{8}$    11          $10\frac{4}{8}$ rounds to 11.

  $\frac{4}{8} = \frac{1}{2}$
  Round **up**.

---

**Round each mixed number to the nearest whole number.**

16. $3\frac{1}{3}$    17. $9\frac{5}{7}$    18. $6\frac{4}{8}$    19. $18\frac{1}{5}$    20. $19\frac{10}{13}$    21. $12\frac{4}{9}$

22. $7\frac{1}{2}$    23. $10\frac{3}{8}$    24. $5\frac{13}{15}$    25. $11\frac{4}{9}$    26. $8\frac{5}{8}$    27. $13\frac{6}{12}$

**PROBLEM SOLVING**

28. A recipe calls for $2\frac{1}{3}$ cups of flour. About how many cups of flour will be needed for the recipe?

29. Sabina studied for $3\frac{3}{8}$ hours. About how many hours did she study?

# Improper Fractions

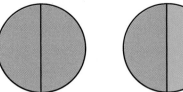
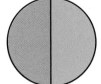

An **improper fraction** has a numerator that is *equal to* or *greater than* its denominator. It has a value that is *equal to* or *greater than 1*.

$\frac{2}{2}$ ⟩ $2 = 2$    So $\frac{2}{2} = 1$ and $\frac{2}{2}$ is an improper fraction.

$\frac{3}{2}$ ⟩ $3 > 2$    So $\frac{3}{2} > 1$ and $\frac{3}{2}$ is an improper fraction.

▶ You can express an improper fraction as a whole number or a mixed number. The number line shows that:

$\frac{2}{2} = 1$     $\frac{3}{2} = 1\frac{1}{2}$     $\frac{4}{2} = 2$

whole number    mixed number    whole number

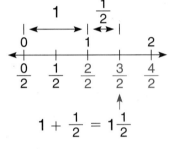

$1 + \frac{1}{2} = 1\frac{1}{2}$

▶ To **rename an improper fraction** as a whole number or a mixed number in simplest form:
- Divide the numerator by the denominator.
- Write the quotient as the whole number part.
- Write the remainder as the numerator and the divisor as the denominator of the fraction part.
- Express the fraction in simplest form.

$\frac{22}{6} = \underline{\ ?\ }$

$\frac{22}{6} \longrightarrow 6\overline{)22}$   $3$ R4

$\frac{22}{6} = 3\frac{4}{6}$

$3\frac{4}{6} = 3\frac{2}{3}$ ← simplest form

**Study these examples.**

$\frac{18}{9} \longrightarrow 9\overline{)18}$   $2$   $= 2$

$\frac{39}{7} \longrightarrow 7\overline{)39}$   $5$ R4   $= 5\frac{4}{7}$

**Find the improper fractions in each set.**

**1. a.** $\frac{9}{8}$    **b.** $\frac{7}{7}$    **c.** $\frac{3}{5}$    **d.** $\frac{6}{7}$    **e.** $\frac{10}{7}$    **f.** $\frac{8}{4}$

**2. a.** $\frac{5}{11}$    **b.** $\frac{17}{4}$    **c.** $\frac{25}{5}$    **d.** $\frac{5}{8}$    **e.** $\frac{9}{2}$    **f.** $\frac{36}{6}$

**Write a numerator to give each improper fraction a value equal to 1.**

3. $\frac{?}{4}$
4. $\frac{?}{6}$
5. $\frac{?}{3}$
6. $\frac{?}{8}$
7. $\frac{?}{10}$
8. $\frac{?}{7}$

9. $\frac{?}{12}$
10. $\frac{?}{9}$
11. $\frac{?}{15}$
12. $\frac{?}{11}$
13. $\frac{?}{13}$
14. $\frac{?}{5}$

**Write a numerator to give each improper fraction a value greater than 1.**

15. $\frac{?}{4}$
16. $\frac{?}{9}$
17. $\frac{?}{5}$
18. $\frac{?}{7}$
19. $\frac{?}{10}$
20. $\frac{?}{6}$

21. $\frac{?}{8}$
22. $\frac{?}{11}$
23. $\frac{?}{19}$
24. $\frac{?}{3}$
25. $\frac{?}{15}$
26. $\frac{?}{12}$

**Write each as a whole number or a mixed number in simplest form.**

27. $\frac{10}{9}$
28. $\frac{44}{7}$
29. $\frac{24}{8}$
30. $\frac{18}{3}$
31. $\frac{6}{4}$
32. $\frac{50}{6}$

33. $\frac{42}{10}$
34. $\frac{37}{7}$
35. $\frac{53}{6}$
36. $\frac{41}{3}$
37. $\frac{30}{8}$
38. $\frac{65}{7}$

39. $\frac{75}{9}$
40. $\frac{45}{8}$
41. $\frac{26}{2}$
42. $\frac{110}{5}$
43. $\frac{192}{9}$
44. $\frac{210}{8}$

**Tell which whole number each improper fraction is closer to.**
You may use a number line.

45. $\frac{9}{2}$
46. $\frac{13}{3}$
47. $\frac{19}{5}$
48. $\frac{40}{7}$
49. $\frac{65}{9}$
50. $\frac{88}{6}$

**PROBLEM SOLVING**
**Write the answer as a mixed number.**

51. A piece of lumber is 43 inches long. If it is cut into 6 equal pieces, how long is each piece?

52. If 6 identical items weigh a total of 23 pounds, how much does each item weigh?

 **Finding Together**

*Discuss* ✓

53. Ms. Rill served 4 different pies for the party: apple, blueberry, cherry, and banana. She cut each pie into eighths. After the party, she found that there were 3 slices of apple pie, 1 slice of blueberry pie, 2 slices of cherry pie, and 5 slices of banana pie left. Write an improper fraction and a mixed number to express the number of pie slices eaten. Explain the method you used to find your answer.

# Comparing and Ordering Fractions

Compare: $\dfrac{5}{8}$ ? $\dfrac{7}{8}$

▶ To **compare fractions** with *like denominators*, compare the numerators.

Compare: $\dfrac{5}{6}$ ? $\dfrac{1}{2}$

▶ To **compare fractions** with *unlike denominators*:
  • Find the least common denominator (LCD) of the fractions.
  • Use the LCD to rename the fractions as equivalent fractions with the same denominator.

  • Compare the numerators.

Compare: $1\dfrac{1}{4}$ ? $1\dfrac{7}{8}$

▶ To **compare mixed numbers**:
  • Compare the whole numbers.
  • Compare the fractions.

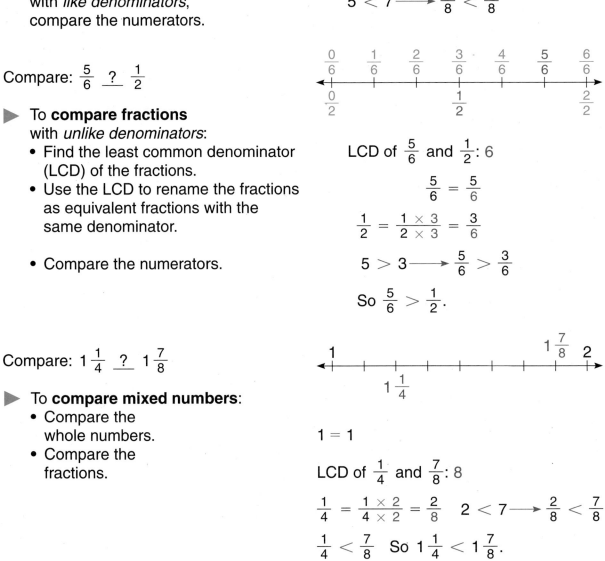

$5 < 7 \longrightarrow \dfrac{5}{8} < \dfrac{7}{8}$

LCD of $\dfrac{5}{6}$ and $\dfrac{1}{2}$: 6

$$\dfrac{5}{6} = \dfrac{5}{6}$$

$$\dfrac{1}{2} = \dfrac{1 \times 3}{2 \times 3} = \dfrac{3}{6}$$

$5 > 3 \longrightarrow \dfrac{5}{6} > \dfrac{3}{6}$

So $\dfrac{5}{6} > \dfrac{1}{2}$.

$1 = 1$

LCD of $\dfrac{1}{4}$ and $\dfrac{7}{8}$: 8

$\dfrac{1}{4} = \dfrac{1 \times 2}{4 \times 2} = \dfrac{2}{8}$    $2 < 7 \longrightarrow \dfrac{2}{8} < \dfrac{7}{8}$

$\dfrac{1}{4} < \dfrac{7}{8}$   So $1\dfrac{1}{4} < 1\dfrac{7}{8}$.

**Study these examples.**

$\dfrac{19}{6}$ ? $\dfrac{17}{6}$    Think: $19 > 17$

$\dfrac{19}{6} > \dfrac{17}{6}$

$3\dfrac{4}{5}$ ? $5\dfrac{4}{5}$    Think: $3 < 5$

$3\dfrac{4}{5} < 5\dfrac{4}{5}$

$\dfrac{21}{4}$ ? $5\dfrac{3}{4}$    Think: $\dfrac{21}{4} = 5\dfrac{1}{4}$

$\dfrac{21}{4} < 5\dfrac{3}{4}$

**Compare. Write $<$, $=$, or $>$.**

1. $\frac{3}{4}$ ___?___ $\frac{2}{4}$

2. $\frac{4}{9}$ ___?___ $\frac{7}{9}$

3. $\frac{5}{6}$ ___?___ $\frac{11}{12}$

4. $\frac{4}{5}$ ___?___ $\frac{12}{15}$

5. $\frac{5}{5}$ ___?___ $\frac{10}{10}$

6. $\frac{8}{8}$ ___?___ $\frac{3}{4}$

7. $3\frac{2}{5}$ ___?___ $3\frac{4}{5}$

8. $5\frac{1}{9}$ ___?___ $2\frac{1}{9}$

9. $3\frac{7}{8}$ ___?___ $3\frac{5}{6}$

10. $1\frac{5}{9}$ ___?___ $1\frac{2}{3}$

11. $4\frac{1}{3}$ ___?___ $4\frac{3}{4}$

12. $\frac{15}{4}$ ___?___ $4$

---

## Ordering Fractions

To **order fractions**:

Order: $\frac{1}{3}$, $\frac{2}{9}$, $\frac{1}{4}$

- Use the LCD to rename the fractions as equivalent fractions with the same denominator.

LCD of $\frac{1}{3}$, $\frac{2}{9}$, and $\frac{1}{4}$: 36

$$\frac{1}{3} = \frac{1 \times 12}{3 \times 12} = \frac{12}{36}$$

$$\frac{2}{9} = \frac{2 \times 4}{9 \times 4} = \frac{8}{36}$$

$$\frac{1}{4} = \frac{1 \times 9}{4 \times 9} = \frac{9}{36}$$

- Compare the fractions.

$$\frac{8}{36} < \frac{9}{36} < \frac{12}{36}$$

So $\frac{2}{9} < \frac{1}{4} < \frac{1}{3}$.

Think:
$8 < 9 < 12$

- Arrange the fractions in order from *least to greatest* or from *greatest to least*.

From least to greatest: $\frac{2}{9}$, $\frac{1}{4}$, $\frac{1}{3}$

From greatest to least: $\frac{1}{3}$, $\frac{1}{4}$, $\frac{2}{9}$

---

**Write in order from least to greatest.**

13. $\frac{2}{7}$, $\frac{4}{7}$, $\frac{3}{7}$

14. $\frac{5}{13}$, $\frac{12}{13}$, $\frac{8}{13}$

15. $\frac{1}{2}$, $\frac{1}{3}$, $\frac{1}{6}$

16. $\frac{4}{5}$, $\frac{1}{4}$, $\frac{7}{8}$

**Write in order from greatest to least.**

17. $\frac{4}{5}$, $\frac{7}{10}$, $\frac{3}{4}$

18. $\frac{11}{12}$, $\frac{3}{8}$, $\frac{5}{6}$

19. $2\frac{7}{9}$, $2\frac{5}{6}$, $2\frac{2}{3}$

20. $1\frac{4}{5}$, $1\frac{7}{10}$, $1\frac{3}{4}$

## PROBLEM SOLVING

21. Teams A, B, and C each played the same number of games in a tournament. Team A won $\frac{7}{10}$ of its games, Team B won $\frac{2}{3}$ of its games, and Team C won $\frac{4}{5}$ of its games. Which team won the fewest games?

22. Lily made three jumps in a broad-jump contest. She jumped $3\frac{1}{2}$ ft in her first jump, $3\frac{2}{5}$ ft in her second jump, and $3\frac{5}{6}$ ft in her third jump. Which was her longest jump?

# TECHNOLOGY

## Primes and Patterns

You can use a calculator to find
the prime factorization of a number.

Remember: Use divisibility
rules to find divisors.

To find the prime factorization of 180,

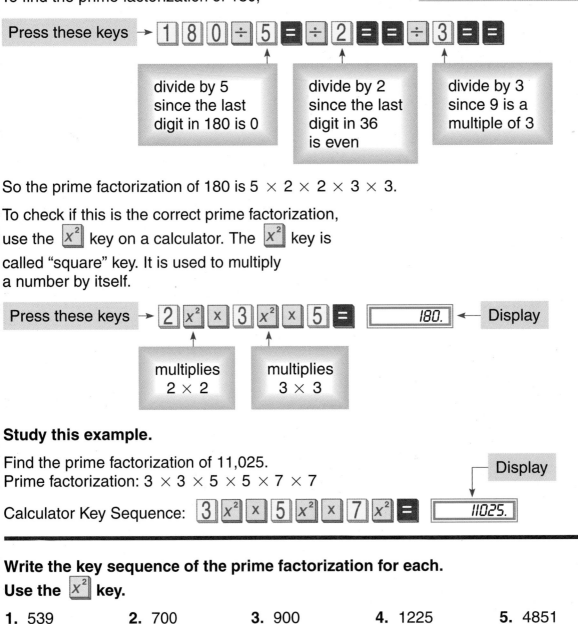

Press these keys → $1$ $8$ $0$ $\div$ $5$ $=$ $\div$ $2$ $=$ $=$ $\div$ $3$ $=$ $=$

| divide by 5 since the last digit in 180 is 0 | divide by 2 since the last digit in 36 is even | divide by 3 since 9 is a multiple of 3 |

So the prime factorization of 180 is $5 \times 2 \times 2 \times 3 \times 3$.

To check if this is the correct prime factorization,
use the $x^2$ key on a calculator. The $x^2$ key is
called "square" key. It is used to multiply
a number by itself.

Press these keys → $2$ $x^2$ $\times$ $3$ $x^2$ $\times$ $5$ $=$ | 180. | ← Display

| multiplies $2 \times 2$ | multiplies $3 \times 3$ |

**Study this example.**

Find the prime factorization of 11,025.
Prime factorization: $3 \times 3 \times 5 \times 5 \times 7 \times 7$

Display

Calculator Key Sequence: $3$ $x^2$ $\times$ $5$ $x^2$ $\times$ $7$ $x^2$ $=$ | 11025. |

---

**Write the key sequence of the prime factorization for each.**
**Use the $x^2$ key.**

**1.** 539     **2.** 700     **3.** 900     **4.** 1225     **5.** 4851

**6.** 3751     **7.** 1694     **8.** 2205     **9.** 10,571     **10.** 20,449

154

Use the $x^2$ key to multiply each number by itself.
Make a conjecture about the products of each.

**11.** 11; 111; 1111  **12.** 34; 334; 3334  **13.** 67; 667; 6667

## PROBLEM SOLVING

Algebra ✓

**14.** Enter each key sequence.

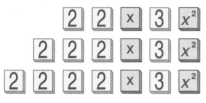

**15.** Predict the product of 22,222 × 9. Check using the $x^2$ key on a calculator.

**16.** Find each product. Look for a pattern.

37 ×  3 =  ?
37 ×  6 =  ?
37 ×  9 =  ?
37 × 12 =  ?

**17.** Predict the product of 37 × 21. Check your results with a calculator.

**18.** Find each product. Look for a pattern.

37 × 30 =  ?
37 × 33 =  ?
37 × 36 =  ?
37 × 39 =  ?

**19.** Predict the product of 37 × 45. Check your results with a calculator.

**20.** Describe the pattern in the products when 37 is multiplied by a multiple of 3.

**21.** Find each product. Look for a pattern.

99 × 11 =  ?
99 × 12 =  ?
99 × 13 =  ?
99 × 14 =  ?

**22.** Predict the product of 99 × 17. Check your results with a calculator.

**23.** Describe the pattern in the products in exercise 21.

**24.** Find and describe a pattern in the products when 999,999 is multiplied by 2, 3, 4, 5, and so on.

## 4-12 Problem Solving: Organized List

**Problem:** Elliot Pet Shop houses a pair of puppies in each dog cage. If there are 6 different puppies: a shepherd, a collie, a poodle, a retriever, a terrier, and a dalmatian, how many possible pairs can be formed?

**1 IMAGINE** Put yourself in the problem.

**2 NAME**

*Facts:* a pair of puppies in each cage
6 different puppies: a shepherd, a collie, a poodle, a retriever, a terrier, and a dalmatian

*Question:* How many possible pairs can be formed?

**3 THINK** Make a list of the possible pairs. Let the first letters of the puppies' names stand for each pair.

**4 COMPUTE**

A shepherd can be housed with any of the 5 other puppies.

| | | |
|---|---|---|
| S | and | C |
| S | and | P |
| S | and | R |
| S | and | T |
| S | and | D |

A collie can be housed with any of the 4 other puppies.

| | | |
|---|---|---|
| C | and | P |
| C | and | R |
| C | and | T |
| C | and | D |

A poodle can be housed with any of the 3 other puppies.

| | | |
|---|---|---|
| P | and | R |
| P | and | T |
| P | and | D |

A retriever can be housed with any of the 2 other puppies.

| | | |
|---|---|---|
| R | and | T |
| R | and | D |

A terrier can be housed with the other puppy that is left.

| | | |
|---|---|---|
| T | and | D |

Count the number of pairs.
$5 + 4 + 3 + 2 + 1 = 15$

So 15 pairs can be formed from the 6 different puppies.

**5 CHECK** Make a second list that begins with a different choice of puppy. Both lists should have the same number of pairs.

**Make an organized list to solve each problem.**

1. Tamisha has 3 shirts: one white, one green, and one purple; 2 pairs of shorts: one white and one black; and 2 vests: one plaid and one flowered. How many different three-piece outfits can she make?

**IMAGINE**  Picture Tamisha arranging the clothes to make different outfits.

**NAME**  *Facts:*  3 shirts—1 white, 1 green, 1 purple
2 pairs of shorts—1 white, 1 black
2 vests—1 plaid, 1 flowered

*Question:*  How many three-piece outfits can she make?

**THINK**  To find how many outfits Tamisha can make, make an organized list showing the possible combinations she can use.

| Shirts | Shorts | Vests |
|--------|--------|-------|
| white | white | plaid |
| white | white | flowered |
| white | black | plaid |
| white | black | flowered |

**COMPUTE** ⟶ **CHECK**

2. How many different 3-digit numbers can be made using the digits 6, 7, and 8 if *no* digit is repeated? if *one* digit is repeated?

3. The juice in a machine costs 60¢ a bottle. The machine will accept only exact change, it cannot give change, and it will not accept pennies or half dollars. How many different combinations of coins can you use to buy a bottle of juice?

**Make Up Your Own**

Communicate

4. Write a problem using the Organized List strategy. Have someone solve it.

**Connections: Physical Education**

Imagine

**Solve each problem and explain the method you used.**

1. At last week's track meet, Stacy ran $\frac{9}{12}$ of a mile, Jules ran $\frac{4}{5}$ of a mile, and Raul ran $\frac{3}{4}$ of a mile. Which two students ran the same distance?

2. Regina ran $2\frac{8}{20}$ miles. Write this number in lowest terms.

Name

3. There were 63 students at the track meet and 9 of them ran in the 100-meter race. What fractional part of the students ran in the race?

4. Ashlee ran $\frac{1}{4}$ of the race before tagging Adam. Then Adam ran $\frac{8}{32}$ of the race. Who ran farther? Explain.

Think

5. Ruby ran $\frac{12}{3}$ miles. Then she ran 3 more miles. How far did she run?

6. Jake ran $\frac{5}{6}$ of a mile. Frank ran $\frac{15}{20}$ of a mile. Who ran farther?

Compute

7. Of the 63 students at the track meet, 34 are girls. What fractional part of the students are boys?

8. Students at the track meet drank fresh apple cider. There were 14 gallons of cider and 4 teams. Each team drank the same amount of cider. At most, how many gallons did Roxanne's team drink?

Check

9. $\frac{32}{8}$ is to 4 as $\frac{20}{4}$ is to  ?

10. $\frac{6}{9}$ is to $\frac{2}{3}$ as $\frac{15}{18}$ is to  ?

11. $1\frac{1}{2}$ is to $\frac{3}{2}$ as $2\frac{3}{5}$ is to  ?

12. $\frac{4}{5}$ is to $\frac{16}{20}$ as $\frac{7}{8}$ is to  ?

**Complete.** Write *True* or *False*.

13. Some improper fractions equal whole numbers.

14. A fraction whose denominator is 1 more than its numerator is sometimes in lowest terms.

**Choose a strategy from the list or use another strategy you know to solve each problem.**

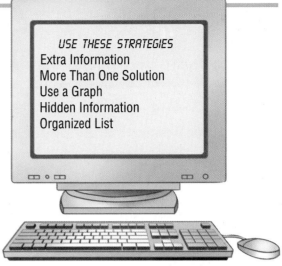

**15.** The judges at the track meet will award prizes to the top 4 teams. How many different ways can the top 4 teams place?

**16.** The long-jump winner jumped $8\frac{1}{2}$ ft. Did the winner jump more than 100 in.?

**17.** There were 12 students at last week's track meet. A little less than half were girls. Write a fraction that might represent the part of the team that were girls.

**18.** The team from Dellmont won $\frac{1}{5}$ of the medals, the team from Edgarton won $\frac{1}{3}$ of the medals, and the team from Fredonia won 11 of the 30 medals given at the meet. Five girls were on the teams. Which team won the most medals?

**Use the chart for problems 19–21.**

**19.** All teams scored at least 82 points. What team scored closest to 90 points?

**20.** The average team score was 84 points. Which team scored the most points? the fewest points?

**21.** Which teams scored between 80 and 85 points?

**Use the graph for problems 22 and 23.**

**22.** How many students participated in the meet?

**23.** In which two events did a total of $\frac{1}{4}$ of the students participate?

### Track Meet Scores

| Team | Points |
|------|--------|
| Spartans | $82\frac{4}{6}$ |
| Lions | $85\frac{2}{3}$ |
| Eagles | $85\frac{6}{8}$ |
| Vikings | $83\frac{5}{20}$ |

### Track Meet Participants

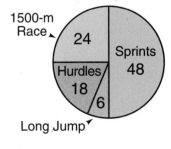

**Share Your Thinking**

Math Journal ✓

**24.** Write in your Math Journal which problems you solved using the same strategy, and explain why. Write a problem modeled on one of these problems and have a classmate solve it.

# Chapter Review and Practice

**Write whether each number is prime or composite.** *(See pp. 134–137.)*

**1.** 43      **2.** 39      **3.** 24      **4.** 57      **5.** 18      **6.** 101

**Use a factor tree to find the prime factorization of each.**

**7.** 16      **8.** 27      **9.** 32      **10.** 44      **11.** 56

**Find the greatest common factor (GCF) of each set of numbers.** *(See pp. 138–139.)*

**12.** 6 and 15      **13.** 9 and 21      **14.** 8 and 12      **15.** 2, 6, and 18

**Copy and complete.** *(See pp. 140–145.)*

**16.** $\frac{1}{4} = \frac{?}{8} = \frac{?}{12}$      **17.** $\frac{3}{7} = \frac{6}{?} = \frac{9}{?}$      **18.** $\frac{5}{9} = \frac{?}{18} = \frac{15}{?}$

**Write each fraction in lowest terms.** *(See pp. 142–143.)*

**19.** $\frac{9}{21}$      **20.** $\frac{16}{24}$      **21.** $\frac{24}{30}$      **22.** $\frac{4}{12}$      **23.** $\frac{14}{35}$      **24.** $\frac{21}{42}$

**Find the least common denominator (LCD) of each set of fractions.** *(See pp. 146–147.)*

**25.** $\frac{1}{4}, \frac{1}{8}$      **26.** $\frac{1}{3}, \frac{3}{10}$      **27.** $\frac{4}{5}, \frac{1}{2}$      **28.** $\frac{5}{9}, \frac{2}{3}, \frac{7}{27}$

**Draw a picture or a number line to show each mixed number.** *(See pp. 148–149.)*

**29.** $2\frac{1}{4}$      **30.** $3\frac{2}{3}$      **31.** $4\frac{3}{5}$      **32.** $6\frac{4}{7}$

**Write as a whole number or a mixed number in simplest form.** *(See pp. 150–151.)*

**33.** $\frac{11}{6}$      **34.** $\frac{36}{9}$      **35.** $\frac{22}{3}$      **36.** $\frac{24}{5}$      **37.** $\frac{47}{7}$

**Compare. Write <, =, or >.** *(See pp. 152–153.)*

**38.** $\frac{5}{9} \underline{\ ?\ } \frac{7}{9}$      **39.** $\frac{5}{9} \underline{\ ?\ } \frac{10}{18}$      **40.** $\frac{2}{3} \underline{\ ?\ } \frac{1}{2}$      **41.** $2\frac{3}{8} \underline{\ ?\ } 2\frac{5}{16}$

## PROBLEM SOLVING

*(See pp. 156–158.)*

**42.** Tom uses three 1–6 number cubes. He is looking for different ways to roll the sum of 12. How many ways will he find?

(See *Still More Practice*, p. 480.)

## DENSITY OF FRACTIONS

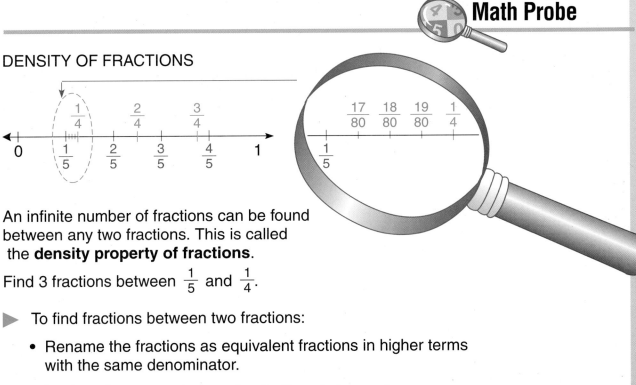

An infinite number of fractions can be found between any two fractions. This is called the **density property of fractions**.

Find 3 fractions between $\frac{1}{5}$ and $\frac{1}{4}$.

▶ To find fractions between two fractions:

- Rename the fractions as equivalent fractions in higher terms with the same denominator.

- Look at the numerators and write the whole numbers between them.

- Write the new fractions. Use the whole numbers as the numerators and the common denominator as the denominators.

- Repeat the steps until the desired number of fractions is found.

$$\frac{1}{5} = \frac{1 \times 4}{5 \times 4} = \frac{4}{20} \longrightarrow \frac{4 \times 2}{20 \times 2} = \frac{8}{40} \longrightarrow \frac{8 \times 2}{40 \times 2} = \frac{16}{80}$$

$$\frac{1}{4} = \frac{1 \times 5}{4 \times 5} = \frac{5}{20} \longrightarrow \frac{5 \times 2}{20 \times 2} = \frac{10}{40} \longrightarrow \frac{10 \times 2}{40 \times 2} = \frac{20}{80}$$

| No whole numbers between 4 and 5: continue renaming. | One whole number between 8 and 10: continue renaming. | Three whole numbers between 16 and 20: 17, 18, 19 |
| --- | --- | --- |

Three fractions between $\frac{1}{5}$ and $\frac{1}{4}$: $\frac{17}{80}, \frac{18}{80}, \frac{19}{80}$

## Find three fractions between each pair of fractions.

**1.** $\frac{1}{10}, \frac{1}{6}$     **2.** $\frac{1}{3}, \frac{2}{5}$     **3.** $\frac{1}{2}, \frac{3}{5}$     **4.** $\frac{7}{10}, \frac{3}{4}$     **5.** $\frac{4}{5}, \frac{5}{6}$

**6.** $\frac{1}{2}, \frac{5}{6}$     **7.** $\frac{7}{15}, \frac{3}{5}$     **8.** $\frac{1}{3}, \frac{3}{8}$     **9.** $\frac{1}{4}, \frac{2}{7}$     **10.** $\frac{3}{4}, \frac{5}{6}$

# Check Your Mastery

## Performance Assessment

**Use a number line.**

Tina cut 3 different lengths of ribbon: $1\frac{1}{2}$ yd, $\frac{2}{3}$ yd, and $1\frac{5}{9}$ yd.

1. Show each on a number line.

2. Use $<$ and $>$ to compare the lengths in 2 ways.

3. Order the lengths from greatest to least.

**Write whether each number is prime or composite.**

**4.** 2        **5.** 13        **6.** 49        **7.** 37        **8.** 111

**Find the greatest common factor (GCF) for each set of numbers.**

**9.** 6 and 21        **10.** 9 and 15        **11.** 12, 16, and 24

**Write whether each fraction is *closer to 0, closer to $\frac{1}{2}$*, or *closer to 1*.**

**12.** $\frac{13}{27}$      **13.** $\frac{39}{40}$      **14.** $\frac{5}{61}$      **15.** $\frac{17}{28}$      **16.** $\frac{197}{200}$

**Find the missing term.**

**17.** $\frac{9}{10} = \frac{?}{100}$      **18.** $\frac{4}{5} = \frac{?}{60}$      **19.** $\frac{2}{9} = \frac{10}{?}$      **20.** $\frac{3}{4} = \frac{24}{?}$

**Write each fraction in lowest terms.**

**21.** $\frac{8}{12}$      **22.** $\frac{4}{8}$      **23.** $\frac{12}{15}$      **24.** $\frac{18}{27}$      **25.** $\frac{36}{54}$

**Find the least common denominator (LCD) of each set of fractions.**

**26.** $\frac{4}{5}, \frac{1}{2}$      **27.** $\frac{2}{3}, \frac{4}{7}$      **28.** $\frac{3}{8}, \frac{1}{4}$      **29.** $\frac{1}{2}, \frac{5}{6}, \frac{7}{18}$

**Write each as a whole number or mixed number in simplest form.**

**30.** $\frac{19}{4}$      **31.** $\frac{37}{8}$      **32.** $\frac{48}{8}$      **33.** $\frac{57}{9}$      **34.** $\frac{84}{12}$

**PROBLEM SOLVING**    *Use a strategy you have learned.*

**35.** Football practice lasted $2\frac{1}{6}$ hours yesterday.
About how many hours was the football practice?

# Fractions: Addition and Subtraction 5

## Grandmother's Almond Cookies

No need cookbook, measuring cup.
Stand close. Watch me. No mess up.

One hand sugar, one hand lard
(cut in pieces when still hard),

two hands flour, more or less,
one pinch baking powder. Guess.

One hand almond, finely crushed.
Mix it with both hands. No rush.

Put two eggs. Brown is better.
Keep on mixing. Should be wetter.

Sprinkle water in it. Make
cookies round and flat. Now bake

one big sheet at three-seven-five.
When they done, they come alive.

*Janet S. Wong*

**In this chapter you will:**

Learn to add or subtract with renaming
Estimate sums and differences of
  mixed numbers
Use the Working Backwards strategy

**Critical Thinking/Finding Together**

One cup of condensed milk weighs 11 oz.
How many ounces of milk will remain
unused after a grandmother opens three
6-oz cans for a recipe that requires $1\frac{1}{2}$
cups of milk?

*Update your skills. See page 10.*

# 5-1

# Renaming Fraction Sums: Like Denominators

Some fifth graders experimented with plant growth in three different types of soil. They recorded their results in a table. What was the total amount of plant growth over a two-week period for each type of soil?

| Period | Plant Growth | | |
|--------|--------|--------|--------|
| | Soil A | Soil B | Soil C |
| Week 1 | $\frac{6}{8}$ in. | $\frac{5}{8}$ in. | $\frac{3}{8}$ in. |
| Week 2 | $\frac{7}{8}$ in. | $\frac{3}{8}$ in. | $\frac{3}{8}$ in. |

 **Hands-On Understanding**

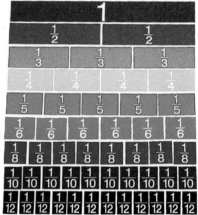

**Materials Needed:** fraction strips, colored pencils or crayons, scissors

**Step 1** Shade fraction strips to show $\frac{6}{8}$ and $\frac{7}{8}$ for the amount of plant growth in Week 1 and Week 2 for Soil A. Then place the shaded fraction strips as shown below.

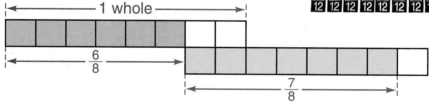

How many eighths are shaded altogether?

**Step 2** Write an addition sentence for the shaded fraction strips. What do you notice about the sum?

**Step 3** Rename the sum as a mixed number in simplest form that tells the total amount of plant growth over a two-week period for Soil A.

1. Repeat the steps to find the total amount of plant growth over a two-week period for Soil B; for Soil C.

2. Can the sum of fractions with like denominators be greater than 1? equal to 1? less than 1? Explain your answers.

**Use fraction strips to model each sum. Write an addition sentence with the sum in simplest form.**

3. $\frac{4}{5} + \frac{3}{5}$

4. $\frac{3}{10} + \frac{9}{10}$

5. $\frac{7}{12} + \frac{11}{12}$

6. $\frac{8}{9} + \frac{8}{9}$

7. $\frac{1}{4} + \frac{3}{4}$

8. $\frac{3}{5} + \frac{2}{5}$

9. $\frac{5}{9} + \frac{4}{9}$

10. $\frac{7}{10} + \frac{3}{10}$

11. $\frac{7}{12} + \frac{1}{12}$

12. $\frac{3}{8} + \frac{1}{8}$

13. $\frac{4}{9} + \frac{2}{9}$

14. $\frac{3}{10} + \frac{3}{10}$

**Write an addition sentence with the sum in simplest form to show how much is shaded.**

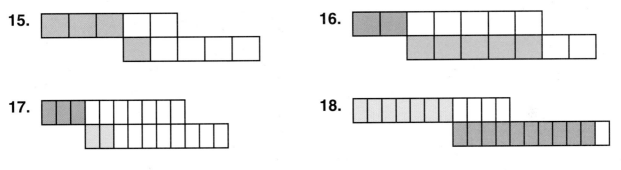

15.

16.

17.

18.

## Communicate

Discuss

19. When adding two fractions with like denominators, when would the sum be a mixed number? 1? a fraction? Explain your answers.

20. Write in your Math Journal the different types of answers you get when adding fractions with like denominators. Give an example for each type.

Math Journal

## Skills to Remember

**Find the least common multiple (LCM) of each set of numbers.**

21. 9, 12

22. 8, 10

23. 5, 7

24. 4, 10

25. 10, 15

26. 4, 6, and 12

27. 5, 6, and 15

28. 15, 20, and 30

# 5-2  Adding Fractions: Unlike Denominators

Dave worked $\frac{3}{4}$ of an hour on his model plane. His dad worked on it for $\frac{2}{3}$ of an hour. How much time did both work on the model plane?

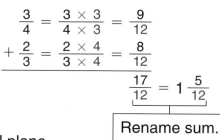

To find the amount of time, add: $\frac{3}{4} + \frac{2}{3} = \underline{\ ?\ }$

$$\frac{3}{4} + \frac{2}{3} = 1\frac{5}{12}$$

▶ To **add fractions** with *unlike* denominators:

- Find the least common denominator (LCD) of the fractions.

- Use the LCD to rename the fractions as equivalent fractions with the same denominator.

- Add. Then write the sum in simplest form.

Multiples of 4: 4, 8, 12, 16, . . .
Multiples of 3: 3, 6, 9, 12, . . .
LCD of $\frac{3}{4}$ and $\frac{2}{3}$: 12

$$\frac{3}{4} = \frac{3 \times 3}{4 \times 3} = \frac{9}{12}$$
$$+\frac{2}{3} = \frac{2 \times 4}{3 \times 4} = \frac{8}{12}$$
$$\overline{\hspace{3cm}}$$
$$\frac{17}{12} = 1\frac{5}{12}$$

Rename sum.

Dave and his dad worked $1\frac{5}{12}$ h on the model plane.

▶ The properties of addition for whole numbers also apply to fractions.

**Commutative Property**

Think: "order."

$$\frac{3}{4} + \frac{2}{3} = \frac{2}{3} + \frac{3}{4}$$
$$1\frac{5}{12} = 1\frac{5}{12}$$

You can check addition by applying the *commutative property.*

**Identity Property**

Think: "same."

$$\frac{3}{4} + 0 = \frac{3}{4}$$
$$0 + \frac{3}{4} = \frac{3}{4}$$

**Study these examples.**

$$\begin{aligned}\frac{3}{16} &= \frac{3}{16}\\ +\frac{5}{8} &= \frac{5 \times 2}{8 \times 2} = \frac{10}{16}\\ \hline &\quad\ \ \frac{13}{16}\end{aligned}$$

$$\frac{5}{6} + \frac{2}{3} = \frac{5}{6} + \frac{2 \times 2}{3 \times 2}$$
$$= \frac{5}{6} + \frac{4}{6}$$
$$= \frac{9}{6} = 1\frac{3}{6} = 1\frac{1}{2} \longleftarrow \text{Simplest form}$$

**Add.**

1. $\frac{2}{3}$
   $+ \frac{1}{6}$

2. $\frac{2}{5}$
   $+ \frac{3}{10}$

3. $\frac{1}{6}$
   $+ \frac{2}{5}$

4. $\frac{1}{3}$
   $+ \frac{5}{9}$

5. $\frac{1}{3}$
   $+ \frac{7}{12}$

6. $\frac{2}{3}$
   $+ \frac{1}{5}$

7. $\frac{2}{3}$
   $+ \frac{1}{12}$

8. $\frac{4}{15}$
   $+ \frac{2}{5}$

9. $\frac{1}{2}$
   $+ \frac{3}{14}$

10. $\frac{3}{5}$
    $+ \frac{2}{3}$

11. $\frac{4}{5}$
    $+ \frac{3}{4}$

12. $\frac{7}{16}$
    $+ \frac{3}{4}$

**Find the sum.** Use the commutative property to check your answers.

13. $\frac{1}{2} + \frac{1}{7}$

14. $\frac{4}{15} + \frac{2}{3}$

15. $\frac{3}{10} + \frac{1}{4}$

16. $\frac{2}{3} + \frac{1}{8}$

17. $\frac{2}{3} + \frac{4}{9}$

18. $\frac{5}{8} + \frac{1}{2}$

19. $\frac{4}{5} + \frac{9}{10}$

20. $\frac{5}{6} + \frac{5}{9}$

**Find the missing fraction. Name the property of addition that is used.**

21. $\frac{4}{7} + \frac{3}{14} = \underline{\ ?\ } + \frac{4}{7}$

22. $0 + \underline{\ ?\ } = \frac{3}{10}$

**PROBLEM SOLVING**

23. Find the sum of three fourths and seven twelfths.

24. What is one third plus eight ninths?

25. Lin spent $\frac{1}{10}$ of her allowance for a gift and $\frac{2}{5}$ for a movie ticket. What part of her allowance did she spend in all?

**Finding Together**

June and Paul recorded the distances they swam each day.

26. On which day did they swim a total of half a mile?

27. Who swam farther on Wednesday and Thursday?

28. On which day did they swim the shortest combined distance?

| Day | Distance in Miles | |
|---|---|---|
| | June | Paul |
| Monday | $\frac{1}{10}$ | $\frac{1}{5}$ |
| Tuesday | $\frac{1}{6}$ | $\frac{1}{3}$ |
| Wednesday | $\frac{1}{2}$ | $\frac{3}{8}$ |
| Thursday | $\frac{3}{8}$ | $\frac{1}{4}$ |

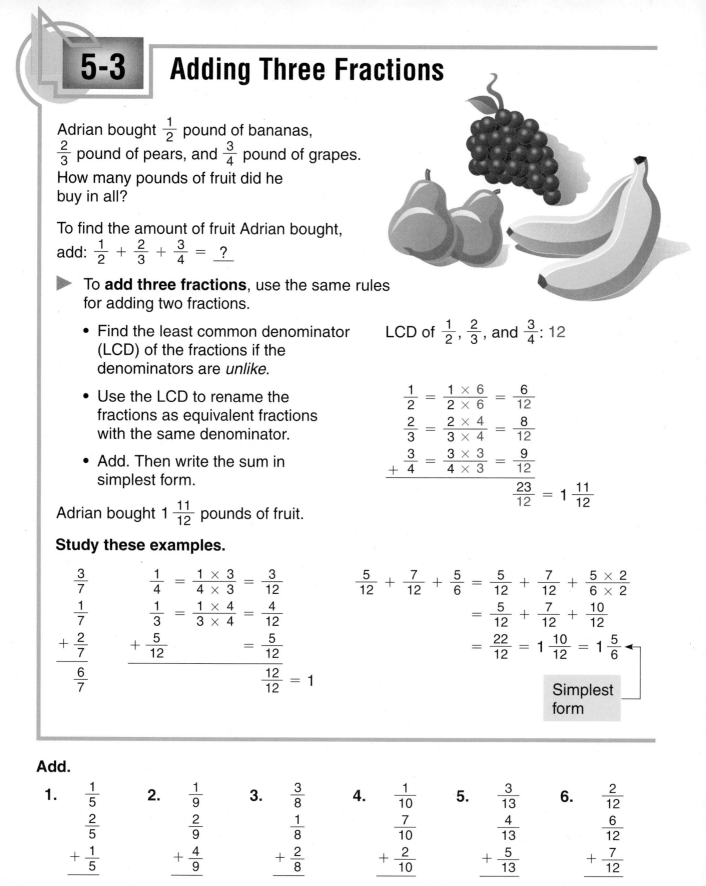

## 5-3 Adding Three Fractions

Adrian bought $\frac{1}{2}$ pound of bananas, $\frac{2}{3}$ pound of pears, and $\frac{3}{4}$ pound of grapes. How many pounds of fruit did he buy in all?

To find the amount of fruit Adrian bought, add: $\frac{1}{2} + \frac{2}{3} + \frac{3}{4} = \underline{\ ?\ }$

▶ To **add three fractions**, use the same rules for adding two fractions.

- Find the least common denominator (LCD) of the fractions if the denominators are *unlike*.

LCD of $\frac{1}{2}$, $\frac{2}{3}$, and $\frac{3}{4}$: 12

- Use the LCD to rename the fractions as equivalent fractions with the same denominator.

- Add. Then write the sum in simplest form.

$$\frac{1}{2} = \frac{1 \times 6}{2 \times 6} = \frac{6}{12}$$
$$\frac{2}{3} = \frac{2 \times 4}{3 \times 4} = \frac{8}{12}$$
$$+\ \frac{3}{4} = \frac{3 \times 3}{4 \times 3} = \frac{9}{12}$$
$$\frac{23}{12} = 1\frac{11}{12}$$

Adrian bought $1\frac{11}{12}$ pounds of fruit.

**Study these examples.**

$$\begin{array}{r} \frac{3}{7} \\ \frac{1}{7} \\ +\ \frac{2}{7} \\ \hline \frac{6}{7} \end{array}$$

$$\frac{1}{4} = \frac{1 \times 3}{4 \times 3} = \frac{3}{12}$$
$$\frac{1}{3} = \frac{1 \times 4}{3 \times 4} = \frac{4}{12}$$
$$+\ \frac{5}{12} \qquad\qquad = \frac{5}{12}$$
$$\frac{12}{12} = 1$$

$$\frac{5}{12} + \frac{7}{12} + \frac{5}{6} = \frac{5}{12} + \frac{7}{12} + \frac{5 \times 2}{6 \times 2}$$
$$= \frac{5}{12} + \frac{7}{12} + \frac{10}{12}$$
$$= \frac{22}{12} = 1\frac{10}{12} = 1\frac{5}{6}$$

Simplest form

**Add.**

1.  $\frac{1}{5}$
    $\frac{2}{5}$
    $+\ \frac{1}{5}$

2.  $\frac{1}{9}$
    $\frac{2}{9}$
    $+\ \frac{4}{9}$

3.  $\frac{3}{8}$
    $\frac{1}{8}$
    $+\ \frac{2}{8}$

4.  $\frac{1}{10}$
    $\frac{7}{10}$
    $+\ \frac{2}{10}$

5.  $\frac{3}{13}$
    $\frac{4}{13}$
    $+\ \frac{5}{13}$

6.  $\frac{2}{12}$
    $\frac{6}{12}$
    $+\ \frac{7}{12}$

**Find the sum.**

7. $\dfrac{1}{3}$
$\dfrac{2}{3}$
$+\dfrac{5}{9}$

8. $\dfrac{1}{5}$
$\dfrac{1}{10}$
$+\dfrac{3}{5}$

9. $\dfrac{3}{4}$
$\dfrac{3}{8}$
$+\dfrac{1}{8}$

10. $\dfrac{1}{4}$
$\dfrac{1}{12}$
$+\dfrac{1}{3}$

11. $\dfrac{1}{6}$
$\dfrac{2}{9}$
$+\dfrac{1}{18}$

12. $\dfrac{2}{5}$
$\dfrac{1}{4}$
$+\dfrac{9}{20}$

13. $\dfrac{4}{5} + \dfrac{3}{10} + \dfrac{1}{4}$

14. $\dfrac{2}{3} + \dfrac{1}{5} + \dfrac{3}{10}$

15. $\dfrac{5}{6} + \dfrac{7}{8} + \dfrac{1}{4}$

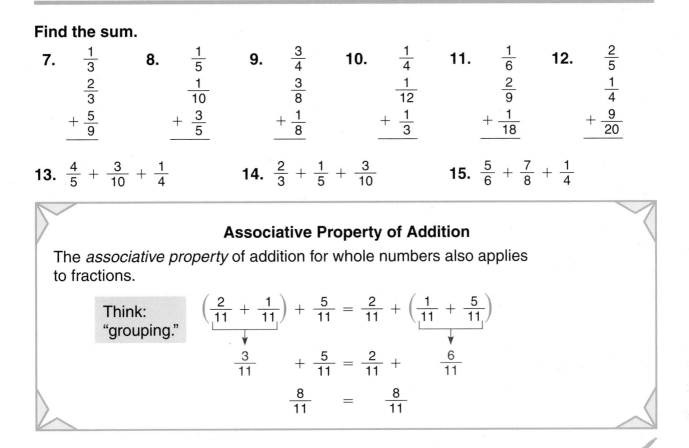

**Associative Property of Addition**

The *associative property* of addition for whole numbers also applies to fractions.

Think: "grouping."

$$\left(\dfrac{2}{11} + \dfrac{1}{11}\right) + \dfrac{5}{11} = \dfrac{2}{11} + \left(\dfrac{1}{11} + \dfrac{5}{11}\right)$$

$$\dfrac{3}{11} + \dfrac{5}{11} = \dfrac{2}{11} + \dfrac{6}{11}$$

$$\dfrac{8}{11} = \dfrac{8}{11}$$

*Algebra* ✓

**Find the missing fraction. Then check by adding.**

16. $\left(\dfrac{2}{9} + \dfrac{1}{9}\right) + \dfrac{4}{9} = \dfrac{2}{9} + \left(\dfrac{1}{9} + \underline{\ ?\ }\right)$

17. $\dfrac{3}{10} + \left(\dfrac{2}{10} + \dfrac{1}{10}\right) = \left(\dfrac{3}{10} + \underline{\ ?\ }\right) + \dfrac{1}{10}$

18. $\left(\dfrac{3}{4} + \underline{\ ?\ }\right) + \dfrac{5}{6} = \dfrac{3}{4} + \left(\dfrac{2}{3} + \dfrac{5}{6}\right)$

19. $\underline{\ ?\ } + \left(\dfrac{1}{2} + \dfrac{1}{6}\right) = \left(\dfrac{2}{5} + \dfrac{1}{2}\right) + \dfrac{1}{6}$

**PROBLEM SOLVING**

20. Zaffar bought $\dfrac{2}{3}$ qt of fresh orange juice, $\dfrac{3}{4}$ qt of fresh mango juice, and $\dfrac{1}{2}$ qt of fresh grape juice. How many quarts of fruit juice did he buy?

21. Yvonne sifted together $\dfrac{3}{4}$ cup of rye flour, $\dfrac{3}{5}$ cup of wheat flour, and $\dfrac{7}{10}$ cup of white flour. How many cups of flour did she sift?

22. Ms. Russell added $\dfrac{1}{8}$ teaspoon of pepper, $\dfrac{1}{3}$ teaspoon of salt, and $\dfrac{1}{4}$ teaspoon of curry powder to the stew. How many teaspoons of seasoning did she add to the stew?

23. Mr. Clarke bought $\dfrac{3}{8}$ pound of peanuts, $\dfrac{3}{4}$ pound of pecans, and $\dfrac{5}{6}$ pound of walnuts. How many pounds of nuts did he buy?

169

# 5-4  Adding Mixed Numbers

Esther used $2\frac{1}{4}$ yd of gold ribbon and $1\frac{1}{4}$ yd of blue ribbon to make certificates. How many yards of ribbon did she use for the certificates?

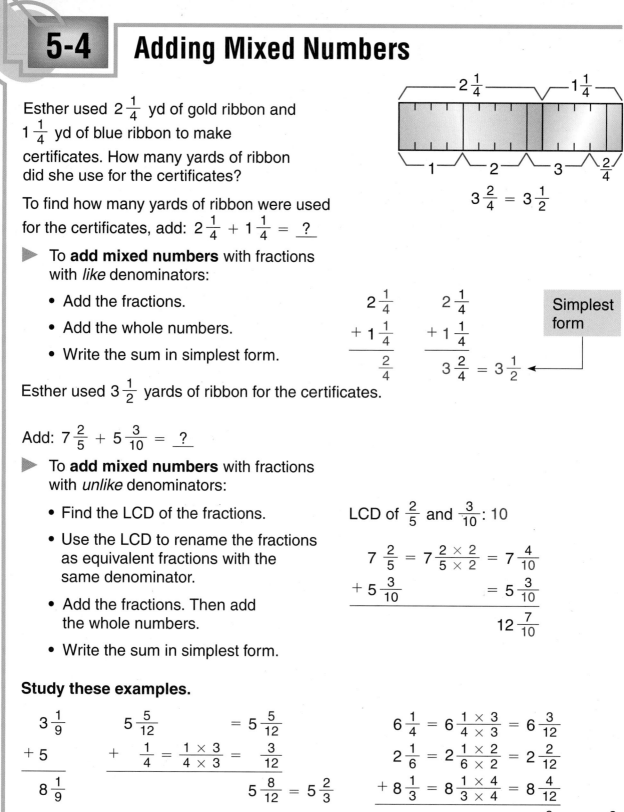

$$3\frac{2}{4} = 3\frac{1}{2}$$

To find how many yards of ribbon were used for the certificates, add: $2\frac{1}{4} + 1\frac{1}{4} = \underline{\ ?\ }$

▶ To **add mixed numbers** with fractions with *like* denominators:

- Add the fractions.

- Add the whole numbers.

- Write the sum in simplest form.

$$\begin{array}{r} 2\frac{1}{4} \\ +\,1\frac{1}{4} \\ \hline \frac{2}{4} \end{array} \qquad \begin{array}{r} 2\frac{1}{4} \\ +\,1\frac{1}{4} \\ \hline 3\frac{2}{4} = 3\frac{1}{2} \end{array}$$

Simplest form

Esther used $3\frac{1}{2}$ yards of ribbon for the certificates.

Add: $7\frac{2}{5} + 5\frac{3}{10} = \underline{\ ?\ }$

▶ To **add mixed numbers** with fractions with *unlike* denominators:

- Find the LCD of the fractions.

- Use the LCD to rename the fractions as equivalent fractions with the same denominator.

- Add the fractions. Then add the whole numbers.

- Write the sum in simplest form.

LCD of $\frac{2}{5}$ and $\frac{3}{10}$: 10

$$\begin{aligned} 7\frac{2}{5} &= 7\frac{2\times 2}{5\times 2} = 7\frac{4}{10} \\ +\,5\frac{3}{10} &\phantom{= 7\frac{2\times 2}{5\times 2}} = 5\frac{3}{10} \\ \hline &\phantom{= 7\frac{2\times 2}{5\times 2}} 12\frac{7}{10} \end{aligned}$$

**Study these examples.**

$$\begin{array}{r} 3\frac{1}{9} \\ +\,5\phantom{\frac{1}{9}} \\ \hline 8\frac{1}{9} \end{array}$$

$$\begin{array}{r} 5\frac{5}{12} \\ +\ \ \frac{1}{4} = \frac{1\times 3}{4\times 3} = \frac{3}{12} \\ \hline \end{array} \begin{array}{l} = 5\frac{5}{12} \\ \\ 5\frac{8}{12} = 5\frac{2}{3} \end{array}$$

$$\begin{array}{r} 6\frac{1}{4} = 6\frac{1\times 3}{4\times 3} = 6\frac{3}{12} \\ 2\frac{1}{6} = 2\frac{1\times 2}{6\times 2} = 2\frac{2}{12} \\ +\,8\frac{1}{3} = 8\frac{1\times 4}{3\times 4} = 8\frac{4}{12} \\ \hline 16\frac{9}{12} = 16\frac{3}{4} \end{array}$$

**Add.**

1. $3\frac{4}{9}$ $+2\frac{1}{9}$

2. $8\frac{5}{12}$ $+9\frac{1}{12}$

3. $9\frac{1}{6}$ $+2\frac{3}{4}$

4. $10\frac{3}{5}$ $+3$

5. $4\frac{1}{3}$ $+7\frac{1}{6}$

6. $\frac{4}{5}$ $+8\frac{1}{6}$

7. $6\frac{3}{7}$ $3\frac{1}{7}$ $+2\frac{2}{7}$

8. $5\frac{1}{9}$ $3\frac{4}{9}$ $+4\frac{1}{9}$

9. $9\frac{1}{3}$ $2\frac{1}{4}$ $+3\frac{1}{12}$

10. $2\frac{2}{5}$ $6\frac{1}{3}$ $+4\frac{1}{15}$

11. $8\frac{1}{4}$ $2\frac{2}{5}$ $+5\frac{3}{20}$

12. $2\frac{1}{3}$ $5\frac{3}{8}$ $+\frac{1}{4}$

13. $6\frac{1}{4} + 5\frac{2}{4}$

14. $3\frac{5}{12} + \frac{1}{3}$

15. $8\frac{2}{5} + 5$

16. $7\frac{1}{6} + 3\frac{1}{6} + 5\frac{1}{6}$

17. $8\frac{2}{5} + 7\frac{1}{4} + \frac{1}{10}$

18. $9 + 8\frac{1}{3} + 3\frac{1}{12}$

19. Explain in your Math Journal how the properties of addition can be used to solve $3\frac{1}{3} + 6\frac{1}{4} + 1\frac{2}{3}$ mentally.

*Math Journal* ✓

## PROBLEM SOLVING

20. Ethel bought $1\frac{5}{12}$ yd of white fabric and $2\frac{1}{2}$ yd of yellow fabric to make curtains. How many yards of fabric did she buy?

21. The chef spent $4\frac{1}{4}$ h cooking dinner and $1\frac{2}{3}$ h cooking breakfast and lunch. How many hours did he spend cooking?

22. In the long-jump competition, Mac's first jump was $22\frac{1}{8}$ ft. His second jump was $21\frac{2}{3}$ ft, and his third jump was $20\frac{3}{4}$ ft. Find the sum of his jumps.

## Connections: Art

23. Arthur mixed $1\frac{3}{4}$ pints of red paint and $3\frac{1}{2}$ pints of yellow paint to make orange paint. How many pints of orange paint did he make?

24. To make purple paint, Lauren mixed $1\frac{2}{3}$ pints of red paint with $1\frac{3}{5}$ pints of blue paint. How much purple paint did she make?

171

# 5-5 Renaming Mixed Number Sums

When the *sum* of two or more mixed numbers contains an *improper fraction*, **rename** the fraction as a whole or mixed number. Then add the whole numbers.

▶ Add: $5\frac{3}{4} + 3\frac{5}{6} = \underline{\ ?\ }$

$$5\frac{3}{4} = 5\frac{3 \times 3}{4 \times 3} = 5\frac{9}{12}$$
$$+\ 3\frac{5}{6} = 3\frac{5 \times 2}{6 \times 2} = 3\frac{10}{12}$$
$$8\frac{19}{12} = 8 + 1\frac{7}{12}$$

Rename $\frac{19}{12}$ as $1\frac{7}{12}$.
$$= 9\frac{7}{12}$$

Add: $9\frac{1}{6} + 1\frac{1}{3} + 3\frac{1}{2} = \underline{\ ?\ }$

$$9\frac{1}{6} \qquad\qquad = 9\frac{1}{6}$$
$$1\frac{1}{3} = 1\frac{1 \times 2}{3 \times 2} = 1\frac{2}{6}$$
$$+\ 3\frac{1}{2} = 3\frac{1 \times 3}{2 \times 3} = 3\frac{3}{6}$$
$$13\frac{6}{6} = 13 + 1$$

Rename $\frac{6}{6}$ as 1.
$$= 14$$

**Study this example.**

$$6\frac{3}{5} = 6\frac{3 \times 3}{5 \times 3} = 6\frac{9}{15}$$
$$+\ \frac{13}{15} \qquad\qquad = \frac{13}{15}$$
$$6\frac{22}{15} = 6 + 1\frac{7}{15} = 7\frac{7}{15}$$

**Rename each as a mixed number in simplest form.**

**1.** $6\frac{11}{9}$  **2.** $10\frac{5}{5}$  **3.** $14\frac{7}{7}$  **4.** $9\frac{10}{8}$  **5.** $8\frac{6}{4}$  **6.** $11\frac{9}{6}$

**7.** $3\frac{20}{15}$  **8.** $21\frac{14}{12}$  **9.** $32\frac{16}{14}$  **10.** $17\frac{18}{15}$  **11.** $19\frac{24}{18}$  **12.** $25\frac{15}{10}$

**13.** $36\frac{8}{8}$  **14.** $42\frac{16}{15}$  **15.** $53\frac{17}{14}$  **16.** $83\frac{12}{8}$  **17.** $75\frac{19}{17}$  **18.** $41\frac{13}{11}$

**Add.**

**19.** $\quad 4\frac{5}{7}$  **20.** $\quad 4\frac{1}{8}$  **21.** $\quad 6\frac{2}{6}$  **22.** $\quad 3\frac{7}{9}$  **23.** $\quad 5\frac{3}{8}$  **24.** $\quad 5\frac{2}{4}$

$\quad\ +\ 2\frac{3}{7}$ $\qquad +\ 5\frac{7}{8}$ $\qquad +\ 4\frac{4}{6}$ $\qquad +\ 6\frac{2}{9}$ $\qquad +\ 3\frac{7}{8}$ $\qquad +\ 3\frac{4}{4}$

**Find the sum.**

**25.** $4\frac{3}{4}$  **26.** $8\frac{5}{6}$  **27.** $7\frac{5}{9}$  **28.** $\frac{3}{5}$  **29.** $6\frac{2}{5}$  **30.** $3\frac{5}{12}$
$\underline{+\ 2\frac{7}{20}}$  $\underline{+\ 2\frac{5}{12}}$  $\underline{+\ 4\frac{8}{18}}$  $\underline{+\ 9\frac{8}{20}}$  $\underline{+\ \ \frac{2}{3}}$  $\underline{+\ 9\frac{7}{8}}$

**31.** $4\frac{1}{5}$  **32.** $3\frac{4}{9}$  **33.** $3\frac{3}{4}$  **34.** $5\frac{1}{8}$  **35.** $6\frac{1}{2}$  **36.** $2\frac{5}{6}$
$\quad 6\frac{9}{10}$  $\quad 6\frac{2}{3}$  $\quad 5\frac{3}{8}$  $\quad \frac{3}{4}$  $\quad 9\frac{1}{4}$  $\quad 9\frac{1}{3}$
$\underline{+\ 2\frac{2}{5}}$  $\underline{+\ 4\frac{2}{9}}$  $\underline{+\ 7\frac{5}{8}}$  $\underline{+\ 6\frac{1}{2}}$  $\underline{+\ 3\frac{2}{3}}$  $\underline{+\ \ \frac{1}{12}}$

**37.** $6\frac{5}{9} + 4\frac{2}{3}$  **38.** $\frac{3}{4} + 2\frac{4}{5}$  **39.** $3\frac{5}{8} + 7\frac{2}{3}$

**40.** $8\frac{1}{2} + 5\frac{7}{12} + 3\frac{2}{3}$  **41.** $3\frac{3}{10} + 2\frac{3}{4} + 6\frac{1}{5}$  **42.** $4\frac{1}{3} + \frac{5}{8} + 1\frac{1}{4}$

**Write *always*, *sometimes*, or *never*.**

*Communicate* ✓

**43.** When you add two mixed numbers, the fractional part of the sum is more than 1. Give examples to support your answer.

## PROBLEM SOLVING

**44.** A $10\frac{1}{2}$-ft ladder has a $4\frac{3}{4}$-ft extension. What is the height of the ladder when totally extended?

**45.** Harriet exercised $14\frac{2}{3}$ min in the morning and $23\frac{5}{6}$ min in the afternoon. How long did she exercise in all?

**46.** The Madrigal family drank $2\frac{2}{3}$ bottles of spring water for breakfast, $2\frac{1}{8}$ bottles for lunch, and $1\frac{3}{4}$ bottles for dinner. How many bottles of spring water did the family drink for their three meals?

### Mental Math

**Add.** Look for sums of 1.

**47.** $10 + 3\frac{1}{2} + 4\frac{1}{2}$  **48.** $6\frac{1}{4} + 11 + 5\frac{3}{4}$  **49.** $9\frac{4}{5} + 7\frac{1}{5} + 3$

**50.** $6\frac{1}{2} + 4\frac{1}{2} + 5\frac{3}{4}$  **51.** $3\frac{1}{3} + 6\frac{1}{5} + 10\frac{2}{3}$  **52.** $8\frac{3}{4} + 2\frac{2}{7} + 9\frac{1}{4}$

# 5-6 Renaming Differences: Like Denominators

How much farther did Mary run than Ellen on each day over the long holiday weekend?

$$\frac{8}{9} - \frac{5}{9} = \underline{?}$$

Mary's distance    Ellen's distance

| | Distance in Miles | |
|---|---|---|
| **Day** | **Mary** | **Ellen** |
| Saturday | $\frac{8}{9}$ | $\frac{5}{9}$ |
| Sunday | $\frac{14}{9}$ | $\frac{5}{9}$ |
| Monday | $\frac{17}{9}$ | $\frac{7}{9}$ |

## Hands-On Understanding

**Materials Needed:** fraction strips, colored pencils or crayons, scissors

**Step 1**   Shade $\frac{8}{9}$ for the distance Mary ran on Saturday.

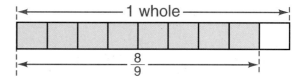

**Step 2**   Mark off $\frac{5}{9}$ for the distance Ellen ran on Saturday.

How many of the shaded ninths are left?

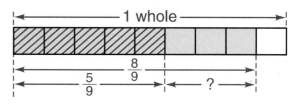

**Step 3**   Write a subtraction sentence that represents your subtraction model.

Is the difference in simplest form?

**Step 4**   Rename the difference as an equivalent fraction in simplest form that tells how much farther Mary ran than Ellen on Saturday.

1. Repeat the steps to find how much farther Mary ran than Ellen on Sunday; on Monday.

2. Can the difference of fractions with like denominators be less than 1? equal to 1? greater than 1? Explain your answers.

**Use fraction strips to model each difference. Write a subtraction
sentence with the difference in simplest form.**

**3.** $\dfrac{5}{8} - \dfrac{1}{8}$    **4.** $\dfrac{5}{6} - \dfrac{1}{6}$    **5.** $\dfrac{11}{12} - \dfrac{5}{12}$    **6.** $\dfrac{7}{10} - \dfrac{3}{10}$

**7.** $\dfrac{15}{8} - \dfrac{7}{8}$    **8.** $\dfrac{7}{6} - \dfrac{1}{6}$    **9.** $\dfrac{19}{12} - \dfrac{7}{12}$    **10.** $\dfrac{17}{10} - \dfrac{7}{10}$

**11.** $\dfrac{13}{8} - \dfrac{2}{8}$    **12.** $\dfrac{19}{6} - \dfrac{5}{6}$    **13.** $\dfrac{17}{12} - \dfrac{4}{12}$    **14.** $\dfrac{13}{10} - \dfrac{2}{10}$

**Write a subtraction sentence with the difference in simplest form
to show how much of the shaded part is left.**

**15.** 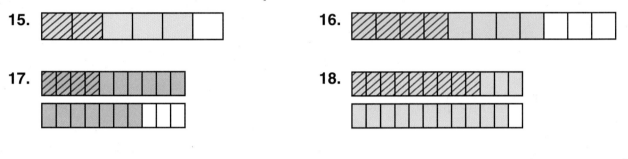    **16.**

**17.**    **18.**

## Communicate

Discuss ✓

**19.** When subtracting two fractions with like denominators,
when would the difference be a fraction? 1? a mixed number?
Explain your answers.

**20.** Use fraction strips to show and explain why
$\dfrac{2}{3} - \dfrac{2}{3} = 0$; $\dfrac{2}{3} - \dfrac{0}{3} = \dfrac{2}{3}$.

**21.** Write in your Math Journal the different types of
answers you get when subtracting fractions with like
denominators. Give an example for each type.

Math
Journal ✓

## Project

**22.** Research how Egyptians worked with fractions.
Show one of their fraction problems and how
it was solved.

# Subtracting: Unlike Denominators

A piece of ribbon $\frac{1}{6}$ yd long is cut from a ribbon that is $\frac{2}{3}$ yd long. How much of the ribbon is left?

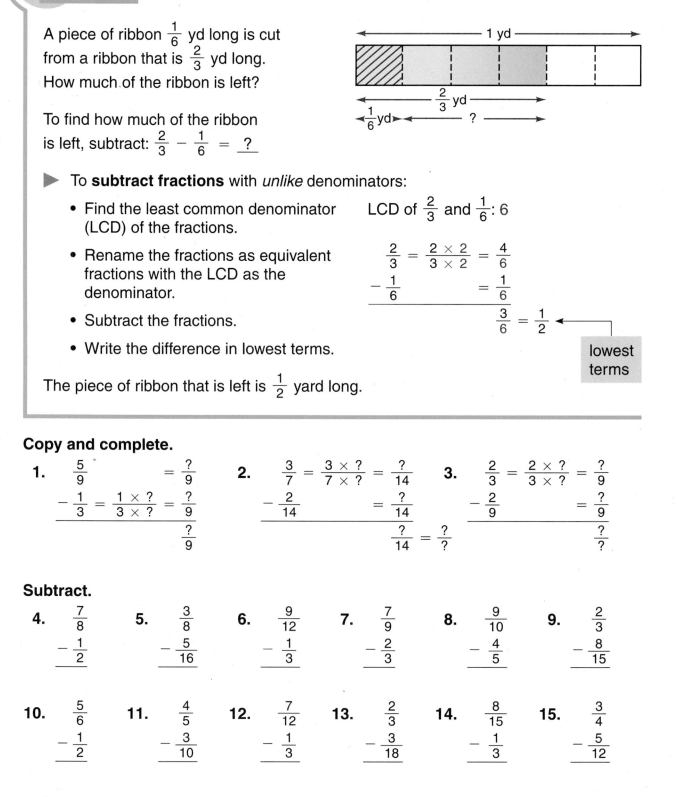

To find how much of the ribbon is left, subtract: $\frac{2}{3} - \frac{1}{6} = \underline{?}$

▶ To **subtract fractions** with *unlike* denominators:

- Find the least common denominator (LCD) of the fractions.

- Rename the fractions as equivalent fractions with the LCD as the denominator.

- Subtract the fractions.

- Write the difference in lowest terms.

LCD of $\frac{2}{3}$ and $\frac{1}{6}$: 6

$$\frac{2}{3} = \frac{2 \times 2}{3 \times 2} = \frac{4}{6}$$
$$-\frac{1}{6} \qquad\qquad = \frac{1}{6}$$
$$\overline{\qquad\qquad\qquad \frac{3}{6} = \frac{1}{2}} \longleftarrow$$

lowest terms

The piece of ribbon that is left is $\frac{1}{2}$ yard long.

## Copy and complete.

1. $\frac{5}{9} = \frac{?}{9}$
$-\frac{1}{3} = \frac{1 \times ?}{3 \times ?} = \frac{?}{9}$
$\overline{\qquad \frac{?}{9}}$

2. $\frac{3}{7} = \frac{3 \times ?}{7 \times ?} = \frac{?}{14}$
$-\frac{2}{14} \qquad\qquad = \frac{?}{14}$
$\overline{\qquad \frac{?}{14} = \frac{?}{?}}$

3. $\frac{2}{3} = \frac{2 \times ?}{3 \times ?} = \frac{?}{9}$
$-\frac{2}{9} \qquad\qquad = \frac{?}{9}$
$\overline{\qquad \frac{?}{?}}$

## Subtract.

4. $\frac{7}{8}$
$-\frac{1}{2}$

5. $\frac{3}{8}$
$-\frac{5}{16}$

6. $\frac{9}{12}$
$-\frac{1}{3}$

7. $\frac{7}{9}$
$-\frac{2}{3}$

8. $\frac{9}{10}$
$-\frac{4}{5}$

9. $\frac{2}{3}$
$-\frac{8}{15}$

10. $\frac{5}{6}$
$-\frac{1}{2}$

11. $\frac{4}{5}$
$-\frac{3}{10}$

12. $\frac{7}{12}$
$-\frac{1}{3}$

13. $\frac{2}{3}$
$-\frac{3}{18}$

14. $\frac{8}{15}$
$-\frac{1}{3}$

15. $\frac{3}{4}$
$-\frac{5}{12}$

**Find the difference.**

16. $\frac{2}{3}$ $-\frac{7}{12}$

17. $\frac{5}{6}$ $-\frac{5}{18}$

18. $\frac{3}{4}$ $-\frac{7}{16}$

19. $\frac{5}{8}$ $-\frac{7}{24}$

20. $\frac{4}{5}$ $-\frac{4}{15}$

21. $\frac{3}{4}$ $-\frac{5}{20}$

22. $\frac{4}{11}$ $-\frac{5}{22}$

23. $\frac{7}{8}$ $-\frac{5}{24}$

24. $\frac{8}{9}$ $-\frac{5}{18}$

25. $\frac{6}{7}$ $-\frac{5}{21}$

26. $\frac{9}{10}$ $-\frac{3}{20}$

27. $\frac{5}{6}$ $-\frac{7}{30}$

**Subtract.**

28. $\frac{4}{6} - \frac{2}{12}$

29. $\frac{11}{18} - \frac{1}{6}$

30. $\frac{2}{3} - \frac{5}{12}$

31. $\frac{14}{16} - \frac{1}{4}$

32. $\frac{5}{9} - \frac{5}{18}$

33. $\frac{15}{16} - \frac{1}{8}$

34. $\frac{7}{15} - \frac{1}{3}$

35. $\frac{4}{5} - \frac{3}{20}$

## PROBLEM SOLVING

36. Nelia had $\frac{2}{3}$ cup of fruit. She put $\frac{3}{6}$ cup into the salad she was making. What fractional part of a cup of fruit was left?

37. Marsha needs $\frac{2}{3}$ qt of paint for a project. She has $\frac{7}{12}$ qt of paint. How much more paint does she need for the project?

38. Chris had $\frac{3}{4}$ yd of ribbon. He used $\frac{3}{8}$ yd for a bow. How much of the ribbon was *not* used for the bow?

39. Juan ran $\frac{6}{8}$ of a mile and Charles ran $\frac{1}{4}$ of a mile. How much farther did Juan run than Charles?

40. Naty had $\frac{11}{12}$ of a tank of gas. She used some and had $\frac{1}{3}$ of a tank left. How much gas did she use?

41. Denroy walked $\frac{7}{8}$ mile on Monday. He walked $\frac{1}{4}$ mile less on Tuesday. How far did he walk on Tuesday?

## Challenge

How much greater than 1 is the sum of:

42. $\frac{2}{3}$ and $\frac{1}{2}$

43. $\frac{2}{3}$ and $\frac{4}{9}$

44. $\frac{5}{6}$ and $\frac{1}{4}$

45. $\frac{4}{5}$ and $\frac{3}{10}$

46. $\frac{3}{4}$ and $\frac{5}{8}$

47. $\frac{5}{6}$ and $\frac{5}{12}$

48. $\frac{3}{7}$ and $\frac{17}{21}$

49. $\frac{7}{8}$ and $\frac{1}{2}$

# More Subtraction of Fractions

Flor uses $\frac{2}{3}$ yd of a $\frac{3}{4}$-yd strip of wood to make a name plate. How long is the piece of wood that is left?

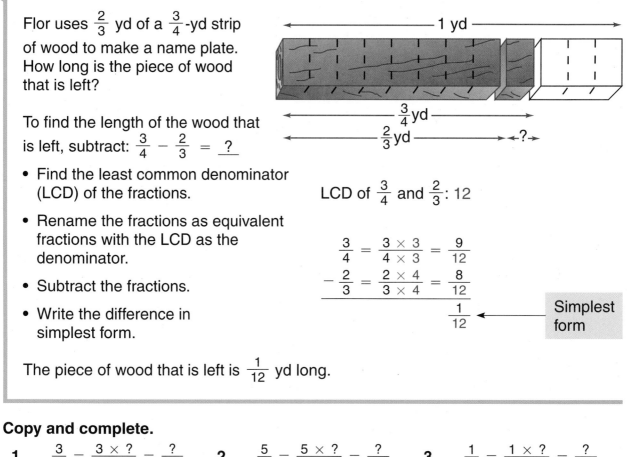

To find the length of the wood that is left, subtract: $\frac{3}{4} - \frac{2}{3} = \underline{?}$

- Find the least common denominator (LCD) of the fractions.

LCD of $\frac{3}{4}$ and $\frac{2}{3}$: 12

- Rename the fractions as equivalent fractions with the LCD as the denominator.

- Subtract the fractions.

$$
\begin{array}{r}
\frac{3}{4} = \frac{3 \times 3}{4 \times 3} = \frac{9}{12} \\
-\frac{2}{3} = \frac{2 \times 4}{3 \times 4} = \frac{8}{12} \\
\hline
\frac{1}{12}
\end{array}
$$

- Write the difference in simplest form.

Simplest form

The piece of wood that is left is $\frac{1}{12}$ yd long.

## Copy and complete.

**1.** $\frac{3}{5} = \frac{3 \times ?}{5 \times ?} = \frac{?}{15}$
$-\frac{1}{3} = \frac{1 \times ?}{3 \times ?} = \frac{?}{15}$
$\overline{\quad\quad\quad\quad\quad \frac{?}{15}}$

**2.** $\frac{5}{6} = \frac{5 \times ?}{6 \times ?} = \frac{?}{24}$
$-\frac{3}{8} = \frac{3 \times ?}{8 \times ?} = \frac{?}{24}$
$\overline{\quad\quad\quad\quad\quad \frac{?}{24}}$

**3.** $\frac{1}{2} = \frac{1 \times ?}{2 \times ?} = \frac{?}{18}$
$-\frac{2}{9} = \frac{2 \times ?}{9 \times ?} = \frac{?}{18}$
$\overline{\quad\quad\quad\quad\quad \frac{?}{?}}$

## Subtract.

**4.** $\frac{1}{3}$
$-\frac{1}{4}$

**5.** $\frac{4}{5}$
$-\frac{3}{4}$

**6.** $\frac{7}{9}$
$-\frac{1}{2}$

**7.** $\frac{2}{5}$
$-\frac{1}{3}$

**8.** $\frac{4}{5}$
$-\frac{1}{2}$

**9.** $\frac{3}{4}$
$-\frac{1}{6}$

**10.** $\frac{6}{7}$
$-\frac{2}{3}$

**11.** $\frac{3}{5}$
$-\frac{1}{8}$

**12.** $\frac{7}{10}$
$-\frac{2}{3}$

**13.** $\frac{5}{6}$
$-\frac{5}{8}$

**14.** $\frac{3}{7}$
$-\frac{1}{3}$

**15.** $\frac{9}{10}$
$-\frac{1}{4}$

**Find the difference.**

16. $\dfrac{5}{6}$  $-\dfrac{2}{9}$

17. $\dfrac{4}{5}$  $-\dfrac{1}{3}$

18. $\dfrac{8}{9}$  $-\dfrac{5}{12}$

19. $\dfrac{13}{15}$  $-\dfrac{4}{9}$

20. $\dfrac{6}{7}$  $-\dfrac{3}{4}$

21. $\dfrac{9}{10}$  $-\dfrac{2}{3}$

22. $\dfrac{5}{7}$  $-\dfrac{3}{5}$

23. $\dfrac{7}{9}$  $-\dfrac{2}{3}$

24. $\dfrac{5}{6}$  $-\dfrac{4}{5}$

25. $\dfrac{7}{8}$  $-\dfrac{2}{3}$

26. $\dfrac{4}{5}$  $-\dfrac{3}{7}$

27. $\dfrac{1}{2}$  $-\dfrac{2}{11}$

**Subtract.**

28. $\dfrac{1}{2} - \dfrac{1}{3}$

29. $\dfrac{3}{4} - \dfrac{2}{5}$

30. $\dfrac{4}{5} - \dfrac{1}{6}$

31. $\dfrac{5}{6} - \dfrac{4}{9}$

32. $\dfrac{2}{3} - \dfrac{1}{4}$

33. $\dfrac{7}{8} - \dfrac{5}{6}$

34. $\dfrac{8}{9} - \dfrac{3}{4}$

35. $\dfrac{14}{15} - \dfrac{2}{9}$

**Compare. Write $<$, $=$, or $>$.**

36. $\dfrac{7}{8} - \dfrac{1}{6}$ ? $\dfrac{2}{3} + \dfrac{1}{5}$

37. $\dfrac{1}{4} + \dfrac{2}{9}$ ? $\dfrac{9}{10} - \dfrac{1}{6}$

38. $\dfrac{5}{6} - \dfrac{1}{3}$ ? $\dfrac{1}{6} + \dfrac{1}{3}$

39. $\dfrac{1}{3} + \dfrac{1}{5}$ ? $\dfrac{2}{3} - \dfrac{1}{4}$

40. $\dfrac{4}{5} - \dfrac{1}{10}$ ? $\dfrac{1}{5} + \dfrac{1}{2}$

41. $\dfrac{2}{5} + \dfrac{1}{7}$ ? $\dfrac{2}{3} - \dfrac{3}{7}$

**PROBLEM SOLVING**

42. How much less than $\dfrac{5}{7}$ is $\dfrac{1}{2}$?

43. How much greater than $\dfrac{6}{7}$ is $\dfrac{5}{6}$?

44. On Tuesday $\dfrac{3}{4}$ inch of snow fell. On Thursday $\dfrac{1}{5}$ inch of snow fell. How much more snow fell on Tuesday than on Thursday?

45. Tess has $\dfrac{5}{8}$ of an inch of loose-leaf paper in her binder. Cal has $\dfrac{2}{3}$ of an inch in his. Who has less loose-leaf paper? How much less?

46. Pat, Jett, and Vic went to the library during their break. Jett stayed in the library for $\dfrac{1}{10}$ hour less than Pat. Vic stayed in the library for $\dfrac{1}{4}$ hour more than Jett. If Vic stayed in the library for $\dfrac{4}{5}$ hour, how much time did each one stay in the library?

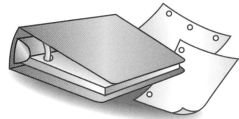

# 5-9 Subtracting Mixed Numbers

Sylvia had $7\frac{3}{4}$ yards of fabric. She used some of the fabric to make curtains and had $2\frac{1}{4}$ yards left. How much fabric did she use for the curtains?

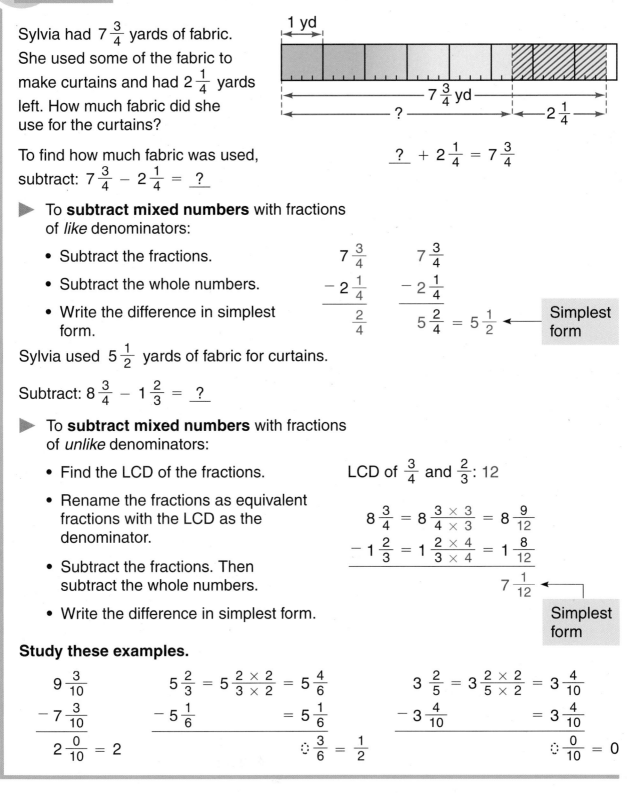

1 yd

$7\frac{3}{4}$ yd

?

$2\frac{1}{4}$

To find how much fabric was used, subtract: $7\frac{3}{4} - 2\frac{1}{4} = \underline{\ ?\ }$

$\underline{\ ?\ } + 2\frac{1}{4} = 7\frac{3}{4}$

▶ To **subtract mixed numbers** with fractions of *like* denominators:

- Subtract the fractions.
- Subtract the whole numbers.
- Write the difference in simplest form.

$$\begin{array}{r} 7\frac{3}{4} \\ -\ 2\frac{1}{4} \\ \hline \frac{2}{4} \end{array} \qquad \begin{array}{r} 7\frac{3}{4} \\ -\ 2\frac{1}{4} \\ \hline 5\frac{2}{4} = 5\frac{1}{2} \end{array}$$

← Simplest form

Sylvia used $5\frac{1}{2}$ yards of fabric for curtains.

Subtract: $8\frac{3}{4} - 1\frac{2}{3} = \underline{\ ?\ }$

▶ To **subtract mixed numbers** with fractions of *unlike* denominators:

- Find the LCD of the fractions.

LCD of $\frac{3}{4}$ and $\frac{2}{3}$: 12

- Rename the fractions as equivalent fractions with the LCD as the denominator.

- Subtract the fractions. Then subtract the whole numbers.

- Write the difference in simplest form.

$$\begin{array}{r} 8\frac{3}{4} = 8\frac{3 \times 3}{4 \times 3} = 8\frac{9}{12} \\ -\ 1\frac{2}{3} = 1\frac{2 \times 4}{3 \times 4} = 1\frac{8}{12} \\ \hline 7\frac{1}{12} \end{array}$$

← Simplest form

**Study these examples.**

$$\begin{array}{r} 9\frac{3}{10} \\ -\ 7\frac{3}{10} \\ \hline 2\frac{0}{10} = 2 \end{array}$$

$$\begin{array}{r} 5\frac{2}{3} = 5\frac{2 \times 2}{3 \times 2} = 5\frac{4}{6} \\ -\ 5\frac{1}{6} = \qquad\quad\ \, = 5\frac{1}{6} \\ \hline \frac{3}{6} = \frac{1}{2} \end{array}$$

$$\begin{array}{r} 3\frac{2}{5} = 3\frac{2 \times 2}{5 \times 2} = 3\frac{4}{10} \\ -\ 3\frac{4}{10} = \qquad\quad\ \, = 3\frac{4}{10} \\ \hline \frac{0}{10} = 0 \end{array}$$

180

**Subtract.**

1. $3\frac{2}{5}$
   $-2\frac{1}{5}$

2. $2\frac{4}{7}$
   $-1\frac{3}{7}$

3. $4\frac{7}{8}$
   $-2\frac{3}{8}$

4. $5\frac{5}{6}$
   $-3\frac{3}{6}$

5. $5\frac{11}{16}$
   $-5\frac{3}{16}$

6. $6\frac{8}{9}$
   $-3\frac{8}{9}$

7. $5\frac{4}{12}$
   $-5\frac{1}{3}$

8. $6\frac{5}{9}$
   $-4\frac{1}{2}$

9. $6\frac{4}{5}$
   $-2\frac{1}{3}$

10. $8\frac{2}{3}$
    $-3\frac{1}{5}$

11. $8\frac{3}{4}$
    $-7\frac{1}{6}$

12. $8\frac{5}{6}$
    $-8\frac{4}{9}$

13. $9\frac{3}{8} - 4\frac{5}{16}$

14. $6\frac{3}{7} - 2\frac{5}{21}$

15. $2\frac{1}{5} - 1\frac{1}{20}$

16. $8\frac{5}{6} - 5\frac{1}{3}$

17. $5\frac{2}{3} - 5\frac{2}{9}$

18. $3\frac{4}{12} - 3\frac{1}{3}$

19. $7\frac{15}{20} - 4\frac{3}{5}$

20. $2\frac{4}{7} - 1\frac{1}{2}$

## PROBLEM SOLVING

21. A motorcyclist rode $9\frac{5}{7}$ miles on flat and hilly roads. If he rode $2\frac{1}{21}$ miles on hilly roads, how many miles did he ride on flat roads?

22. A recipe calls for $2\frac{5}{9}$ cups of flour. Lou has only $1\frac{1}{3}$ cups of flour on hand. How many more cups of flour does she need to make the recipe?

23. From a $5\frac{5}{6}$-ft piece of rope, Val cut off $2\frac{1}{3}$ ft. How much rope was left?

24. Cindy ran the 60-yd hurdles in $11\frac{2}{3}$ s. Elsie ran the same race in $1\frac{1}{2}$ s less than Cindy. What was Elsie's time?

25. In your Math Journal write when the fractional part of the difference of two mixed numbers is equal to zero; the whole-number part of the difference is equal to zero. Use models to explain your answers.

*Math Journal*

## Challenge

*Communicate*

**Compare. Write $<$, $=$, or $>$. Explain the method you used.**

26. $9\frac{5}{10} - 6\frac{3}{10}$ ___?___ $5\frac{2}{5} - 2\frac{1}{5}$

27. $6\frac{3}{4} - 2\frac{1}{4}$ ___?___ $10\frac{2}{3} - 6\frac{1}{3}$

28. $8\frac{4}{7} - 5\frac{3}{14}$ ___?___ $7\frac{4}{5} - 4\frac{3}{10}$

29. $9\frac{12}{20} - 2\frac{3}{10}$ ___?___ $8\frac{1}{4} - 1\frac{1}{8}$

# Subtraction with Renaming

Susan had 3 yards of ribbon.
She used $1\frac{2}{6}$ yards for edging.
How many yards of ribbon
did she have left?

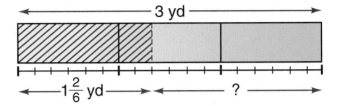

To find the number of yards left,
subtract: $3 - 1\frac{2}{6} = \underline{\ ?\ }$

▶ To **subtract** a *mixed number* from
a *whole number:*

- Rename the whole number as
a mixed number.

- Subtract the mixed numbers.

- Write the difference in simplest form.

$$3 = 2\frac{6}{6}$$
$$-1\frac{2}{6} = 1\frac{2}{6}$$
$$\overline{\qquad 1\frac{4}{6} = 1\frac{2}{3}} \leftarrow$$

$$3 = 2 + 1$$
$$= 2 + \frac{6}{6} = 2\frac{6}{6}$$

Simplest
form

Susan had $1\frac{2}{3}$ yards of ribbon left.

## Study these examples.

$$7 = 6\frac{4}{4}$$
$$-6\frac{1}{4} = 6\frac{1}{4}$$
$$\overline{\qquad\ \frac{3}{4}}$$

$$7 = 6 + 1$$
$$= 6 + \frac{4}{4}$$
$$= 6\frac{4}{4}$$

$$9 - \frac{2}{3} = 8\frac{3}{3} - \frac{2}{3}$$
$$= 8\frac{1}{3}$$

$$9 = 8 + 1$$
$$= 8 + \frac{3}{3}$$
$$= 8\frac{3}{3}$$

## Copy and complete.

**1.** $2 = 1\frac{?}{2}$

**2.** $5 = 4\frac{?}{3}$

**3.** $7 = 6\frac{?}{8}$

**4.** $9 = 8\frac{?}{5}$

**5.** $6 = 5\frac{?}{4}$

**6.** $3 = 2\frac{?}{3}$

**7.** $8 = 7\frac{?}{7}$

**8.** $1 = \frac{?}{9}$

**9.** $8 = \underline{\ ?\ }\frac{?}{4}$

**10.** $6 = \underline{\ ?\ }\frac{?}{6}$

**11.** $5 = \underline{\ ?\ }\frac{?}{2}$

**12.** $4 = \underline{\ ?\ }\frac{?}{3}$

**13.** $10 = \underline{\ ?\ }\frac{?}{5}$

**14.** $12 = \underline{\ ?\ }\frac{?}{7}$

**15.** $14 = \underline{\ ?\ }\frac{?}{11}$

**16.** $11 = \underline{\ ?\ }\frac{?}{9}$

**Subtract.**

**17.** $7$
$-3\frac{2}{3}$

**18.** $6$
$-2\frac{1}{2}$

**19.** $4$
$-1\frac{3}{8}$

**20.** $5$
$-1\frac{1}{4}$

**21.** $10$
$-7\frac{3}{5}$

**22.** $7$
$-2\frac{2}{7}$

**23.** $9$
$-2\frac{1}{6}$

**24.** $6$
$-4\frac{1}{5}$

**25.** $4$
$-1\frac{1}{2}$

**26.** $6$
$-2\frac{2}{3}$

**27.** $7$
$-3\frac{4}{9}$

**28.** $3$
$-1\frac{6}{10}$

**29.** $3$
$-2\frac{2}{5}$

**30.** $7$
$-6\frac{1}{8}$

**31.** $4$
$-2\frac{6}{9}$

**32.** $8$
$-5\frac{2}{4}$

**33.** $10$
$-5\frac{5}{6}$

**34.** $4$
$-1\frac{1}{9}$

**35.** $3$
$-2\frac{1}{6}$

**36.** $7$
$-6\frac{8}{12}$

**37.** $8$
$-4\frac{7}{12}$

**38.** $4$
$-1\frac{2}{3}$

**39.** $16$
$-9\frac{5}{8}$

**40.** $3$
$-1\frac{9}{10}$

**Find the difference.**

**41.** $6 - 2\frac{3}{5}$

**42.** $8 - \frac{1}{4}$

**43.** $9 - \frac{3}{5}$

**44.** $7 - 4\frac{3}{10}$

**45.** $5 - 3\frac{2}{9}$

**46.** $7 - 6\frac{1}{6}$

**47.** $4 - \frac{3}{4}$

**48.** $2 - \frac{1}{5}$

**PROBLEM SOLVING**

**49.** A piece of tin $2\frac{3}{8}$ ft long was cut from a 4-ft sheet of tin. How much of the sheet was left?

**50.** Max lives $4\frac{5}{6}$ miles from school. Don lives 6 miles from school. How much farther away from school does Don live than Max?

**51.** Explain in your Math Journal why renaming is needed when a mixed number is subtracted from a whole number.

*Math Journal*

### Critical Thinking

**Write the next two terms to complete the pattern. Explain the method you used.**

*Algebra*

**52.** $6, 5\frac{1}{2}, 5, 4\frac{1}{2}, \underline{\ ?\ }, \underline{\ ?\ }$

**53.** $8, 6\frac{1}{2}, 5, 3\frac{1}{2}, \underline{\ ?\ }, \underline{\ ?\ }$

**54.** $7, 5\frac{2}{3}, 4\frac{1}{3}, 3, \underline{\ ?\ }, \underline{\ ?\ }$

**55.** $9, 7\frac{3}{4}, 6\frac{1}{2}, 5\frac{1}{4}, \underline{\ ?\ }, \underline{\ ?\ }$

# More Renaming in Subtraction

Elise is biking to the shopping mall, $4\frac{1}{2}$ miles from her home. She has already gone $2\frac{5}{6}$ miles. How much farther does she have to go to reach the mall?

To find how much farther Elise has to go, subtract: $4\frac{1}{2} - 2\frac{5}{6} = \underline{\ ?\ }$

- Find the LCD of the fractions.

LCD of $\frac{1}{2}$ and $\frac{5}{6}$: 6

- Express the fractions as equivalent fractions with the LCD as the denominator.

$$4\frac{1}{2} = 4\frac{1 \times 3}{2 \times 3} = 4\frac{3}{6}$$
$$-\,2\frac{5}{6} \qquad\qquad = 2\frac{5}{6}$$

$\frac{3}{6} < \frac{5}{6}$

- Rename the *minuend* if the fraction in the minuend is less than the fraction in the subtrahend.

$$4\frac{1}{2} = 4\frac{3}{6} = 3\frac{9}{6}$$
$$-\,2\frac{5}{6} = 2\frac{5}{6} = 2\frac{5}{6}$$
$$\rule{2cm}{0.4pt}$$
$$1\frac{4}{6} = 1\frac{2}{3}$$

$$4\frac{3}{6} = 3 + 1 + \frac{3}{6}$$
$$= 3 + \frac{6}{6} + \frac{3}{6}$$
$$= 3 + \frac{9}{6}$$

- Subtract. Write the difference in simplest form.

↑ Simplest form

Elise has to go $1\frac{2}{3}$ miles more to reach the mall.

### Study this example.

$$5\frac{1}{3} = 5\frac{1 \times 8}{3 \times 8} = 5\frac{8}{24}$$
$$-\,4\frac{7}{8} = 4\frac{7 \times 3}{8 \times 3} = 4\frac{21}{24}$$
$$\rule{3cm}{0.4pt}$$
$$\frac{8}{24} < \frac{21}{24}$$

$$5\frac{8}{24} = 4\frac{32}{24}$$
$$-\,4\frac{21}{24} = 4\frac{21}{24}$$
$$\rule{2.5cm}{0.4pt}$$
$$\frac{11}{24}$$

$$5\frac{8}{24} = 4 + 1 + \frac{8}{24}$$
$$= 4 + \frac{24}{24} + \frac{8}{24}$$
$$= 4\frac{32}{24}$$

### Copy and complete.

**1.** $5\frac{1}{5} = 4 + 1 + \frac{1}{5}$
$\qquad = 4 + \frac{5}{5} + \frac{1}{5}$
$\qquad = 4\frac{?}{5}$

**2.** $8\frac{2}{3} = 7 + 1 + \frac{2}{3}$
$\qquad = 7 + \frac{?}{3} + \frac{2}{3}$
$\qquad = 7\frac{?}{3}$

**3.** $4\frac{3}{7} = 3 + 1 + \frac{3}{7}$
$\qquad = 3 + \frac{?}{7} + \frac{?}{7}$
$\qquad = 3\frac{?}{7}$

**Subtract.**

4.  $6\frac{1}{2}$
    $-3\frac{3}{4}$

5.  $10\frac{1}{4}$
    $-\ 9\frac{3}{8}$

6.  $4\frac{1}{6}$
    $-2\frac{2}{3}$

7.  $8\frac{1}{5}$
    $-2\frac{5}{10}$

8.  $8\frac{1}{3}$
    $-4\frac{5}{12}$

9.  $8\frac{1}{3}$
    $-2\frac{4}{15}$

10. $7\frac{3}{4}$
    $-2\frac{7}{8}$

11. $6\frac{1}{3}$
    $-4\frac{4}{9}$

12. $12\frac{1}{6}$
    $-\ 7\frac{7}{12}$

13. $10\frac{3}{10}$
    $-\ 4\frac{3}{5}$

14. $8\frac{1}{3}$
    $-3\frac{7}{15}$

15. $6\frac{1}{2}$
    $-5\frac{9}{10}$

16. $9\frac{1}{4}$
    $-2\frac{3}{7}$

17. $12\frac{1}{4}$
    $-\ 8\frac{2}{3}$

18. $2\frac{1}{5}$
    $-1\frac{2}{3}$

19. $5\frac{1}{4}$
    $-2\frac{5}{6}$

20. $6\frac{1}{9}$
    $-4\frac{1}{2}$

21. $2\frac{1}{4}$
    $-\ \frac{3}{5}$

**Find the difference.**

22. $8\frac{3}{8} - 5\frac{3}{4}$

23. $7\frac{1}{2} - 4\frac{7}{10}$

24. $9\frac{1}{3} - 8\frac{5}{6}$

25. $6\frac{1}{4} - \frac{3}{8}$

26. $5\frac{1}{4} - 4\frac{2}{3}$

27. $4\frac{3}{4} - 2\frac{5}{6}$

28. $10\frac{1}{5} - \frac{1}{3}$

29. $3\frac{1}{8} - \frac{3}{5}$

30. $11\frac{3}{8} - 8\frac{2}{3}$

31. $5\frac{1}{5} - \frac{7}{9}$

32. $8\frac{2}{3} - 4\frac{4}{5}$

33. $7\frac{4}{7} - \frac{3}{4}$

**PROBLEM SOLVING**

34. What number is $\frac{5}{7}$ less than $3\frac{1}{2}$?

35. Find the difference between $7\frac{3}{8}$ and $5\frac{2}{3}$.

36. Chuck roller-skates $4\frac{1}{3}$ miles from his home to school. After he goes $2\frac{7}{8}$ miles from his home, he passes Arnie's house. How far from school is Arnie's house?

37. Dad caught a trout that weighed $7\frac{3}{8}$ pounds. Tom caught one that weighed $3\frac{3}{4}$ pounds. How many pounds heavier was Dad's trout than Tom's trout?

38. From a $10\frac{1}{3}$-ft piece of rope, a piece of $5\frac{5}{6}$ ft was cut off. How much rope was left?

39. Owen needs $6\frac{2}{5}$ yd of wire. He has $4\frac{3}{4}$ yd. How much more wire does he need?

 **Share Your Thinking**

Communicate ✓

40. Explain to your teacher how to rename $5\frac{1}{6}$ so that you could subtract $3\frac{2}{9}$ from it.

# Estimate to Compute

About how many miles in all did Jack hike on his three hiking trips?

To find about how many miles in all, estimate the sum: $12\frac{1}{3} + 11\frac{5}{6} + 14\frac{4}{9}$

**Jack's Hiking Trips**

| Trip | Distance Hiked |
|------|----------------|
| 1 | $12\frac{1}{3}$ mi |
| 2 | $11\frac{5}{6}$ mi |
| 3 | $14\frac{4}{9}$ mi |

▶ To **estimate the sum** of two or more mixed numbers:

- Round each mixed number to the nearest whole number.

- Add the rounded numbers.

$$12\frac{1}{3} + 11\frac{5}{6} + 14\frac{4}{9}$$
$$12 \ + \ 12 \ + \ 14 \ = \ 38 \quad \text{estimated sum}$$

Jack hiked about 38 miles on the three trips.

About how many miles more did Jack hike on the third trip than on the second trip?

To find about how many miles more, estimate the difference: $14\frac{4}{9} - 11\frac{5}{6}$

▶ To **estimate the difference** of two mixed numbers:

- Round each mixed number to the nearest whole number.

- Subtract the rounded numbers.

$$14\frac{4}{9} - 11\frac{5}{6}$$
$$14 \ - \ 12 \ = \ 2 \quad \text{estimated difference}$$

Jack hiked about 2 miles more on the third trip than on the second trip.

## Estimate the sum.

**1.** $9\frac{1}{3} + 2\frac{3}{8}$     **2.** $8\frac{2}{3} + 3\frac{3}{4}$     **3.** $14\frac{1}{3} + 12\frac{1}{2}$     **4.** $16\frac{2}{7} + 13\frac{5}{9}$

**5.** $11\frac{3}{5} + 4\frac{7}{8}$     **6.** $16\frac{1}{4} + 4\frac{3}{8}$     **7.** $19\frac{2}{9} + 15\frac{3}{4}$     **8.** $15\frac{1}{8} + 14\frac{8}{9}$

**9.** $7\frac{1}{5} + 3\frac{4}{9} + 5\frac{1}{3}$     **10.** $4\frac{2}{11} + 7\frac{1}{8} + 9\frac{3}{10}$     **11.** $8\frac{3}{5} + 9\frac{4}{7} + 3\frac{5}{6}$

**Estimate the difference.**

**12.** $6\frac{4}{7} - 2\frac{1}{3}$     **13.** $5\frac{7}{10} - 2\frac{3}{5}$     **14.** $9\frac{2}{3} - 2\frac{5}{6}$     **15.** $8\frac{3}{4} - 3\frac{2}{7}$

**16.** $8\frac{7}{12} - 4\frac{3}{4}$     **17.** $10\frac{1}{5} - 2\frac{3}{10}$     **18.** $18\frac{2}{9} - 4\frac{1}{2}$     **19.** $15\frac{2}{3} - 4\frac{7}{8}$

---

### Front-End Estimation

▶ To estimate sums using front-end estimation:

- Add the whole number parts.     $9\frac{1}{3} + 4\frac{1}{5} + 5\frac{7}{9} \rightarrow 18$

- Adjust the estimate with the fraction parts.     $9\frac{1}{3} + 4\frac{1}{5} + 5\frac{7}{9}$

$\downarrow$

about 1

Adjusted estimate: $18 + 1 = 19$

▶ To estimate differences using front-end estimation:

- Subtract the whole number parts.     $15\frac{5}{9} - 6\frac{1}{4} = 9 \longleftarrow$ estimated difference

---

**Estimate. Use front-end estimation.**

**20.** $12\frac{1}{8} + 3\frac{2}{3}$     **21.** $9\frac{8}{11} + 7\frac{2}{9} + 6\frac{1}{10}$     **22.** $8\frac{3}{4} + 9\frac{4}{5} + 4\frac{1}{3}$

**23.** $9\frac{5}{16} - 6\frac{1}{5}$     **24.** $10\frac{3}{5} - 4\frac{2}{3}$     **25.** $18\frac{7}{12} - 5\frac{2}{7}$     **26.** $25\frac{1}{8} - 13\frac{11}{15}$

### PROBLEM SOLVING

**27.** Jan ran $2\frac{3}{8}$ mi on Saturday and $6\frac{5}{6}$ mi on Sunday. About how many miles did he run that weekend?

**28.** Which estimation method would give a more reasonable estimate for $10\frac{1}{3} - 9\frac{5}{9}$? Why?

### Critical Thinking

Communicate ✓

**Write a mixed number that will give a sum or difference in the given range. Explain the method you used.**

**29.** $6\frac{2}{7} + \underline{\ ?\ }$ is between 8 and 9.

**30.** $8\frac{7}{12} + \underline{\ ?\ }$ is between 12 and 13.

**31.** $9\frac{2}{9} - \underline{\ ?\ }$ is between 4 and 5.

**32.** $10\frac{3}{5} - \underline{\ ?\ }$ is between 6 and 7.

# 5-13 Problem Solving: Working Backwards

**Problem:** At a bake sale Ms. Talbot sold $6\frac{1}{3}$ dozen muffins before lunch. After lunch, Mr. Vaner donated 2 dozen muffins. Ms. Talbot sold another $7\frac{1}{2}$ dozen. Then she had $1\frac{1}{2}$ dozen muffins left. How many muffins did she have at the start of the sale?

**1 IMAGINE** Put yourself in the problem.

**2 NAME**

*Facts:*

before lunch — $6\frac{1}{3}$ doz sold

after lunch — 2 doz donated

$7\frac{1}{2}$ doz sold

$1\frac{1}{2}$ doz left

*Question:* How many muffins did she have at the start of the sale?

**3 THINK**

First write a number sentence to show what happened.

Total − doz sold + doz donated − doz sold = doz left

$\underline{\ ?\ }$ − $6\frac{1}{3}$ + 2 − $7\frac{1}{2}$ = $1\frac{1}{2}$

To find the original number, start with the number left and work backwards. Add the muffins that were sold and subtract the muffins that were donated.

$1\frac{1}{2} + 7\frac{1}{2} - 2 + 6\frac{1}{3} = \underline{\ ?\ }$ Total

**4 COMPUTE**

$1\frac{1}{2} + 7\frac{1}{2} - 2 + 6\frac{1}{3} = \underline{\ ?\ }$

$9 \quad\ - 2 + 6\frac{1}{3} = \underline{\ ?\ }$

$7 \quad + 6\frac{1}{3} = 13\frac{1}{3}$

Use the order of operations.

Ms. Talbot had $13\frac{1}{3}$ doz muffins at the start of the sale.

**5 CHECK**

Begin with the total and work *forward*.

$13\frac{1}{3} - 6\frac{1}{3} + \quad 2 \quad - 7\frac{1}{2} \overset{?}{=} 1\frac{1}{2}$

$7 \quad + \quad 2 \quad - 7\frac{1}{2} \overset{?}{=} 1\frac{1}{2}$

$9 \quad\quad - 7\frac{1}{2} \overset{?}{=} 1\frac{1}{2}$

$1\frac{1}{2} = 1\frac{1}{2}$ The answer checks.

**Use the Working Backwards strategy to solve each problem.**

1. The final cost of Elihu's bicycle was $94.00.
   This included a discount of $10.25 and tax
   of $5.50. What was the original price of the
   bicycle without the tax and discount?

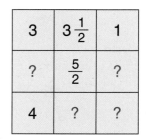

| IMAGINE | Picture yourself buying a bicycle. |
|---|---|

| NAME | *Facts:* | final cost—$94.00 |
|---|---|---|
| | | discount —$10.25 |
| | | tax       —$5.50 |
| | | |
| | *Question:* | What was the original price of the bicycle without the tax and discount? |

| THINK | First write a number sentence to show what happened. Then work backwards. |
|---|---|

Cost    = price  −  discount + tax
$94.00 =  _?_   −   $10.25 + $5.50
$94.00 − $5.50 +   $10.25 =   _?_

**COMPUTE** ⟶ **CHECK**

2. Find the missing addends in the magic square.
   (*Hint:* Find the sums first. Remember: All the
   sums are the *same.*)

| 3 | $3\frac{1}{2}$ | 1 |
|---|---|---|
| ? | $\frac{5}{2}$ | ? |
| 4 | ? | ? |

3. Nick ordered 2 suits for $249.95 each and a pair of slacks.
   The total cost was $554.85. What was the cost of the slacks?

4. After Dad cut fencing to put around his garden, he had
   $\frac{3}{4}$ ft of fencing left over. He had already cut three $3\frac{1}{4}$-ft
   pieces, one $2\frac{1}{3}$-ft piece, and one $3\frac{1}{2}$-ft piece. How long
   was the fencing originally?

5. The Dinger Catering Service prepared punch for 3 wedding
   receptions on one Saturday. If they served $10\frac{1}{3}$ gal of punch
   at the first, $13\frac{1}{2}$ gal at the second, $13\frac{2}{3}$ gal at the third, and
   had $2\frac{1}{2}$ gal left over, how much punch did they prepare for
   the day?

# 5-14 Problem-Solving Applications

**Solve each problem and explain the method you used.**

1. At Pet Palace, Meg spent $\frac{1}{5}$ h bathing a terrier and $\frac{3}{5}$ h cutting its hair. How long did Meg spend grooming the terrier?

2. Meg opened a new bottle of dog shampoo in the morning. She used $\frac{1}{4}$ of the bottle before noon and $\frac{2}{5}$ of the bottle after noon. How much of the bottle of shampoo did she use in all?

3. A sheepdog's hair was $4\frac{3}{4}$ in. long. Meg trimmed off $1\frac{3}{8}$ in. How long was the dog's hair after cutting?

4. A bottle of flea spray was $\frac{5}{6}$ full at the beginning of the day. At the end of the day, the bottle was $\frac{1}{3}$ full. How much of the bottle was used that day?

5. The tallest client at Pet Palace, Hercules, is $30\frac{1}{8}$ inches tall. The shortest, Muffin, is $11\frac{3}{16}$ inches tall. How much taller is Hercules than Muffin?

6. Koji worked for $3\frac{1}{4}$ h before lunch and $3\frac{1}{4}$ h after lunch. How long did he work in all?

7. Koji gave a dalmatian $2\frac{1}{2}$ dog biscuits. He gave a poodle $1\frac{1}{2}$ biscuits, and he gave a collie $2\frac{3}{4}$ biscuits. How many biscuits did Koji give to the dogs?

8. In a 50-lb bag of dog food, $19\frac{1}{4}$ lb are meat protein, and $18\frac{7}{8}$ lb are vitamin compound. To fill the bag, how many pounds of cereal compound are needed?

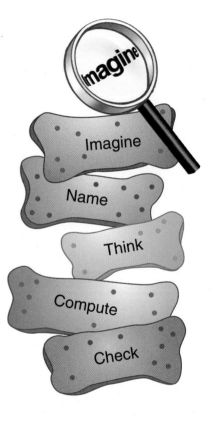

Imagine

Name

Think

Compute

Check

**Choose a strategy from the list or use another strategy you know to solve each problem.**

USE THESE STRATEGIES
Working Backwards
Guess and Test
Hidden Information
Logical Reasoning
Find a Pattern
Use a Graph

9. Avi cut $\frac{2}{3}$ in. off a poodle's hair, but it was not short enough, so he cut another $\frac{1}{4}$ in. Then the dog's hair was perfect at $5\frac{1}{2}$ in. How long was the poodle's hair before cutting?

10. Loxy and Foxy are cats. Together they weigh 16 pounds. Loxy weighs $\frac{1}{2}$ lb more than Foxy, and each cat weighs more than 7 pounds. How much could each cat weigh?

11. Ace, Champ, and Ruffy are dogs that weighed $23\frac{3}{4}$ lb, $23\frac{5}{6}$ lb, and $23\frac{5}{8}$ lb. Ace and Ruffy together weigh more than double Champ's weight. Ace weighs more than Ruffy. Place the dogs in order from lightest to heaviest.

12. A puppy weighed $1\frac{3}{4}$ lb at birth. Each day it gained $\frac{1}{8}$ lb. What will its weight be after one week?

**Use the circle graph for problems 13–15.**

13. What fractional part of the clients were dogs?

14. What fractional part of the clients were cats?

15. From which group of pets does Pet Palace obtain most of its clients?
Group A: large and miniature dogs
Group B: miniature and small dogs
How much greater is this group than the other?

**Pet Palace Clients**

Short-Hair Cats
Long-Hair Cats
$\frac{1}{5}$ $\frac{1}{8}$ $\frac{1}{8}$
Small Dogs
$\frac{1}{4}$ $\frac{3}{10}$
Large Dogs
Miniature Dogs

**Share Your Thinking**

Math Journal

16. Write in your Math Journal which problem you solved using more than one strategy and explain why. Then write a problem modeled on that problem and have a classmate solve it.

# Chapter Review and Practice

**Add.**                                                                        *(See pp. 164–167, 170–173.)*

1. $\dfrac{9}{8}$ $+\dfrac{1}{8}$

2. $\dfrac{5}{6}$ $+\dfrac{1}{6}$

3. $\dfrac{7}{8}$ $+\dfrac{5}{8}$

4. $\dfrac{2}{5}$ $+\dfrac{1}{10}$

5. $\dfrac{1}{4}$ $+\dfrac{1}{12}$

6. $\dfrac{7}{10}$ $+\dfrac{1}{2}$

7. $6\dfrac{7}{9}$ $+1\dfrac{4}{9}$

8. $9\dfrac{7}{16}$ $+2\dfrac{5}{16}$

9. $7\dfrac{1}{2}$ $+1\dfrac{1}{4}$

10. $6\dfrac{1}{5}$ $+4\dfrac{3}{10}$

11. $8\dfrac{5}{12}$ $+2\dfrac{1}{2}$

12. $5\dfrac{3}{5}$ $+2\dfrac{2}{3}$

**Subtract.**                                                                   *(See pp. 174–185.)*

13. $\dfrac{15}{11}$ $-\dfrac{1}{11}$

14. $\dfrac{13}{8}$ $-\dfrac{5}{8}$

15. $\dfrac{7}{12}$ $-\dfrac{1}{3}$

16. $\dfrac{1}{2}$ $-\dfrac{3}{8}$

17. $\dfrac{2}{3}$ $-\dfrac{1}{9}$

18. $\dfrac{5}{6}$ $-\dfrac{1}{2}$

19. $9\dfrac{3}{5}$ $-3\dfrac{1}{4}$

20. $4\dfrac{2}{3}$ $-2\dfrac{1}{2}$

21. $6\dfrac{1}{8}$ $-3\dfrac{1}{2}$

22. $4\dfrac{1}{4}$ $-3\dfrac{3}{8}$

23. $2\dfrac{7}{16}$ $-1\dfrac{3}{4}$

24. $14\dfrac{1}{3}$ $-9\dfrac{3}{5}$

**Add or subtract.**                                                            *(See pp. 164–185.)*

25. $\dfrac{7}{13}+\dfrac{4}{13}$

26. $\dfrac{15}{16}-\dfrac{3}{16}$

27. $\dfrac{3}{5}+\dfrac{1}{3}+\dfrac{4}{15}$

28. $13\dfrac{1}{8}-11\dfrac{1}{2}$

29. $7-2\dfrac{3}{5}$

30. $9\dfrac{7}{8}+2\dfrac{5}{16}+4\dfrac{1}{2}$

**Estimate.** Use front-end estimation.                                         *(See pp. 186–187.)*

31. $10\dfrac{3}{5}+14\dfrac{2}{3}$

32. $2\dfrac{5}{6}-1\dfrac{7}{12}$

33. $1\dfrac{1}{2}+3\dfrac{3}{8}$

34. $3\dfrac{5}{8}-1\dfrac{1}{2}$

## PROBLEM SOLVING                                                              *(See pp. 180–185, 188–190.)*

35. Vicky came home from the matinee at 5:45 P.M. The travel time to and from the cinema was $\dfrac{1}{2}$ hour each way. She spent $2\dfrac{1}{4}$ hours at the cinema. What time did she leave home?

36. Elsie bought $3\dfrac{1}{8}$ lb of peaches and $2\dfrac{5}{6}$ lb of grapes. How many more pounds of peaches than grapes did she buy?

(See *Still More Practice*, p. 481.)

## UNIT FRACTIONS

A **unit fraction** is a fraction with a numerator of 1.

$\frac{1}{2}, \frac{1}{3}, \frac{1}{4}, \frac{1}{5}, \frac{1}{11}, \frac{1}{20}$ are unit fractions.

▶ To express a nonunit fraction as the sum of two or more different unit fractions:

$$\frac{3}{4} = \frac{1}{?} + \frac{1}{?}$$

- Find the unit fractions that have a least common denominator (LCD) equal to the denominator of the nonunit fraction.

What unit fractions have an LCD of 4?
$\frac{1}{2}$ and $\frac{1}{4}$

$$\frac{3}{4} \overset{?}{=} \frac{1}{2} + \frac{1}{4}$$

- Check if the sum of the unit fractions is equal to the given nonunit fraction.

$$\frac{1}{2} = \frac{1 \times 2}{2 \times 2} = \frac{2}{4}$$
$$+ \frac{1}{4} \qquad\qquad = \frac{1}{4}$$
$$\overline{\qquad\qquad\qquad \frac{3}{4}}$$

So $\frac{3}{4} = \frac{1}{2} + \frac{1}{4}$ ⎯ unit fractions

**Write each fraction as the sum of different unit fractions.**

1. $\frac{5}{6} = \frac{1}{?} + \frac{1}{?}$ 

2. $\frac{7}{10} = \frac{1}{?} + \frac{1}{?}$ 

3. $\frac{5}{8} = \frac{1}{?} + \frac{1}{?}$

4. $\frac{4}{9} = \frac{1}{?} + \frac{1}{?}$ 

5. $\frac{2}{3} = \frac{1}{?} + \frac{1}{?}$ 

6. $\frac{7}{12} = \frac{1}{?} + \frac{1}{?}$

7. $\frac{8}{15} = \frac{1}{?} + \frac{1}{?}$ 

8. $\frac{9}{20} = \frac{1}{?} + \frac{1}{?}$ 

9. $\frac{3}{5} = \frac{1}{?} + \frac{1}{?}$

10. $\frac{9}{14} = \frac{1}{?} + \frac{1}{?}$ 

11. $\frac{10}{21} = \frac{1}{?} + \frac{1}{?}$ 

12. $\frac{7}{24} = \frac{1}{?} + \frac{1}{?}$

## PROJECT

13. Research the Italian mathematician Leonardo of Pisa, known as *Fibonacci*. Report his work on unit fractions to the class.

Communicate ✓

# Check Your Mastery

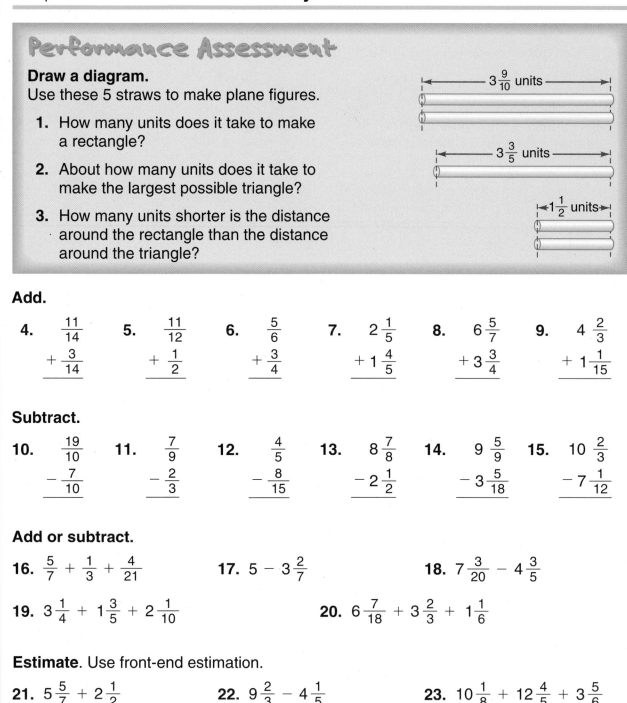

**Performance Assessment**

**Draw a diagram.**
Use these 5 straws to make plane figures.

1. How many units does it take to make a rectangle?

2. About how many units does it take to make the largest possible triangle?

3. How many units shorter is the distance around the rectangle than the distance around the triangle?

$3\frac{9}{10}$ units

$3\frac{3}{5}$ units

$1\frac{1}{2}$ units

**Add.**

4. $\frac{11}{14}$
$+\frac{3}{14}$

5. $\frac{11}{12}$
$+\frac{1}{2}$

6. $\frac{5}{6}$
$+\frac{3}{4}$

7. $2\frac{1}{5}$
$+1\frac{4}{5}$

8. $6\frac{5}{7}$
$+3\frac{3}{4}$

9. $4\frac{2}{3}$
$+1\frac{1}{15}$

**Subtract.**

10. $\frac{19}{10}$
$-\frac{7}{10}$

11. $\frac{7}{9}$
$-\frac{2}{3}$

12. $\frac{4}{5}$
$-\frac{8}{15}$

13. $8\frac{7}{8}$
$-2\frac{1}{2}$

14. $9\frac{5}{9}$
$-3\frac{5}{18}$

15. $10\frac{2}{3}$
$-7\frac{1}{12}$

**Add or subtract.**

16. $\frac{5}{7} + \frac{1}{3} + \frac{4}{21}$

17. $5 - 3\frac{2}{7}$

18. $7\frac{3}{20} - 4\frac{3}{5}$

19. $3\frac{1}{4} + 1\frac{3}{5} + 2\frac{1}{10}$

20. $6\frac{7}{18} + 3\frac{2}{3} + 1\frac{1}{6}$

**Estimate.** Use front-end estimation.

21. $5\frac{5}{7} + 2\frac{1}{2}$

22. $9\frac{2}{3} - 4\frac{1}{5}$

23. $10\frac{1}{8} + 12\frac{4}{5} + 3\frac{5}{6}$

**PROBLEM SOLVING** *Use a strategy you have learned.*

24. Anthony needs $6\frac{7}{8}$ yd of wire. He has $4\frac{1}{3}$ yd. How much more wire does he need?

# Cumulative Review II

**Choose the best answer.**

---

**1.** What is the value of 7 in 376,148,206?

    **a.** 7000
    **b.** 70,000
    **c.** 70,000,000
    **d.** 80,000,000

**2.** Round 592,067,208 to its greatest place.

    **a.** 550,000,000
    **b.** 592,000,000
    **c.** 600,000,000
    **d.** none of these

---

**3.** $900 − $46.54

    **a.** $854.54
    **b.** $864.56
    **c.** $946.54
    **d.** $853.46

**4.** Estimate.

4632 × 221

    **a.** 80,000
    **b.** 100,000
    **c.** 1,000,000
    **d.** 10,000,000

---

**5.** How many times greater than

$30 \times 20$ is $30 \times 2000$?

    **a.** 10   **b.** 100   **c.** 200   **d.** 1000

**6.**    946
    × 608

    **a.** 264,109   **b.** 575,168
    **c.** 755,618   **d.** 576,168

---

**7.** Find the missing dividend.

$\underline{\ ?\ } \div 4 = 12$

    **a.** 3    **b.** 16
    **c.** 36   **d.** 48

**8.** $34\overline{)26{,}588}$

    **a.** 782    **b.** 799 R22
    **c.** 882   **d.** 881 R29

---

**9.** Which is a prime number?

    **a.** 9    **b.** 13
    **c.** 15   **d.** 25

**10.** What is the GCF of 12 and 24?

    **a.** 6    **b.** 8
    **c.** 12   **d.** 24

---

**11.** Which shows $\frac{18}{54}$ in lowest terms?

    **a.** $\frac{1}{4}$   **b.** $\frac{1}{3}$   **c.** $\frac{1}{2}$   **d.** none of these

**12.** Name the mixed number

    **a.** $2\frac{4}{5}$   **b.** $2\frac{5}{6}$   **c.** $2\frac{7}{8}$   **d.** $3\frac{5}{6}$

---

**13.** Find the missing number.

$7 = 6\frac{?}{11}$

    **a.** 6    **b.** 7
    **c.** 11   **d.** 33

**14.** $\frac{1}{2} + \frac{5}{6} + \frac{3}{4}$

    **a.** $\frac{9}{12}$   **b.** $2\frac{1}{12}$
    **c.** $3\frac{1}{4}$   **d.** $1\frac{1}{12}$

---

**15.** $8\frac{6}{7} - 2\frac{1}{7}$

    **a.** $5\frac{5}{7}$   **b.** $8\frac{3}{7}$

    **c.** $10\frac{1}{7}$   **d.** not given

**16.** $12 - 2\frac{1}{3}$

    **a.** $9\frac{2}{3}$   **b.** $10\frac{1}{3}$

    **c.** $14\frac{1}{3}$   **d.** $10\frac{2}{3}$

---

## For Your Portfolio

**Solve each problem. Explain the steps and the strategy or strategies you used for each. Then choose one from problems 1–4 for your Portfolio.**

1. Mr. Diaz needs $8\frac{5}{16}$ feet of molding to finish a closet. He has $7\frac{1}{8}$ feet of molding. How many more feet of molding does Mr. Diaz need?

2. Ilka started with a number, added $\frac{2}{5}$ to it, and then subtracted $1\frac{1}{10}$. She ended up with the number $2\frac{3}{10}$. What was Ilka's original number?

3. Ellen collected $1072.61 from 49 customers at her booth at the antiques fair. About how much did she receive from each person if each person gave her approximately the same amount?

4. Natasha has completed 7 miles of her 13-mile trip. Tyrone has completed 6 miles of his 9-mile trip. Which person has completed the greater part of her/his trip?

**Tell about it.**

5. How can you use estimation to help you solve problem 3? Explain.

*Communicate* ✓

6. Describe the strategy you used to solve problem 4. Name another strategy you could have used.

---

## For Rubric Scoring

**Listen for information on how your work will be scored.**

7. Use each of the digits 1, 2, 3, 4, 5, 6 one time to fill in the boxes. The subtraction example contains mixed numbers and must satisfy these four clues:

   *Clue #1:* Difference is between 4 and 7.

   *Clue #2:* LCD of the fractions is greater than 12.

   *Clue #3:* No regrouping is needed to subtract.

   *Clue #4:* The whole-number part of the answer is an odd number.

# Fractions: Multiplication and Division

## Arithmetic

Multiplication is vexation.
Division is as bad;
The Rule of Three it puzzles me,
And fractions drive me mad.

*Anonymous*

**In this chapter you will:**

Multiply fractions and mixed
   numbers with cancellation
Explore division with models
Learn about reciprocals,
   dividing fractions, and
   mixed numbers
Estimate mixed-number
   products and quotients
Study order of operations
   using a calculator
Solve problems using
   simpler numbers

**Critical Thinking/
Finding Together**

By how many sixteenths
is $\frac{1}{3}$ of $\frac{3}{4}$ more than
$\frac{1}{4}$ of $\frac{3}{4}$ ?

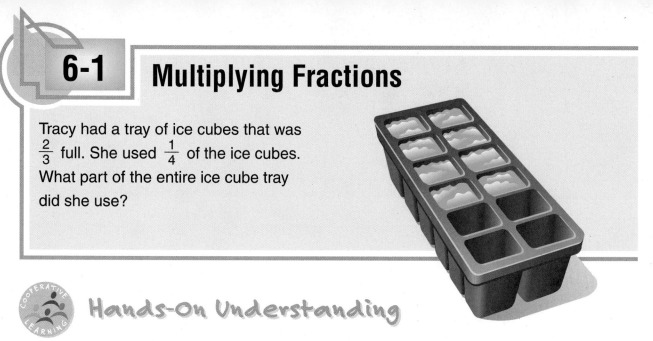

# 6-1 Multiplying Fractions

Tracy had a tray of ice cubes that was $\frac{2}{3}$ full. She used $\frac{1}{4}$ of the ice cubes. What part of the entire ice cube tray did she use?

## Hands-On Understanding

**Materials Needed:** paper, ruler, colored pencils or crayons

**Step 1**   Fold a rectangular piece of paper in thirds *horizontally*. Open it up and then draw a line along each fold. Shade two of the horizontal sections to show $\frac{2}{3}$, which is how much of the tray is full.

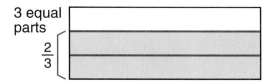

**Step 2**   Fold the paper in fourths *vertically*. Open it up and then draw a line along each fold. Mark off $\frac{1}{4}$ of the shaded vertical sections, which is how much of the entire ice cube tray Tracy used.

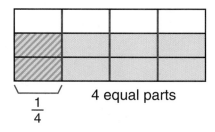

Into how many sections did you finally divide the rectangle? What part of the rectangle is marked off?

**Step 3**   Write a multiplication sentence that tells what part of the entire tray Tracy used.

**1.** Is the product $\frac{1}{4} \times \frac{2}{3}$ less than 1?

**2.** Explain what $\frac{1}{4}$ of $\frac{2}{3}$ means.

198

**Use the pair of diagrams to complete the statement.**

3.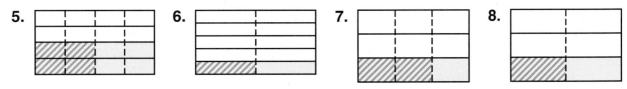

$\frac{2}{4}$ of $\frac{1}{2}$ = ___?___

4. $\frac{1}{3} \times \frac{3}{4}$ = ___?___

**Write a multiplication sentence for each diagram.**

5.      6.      7.      8.

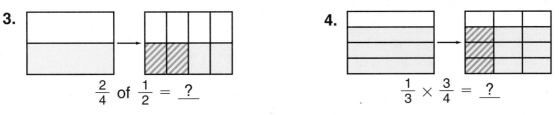

**Draw a diagram to show each product. Then write a multiplication sentence.**

9. $\frac{4}{5} \times \frac{1}{4}$      10. $\frac{2}{9} \times \frac{1}{3}$      11. $\frac{1}{5} \times \frac{2}{3}$      12. $\frac{1}{4} \times \frac{3}{8}$

13. $\frac{5}{6} \times \frac{1}{3}$      14. $\frac{2}{3} \times \frac{4}{5}$      15. $\frac{3}{4} \times \frac{1}{2}$      16. $\frac{3}{8} \times \frac{1}{3}$

**Find the diagram that matches each statement. Then complete.**

17. $\frac{1}{3}$ of $\frac{1}{3}$ = ___?___    18. $\frac{3}{4}$ of $\frac{1}{3}$ = ___?___    19. $\frac{1}{2}$ of $\frac{1}{4}$ = ___?___    20. $\frac{2}{3}$ of $\frac{1}{5}$ = ___?___

a.      b.      c.      d.

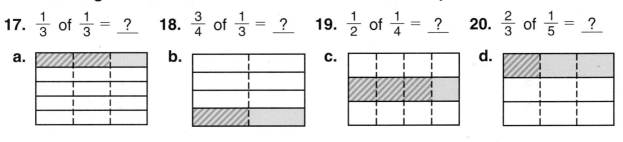

## Communicate

21. How does shading and marking off help you find the product of two fractions?

22. Study the relationship between the numerators of the factors and the numerator of the product. What do you observe? Explain your answer. Is the same true for the denominator?

23. Use what you have observed to write a rule in your Math Journal on how you multiply fractions. Compare your rule to those of your classmates.

*Discuss*

*Math Journal*

# 6-2    Multiplying Fractions by Fractions

One third of a swimming pool is roped off for nonswimmers. Three fourths of this space is used for swimming lessons. What part of the pool is used for swimming lessons?

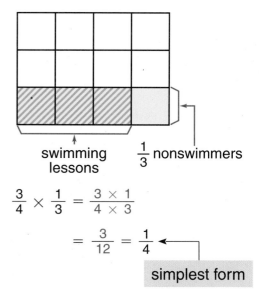

swimming lessons    $\frac{1}{3}$ nonswimmers

To find what part of the pool is used for swimming lessons, multiply: $\frac{3}{4} \times \frac{1}{3} = \underline{\ ?\ }$

▶ To **multiply** a *fraction* by a *fraction*:

- Multiply the numerators.
- Multiply the denominators.
- Write the product in simplest form.

$$\frac{3}{4} \times \frac{1}{3} = \frac{3 \times 1}{4 \times 3}$$

$$= \frac{3}{12} = \frac{1}{4}$$

simplest form

One fourth of the pool is used for swimming lessons.

▶ To check multiplication use the *commutative property*.

Think: "order."

$$\frac{1}{2} \times \frac{3}{4} = \frac{1 \times 3}{2 \times 4} \qquad \frac{3}{4} \times \frac{1}{2} = \frac{3 \times 1}{4 \times 2}$$
$$= \frac{3}{8} \qquad\qquad = \frac{3}{8}$$

**Study these examples.**

$$\frac{3}{5} \text{ of } \frac{1}{6} = \frac{3}{5} \times \frac{1}{6}$$
$$= \frac{3 \times 1}{5 \times 6}$$
$$= \frac{3}{30} = \frac{1}{10}$$

Compare: $\frac{1}{2} \times \frac{1}{4} \ \underline{\ ?\ } \ \frac{1}{2} \times \frac{1}{5}$

$$\frac{1 \times 1}{2 \times 4} \ \underline{\ ?\ } \ \frac{1 \times 1}{2 \times 5}$$

$$\frac{1}{8} \ > \ \frac{1}{10}$$

**Copy and complete.**

**1.** $\frac{2}{3} \times \frac{4}{5} = \frac{2 \times ?}{3 \times ?}$

$\qquad = \frac{8}{?}$

**2.** $\frac{3}{5} \times \frac{1}{2} = \frac{? \times 1}{? \times 2}$

$\qquad = \frac{?}{10}$

**3.** $\frac{5}{7} \times \frac{1}{4} = \frac{? \times ?}{? \times ?}$

$\qquad = \frac{?}{?}$

**Multiply.**

**4.** $\frac{1}{3} \times \frac{1}{8}$

**5.** $\frac{1}{4} \times \frac{3}{5}$

**6.** $\frac{4}{5} \times \frac{1}{7}$

**7.** $\frac{1}{3} \times \frac{2}{9}$

**Find the product.** Use the commutative property to check your answers.

**8.** $\frac{7}{10} \times \frac{1}{3}$  **9.** $\frac{3}{4} \times \frac{3}{5}$  **10.** $\frac{3}{8} \times \frac{5}{7}$  **11.** $\frac{5}{6} \times \frac{2}{9}$

**12.** $\frac{3}{4}$ of $\frac{2}{9}$  **13.** $\frac{4}{5}$ of $\frac{4}{7}$  **14.** $\frac{3}{10}$ of $\frac{2}{5}$  **15.** $\frac{5}{8}$ of $\frac{4}{9}$

**Find the missing fraction. Then check by multiplying.**

Algebra

**16.** $\frac{3}{4} \times \underline{\ ?\ } = \frac{5}{6} \times \frac{3}{4}$  **17.** $\frac{6}{7} \times \frac{1}{4} = \underline{\ ?\ } \times \frac{6}{7}$  **18.** $\underline{\ ?\ } \times \frac{2}{9} = \frac{2}{9} \times \frac{4}{5}$

**Compare. Write <, =, or >.**

**19.** $\frac{2}{5} \times \frac{1}{4} \underline{\ ?\ } \frac{1}{4} \times \frac{2}{3}$  **20.** $\frac{5}{9} \times \frac{3}{5} \underline{\ ?\ } \frac{5}{6} \times \frac{3}{4}$  **21.** $\frac{1}{4} \times \frac{3}{8} \underline{\ ?\ } \frac{3}{16} \times \frac{1}{2}$

**22.** $\frac{3}{5} \times \frac{1}{6} \underline{\ ?\ } \frac{3}{9} \times \frac{1}{2}$  **23.** $\frac{3}{5} \times \frac{2}{3} \underline{\ ?\ } \frac{7}{8} \times \frac{2}{5}$  **24.** $\frac{5}{6} \times \frac{9}{10} \underline{\ ?\ } \frac{1}{2} \times \frac{4}{5}$

**25.** In your Math Journal write how the product of two fractions compares with 1; with each fraction.

Math Journal

## PROBLEM SOLVING

**26.** It took Peter $\frac{3}{4}$ of the morning to do yard work. He spent $\frac{2}{3}$ of this time pulling weeds. What part of the morning did he pull weeds?

**27.** Five sixths of the books on the shelf are nonfiction. Three fourths of these books are science books. What part of the books on the shelf are science books?

## Finding Together

Discuss

**Using each of the digits 2, 3, 4, and 5 only once, find two fractions that will have:**

**28.** a product close to 1.

**29.** the greatest product possible.

**30.** the least product possible.

Think:
$\frac{?}{?} \times \frac{?}{?} = \underline{\ ?\ }$

# Multiplying Fractions and Whole Numbers

Cara walks $\frac{1}{4}$ mile to the library. Kareem walks three times this distance. What part of a mile does Kareem walk?

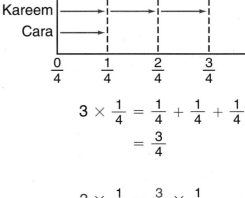

To find what part of a mile Kareem walks, multiply: $3 \times \frac{1}{4} = \underline{\ ?\ }$

$$3 \times \frac{1}{4} = \frac{1}{4} + \frac{1}{4} + \frac{1}{4}$$
$$= \frac{3}{4}$$

▶ To **multiply** a *fraction* and a *whole number*:

- Rename the whole number as an improper fraction with a denominator of 1.

- Multiply the numerators.

- Multiply the denominators.

- Write the product in simplest form.

$$3 \times \frac{1}{4} = \frac{3}{1} \times \frac{1}{4}$$
$$= \frac{3 \times 1}{1 \times 4}$$
$$= \frac{3}{4} \longleftarrow$$

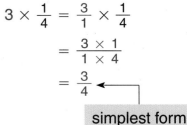

simplest form

Kareem walks $\frac{3}{4}$ mile to the library.

▶ The properties of multiplication for whole numbers also apply to fractions.

**Identity Property**

$$1 \times \frac{5}{6} = \frac{5}{6}$$

Think: "same."

**Zero Property**

$$0 \times \frac{2}{3} = 0$$

Think: "0 product."

**Study these examples.**

$$5 \times \frac{4}{5} = \frac{5}{1} \times \frac{4}{5}$$
$$= \frac{5 \times 4}{1 \times 5}$$
$$= \frac{20}{5} = 4$$

$$7 \times \frac{5}{21} = \frac{7}{1} \times \frac{5}{21}$$
$$= \frac{7 \times 5}{1 \times 21}$$
$$= \frac{35}{21} = 1\frac{14}{21} = 1\frac{2}{3} \longleftarrow \text{simplest form}$$

**Multiply.**

**1.** $16 \times \frac{1}{8}$

**2.** $20 \times \frac{1}{4}$

**3.** $18 \times \frac{1}{6}$

**4.** $24 \times \frac{1}{3}$

**Find the product.**

**5.** $22 \times \frac{1}{2}$      **6.** $30 \times \frac{1}{10}$      **7.** $0 \times \frac{1}{5}$      **8.** $15 \times \frac{2}{3}$

**9.** $2 \times \frac{3}{8}$      **10.** $10 \times \frac{3}{50}$      **11.** $2 \times \frac{3}{7}$      **12.** $2 \times \frac{4}{11}$

**13.** $40 \times \frac{7}{16}$      **14.** $15 \times \frac{4}{25}$      **15.** $45 \times \frac{5}{27}$      **16.** $24 \times \frac{5}{16}$

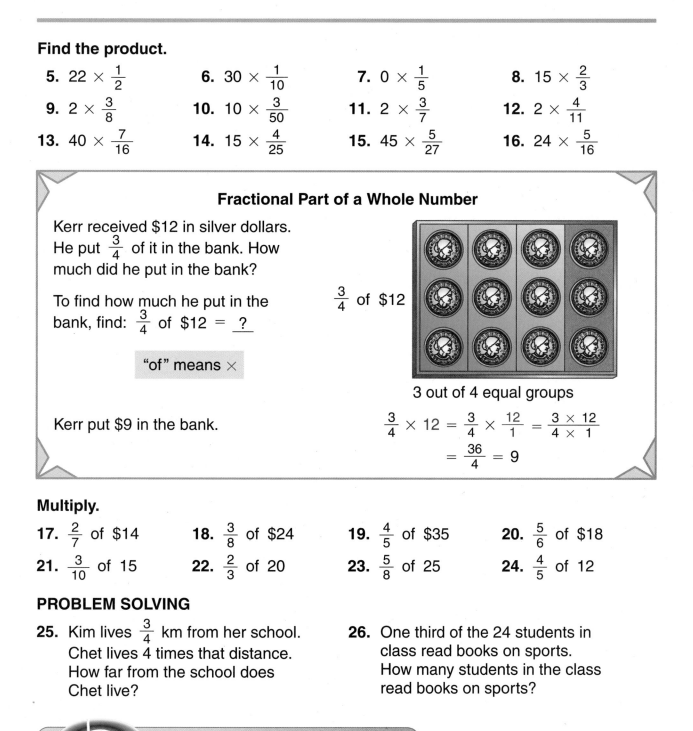

### Fractional Part of a Whole Number

Kerr received $12 in silver dollars. He put $\frac{3}{4}$ of it in the bank. How much did he put in the bank?

To find how much he put in the bank, find: $\frac{3}{4}$ of $12 = $ __?__

$\frac{3}{4}$ of $12

"of" means $\times$

3 out of 4 equal groups

Kerr put $9 in the bank.

$$\frac{3}{4} \times 12 = \frac{3}{4} \times \frac{12}{1} = \frac{3 \times 12}{4 \times 1}$$
$$= \frac{36}{4} = 9$$

**Multiply.**

**17.** $\frac{2}{7}$ of $14      **18.** $\frac{3}{8}$ of $24      **19.** $\frac{4}{5}$ of $35      **20.** $\frac{5}{6}$ of $18

**21.** $\frac{3}{10}$ of 15      **22.** $\frac{2}{3}$ of 20      **23.** $\frac{5}{8}$ of 25      **24.** $\frac{4}{5}$ of 12

### PROBLEM SOLVING

**25.** Kim lives $\frac{3}{4}$ km from her school. Chet lives 4 times that distance. How far from the school does Chet live?

**26.** One third of the 24 students in class read books on sports. How many students in the class read books on sports?

## Skills to Remember

**Find the greatest common factor (GCF) of each set of numbers.**

**27.** 4 and 8      **28.** 6 and 12      **29.** 9 and 18      **30.** 5 and 20

**31.** 7 and 21      **32.** 8 and 18      **33.** 16 and 20      **34.** 10 and 25

# 6-4 Multiplying Using Cancellation

To **multiply fractions using cancellation** divide any numerator and denominator by their greatest common factor (GCF).

Multiply: $\frac{20}{21} \times \frac{7}{8} = \underline{\ ?\ }$

**GCF:**

$20 = 1, 2, 4, \mathbf{5}, 10, 20$

$8 = 1, 2, \mathbf{4}, 8$

▶ You can **multiply fractions** quickly by using *cancellation*.

- Divide *any* numerator and denominator by their GCF.

- Cancel until the numerators and denominators have no common factor other than 1.

- Multiply the numerators. Then multiply the denominators. The product will be in simplest form.

- If the product is an improper fraction, rename it as a whole or mixed number.

$$\frac{20}{21} \times \frac{7}{8} = \frac{\overset{5}{\cancel{20}} \times \overset{1}{\cancel{7}}}{\underset{3}{\cancel{21}} \times \underset{2}{\cancel{8}}}$$

GCF of 20 and 8: 4
GCF of 7 and 21: 7

$$= \frac{5 \times 1}{3 \times 2} = \frac{5}{6}$$

simplest form

**Study these examples.**

$$\frac{2}{5} \times 25 = \frac{2}{5} \times \frac{25}{1}$$

$$= \frac{2 \times \overset{5}{\cancel{25}}}{\underset{1}{\cancel{5}} \times 1}$$

GCF of 25 and 5: 5

$$= \frac{2 \times 5}{1 \times 1} = \frac{10}{1} = 10$$

whole number

$$49 \times \frac{5}{14} = \frac{49}{1} \times \frac{5}{14}$$

$$= \frac{\overset{7}{\cancel{49}} \times 5}{1 \times \underset{2}{\cancel{14}}}$$

GCF of 49 and 14: 7

$$= \frac{7 \times 5}{1 \times 2} = \frac{35}{2} = 17\frac{1}{2}$$

mixed number

**Copy and complete.**

1. $\dfrac{4}{7} \times \dfrac{35}{36} = \dfrac{\overset{1}{\cancel{4}} \times \overset{?}{\cancel{35}}}{\underset{1}{\cancel{7}} \times \underset{?}{\cancel{36}}}$

   $= \dfrac{1 \times ?}{1 \times ?} = \dfrac{?}{?}$

2. $\dfrac{3}{8} \times 16 = \dfrac{3}{8} \times \dfrac{16}{?} = \dfrac{3 \times \overset{?}{\cancel{16}}}{\underset{?}{\cancel{8}} \times ?}$

   $= \dfrac{3 \times ?}{? \times ?} = \dfrac{?}{?} = ?$

**Multiply using cancellation.**

3. $\frac{1}{2} \times \frac{2}{3}$
4. $\frac{1}{4} \times \frac{2}{7}$
5. $\frac{2}{9} \times \frac{1}{6}$
6. $\frac{3}{4} \times \frac{1}{9}$
7. $\frac{4}{9} \times \frac{3}{5}$

8. $\frac{4}{7} \times \frac{3}{8}$
9. $\frac{4}{15} \times \frac{5}{9}$
10. $\frac{2}{3} \times \frac{3}{8}$
11. $\frac{6}{7} \times \frac{7}{8}$
12. $\frac{3}{10} \times \frac{7}{9}$

13. $\frac{3}{4} \times 16$
14. $\frac{4}{25} \times 10$
15. $\frac{7}{8} \times 24$
16. $\frac{4}{21} \times 49$
17. $\frac{5}{8} \times 32$

18. $32 \times \frac{5}{6}$
19. $33 \times \frac{4}{11}$
20. $35 \times \frac{5}{42}$
21. $24 \times \frac{3}{8}$
22. $25 \times \frac{2}{15}$

**Find the product using cancellation.**

23. $\frac{3}{10} \times \frac{25}{27}$
24. $\frac{8}{27} \times \frac{9}{20}$
25. $\frac{9}{14} \times \frac{7}{15}$
26. $\frac{7}{8} \times \frac{6}{21}$
27. $\frac{2}{9} \times \frac{21}{26}$

28. $14 \times \frac{3}{7}$
29. $36 \times \frac{7}{8}$
30. $20 \times \frac{3}{25}$
31. $\frac{5}{12} \times 8$
32. $\frac{3}{8} \times 30$

33. $\frac{5}{8} \times \frac{4}{15}$
34. $\frac{3}{4} \times 18$
35. $\frac{5}{7} \times \frac{8}{15}$
36. $72 \times \frac{5}{12}$
37. $\frac{5}{6} \times 54$

38. Explain in your Math Journal why your answer is already in lowest terms when you multiply fractions using cancellation.

Math Journal ✓

**PROBLEM SOLVING**

39. There are 20 members of the basketball team. Three fifths are fifth-grade students. How many members of the basketball team are fifth-grade students?

40. Mary Ellen had $\frac{2}{3}$ of a pie left. She ate $\frac{3}{8}$ of it at lunchtime. How much of the pie did she eat at lunchtime? How much of the pie was left?

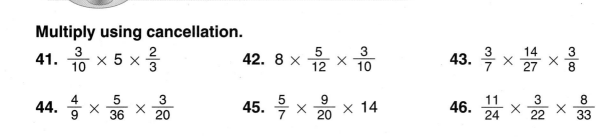

**Challenge**

**Multiply using cancellation.**

41. $\frac{3}{10} \times 5 \times \frac{2}{3}$
42. $8 \times \frac{5}{12} \times \frac{3}{10}$
43. $\frac{3}{7} \times \frac{14}{27} \times \frac{3}{8}$

44. $\frac{4}{9} \times \frac{5}{36} \times \frac{3}{20}$
45. $\frac{5}{7} \times \frac{9}{20} \times 14$
46. $\frac{11}{24} \times \frac{3}{22} \times \frac{8}{33}$

# Mixed Numbers to Improper Fractions

Rename $2\frac{3}{8}$ as an improper fraction.

$$2\frac{3}{8} = 1 + 1 + \frac{3}{8}$$

$$= \frac{8}{8} + \frac{8}{8} + \frac{3}{8}$$

$$= \frac{19}{8}$$

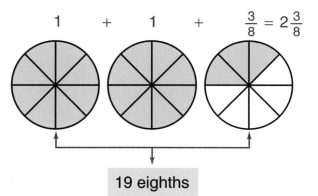

$$1 \quad + \quad 1 \quad + \quad \frac{3}{8} = 2\frac{3}{8}$$

19 eighths

▶ To **rename** a *mixed number* as an *improper fraction*:

- Multiply the whole number by the denominator.

- Add the product to the numerator.

- Write the sum as the numerator and the given denominator as the denominator.

Multiply:
$8 \times 2 = 16$

Then add:
$16 + 3 = 19$

$$2\frac{3}{8} = \frac{(8 \times 2) + 3}{8}$$

$$= \frac{16 + 3}{8}$$

$$= \frac{19}{8}$$

## Copy and complete.

**1.** $2\frac{1}{2} = \dfrac{(2 \times ?) + ?}{2}$

$\quad\quad = \dfrac{?}{2}$

**2.** $6\frac{3}{4} = \dfrac{(4 \times ?) + ?}{4}$

$\quad\quad = \dfrac{?}{4}$

**3.** $10\frac{3}{5} = \dfrac{(? \times 10) + ?}{?}$

$\quad\quad = \dfrac{?}{?}$

## Write the letter of the correct answer.

**4.** $11\frac{3}{5}$
  **a.** $\frac{55}{3}$
  **b.** $\frac{5}{58}$
  **c.** $\frac{58}{5}$
  **d.** $\frac{3}{55}$

**5.** $12\frac{4}{7}$
  **a.** $\frac{7}{88}$
  **b.** $\frac{84}{7}$
  **c.** $\frac{7}{84}$
  **d.** $\frac{88}{7}$

**6.** $15\frac{5}{6}$
  **a.** $\frac{95}{6}$
  **b.** $\frac{85}{6}$
  **c.** $\frac{6}{95}$
  **d.** $\frac{6}{85}$

**Rename each as an improper fraction.**

**7.** $2\frac{1}{8}$     **8.** $5\frac{3}{4}$     **9.** $3\frac{1}{7}$     **10.** $6\frac{7}{8}$     **11.** $11\frac{2}{3}$     **12.** $4\frac{3}{5}$

**13.** $6\frac{1}{5}$     **14.** $7\frac{5}{8}$     **15.** $3\frac{3}{8}$     **16.** $10\frac{1}{2}$     **17.** $9\frac{5}{6}$     **18.** $8\frac{7}{9}$

**19.** $2\frac{4}{7}$     **20.** $3\frac{7}{8}$     **21.** $10\frac{2}{5}$     **22.** $12\frac{1}{6}$     **23.** $5\frac{2}{9}$     **24.** $7\frac{4}{11}$

**25.** $15\frac{1}{4}$     **26.** $8\frac{2}{3}$     **27.** $14\frac{4}{5}$     **28.** $10\frac{5}{8}$     **29.** $12\frac{3}{4}$     **30.** $5\frac{9}{10}$

**31.** $4\frac{5}{12}$     **32.** $3\frac{5}{7}$     **33.** $5\frac{9}{11}$     **34.** $3\frac{4}{13}$     **35.** $4\frac{2}{17}$     **36.** $2\frac{17}{19}$

**Rename the mixed number as an improper fraction.**

**37.** A sheet of tin is $4\frac{5}{9}$ ft long.

**38.** A book page is $7\frac{1}{10}$ in. wide.

**39.** A bag of fertilizer weighs $31\frac{3}{8}$ lb.

**40.** A gasoline tank contains $20\frac{3}{4}$ gal.

**41.** Explain in your Math Journal:

- how to use drawings or models to prove that $2\frac{3}{4} = \frac{11}{4}$.

- why $2\frac{5}{6}$, $1\frac{11}{6}$, and $\frac{17}{6}$ are equivalent or are not equivalent.

Math Journal ✓

**PROBLEM SOLVING**

**42.** A piece of lumber that is 40 in. long has been cut into 7 equal pieces. How long is each piece? Write the length as a mixed number.

**43.** The flying time from New York to Los Angeles is $5\frac{2}{3}$ h. Write this time as an improper fraction.

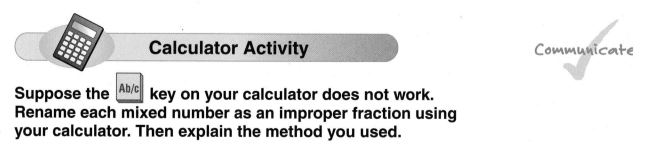

**Calculator Activity**

Communicate ✓

Suppose the Ab/c key on your calculator does not work. Rename each mixed number as an improper fraction using your calculator. Then explain the method you used.

**44.** $5\frac{17}{36}$     **45.** $24\frac{23}{24}$     **46.** $42\frac{28}{41}$     **47.** $76\frac{23}{25}$     **48.** $86\frac{85}{86}$

# Multiplying Fractions and Mixed Numbers

Arlene bought $4\frac{2}{3}$ yards of material. She used $\frac{6}{7}$ of it for a dress. How much material did she use for the dress?

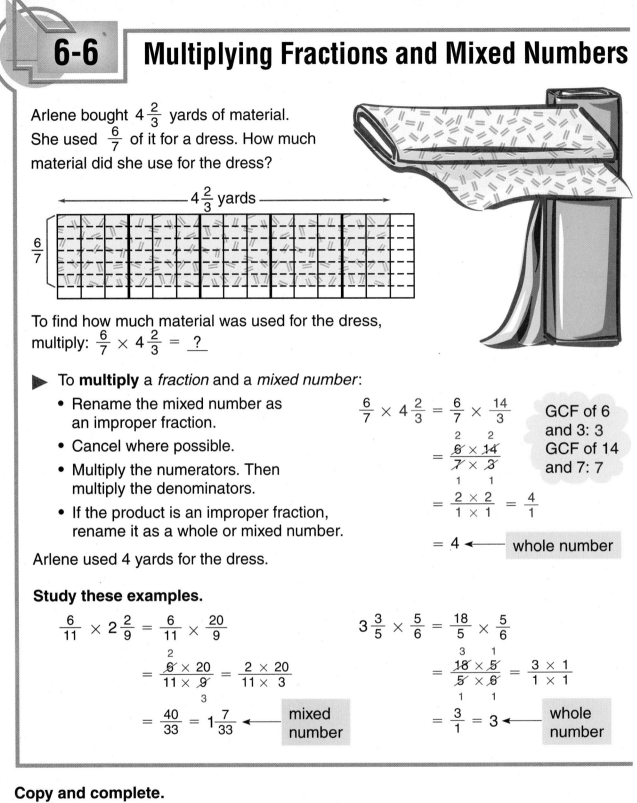

$4\frac{2}{3}$ yards

$\frac{6}{7}$

To find how much material was used for the dress, multiply: $\frac{6}{7} \times 4\frac{2}{3} = \underline{\ ?\ }$

▶ To **multiply** a *fraction* and a *mixed number*:

- Rename the mixed number as an improper fraction.
- Cancel where possible.
- Multiply the numerators. Then multiply the denominators.
- If the product is an improper fraction, rename it as a whole or mixed number.

Arlene used 4 yards for the dress.

$$\frac{6}{7} \times 4\frac{2}{3} = \frac{6}{7} \times \frac{14}{3}$$

$$= \frac{\overset{2}{\cancel{6}} \times \overset{2}{\cancel{14}}}{\underset{1}{\cancel{7}} \times \underset{1}{\cancel{3}}}$$

$$= \frac{2 \times 2}{1 \times 1} = \frac{4}{1}$$

$$= 4 \longleftarrow \text{ whole number}$$

GCF of 6 and 3: 3
GCF of 14 and 7: 7

**Study these examples.**

$$\frac{6}{11} \times 2\frac{2}{9} = \frac{6}{11} \times \frac{20}{9}$$

$$= \frac{\overset{2}{\cancel{6}} \times 20}{11 \times \underset{3}{\cancel{9}}} = \frac{2 \times 20}{11 \times 3}$$

$$= \frac{40}{33} = 1\frac{7}{33} \longleftarrow \begin{array}{c}\text{mixed} \\ \text{number}\end{array}$$

$$3\frac{3}{5} \times \frac{5}{6} = \frac{18}{5} \times \frac{5}{6}$$

$$= \frac{\overset{3}{\cancel{18}} \times \overset{1}{\cancel{5}}}{\underset{1}{\cancel{5}} \times \underset{1}{\cancel{6}}} = \frac{3 \times 1}{1 \times 1}$$

$$= \frac{3}{1} = 3 \longleftarrow \begin{array}{c}\text{whole} \\ \text{number}\end{array}$$

**Copy and complete.**

1. $\frac{4}{7} \times 3\frac{1}{2} = \frac{4}{7} \times \frac{?}{2}$

   $= \underline{\ ?\ }$

2. $6\frac{2}{5} \times \frac{3}{8} = \frac{?}{5} \times \frac{?}{8}$

   $= \underline{\ ?\ }$

3. $\frac{8}{9} \times 2\frac{3}{4} = \frac{?}{9} \times \frac{?}{?}$

   $= \underline{\ ?\ }$

**Find the product.**

4. $3\frac{1}{2} \times \frac{1}{3}$

5. $2\frac{1}{2} \times \frac{3}{5}$

6. $\frac{5}{14} \times 2\frac{1}{3}$

7. $\frac{1}{2} \times 5\frac{1}{3}$

8. $\frac{2}{3} \times 4\frac{1}{5}$

9. $\frac{3}{7} \times 5\frac{3}{5}$

10. $2\frac{1}{5} \times \frac{4}{11}$

11. $1\frac{5}{7} \times \frac{5}{12}$

12. $6\frac{1}{8} \times \frac{4}{7}$

13. $4\frac{1}{2} \times \frac{2}{3}$

14. $\frac{9}{12} \times 1\frac{1}{3}$

15. $\frac{3}{14} \times 2\frac{1}{3}$

16. $\frac{3}{4} \times 1\frac{2}{6}$

17. $\frac{7}{8} \times 2\frac{2}{7}$

18. $2\frac{1}{2} \times \frac{2}{15}$

19. $2\frac{4}{5} \times \frac{5}{7}$

20. $\frac{6}{7} \times 4\frac{1}{3}$

21. $\frac{6}{7} \times 2\frac{1}{3}$

22. $2\frac{2}{5} \times \frac{5}{6}$

23. $4\frac{1}{5} \times \frac{6}{7}$

---

### Using the Distributive Property

The **distributive property** is sometimes used when multiplying a fraction and a mixed number.

$$\frac{2}{3} \times 9\frac{3}{10} = \frac{2}{3} \times \left(9 + \frac{3}{10}\right) = \left(\frac{2}{3} \times 9\right) + \left(\frac{2}{3} \times \frac{3}{10}\right)$$

$$= \left(\frac{2}{3} \times \frac{9}{1}\right) + \left(\frac{2}{3} \times \frac{3}{10}\right)$$

$$= \left(\frac{2 \times \overset{3}{\cancel{9}}}{\cancel{3} \times 1}\right) + \left(\frac{\overset{1}{\cancel{2}} \times \overset{1}{\cancel{3}}}{\cancel{3} \times \cancel{10}}\right) = \left(\frac{2 \times 3}{1 \times 1}\right) + \left(\frac{1 \times 1}{1 \times 5}\right)$$

$$= \frac{6}{1} + \frac{1}{5} = 6 + \frac{1}{5} = 6\frac{1}{5}$$

---

**Multiply.** Use the distributive property.

24. $\frac{1}{8} \times 8\frac{8}{11}$

25. $\frac{1}{6} \times 12\frac{3}{5}$

26. $\frac{1}{5} \times 10\frac{5}{9}$

27. $\frac{1}{3} \times 15\frac{3}{8}$

28. $\frac{3}{4} \times 4\frac{1}{3}$

29. $\frac{8}{9} \times 18\frac{1}{4}$

30. $\frac{3}{7} \times 14\frac{1}{9}$

31. $\frac{5}{6} \times 18\frac{9}{10}$

### PROBLEM SOLVING

32. Celia had $4\frac{1}{2}$ yards of ribbon. She used $\frac{5}{6}$ of it for her project. How many yards of ribbon did she use for her project? How many yards were *not* used?

33. Arnold lives $8\frac{3}{4}$ miles from the library. Miriam lives $\frac{4}{5}$ of this distance from the library. How far does Miriam live from the library?

# 6-7 Multiplying Mixed Numbers

Stan bought $2\frac{1}{4}$ feet of wood for shelving. Ralph bought $1\frac{2}{3}$ times as much. How many feet of wood did Ralph buy?

To find how many feet of wood, multiply: $1\frac{2}{3} \times 2\frac{1}{4} = \underline{?}$

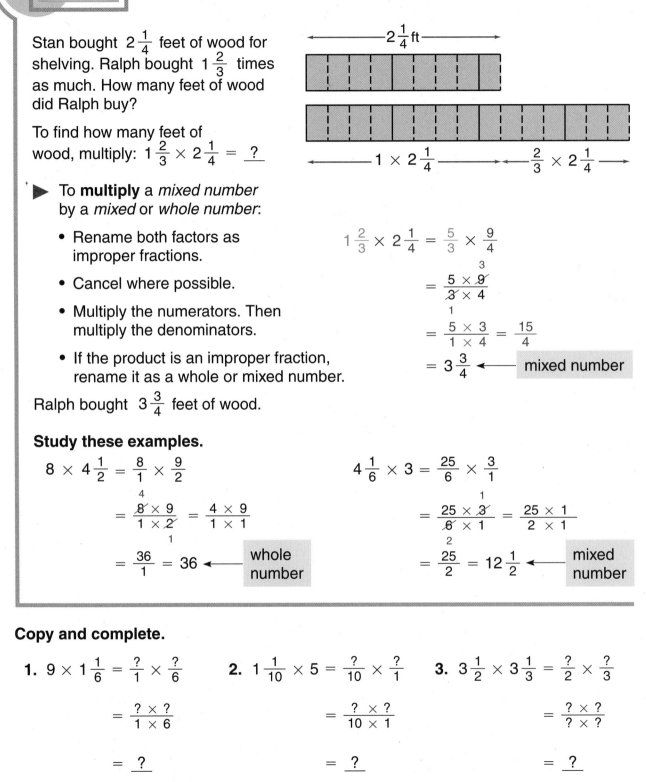

▶ To **multiply** a *mixed number* by a *mixed* or *whole number*:

- Rename both factors as improper fractions.

- Cancel where possible.

- Multiply the numerators. Then multiply the denominators.

- If the product is an improper fraction, rename it as a whole or mixed number.

$$1\frac{2}{3} \times 2\frac{1}{4} = \frac{5}{3} \times \frac{9}{4}$$

$$= \frac{5 \times \overset{3}{\cancel{9}}}{\underset{1}{\cancel{3}} \times 4}$$

$$= \frac{5 \times 3}{1 \times 4} = \frac{15}{4}$$

$$= 3\frac{3}{4} \quad \longleftarrow \quad \boxed{\text{mixed number}}$$

Ralph bought $3\frac{3}{4}$ feet of wood.

## Study these examples.

$$8 \times 4\frac{1}{2} = \frac{8}{1} \times \frac{9}{2}$$

$$= \frac{\overset{4}{\cancel{8}} \times 9}{1 \times \underset{1}{\cancel{2}}} = \frac{4 \times 9}{1 \times 1}$$

$$= \frac{36}{1} = 36 \quad \longleftarrow \quad \boxed{\begin{array}{c}\text{whole}\\\text{number}\end{array}}$$

$$4\frac{1}{6} \times 3 = \frac{25}{6} \times \frac{3}{1}$$

$$= \frac{25 \times \overset{1}{\cancel{3}}}{\underset{2}{\cancel{6}} \times 1} = \frac{25 \times 1}{2 \times 1}$$

$$= \frac{25}{2} = 12\frac{1}{2} \quad \longleftarrow \quad \boxed{\begin{array}{c}\text{mixed}\\\text{number}\end{array}}$$

## Copy and complete.

**1.** $9 \times 1\frac{1}{6} = \frac{?}{1} \times \frac{?}{6}$

$$= \frac{? \times ?}{1 \times 6}$$

$$= \underline{?}$$

**2.** $1\frac{1}{10} \times 5 = \frac{?}{10} \times \frac{?}{1}$

$$= \frac{? \times ?}{10 \times 1}$$

$$= \underline{?}$$

**3.** $3\frac{1}{2} \times 3\frac{1}{3} = \frac{?}{2} \times \frac{?}{3}$

$$= \frac{? \times ?}{? \times ?}$$

$$= \underline{?}$$

**Multiply.**

**4.** $6 \times 3\frac{1}{3}$

**5.** $9 \times 1\frac{2}{3}$

**6.** $4 \times 2\frac{2}{5}$

**7.** $3 \times 4\frac{4}{9}$

**8.** $7\frac{1}{2} \times 2\frac{2}{5}$

**9.** $1\frac{1}{3} \times 5\frac{1}{4}$

**10.** $3\frac{3}{5} \times 1\frac{2}{3}$

**11.** $6\frac{1}{4} \times 2\frac{2}{5}$

**12.** $3\frac{3}{4} \times 3\frac{1}{3}$

**13.** $6\frac{1}{4} \times 1\frac{1}{5}$

**14.** $8\frac{2}{3} \times 2\frac{1}{2}$

**15.** $3\frac{3}{8} \times 3\frac{1}{2}$

**16.** $4\frac{2}{7} \times 3$

**17.** $4\frac{1}{6} \times 12$

**18.** $1\frac{1}{7} \times 7$

**19.** $5\frac{2}{5} \times 15$

**20.** $2\frac{1}{3} \times 1\frac{2}{7}$

**21.** $4\frac{2}{5} \times 2\frac{1}{2}$

**22.** $6\frac{1}{8} \times 2\frac{2}{7}$

**23.** $9\frac{2}{3} \times 1\frac{1}{2}$

**24.** $\frac{7}{8} \times 2\frac{2}{5}$

**25.** $\frac{3}{10} \times 4\frac{4}{9}$

**26.** $7\frac{1}{2} \times 2\frac{4}{5}$

**27.** $1\frac{1}{2} \times 4\frac{2}{3}$

**Compare. Write $<$, $=$, or $>$.**

**28.** $1\frac{6}{7} \times 21$ __?__ $6 \times 1\frac{5}{6}$

**29.** $18 \times 2\frac{2}{9}$ __?__ $2\frac{1}{2} \times 16$

**30.** $2\frac{1}{2} \times 1\frac{2}{3}$ __?__ $2\frac{2}{5} \times 1\frac{1}{2}$

**31.** $3\frac{1}{2} \times 2\frac{1}{4}$ __?__ $3\frac{1}{4} \times 2\frac{1}{2}$

## PROBLEM SOLVING

**32.** Lilia made 2 dresses. Each dress needed $2\frac{3}{4}$ yards of fabric. How many yards of fabric did she use?

**33.** Vito is $4\frac{2}{3}$ feet tall. His father is $1\frac{1}{4}$ times as tall. How tall is Vito's father?

**34.** The hour hand of a clock moves 30 degrees every hour. How many degrees does it move in $2\frac{3}{4}$ hours?

**35.** Cayo uses $9\frac{1}{2}$ ounces of flour to make a loaf of bread. How much flour does he use to make 6 loaves of bread?

### Project

**36.** Keep a daily log for a week. Use mixed numbers and fractions to record, to the nearest quarter of an hour, the time you spend at school, do your homework, watch TV, and play. At the end of the week, find the total time spent for each category.

**37.** Suppose the total time you found in each category is the same every week. How much time do you spend on each activity in a month?

# Division of Fractions

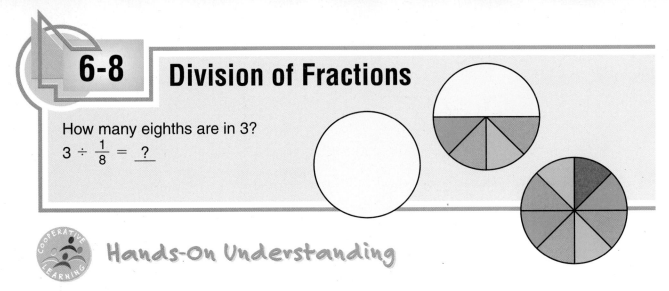

How many eighths are in 3?

$$3 \div \frac{1}{8} = \underline{\ ?\ }$$

## Hands-On Understanding

**Materials Needed:** fraction circles and strips, colored pencils or crayons

| **Step 1** | Find 3 fraction circles that show eighths. Shade all of the eighths. |

8 equal parts    8 equal parts    8 equal parts

| **Step 2** | Count the number of eighths shaded. How many eighths are there altogether? |

| **Step 3** | Write a division sentence that tells how many eighths are in 3. |

1. The diagram at the right shows how to model the number of two thirds in 2.

   Use fraction strips to model $2 \div \frac{2}{3}$ as shown. How many $\frac{2}{3}$s are in 2? What is $2 \div \frac{2}{3}$ ?

   $$\frac{2}{3} \qquad \frac{2}{3} \qquad \frac{2}{3}$$

2. The diagram at the right shows how to model the number of tenths in $\frac{4}{5}$.

   Use fraction strips to model $\frac{4}{5} \div \frac{1}{10}$ as shown. How many $\frac{1}{10}$s are in $\frac{4}{5}$ ? What is $\frac{4}{5} \div \frac{1}{10}$ ?

   $$\frac{1}{10}$$

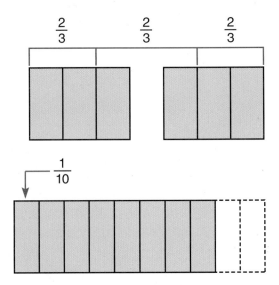

**Use fraction circles or strips to model each quotient.
Then write a division sentence. Look for a pattern.**

**3.** $4 \div \frac{1}{2}$

**4.** $2 \div \frac{1}{8}$

**5.** $3 \div \frac{3}{4}$

**6.** $2 \div \frac{2}{5}$

**7.** $\frac{3}{4} \div \frac{1}{4}$

**8.** $\frac{9}{11} \div \frac{3}{11}$

**9.** $\frac{3}{8} \div \frac{2}{8}$

**10.** $\frac{7}{9} \div \frac{2}{9}$

**Write a division sentence for each diagram.**

**11.**

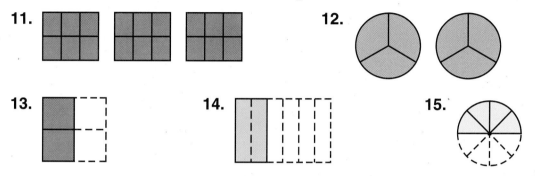

**12.**

**13.**

**14.**

**15.**

## Communicate

**16.** When you divide a whole number by a fraction, how does the quotient compare with the whole number? Explain your answer.

**17.** If you divide fractions with like denominators, when will the quotient be a whole number?

**18.** Explain how you divide fractions with like denominators.

## Challenge

Algebra

**Use the diagram to complete each number sentence.**

**19.**

**20.**

**21.**

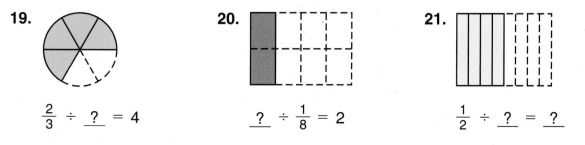

$\frac{2}{3} \div \underline{\ ?\ } = 4$

$\underline{\ ?\ } \div \frac{1}{8} = 2$

$\frac{1}{2} \div \underline{\ ?\ } = \underline{\ ?\ }$

**Reciprocals**

Two numbers with a product of 1 are called
**reciprocals** of each other.

$$\frac{\cancel{2}^{1}}{\cancel{7}_{1}} \times \frac{\cancel{7}^{1}}{\cancel{2}_{1}} = \frac{1}{1} = 1 \quad \text{So } \frac{2}{7} \text{ and } \frac{7}{2} \text{ are reciprocals.}$$

$$5 \times \frac{1}{5} = \frac{\cancel{5}^{1}}{1} \times \frac{1}{\cancel{5}_{1}} = \frac{1}{1} = 1 \quad \text{So 5 and } \frac{1}{5} \text{ are reciprocals.}$$

$$1\frac{1}{4} \times \frac{4}{5} = \frac{\cancel{5}^{1}}{\cancel{4}_{1}} \times \frac{\cancel{4}^{1}}{\cancel{5}_{1}} = \frac{1}{1} = 1 \quad \text{So } 1\frac{1}{4} \text{ and } \frac{4}{5} \text{ are reciprocals.}$$

Find the reciprocal of 2.

▶ To find the **reciprocal of a number:**

- Write the number as a fraction. $\qquad 2 = \frac{2}{1}$

- Invert the fraction by exchanging
  the position of the numerator and $\qquad \frac{2}{1} \diagdown \frac{1}{2}$
  the denominator.

- Check if the product of the
  numbers is 1. $\qquad \frac{\cancel{2}^{1}}{1} \times \frac{1}{\cancel{2}_{1}} = \frac{1}{1} = 1$

> 2 and $\frac{1}{2}$ are
> reciprocals.

**Study these examples.**

$$\frac{5}{9} \diagdown \frac{9}{5} \qquad \frac{\cancel{5}^{1}}{\cancel{9}_{1}} \times \frac{\cancel{9}^{1}}{\cancel{5}_{1}} = \frac{1}{1} = 1 \qquad\qquad 2\frac{1}{3} = \frac{7}{3} \qquad \frac{7}{3} \diagdown \frac{3}{7} \qquad \frac{\cancel{7}^{1}}{\cancel{3}_{1}} \times \frac{\cancel{3}^{1}}{\cancel{7}_{1}} = \frac{1}{1} = 1$$

$\frac{9}{5}$ is the reciprocal of $\frac{5}{9}$. $\qquad\qquad\qquad \frac{3}{7}$ is the reciprocal of $2\frac{1}{3}$.

**Find the missing reciprocal in each statement.**

Algebra ✓

**1.** $7 \times \underline{\ ?\ } = 1$ 
**2.** $2 \times \underline{\ ?\ } = 1$ 
**3.** $\frac{1}{6} \times \underline{\ ?\ } = 1$

**4.** $\frac{7}{11} \times \underline{\ ?\ } = 1$ 
**5.** $\frac{8}{9} \times \underline{\ ?\ } = 1$ 
**6.** $3\frac{1}{2} \times \underline{\ ?\ } = 1$ 
**7.** $4\frac{2}{3} \times \underline{\ ?\ } = 1$

**Write the reciprocal of each number.**

**8.** 11     **9.** 6     **10.** $\frac{1}{2}$     **11.** $\frac{1}{4}$     **12.** $\frac{5}{8}$     **13.** $\frac{8}{13}$

**14.** $\frac{9}{2}$     **15.** $\frac{13}{5}$     **16.** $\frac{15}{7}$     **17.** $3\frac{1}{3}$     **18.** $6\frac{3}{5}$     **19.** $2\frac{5}{6}$

**Write *always*, *sometimes*, or *never* to make each statement true.**

**20.** The reciprocal of a whole number __?__ has a numerator of 1.

**21.** The reciprocal of a mixed number is __?__ an improper fraction.

**22.** The reciprocal of a fraction is __?__ a whole number.

**PROBLEM SOLVING** Use the numbers in the box for problems 23–26.

**23.** Write the fractions that are less than 1.
Then write their reciprocals.

**24.** Write the fractions that are greater than 1.
Then write their reciprocals.

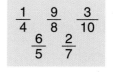

**25.** What numbers have reciprocals less than 1?

**26.** What numbers have reciprocals greater than 1?

**27.** When is the reciprocal of a number greater than the number? less than the number? Give examples.

*Communicate* ✓

**28.** What number is its own reciprocal? Why?

**29.** Is there any number that does *not* have a reciprocal?
Explain your answer.

## Critical Thinking

**Use the numbers in the box.**

**30.** Which number has a whole number as its reciprocal?

**31.** Which numbers have reciprocals close to 1?

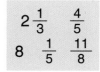

**32.** Which number has its reciprocal close to 0?

**33.** Which number has its reciprocal close to $\frac{1}{2}$?

215

# Dividing Whole Numbers by Fractions

A carpenter cut a 4-ft board into $\frac{2}{3}$-ft boards. How many pieces of board did the carpenter make?

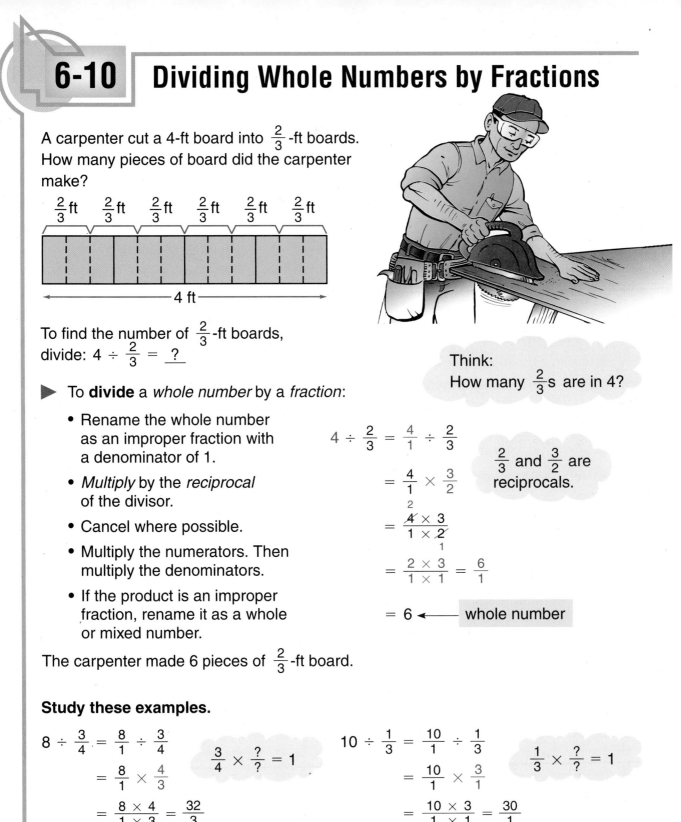

$\frac{2}{3}$ ft    $\frac{2}{3}$ ft    $\frac{2}{3}$ ft    $\frac{2}{3}$ ft    $\frac{2}{3}$ ft    $\frac{2}{3}$ ft

← 4 ft →

To find the number of $\frac{2}{3}$-ft boards, divide: $4 \div \frac{2}{3} = \underline{\ ?\ }$

**Think:**
How many $\frac{2}{3}$s are in 4?

▶ To **divide** a *whole number* by a *fraction*:

- Rename the whole number as an improper fraction with a denominator of 1.

- *Multiply* by the *reciprocal* of the divisor.

- Cancel where possible.

- Multiply the numerators. Then multiply the denominators.

- If the product is an improper fraction, rename it as a whole or mixed number.

$4 \div \frac{2}{3} = \frac{4}{1} \div \frac{2}{3}$

$= \frac{4}{1} \times \frac{3}{2}$

$= \frac{\overset{2}{\cancel{4}} \times 3}{1 \times \underset{1}{\cancel{2}}}$

$= \frac{2 \times 3}{1 \times 1} = \frac{6}{1}$

$= 6 \longleftarrow$ whole number

$\frac{2}{3}$ and $\frac{3}{2}$ are reciprocals.

The carpenter made 6 pieces of $\frac{2}{3}$-ft board.

**Study these examples.**

$8 \div \frac{3}{4} = \frac{8}{1} \div \frac{3}{4}$

$= \frac{8}{1} \times \frac{4}{3}$

$= \frac{8 \times 4}{1 \times 3} = \frac{32}{3}$

$= 10\frac{2}{3} \longleftarrow$ mixed number

$\frac{3}{4} \times \frac{?}{?} = 1$

$10 \div \frac{1}{3} = \frac{10}{1} \div \frac{1}{3}$

$= \frac{10}{1} \times \frac{3}{1}$

$= \frac{10 \times 3}{1 \times 1} = \frac{30}{1}$

$= 30 \longleftarrow$ whole number

$\frac{1}{3} \times \frac{?}{?} = 1$

**Copy and complete the table.**

| | Division Sentence | Reciprocal of Divisor | Multiplication Sentence | Quotient |
|---|---|---|---|---|
| **1.** | $6 \div \frac{3}{8} = \underline{?}$ | $\underline{?}$ | $6 \times \frac{8}{3} = \underline{?}$ | $\underline{?}$ |
| **2.** | $12 \div \frac{4}{5} = \underline{?}$ | $\underline{?}$ | $12 \times \frac{5}{4} = \underline{?}$ | $\underline{?}$ |
| **3.** | $5 \div \frac{7}{8} = \underline{?}$ | $\frac{8}{7}$ | $\underline{?}$ | $\underline{?}$ |
| **4.** | $7 \div \frac{3}{4} = \underline{?}$ | $\frac{4}{3}$ | $\underline{?}$ | $\underline{?}$ |

**Divide.**

**5.** $3 \div \frac{1}{2}$

**6.** $4 \div \frac{1}{3}$

**7.** $18 \div \frac{6}{17}$

**8.** $6 \div \frac{3}{5}$

**9.** $12 \div \frac{3}{4}$

**10.** $8 \div \frac{2}{5}$

**11.** $24 \div \frac{12}{13}$

**12.** $9 \div \frac{3}{7}$

**13.** $7 \div \frac{4}{5}$

**14.** $15 \div \frac{9}{11}$

**15.** $7 \div \frac{2}{7}$

**16.** $5 \div \frac{4}{9}$

**17.** $6 \div \frac{5}{8}$

**18.** $4 \div \frac{3}{10}$

**19.** $20 \div \frac{8}{9}$

**20.** $13 \div \frac{3}{10}$

## PROBLEM SOLVING

**21.** How many pieces of $\frac{1}{4}$-yd copper tubing can be cut from a 10-yd piece of copper tubing?

**22.** Edward jogs $\frac{3}{4}$ mile a day. How many days will it take him to jog 8 miles?

**23.** How many pieces of $\frac{5}{9}$-m board can be cut from a 15-m board?

**24.** How many $\frac{7}{8}$-qt containers can be filled with 14 qt of strawberries?

**25.** Which quotient is greater: $5 \div \frac{1}{10}$ or $5 \div \frac{3}{10}$? Explain your answer.

**26.** Which quotient is less: $10 \div \frac{1}{5}$ or $10 \div \frac{3}{5}$? Explain your answer.

 **Share Your Thinking**

Communicate

**27.** Tell your teacher the two types of answers you get when dividing a whole number by a fraction. Give an example of each type.

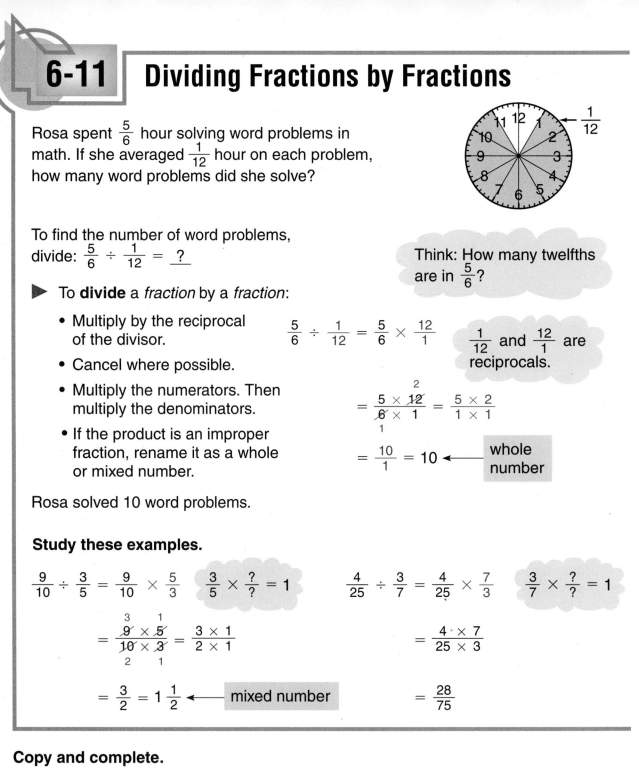

## 6-11 Dividing Fractions by Fractions

Rosa spent $\frac{5}{6}$ hour solving word problems in math. If she averaged $\frac{1}{12}$ hour on each problem, how many word problems did she solve?

To find the number of word problems, divide: $\frac{5}{6} \div \frac{1}{12} = \underline{\ ?\ }$

Think: How many twelfths are in $\frac{5}{6}$?

▶ To **divide** a *fraction* by a *fraction*:

- Multiply by the reciprocal of the divisor.

$$\frac{5}{6} \div \frac{1}{12} = \frac{5}{6} \times \frac{12}{1}$$

$\frac{1}{12}$ and $\frac{12}{1}$ are reciprocals.

- Cancel where possible.
- Multiply the numerators. Then multiply the denominators.

$$= \frac{5 \times \overset{2}{\cancel{12}}}{\underset{1}{\cancel{6}} \times 1} = \frac{5 \times 2}{1 \times 1}$$

- If the product is an improper fraction, rename it as a whole or mixed number.

$$= \frac{10}{1} = 10 \longleftarrow \text{whole number}$$

Rosa solved 10 word problems.

**Study these examples.**

$$\frac{9}{10} \div \frac{3}{5} = \frac{9}{10} \times \frac{5}{3} \qquad \frac{3}{5} \times \frac{?}{?} = 1$$

$$= \frac{\overset{3}{\cancel{9}} \times \overset{1}{\cancel{5}}}{\underset{2}{\cancel{10}} \times \underset{1}{\cancel{3}}} = \frac{3 \times 1}{2 \times 1}$$

$$= \frac{3}{2} = 1\frac{1}{2} \longleftarrow \text{mixed number}$$

$$\frac{4}{25} \div \frac{3}{7} = \frac{4}{25} \times \frac{7}{3} \qquad \frac{3}{7} \times \frac{?}{?} = 1$$

$$= \frac{4 \cdot \times 7}{25 \times 3}$$

$$= \frac{28}{75}$$

**Copy and complete.**

1. $\frac{2}{3} \div \frac{5}{6} = \frac{2}{3} \times \frac{6}{5} = \underline{\ ?\ }$

2. $\frac{4}{5} \div \frac{4}{7} = \frac{4}{5} \times \frac{7}{4} = \underline{\ ?\ }$

3. $\frac{3}{8} \div \frac{15}{16} = \frac{3}{8} \times \frac{?}{?} = \underline{\ ?\ }$

4. $\frac{3}{7} \div \frac{6}{7} = \frac{3}{7} \times \frac{?}{?} = \underline{\ ?\ }$

5. $\frac{4}{25} \div \frac{2}{3} = \frac{?}{?} \times \frac{?}{?} = \underline{\ ?\ }$

6. $\frac{9}{10} \div \frac{3}{5} = \frac{?}{?} \times \frac{?}{?} = \underline{\ ?\ }$

**Divide.**

**7.** $\frac{1}{2} \div \frac{1}{6}$     **8.** $\frac{1}{4} \div \frac{1}{12}$     **9.** $\frac{3}{4} \div \frac{1}{6}$     **10.** $\frac{5}{6} \div \frac{1}{9}$

**11.** $\frac{3}{8} \div \frac{1}{4}$     **12.** $\frac{5}{8} \div \frac{1}{2}$     **13.** $\frac{1}{3} \div \frac{4}{15}$     **14.** $\frac{3}{14} \div \frac{1}{7}$

**15.** $\frac{4}{9} \div \frac{1}{6}$     **16.** $\frac{4}{5} \div \frac{7}{15}$     **17.** $\frac{5}{12} \div \frac{1}{4}$     **18.** $\frac{4}{9} \div \frac{1}{12}$

**19.** $\frac{3}{4} \div \frac{3}{8}$     **20.** $\frac{3}{5} \div \frac{3}{10}$     **21.** $\frac{3}{7} \div \frac{3}{7}$     **22.** $\frac{7}{10} \div \frac{7}{20}$

**23.** $\frac{2}{3} \div \frac{8}{9}$     **24.** $\frac{8}{15} \div \frac{2}{5}$     **25.** $\frac{3}{5} \div \frac{4}{15}$     **26.** $\frac{4}{7} \div \frac{3}{14}$

**27.** $\frac{7}{9} \div \frac{5}{6}$     **28.** $\frac{5}{12} \div \frac{2}{3}$     **29.** $\frac{2}{3} \div \frac{3}{4}$     **30.** $\frac{2}{11} \div \frac{10}{13}$

**31.** $\frac{2}{3} \div \frac{2}{5}$     **32.** $\frac{4}{5} \div \frac{3}{7}$     **33.** $\frac{8}{9} \div \frac{4}{5}$     **34.** $\frac{5}{8} \div \frac{2}{9}$

**Compare. Write <, =, or >.**

**35.** $\frac{1}{2} \div \frac{1}{3}$ ___?___ $\frac{1}{4} \div \frac{1}{6}$     **36.** $\frac{1}{5} \div \frac{1}{7}$ ___?___ $\frac{1}{8} \div \frac{1}{9}$

**37.** $\frac{1}{6} \div \frac{5}{12}$ ___?___ $\frac{1}{5} \div \frac{3}{5}$     **38.** $\frac{1}{8} \div \frac{3}{4}$ ___?___ $\frac{1}{9} \div \frac{2}{3}$

**39.** $\frac{4}{9} \div \frac{2}{3}$ ___?___ $\frac{16}{25} \div \frac{4}{5}$     **40.** $\frac{5}{6} \div \frac{2}{9}$ ___?___ $\frac{4}{7} \div \frac{3}{14}$

**PROBLEM SOLVING**

**41.** Gerald cuts a $\frac{7}{8}$-yd piece of leather into $\frac{1}{16}$-yd strips for key holders. How many strips does he cut?

**42.** Karen divides $\frac{3}{4}$ cup of salad dressing into $\frac{1}{8}$-cup portions. How many portions of salad dressing does she have?

**43.** How many sixths are there in $\frac{1}{3}$?

**44.** How many tenths are there in $\frac{2}{5}$?

**Challenge**

**45.** The reciprocal of a number is the sum of $\frac{1}{3}$ and $\frac{5}{6}$. What is the number?

**46.** The reciprocal of a number is the product of $\frac{1}{3}$ and $\frac{5}{6}$. What is the number?

**Dividing Fractions by Whole Numbers**

Mrs. Kelly divided a half loaf of raisin bread equally among her 6 grandchildren. How much bread did each grandchild receive?

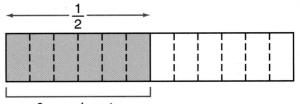

6 equal parts

To find how much bread each received, divide: $\frac{1}{2} \div 6 = \underline{\ ?\ }$

Think: What is $\frac{1}{2}$ divided into 6 equal parts?

▶ To **divide** a *fraction* by a *whole number*:

- Rename the whole number as an improper fraction with a denominator of 1.

- Multiply by the reciprocal of the whole-number divisor.

- Cancel where possible.

- Multiply the numerators. Then multiply the denominators.

- Write the answer in simplest form.

$$\frac{1}{2} \div 6 = \frac{1}{2} \div \frac{6}{1}$$

$$\frac{6}{1} \times \frac{?}{?} = 1$$

$$= \frac{1}{2} \times \frac{1}{6}$$

$$= \frac{1 \times 1}{2 \times 6}$$

$$= \frac{1}{12} \longleftarrow \text{Simplest form}$$

Each grandchild received $\frac{1}{12}$ of the loaf of bread.

**Study this example.**

$$\frac{9}{10} \div 12 = \frac{9}{10} \div \frac{12}{1}$$

Think: $\frac{9}{10}$ divided into 12 equal parts equals what number?

$$= \frac{9}{10} \times \frac{1}{12} = \frac{\overset{3}{\cancel{9}} \times 1}{10 \times \underset{4}{\cancel{12}}}$$

$$= \frac{3 \times 1}{10 \times 4} = \frac{3}{40}$$

**Copy and complete.**

**1.** $\frac{1}{4} \div 3 = \frac{1}{4} \div \frac{3}{?}$

$\qquad = \frac{1}{4} \times \frac{?}{?}$

$\qquad = \underline{\ ?\ }$

**2.** $\frac{2}{3} \div 10 = \frac{2}{3} \div \frac{10}{?}$

$\qquad = \frac{2}{3} \times \frac{?}{?}$

$\qquad = \underline{\ ?\ }$

**3.** $\frac{3}{5} \div 9 = \frac{?}{5} \div \frac{9}{?}$

$\qquad = \frac{?}{5} \times \frac{?}{?}$

$\qquad = \underline{\ ?\ }$

**Divide.**

**4.** $\frac{1}{5} \div 2$    **5.** $\frac{1}{7} \div 4$    **6.** $\frac{5}{8} \div 10$    **7.** $\frac{3}{5} \div 9$

**8.** $\frac{12}{33} \div 4$    **9.** $\frac{9}{10} \div 3$    **10.** $\frac{6}{7} \div 4$    **11.** $\frac{15}{19} \div 6$

**12.** $\frac{4}{5} \div 6$    **13.** $\frac{6}{7} \div 9$    **14.** $\frac{12}{25} \div 6$    **15.** $\frac{6}{7} \div 15$

**16.** $\frac{4}{9} \div 36$    **17.** $\frac{5}{8} \div 40$    **18.** $\frac{3}{5} \div 27$    **19.** $\frac{9}{10} \div 81$

**20.** $\frac{7}{8} \div 49$    **21.** $\frac{6}{7} \div 42$    **22.** $\frac{4}{11} \div 8$    **23.** $\frac{3}{4} \div 9$

**24.** $\frac{3}{5} \div 21$    **25.** $\frac{2}{3} \div 50$    **26.** $\frac{5}{6} \div 20$    **27.** $\frac{7}{8} \div 14$

**28.** $\frac{5}{12} \div 25$    **29.** $\frac{11}{12} \div 22$    **30.** $\frac{9}{10} \div 27$    **31.** $\frac{3}{11} \div 12$

**32.** $\frac{6}{7} \div 8$    **33.** $\frac{4}{5} \div 30$    **34.** $\frac{12}{13} \div 16$    **35.** $\frac{10}{11} \div 15$

**Compare. Write <, =, or >.**

**36.** $\frac{1}{2} \div 10 \; \underline{\;?\;} \; \frac{1}{4} \div 5$    **37.** $\frac{1}{5} \div 8 \; \underline{\;?\;} \; \frac{2}{5} \div 8$    **38.** $\frac{1}{5} \div 6 \; \underline{\;?\;} \; \frac{1}{6} \div 5$

**39.** $\frac{3}{4} \div 6 \; \underline{\;?\;} \; \frac{3}{4} \div 9$    **40.** $\frac{1}{3} \div 4 \; \underline{\;?\;} \; \frac{2}{6} \div 4$    **41.** $\frac{5}{6} \div 25 \; \underline{\;?\;} \; \frac{4}{7} \div 28$

**PROBLEM SOLVING**

**42.** One third of the class is divided into 3 equal groups. What part of the class is each group?

**43.** Three fourths of a squad is divided into 2 teams. What part of the squad is each team?

**44.** Jenny has $\frac{7}{8}$ yard of lace to use for 3 dresses. If the same amount of lace is used for each dress, how many yards of lace are used for one dress?

**45.** Camilo has $\frac{3}{5}$ hour to solve 12 math problems. If he spends the same amount of time on each problem, what part of an hour does he spend on each problem?

**Dividing Mixed Numbers by Fractions**

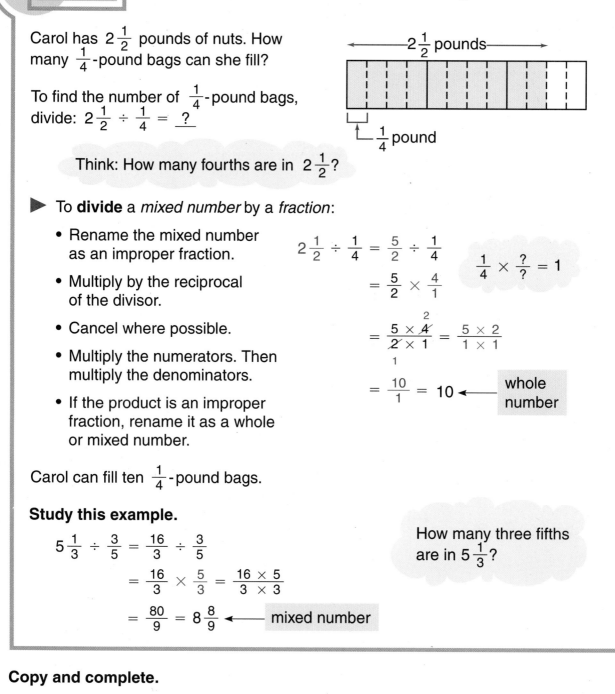

Carol has $2\frac{1}{2}$ pounds of nuts. How many $\frac{1}{4}$-pound bags can she fill?

To find the number of $\frac{1}{4}$-pound bags, divide: $2\frac{1}{2} \div \frac{1}{4} = \underline{\ ?\ }$

$2\frac{1}{2}$ pounds

$\frac{1}{4}$ pound

Think: How many fourths are in $2\frac{1}{2}$?

▶ To **divide** a *mixed number* by a *fraction*:

- Rename the mixed number as an improper fraction.

- Multiply by the reciprocal of the divisor.

- Cancel where possible.

- Multiply the numerators. Then multiply the denominators.

- If the product is an improper fraction, rename it as a whole or mixed number.

$$2\frac{1}{2} \div \frac{1}{4} = \frac{5}{2} \div \frac{1}{4}$$
$$= \frac{5}{2} \times \frac{4}{1}$$
$$= \frac{5 \times \overset{2}{\cancel{4}}}{\cancel{2} \times 1} = \frac{5 \times 2}{1 \times 1}$$
$$= \frac{10}{1} = 10 \longleftarrow \text{ whole number}$$

$\frac{1}{4} \times \frac{?}{?} = 1$

Carol can fill ten $\frac{1}{4}$-pound bags.

**Study this example.**

$$5\frac{1}{3} \div \frac{3}{5} = \frac{16}{3} \div \frac{3}{5}$$
$$= \frac{16}{3} \times \frac{5}{3} = \frac{16 \times 5}{3 \times 3}$$
$$= \frac{80}{9} = 8\frac{8}{9} \longleftarrow \text{ mixed number}$$

How many three fifths are in $5\frac{1}{3}$?

**Copy and complete.**

1. $2\frac{1}{3} \div \frac{1}{6} = \frac{?}{3} \div \frac{1}{6}$
$= \frac{?}{3} \times \frac{?}{?}$
$= \underline{\ ?\ }$

2. $1\frac{1}{2} \div \frac{9}{10} = \frac{?}{2} \div \frac{9}{10}$
$= \frac{?}{2} \times \frac{?}{?}$
$= \underline{\ ?\ }$

3. $1\frac{1}{4} \div \frac{3}{8} = \frac{?}{?} \div \frac{3}{8}$
$= \frac{?}{?} \times \frac{?}{?}$
$= \underline{\ ?\ }$

**Divide.**

4. $2\frac{1}{2} \div \frac{5}{6}$

5. $2\frac{1}{4} \div \frac{3}{4}$

6. $2\frac{11}{12} \div \frac{5}{12}$

7. $6\frac{7}{8} \div \frac{5}{8}$

8. $3\frac{1}{5} \div \frac{4}{15}$

9. $5\frac{1}{4} \div \frac{3}{8}$

10. $3\frac{1}{7} \div \frac{2}{7}$

11. $7\frac{1}{2} \div \frac{5}{6}$

12. $4\frac{4}{5} \div \frac{4}{15}$

13. $3\frac{6}{7} \div \frac{9}{14}$

14. $2\frac{1}{4} \div \frac{9}{10}$

15. $2\frac{8}{9} \div \frac{2}{3}$

16. $6\frac{3}{4} \div \frac{3}{5}$

17. $2\frac{3}{4} \div \frac{5}{12}$

18. $4\frac{1}{8} \div \frac{5}{16}$

19. $4\frac{1}{5} \div \frac{3}{7}$

20. $4\frac{5}{8} \div \frac{3}{4}$

21. $3\frac{7}{8} \div \frac{3}{8}$

22. $6\frac{5}{9} \div \frac{5}{9}$

23. $2\frac{4}{9} \div \frac{5}{6}$

**Compare. Write <, =, or >.**

24. $1\frac{1}{2} \div \frac{3}{4}$ __?__ $1\frac{1}{3} \div \frac{1}{3}$

25. $2\frac{1}{2} \div \frac{1}{8}$ __?__ $3\frac{1}{3} \div \frac{1}{6}$

26. $3\frac{3}{4} \div \frac{3}{4}$ __?__ $3\frac{1}{5} \div \frac{4}{5}$

27. $3\frac{1}{5} \div \frac{4}{15}$ __?__ $8\frac{1}{3} \div \frac{5}{6}$

28. $3\frac{1}{2} \div \frac{3}{4}$ __?__ $1\frac{1}{4} \div \frac{3}{8}$

29. $4\frac{1}{2} \div \frac{1}{4}$ __?__ $2\frac{1}{4} \div \frac{1}{2}$

**PROBLEM SOLVING**

30. Pang has $8\frac{2}{3}$ pounds of coffee beans. How many $\frac{2}{3}$-pound bags can he fill?

31. Eli jogs $\frac{3}{4}$ mile a day. How many days will it take him to jog $6\frac{1}{4}$ miles?

32. Kim had $2\frac{1}{2}$ meters of copper tubing that he cut into $\frac{1}{4}$-meter pieces. How many $\frac{1}{4}$-meter pieces of tubing did he cut?

33. A carpenter cuts a $4\frac{1}{6}$-yard length of board into $\frac{5}{6}$-yard pieces. How many $\frac{5}{6}$-yard pieces of board does he cut?

**Mental Math**

**Divide.**

34. $\frac{1}{2} \div \frac{1}{4}$

35. $\frac{1}{3} \div \frac{1}{9}$

36. $\frac{1}{4} \div \frac{1}{16}$

37. $\frac{1}{5} \div \frac{1}{25}$

38. $\frac{1}{6} \div \frac{1}{36}$

39. $2 \div \frac{1}{4}$

40. $3 \div \frac{1}{5}$

41. $4 \div \frac{1}{6}$

42. $5 \div \frac{1}{7}$

43. $6 \div \frac{1}{8}$

How many boxes are needed to pack $7\frac{1}{2}$ dozen apples if a box holds $2\frac{1}{2}$ dozen?

$$\longleftarrow 7\frac{1}{2} \text{ dozen} \longrightarrow$$

1 box

To find how many boxes are needed, divide: $7\frac{1}{2} \div 2\frac{1}{2} = \underline{\ ?\ }$

Think: How many $2\frac{1}{2}$s are in $7\frac{1}{2}$?

▶ To **divide** a *mixed* or *whole number* by another *mixed* or *whole number*:

- Rename both numbers as improper fractions.

- Multiply by the reciprocal of the divisor.

- Cancel where possible.

- Multiply the numerators. Then multiply the denominators.

- Write the answer in simplest form.

$$7\frac{1}{2} \div 2\frac{1}{2} = \frac{15}{2} \div \frac{5}{2}$$

$$\frac{5}{2} \times \frac{?}{?} = 1$$

$$= \frac{15}{2} \times \frac{2}{5}$$

$$= \frac{\overset{3}{\cancel{15}} \times \overset{1}{\cancel{2}}}{\underset{1}{\cancel{2}} \times \underset{1}{\cancel{5}}} = \frac{3 \times 1}{1 \times 1}$$

$$= \frac{3}{1} = 3$$

Three boxes are needed to pack $7\frac{1}{2}$ dozen apples.

**Study these examples.**

$$7\frac{1}{5} \div 9 = \frac{36}{5} \div \frac{9}{1}$$

Divide $7\frac{1}{5}$ into 9 equal parts.

$$= \frac{36}{5} \times \frac{1}{9} = \frac{\overset{4}{36} \times 1}{5 \times \underset{1}{9}}$$

$$= \frac{4 \times 1}{5 \times 1} = \frac{4}{5}$$

$$16 \div 1\frac{1}{3} = \frac{16}{1} \div \frac{4}{3}$$

How many $1\frac{1}{3}$s are in 16?

$$= \frac{16}{1} \times \frac{3}{4} = \frac{\overset{4}{16} \times 3}{1 \times \underset{1}{4}}$$

$$= \frac{4 \times 3}{1 \times 1} = 12$$

**Copy and complete.**

**1.** $3\frac{1}{3} \div 1\frac{2}{3} = \frac{10}{3} \div \frac{?}{3}$

$$= \frac{?}{?} \times \frac{?}{?} = \underline{\ ?\ }$$

**2.** $7 \div 3\frac{1}{2} = \frac{7}{?} \div \frac{?}{?}$

$$= \frac{?}{?} \times \frac{?}{?} = \underline{\ ?\ }$$

**Divide.**

**3.** $3\frac{1}{2} \div 1\frac{3}{4}$     **4.** $5\frac{1}{3} \div 1\frac{1}{3}$     **5.** $10\frac{1}{2} \div 3\frac{1}{2}$     **6.** $3\frac{6}{7} \div 1\frac{2}{7}$

**7.** $3\frac{1}{5} \div 8$     **8.** $3\frac{1}{3} \div 10$     **9.** $7\frac{1}{3} \div 11$     **10.** $3\frac{2}{5} \div 17$

**11.** $6 \div 1\frac{1}{2}$     **12.** $14 \div 4\frac{2}{3}$     **13.** $5 \div 6\frac{3}{5}$     **14.** $23 \div 3\frac{5}{6}$

**15.** $7\frac{1}{2} \div 1\frac{2}{3}$     **16.** $4\frac{1}{5} \div 1\frac{3}{4}$     **17.** $5\frac{1}{4} \div 2\frac{1}{3}$     **18.** $6\frac{2}{3} \div 1\frac{1}{4}$

**19.** $4\frac{1}{8} \div 2\frac{3}{4}$     **20.** $6\frac{3}{4} \div 1\frac{1}{2}$     **21.** $6\frac{1}{4} \div 5$     **22.** $6\frac{3}{7} \div 9$

**23.** $6\frac{2}{3} \div 10$     **24.** $9\frac{3}{5} \div 8$     **25.** $15 \div 1\frac{2}{3}$     **26.** $56 \div 3\frac{1}{2}$

**27.** $3\frac{3}{4} \div 1\frac{1}{4}$     **28.** $4\frac{4}{5} \div 1\frac{1}{5}$     **29.** $2\frac{2}{7} \div 1\frac{4}{7}$     **30.** $3\frac{3}{5} \div 2\frac{3}{10}$

**31.** $12 \div 2\frac{2}{5}$     **32.** $18 \div 1\frac{2}{7}$     **33.** $32 \div 1\frac{3}{5}$     **34.** $21 \div 2\frac{1}{3}$

## PROBLEM SOLVING

**35.** How many pieces of $1\frac{1}{4}$-ft board can be cut from a board that is $8\frac{3}{4}$ ft long?

**36.** Jorge cut a $5\frac{1}{5}$-yd board into 5 equal pieces. How long was each piece?

**37.** Delia is making name tags that are each $3\frac{3}{4}$ in. long. How many can she make from a 30-in. roll of label paper?

**38.** Subas packed 5 dozen oranges in boxes. If he put $1\frac{3}{4}$ dozen in each box, how many boxes did he pack?

## Share Your Thinking

*Communicate* ✓

**39.** Teach your parent(s) or another adult:

- how to use manipulatives or drawings to divide: $12 \div 1\frac{1}{2}$.

- how to use the division steps to solve the same problem.

# 6-15 Estimating with Mixed Numbers

Benny walks at a rate of $3\frac{1}{4}$ miles per hour. If he walks for $2\frac{4}{5}$ hours, about how far will he have walked?

To find about how far he will have walked, estimate: $3\frac{1}{4} \times 2\frac{4}{5}$

**Remember:**
When rounding mixed numbers:
- equal to or greater than $\frac{1}{2}$, round *up*.
- less than $\frac{1}{2}$, round *down*.

▶ To **estimate products** of mixed numbers:

- Round each factor to the nearest whole number.

- Multiply the rounded numbers.

$3\frac{1}{4} \times 2\frac{4}{5}$

$\frac{1}{4} < \frac{1}{2}$ Round *down* $3\frac{1}{4}$.

$\frac{4}{5} > \frac{1}{2}$ Round *up* $2\frac{4}{5}$.

$3 \times 3 = 9 \leftarrow$ estimated product

Benny will have walked about 9 miles.

**Study this example.**

$5\frac{1}{2} \times 4$

$\frac{1}{2} = \frac{1}{2}$ Round *up* $5\frac{1}{2}$.

$6 \times 4 = 24 \leftarrow$ estimated product

**Estimate each product.**

1. $9 \times 4\frac{5}{8}$
2. $8 \times 1\frac{1}{4}$
3. $7 \times 4\frac{1}{3}$
4. $6 \times 2\frac{1}{2}$

5. $4\frac{1}{4} \times 3\frac{1}{8}$
6. $10\frac{3}{4} \times 1\frac{6}{7}$
7. $4\frac{1}{2} \times 5\frac{1}{4}$
8. $8\frac{1}{5} \times 3\frac{2}{3}$

9. $5\frac{1}{9} \times 3\frac{1}{5}$
10. $2\frac{1}{5} \times 5\frac{1}{7}$
11. $4\frac{3}{7} \times 6\frac{1}{10}$
12. $5\frac{2}{5} \times 2\frac{5}{6}$

13. $3\frac{5}{6} \times 6$
14. $2\frac{3}{5} \times 9$
15. $4\frac{2}{7} \times 12$
16. $6\frac{3}{10} \times 5$

17. $11\frac{1}{4} \times 4\frac{1}{5}$
18. $16\frac{3}{4} \times 7\frac{1}{8}$
19. $14\frac{1}{7} \times 2\frac{3}{8}$
20. $16\frac{5}{7} \times 1\frac{7}{8}$

226

# Using Compatible Numbers

Estimate the product of a fraction and a mixed or whole number by using compatible numbers.

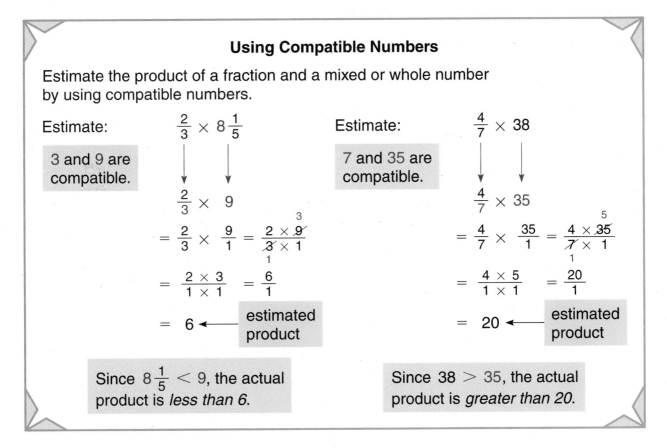

Estimate: $\frac{2}{3} \times 8\frac{1}{5}$

3 and 9 are compatible.

$\frac{2}{3} \times 9$

$= \frac{2}{3} \times \frac{9}{1} = \frac{2 \times \overset{3}{\cancel{9}}}{\underset{1}{\cancel{3}} \times 1}$

$= \frac{2 \times 3}{1 \times 1} = \frac{6}{1}$

$= 6 \leftarrow$ estimated product

Since $8\frac{1}{5} < 9$, the actual product is *less than 6.*

Estimate: $\frac{4}{7} \times 38$

7 and 35 are compatible.

$\frac{4}{7} \times 35$

$= \frac{4}{7} \times \frac{35}{1} = \frac{4 \times \overset{5}{\cancel{35}}}{\underset{1}{\cancel{7}} \times 1}$

$= \frac{4 \times 5}{1 \times 1} = \frac{20}{1}$

$= 20 \leftarrow$ estimated product

Since $38 > 35$, the actual product is *greater than 20.*

**Estimate. Then write whether the actual product is *less than* or *greater than* the estimated product.**

**21.** $\frac{2}{5} \times 9\frac{1}{8}$  **22.** $\frac{7}{9} \times 11\frac{1}{3}$  **23.** $\frac{3}{8} \times 14\frac{1}{4}$  **24.** $\frac{5}{6} \times 7\frac{6}{7}$

**25.** $1\frac{1}{9} \times 28$  **26.** $2\frac{1}{7} \times 15$  **27.** $1\frac{1}{12} \times 35$  **28.** $3\frac{2}{3} \times 17$

## Choose a Computation Method

**Solve and explain the method you used. Write whether you estimated or found an exact answer.**

**29.** Gina uses $3\frac{2}{3}$ cups of flour to make bread. Kate uses $2\frac{1}{2}$ times as much for her recipe. About how much flour does Kate use for her recipe?

**30.** Jane bought $4\frac{2}{3}$ yd of material. She used $\frac{9}{10}$ of it for a dress. How much material did Jane use for her dress?

**31.** If $2\frac{3}{4}$ yards of fabric are needed to make a dress, will 15 yards of fabric be enough for Kathy to make 5 dresses?

**32.** A piece of wood $6\frac{2}{5}$ ft long was purchased for shelving. Only $\frac{7}{8}$ of the wood was used. How much of the entire length was *not* used?

# TECHNOLOGY

## Order of Operations with Fractions

Fractions and mixed numbers can be computed using a fraction calculator.

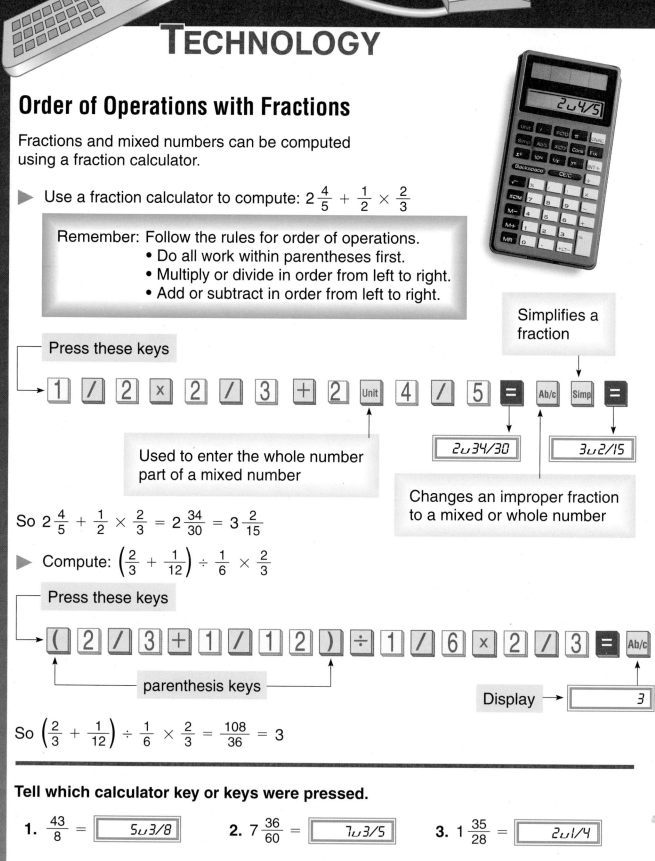

▶ Use a fraction calculator to compute: $2\frac{4}{5} + \frac{1}{2} \times \frac{2}{3}$

Remember: Follow the rules for order of operations.
- Do all work within parentheses first.
- Multiply or divide in order from left to right.
- Add or subtract in order from left to right.

Simplifies a fraction

Press these keys

[1] [/] [2] [×] [2] [/] [3] [+] [2] [Unit] [4] [/] [5] [=] [Ab/c] [Simp] [=]

Used to enter the whole number part of a mixed number

2⌐34/30     3⌐2/15

Changes an improper fraction to a mixed or whole number

So $2\frac{4}{5} + \frac{1}{2} \times \frac{2}{3} = 2\frac{34}{30} = 3\frac{2}{15}$

▶ Compute: $\left(\frac{2}{3} + \frac{1}{12}\right) \div \frac{1}{6} \times \frac{2}{3}$

Press these keys

[(] [2] [/] [3] [+] [1] [/] [1] [2] [)] [÷] [1] [/] [6] [×] [2] [/] [3] [=] [Ab/c]

parenthesis keys

Display → 3

So $\left(\frac{2}{3} + \frac{1}{12}\right) \div \frac{1}{6} \times \frac{2}{3} = \frac{108}{36} = 3$

---

**Tell which calculator key or keys were pressed.**

**1.** $\frac{43}{8} = $ [ 5⌐3/8 ]

**2.** $7\frac{36}{60} = $ [ 7⌐3/5 ]

**3.** $1\frac{35}{28} = $ [ 2⌐1/4 ]

**Tell which operation should be done first. Then compute.
Express your answer in simplest form.**

**4.** $\left(\dfrac{5}{8} + \dfrac{1}{2}\right) \times \dfrac{2}{3}$

**5.** $\dfrac{2}{3} + \dfrac{1}{6} \div \dfrac{4}{5}$

**6.** $3\dfrac{5}{7} - 1\dfrac{3}{8} + \dfrac{1}{2}$

**7.** $\dfrac{1}{8} \times \left(\dfrac{8}{9} - \dfrac{2}{3}\right)$

**8.** $\dfrac{1}{3} + \dfrac{2}{9} \div \dfrac{2}{3}$

**9.** $\left(3\dfrac{5}{6} - 2\dfrac{1}{4}\right) \div \dfrac{3}{8}$

**10.** $6 \times \dfrac{3}{8} \div \dfrac{3}{4}$

**11.** $2\dfrac{5}{9} \times \left(\dfrac{2}{3} \div \dfrac{2}{9}\right)$

**12.** $\dfrac{3}{4} \times \dfrac{5}{6} \div \dfrac{3}{4}$

**13.** $\dfrac{1}{4} \times \dfrac{3}{8} + \left(\dfrac{7}{8} - \dfrac{1}{2}\right)$

**14.** $1\dfrac{1}{2} + \left(\dfrac{3}{5} \times \dfrac{2}{3}\right) \div \dfrac{2}{3}$

**15.** $\dfrac{5}{6} \times \dfrac{4}{5} \div \dfrac{2}{3} + \dfrac{1}{12}$

**16.** $\dfrac{2}{3} \div \dfrac{7}{8} + \left(\dfrac{6}{7} \times \dfrac{2}{3}\right)$

**Place parentheses in the expression to obtain
the given result.**

**17.** $\dfrac{2}{5} \times \dfrac{1}{4} + \dfrac{1}{2} \div \dfrac{2}{3} ; \dfrac{9}{20}$

**18.** $\dfrac{3}{4} - \dfrac{3}{8} \times \dfrac{3}{10} \div \dfrac{3}{8} ; \dfrac{3}{10}$

**Use a calculator to compute.**

**19.** $\dfrac{1}{2}$ of $120

$\dfrac{1}{2}$ of $12.00

$\dfrac{1}{2}$ of $1.20

**20.** Predict the result
of $\dfrac{1}{2}$ of $0.12. Check
using a calculator.

**21.** $\dfrac{1}{3}$ of $0.24

$\dfrac{1}{3}$ of $2.40

$\dfrac{1}{3}$ of $24.00

**22.** Predict the result
of $\dfrac{1}{3}$ of $240. Check
using a calculator.

**23.** $\dfrac{1}{2}$ of $60

$\dfrac{1}{2}$ of $6.00

$\dfrac{1}{2}$ of $0.60

**24.** Predict the result
of $\dfrac{1}{2}$ of $0.06. Check
using a calculator.

**25.** Describe any pattern that you see in
exercises 19, 21, and 23.

Communicate

229

# 6-17 Problem Solving: Using Simpler Numbers

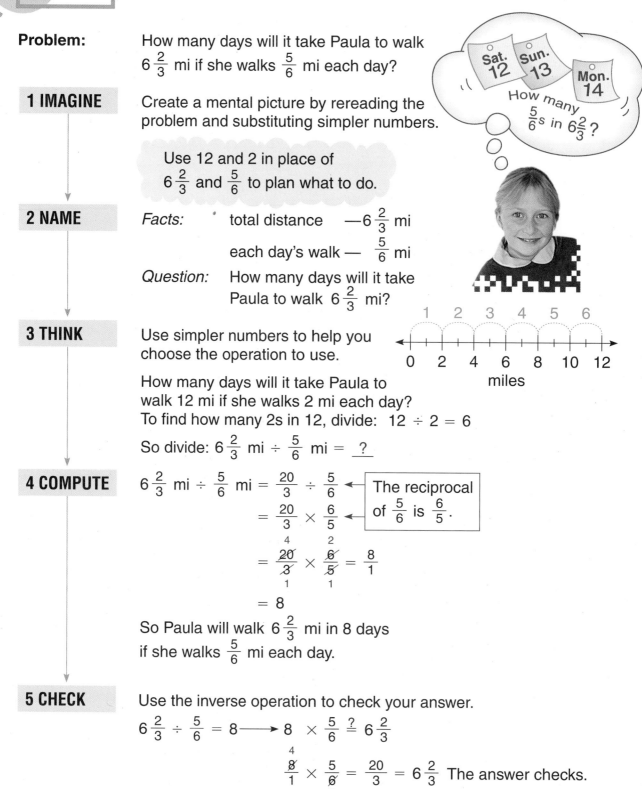

**Problem:** How many days will it take Paula to walk $6\frac{2}{3}$ mi if she walks $\frac{5}{6}$ mi each day?

**1 IMAGINE** Create a mental picture by rereading the problem and substituting simpler numbers.

Use 12 and 2 in place of $6\frac{2}{3}$ and $\frac{5}{6}$ to plan what to do.

How many $\frac{5}{6}$s in $6\frac{2}{3}$?

**2 NAME**

*Facts:*    total distance — $6\frac{2}{3}$ mi

each day's walk — $\frac{5}{6}$ mi

*Question:* How many days will it take Paula to walk $6\frac{2}{3}$ mi?

**3 THINK** Use simpler numbers to help you choose the operation to use.

How many days will it take Paula to walk 12 mi if she walks 2 mi each day?
To find how many 2s in 12, divide: $12 \div 2 = 6$

So divide: $6\frac{2}{3}$ mi $\div \frac{5}{6}$ mi = ___?___

**4 COMPUTE**

$6\frac{2}{3}$ mi $\div \frac{5}{6}$ mi $= \frac{20}{3} \div \frac{5}{6}$ ← The reciprocal of $\frac{5}{6}$ is $\frac{6}{5}$.

$= \frac{20}{3} \times \frac{6}{5}$

$= \frac{\overset{4}{\cancel{20}}}{\underset{1}{\cancel{3}}} \times \frac{\overset{2}{\cancel{6}}}{\underset{1}{\cancel{5}}} = \frac{8}{1}$

$= 8$

So Paula will walk $6\frac{2}{3}$ mi in 8 days if she walks $\frac{5}{6}$ mi each day.

**5 CHECK** Use the inverse operation to check your answer.

$6\frac{2}{3} \div \frac{5}{6} = 8 \longrightarrow 8 \times \frac{5}{6} \overset{?}{=} 6\frac{2}{3}$

$\frac{\overset{4}{\cancel{8}}}{1} \times \frac{5}{\underset{3}{\cancel{6}}} = \frac{20}{3} = 6\frac{2}{3}$ The answer checks.

230

**Use simpler numbers to solve each problem.**

1. Raul reads 38 pages an hour. At that rate, if he reads for $3\frac{1}{4}$ h, how many pages will he read?

30 pages

30 pages

30 pages

**90 pages in all**

| **IMAGINE** | Suppose Raul reads 30 pages an hour. How many pages will he read in 3 h? Draw a picture. |
|---|---|

| **NAME** | *Facts:*     38 pages an hour read     $3\frac{1}{4}$ h total time |
|---|---|

*Question:*    How many pages will Raul read?

| **THINK** | Study the picture. If Raul reads 30 pages each hour, in 3 hours he will read $3 \times 30$ pages or 90 pages. |
|---|---|

Now multiply: $3\frac{1}{4} \times 38 = \underline{\ ?\ }$

**COMPUTE**  →  **CHECK**

2. Each box holds $2\frac{1}{2}$ dozen apples. How many boxes are needed to pack $32\frac{1}{2}$ dozen apples?

3. What is the speed in miles per minute of an airplane that flies $18\frac{3}{4}$ mi in $2\frac{1}{2}$ min?

4. If fourteen children share $9\frac{1}{3}$ lb of a fruit mix, what part of a pound will each receive?

5. Rosa needs $14\frac{1}{2}$ lb of potatoes to make potato salad for the picnic. She has peeled $5\frac{1}{3}$ lb. How many more pounds does she need to peel?

6. Eduardo studies $1\frac{5}{6}$ h each night. How many hours will he study in 5 nights?

**Make Up Your Own**

*Communicate* ✓

7. Write a problem that can be solved using simpler numbers. Have a classmate solve it.

## 6-18 Problem-Solving Applications

**Connections: Home Economics**

**Solve each problem and explain the method you used.**

1. Martin has $\frac{5}{6}$ of a loaf of banana bread. He gives half of what is left to a friend. What part of the loaf does he give to his friend?

2. A recipe calls for $\frac{3}{4}$ c of walnuts. Anna decides to use only $\frac{1}{4}$ of that amount. How much does Anna use?

3. Helen slices 8 carrots into tenths for stew. How many slices are there?

4. Van and Doug make bread. Van uses $\frac{1}{6}$ c of rye flour and Doug uses 4 times as much rye flour. How much rye flour does Doug use?

5. Van's recipe calls for $3\frac{1}{2}$ c of wheat flour. He decides to cut the recipe in half. How much wheat flour should Van use?

6. Holly has $3\frac{1}{4}$ pt of raspberries. She wants to make raspberry muffins. Each muffin uses $\frac{1}{8}$ pt of berries. How many muffins can Holly make?

7. Dorothy buys $10\frac{1}{2}$ lb of apples. She uses $\frac{1}{4}$ of the apples in a pie. How many pounds of apples does she use in the pie?

8. Tom is making burritos. Each burrito uses $\frac{3}{8}$ c of beans and $\frac{1}{4}$ c of rice. How many cups of beans and cups of rice does he need to make 2 dozen burritos?

9. It takes $1\frac{1}{4}$ h to bake a loaf of rye bread. How long will it take to bake a half-dozen loaves if they are baked one at a time?

Imagine

Name

Think

Compute

Check

**Choose a strategy from the list or use another strategy you know to solve each problem.**

USE THESE STRATEGIES
Working Backwards
Use Simpler Numbers
Organized List
Hidden Information
Make a Table/Find a Pattern

10. Jeanine is making her own breakfast cereal. For every cup of oats, she uses $\frac{1}{4}$ c of dates, $\frac{1}{3}$ c of raisins, and $\frac{1}{8}$ c of puffed rice. How many cups of each ingredient will she use for 8 c of oats?

11. Robert is making party mix from raisins, nuts, cereal, and butter. How many different ways can he combine the ingredients if he decides to put the butter in last?

12. Adam decided to divide a carrot cake recipe in half, so he used $\frac{4}{5}$ lb of carrots. How many pounds of carrots did the original recipe require?

13. A recipe calls for $\frac{1}{8}$ lb of pistachio nuts. Heather has 3 oz of pistachios. Does she have enough to make the recipe?

14. Ashlee bakes a loaf of rye bread that weighs $18\frac{1}{3}$ oz. How many $\frac{5}{6}$-oz slices can she cut?

**Use the chart for problems 15–18.**

15. Rosemary makes a double batch of garden salad and a triple batch of cucumber salad. How many pounds of cucumbers does she use?

16. Which uses more tomatoes: three garden salads or six cucumber salads?

17. Which use less oil and vinegar combined: four garden salads or three cucumber salads?

**Salads**

| Item | Garden | Cucumber |
|---|---|---|
| Tomatoes | $\frac{3}{4}$ lb | $\frac{1}{5}$ lb |
| Lettuce | $1\frac{1}{2}$ lb | none |
| Onions | $\frac{1}{6}$ lb | $\frac{1}{3}$ lb |
| Cucumbers | $\frac{1}{6}$ lb | $1\frac{1}{4}$ lb |
| Oil | $\frac{1}{4}$ c | $\frac{1}{3}$ c |
| Vinegar | $\frac{1}{8}$ c | $\frac{1}{3}$ c |

**Make Up Your Own**

Communicate ✓

18. Write a problem that uses the data in the chart. Have someone solve it.

233

# Chapter Review and Practice

**Use the diagram to complete each statement.** *(See pp. 198–199, 212–213.)*

1. $\frac{1}{4}$ of $\frac{1}{3} = $ ___?___

2. $\frac{2}{3} \times \frac{5}{6} = $ ___?___

3. $2 \div \frac{1}{4} = $ ___?___

4. $\frac{2}{3} \div \frac{1}{6} = $ ___?___

**Multiply.** *(See pp. 200–205.)*

5. $\frac{3}{5} \times \frac{1}{2}$

6. $\frac{7}{10} \times \frac{2}{21}$

7. $6 \times \frac{2}{3}$

8. $\frac{4}{9} \times 18$

9. $\frac{2}{3} \times \frac{7}{10}$

10. $\frac{9}{20} \times \frac{24}{45}$

11. $\frac{9}{20} \times 6$

12. $60 \times \frac{3}{5}$

**Rename each as an improper fraction.** *(See pp. 206–207.)*

13. $2\frac{1}{2}$

14. $3\frac{1}{3}$

15. $2\frac{1}{4}$

16. $4\frac{2}{3}$

17. $3\frac{1}{5}$

18. $6\frac{1}{8}$

**Find the product.** *(See pp. 208–211.)*

19. $5\frac{1}{3} \times 3\frac{3}{4}$

20. $2\frac{1}{3} \times \frac{4}{7}$

21. $9\frac{1}{3} \times \frac{1}{7}$

22. $1\frac{7}{8} \times \frac{4}{5}$

**Are the numbers reciprocals? Write *Yes* or *No*.** *(See pp. 214–215.)*

23. $5, \frac{1}{5}$

24. $\frac{2}{3}, 1\frac{1}{2}$

25. $3\frac{1}{4}, \frac{4}{13}$

26. $\frac{4}{5}, \frac{8}{10}$

**Use manipulatives or drawings to divide.** *(See pp. 216–225.)*

27. $9 \div \frac{3}{5}$

28. $\frac{3}{8} \div 6$

29. $\frac{3}{10} \div \frac{3}{5}$

30. $\frac{5}{8} \div \frac{3}{10}$

31. $3\frac{1}{5} \div \frac{1}{3}$

32. $3\frac{1}{2} \div 1\frac{3}{4}$

33. $5 \div 2\frac{2}{3}$

34. $6\frac{1}{8} \div 1\frac{3}{4}$

**Estimate.** *(See pp. 226–227.)*

35. $14\frac{2}{7} \times 4\frac{1}{5}$

36. $4\frac{2}{3} \times 2\frac{1}{2}$

37. $5\frac{1}{4} \times 1\frac{3}{4}$

38. $9\frac{1}{3} \times 5\frac{4}{5}$

## PROBLEM SOLVING
*(See pp. 208–209, 224–225, 230–232.)*

39. Tony ran $3\frac{1}{3}$ times farther than Dot. If Dot ran $\frac{3}{4}$ of a mile, how far did Tony run?

40. Ann uses a $2\frac{1}{2}$-gal container to fill a 20-gal tank with water. How many times must she fill the container?

(See *Still More Practice*, p. 482.)

## LOGIC

In logic, two statements can be combined to form a compound statement using *and* or a compound statement using *or*.

▶ The compound statement using *and* is true only when *both* original statements are true.

A triangle has 3 sides. (true)
A square has 4 angles. (true)
A triangle has 3 sides *and* a square has 4 angles. (true)

A triangle has 4 sides. (false)
A triangle has 4 sides *and* a square has 4 angles. (false)

▶ The compound statement using *or* is true when *both* original statements are true, or *one* of the original statements is true.

A triangle has 3 sides *or* a square has 4 angles. (true)

A square has 5 angles. (false)
A triangle has 3 sides *or* a square has 5 angles. (true)

A triangle has 4 sides *or* a square has 5 angles. (false)

**Write compound statements using *and* and *or*. Then tell whether each compound statement is *true* or *false*.**

**1.** A cat is an animal.
A nickel is a coin.

**2.** Fall follows spring.
December falls in winter.

**3.** Ten is divisible by 2.
Twelve is divisible by 3.

**4.** Four is a prime number.
Five is a composite number.

**5.** $45 \div 9 = 5$
$8 - 3 = 6$

**6.** $4 \times 6 = 20$
$9 + 5 = 14$

**7.** $8 + 2 = 10$
$8 < 9$

**8.** $8 + 20 \div 2 = 18$
$2 + 3 + 5 > 10$

**9.** $\frac{1}{2} + \frac{2}{3} = \frac{3}{5}$
$\frac{6}{7} - \frac{2}{7} = \frac{4}{7}$

**10.** $\frac{1}{2} \times \frac{2}{3} = \frac{1}{3}$
$\frac{6}{7} \div \frac{2}{7} = 3$

# Check Your Mastery

## Performance Assessment

**Use these rule cards.**
Predict the rule for each pattern.
Then tell the next number and rule.

$\div 3$   $\div \frac{2}{3}$   $\times 1\frac{1}{3}$

**1.** $6, 2, \frac{2}{3}, \underline{\ ?\ }$

**2.** $6, 9, 13\frac{1}{2}, \underline{\ ?\ }$

**3.** $6, 8, 10\frac{2}{3}, \underline{\ ?\ }$

**Use the diagram to complete each statement.**

**4.**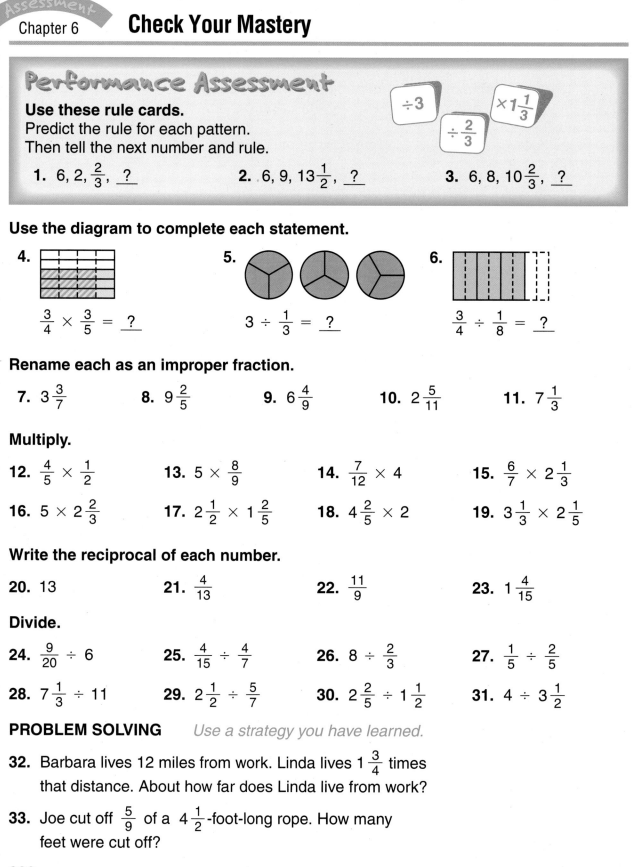

$\frac{3}{4} \times \frac{3}{5} = \underline{\ ?\ }$

**5.**

$3 \div \frac{1}{3} = \underline{\ ?\ }$

**6.**

$\frac{3}{4} \div \frac{1}{8} = \underline{\ ?\ }$

**Rename each as an improper fraction.**

**7.** $3\frac{3}{7}$    **8.** $9\frac{2}{5}$    **9.** $6\frac{4}{9}$    **10.** $2\frac{5}{11}$    **11.** $7\frac{1}{3}$

**Multiply.**

**12.** $\frac{4}{5} \times \frac{1}{2}$    **13.** $5 \times \frac{8}{9}$    **14.** $\frac{7}{12} \times 4$    **15.** $\frac{6}{7} \times 2\frac{1}{3}$

**16.** $5 \times 2\frac{2}{3}$    **17.** $2\frac{1}{2} \times 1\frac{2}{5}$    **18.** $4\frac{2}{5} \times 2$    **19.** $3\frac{1}{3} \times 2\frac{1}{5}$

**Write the reciprocal of each number.**

**20.** $13$    **21.** $\frac{4}{13}$    **22.** $\frac{11}{9}$    **23.** $1\frac{4}{15}$

**Divide.**

**24.** $\frac{9}{20} \div 6$    **25.** $\frac{4}{15} \div \frac{4}{7}$    **26.** $8 \div \frac{2}{3}$    **27.** $\frac{1}{5} \div \frac{2}{5}$

**28.** $7\frac{1}{3} \div 11$    **29.** $2\frac{1}{2} \div \frac{5}{7}$    **30.** $2\frac{2}{5} \div 1\frac{1}{2}$    **31.** $4 \div 3\frac{1}{2}$

**PROBLEM SOLVING**   *Use a strategy you have learned.*

**32.** Barbara lives 12 miles from work. Linda lives $1\frac{3}{4}$ times that distance. About how far does Linda live from work?

**33.** Joe cut off $\frac{5}{9}$ of a $4\frac{1}{2}$-foot-long rope. How many feet were cut off?

# Probability and Statistics 7

## Leaves

The winds that blow—
ask them, which leaf of the tree
will be next to go!

*Soseki*

**In this chapter you will:**

Learn about tree diagrams
and independent and
dependent events
Collect, organize, report,
and interpret data
Interpret and make line plots
and graphs
Use a model or diagram to solve
problems

**Critical Thinking/Finding Together**

On Sunday, leaves start falling into the
swimming pool. The number of leaves
doubles each day, until the whole pool
is covered on the seventh day. On
which day is the pool half-covered?

*Update your skills. See page 20.*

## 7-1 Probability

**Probability** is the chance that a given *event* will occur in an *experiment*.

**Random experiments**, like tossing a coin, rolling a number cube, spinning a spinner, and selecting an item from a set of items without looking, mean you do not know beforehand what the result, or *outcome*, will be.

> Experiment: tossing a coin
> Possible Outcomes: heads (*H*), tails (*T*)
> Event: tossing heads

The experiment of spinning the given spinner involves finding the probability of a spinner landing on different colors. What is the probability of the spinner landing on red? landing on *not* blue?

▶ For *equally likely* outcomes, **the probability of an event, *P* (E)** to occur is given by the formula:

$$P(E) = \frac{\text{number of favorable outcomes}}{\text{number of possible outcomes}}$$

| Probability of spinning red | $\rightarrow P(\text{red}) = \frac{1}{3}$ ← one red ← three possible outcomes |
|---|---|

*not* blue: 1 red, 1 green

$$P(not\ \text{blue}) = \frac{2}{3}$$

▶ You can use probability to make predictions. In 300 spins, predict about how many times the spinner above will land on red.

$$\frac{1}{3} \times 300 = 100$$

The spinner will land on red about 100 out of 300 spins.

$$P(\text{red})$$

---

**Use a coin to find the probability of each event.**

**1.** $P(H)$  **2.** $P(T)$  **3.** $P(not\ H)$  **4.** $P(not\ T)$

**Use the spinner at the right to find the probability of each event.**

**5.** $P(1)$  **6.** $P(2)$  **7.** $P(3)$  **8.** $P(4)$

**9.** In 700 spins, predict how many times the spinner at the right will land on:  **a.** 1  **b.** 2  **c.** 3  **d.** 4

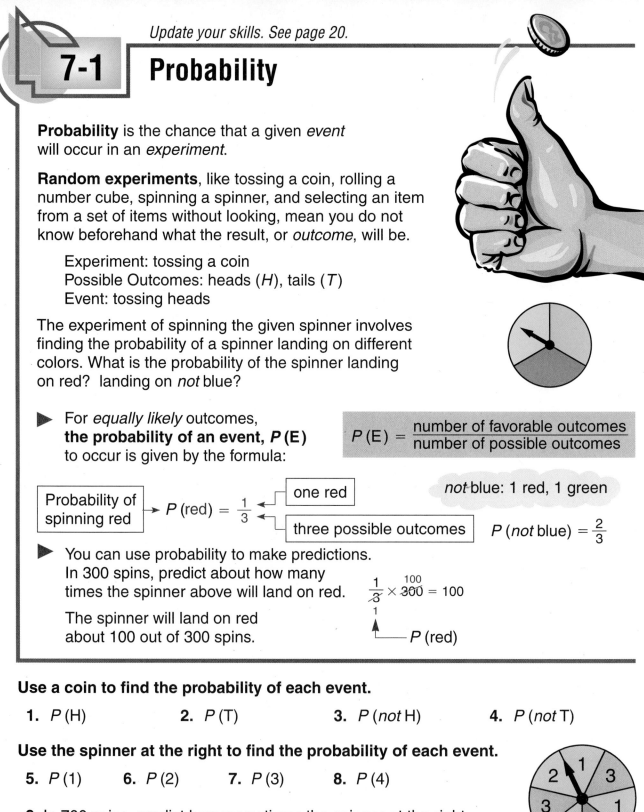

**Use the number cube at the right to find
the probability of each event.**

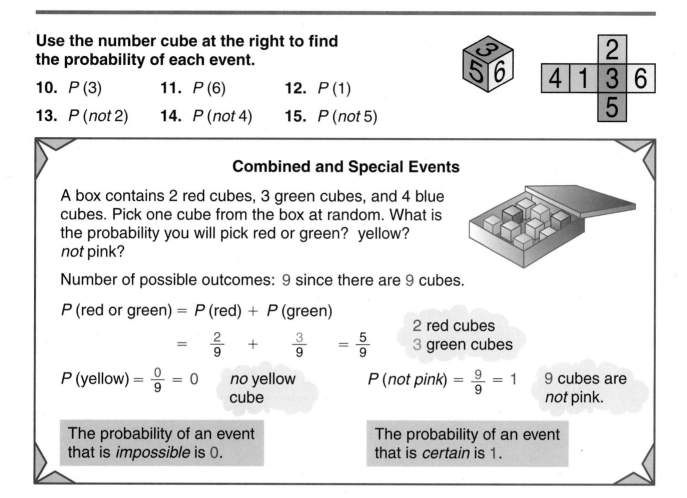

**10.** $P(3)$    **11.** $P(6)$    **12.** $P(1)$

**13.** $P(not\ 2)$    **14.** $P(not\ 4)$    **15.** $P(not\ 5)$

---

### Combined and Special Events

A box contains 2 red cubes, 3 green cubes, and 4 blue
cubes. Pick one cube from the box at random. What is
the probability you will pick red or green? yellow?
*not* pink?

Number of possible outcomes: 9 since there are 9 cubes.

$P(red\ or\ green) = P(red) + P(green)$

$$= \frac{2}{9} + \frac{3}{9} = \frac{5}{9}$$

> 2 red cubes
> 3 green cubes

$P(yellow) = \frac{0}{9} = 0$    *no* yellow cube

$P(not\ pink) = \frac{9}{9} = 1$    9 cubes are *not* pink.

| The probability of an event that is *impossible* is 0. | The probability of an event that is *certain* is 1. |

---

**Find the probability of each event.** Use the box of cubes above.

**16.** $P(red\ or\ blue)$    **17.** $P(blue\ or\ green)$    **18.** $P(not\ red)$

**19.** $P(not\ blue)$    **20.** $P(not\ purple)$    **21.** $P(gray)$

### PROBLEM SOLVING

**22.** A bank contains a nickel, a dime, and a quarter. James selects
one coin at random. What is the probability that the coin is worth:

**a.** exactly 5¢?    **b.** exactly 4¢?    **c.** more than 4¢?

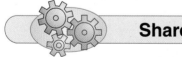

### Share Your Thinking

Communicate ✓

**23.** Tell a classmate that you will put the names of all your
classmates in a bag and then pick a name from it at random.
What is the probability you will pick your classmate's name?
a boy's name? Explain your answers.

# Tree Diagrams

In an experiment, Taylor flips two counters. One side of each counter is green and the other side is red. Find all possible outcomes. What is the probability of both counters landing green side up?

Side 1

Side 2

▶ You can use a **tree diagram** to show all possible outcomes and to determine the *probability of more than one event*.

| **Event 1** | **Event 2** | **Outcomes** | **Write** |
| First Counter | Second Counter | | |

Green (G) →  Green (G) ————→ Green-Green  $(G, G)$
         →  Red (R) ————→ Green-Red  $(G, R)$

Red (R) →  Green (G) ————→ Red-Green  $(R, G)$
       →  Red (R) ————→ Red-Red  $(R, R)$

Probability of both green → $P(G, G) = \dfrac{1}{4}$ ← favorable outcomes
← possible outcomes

There are 4 possible outcomes.
The probability of both counters landing green side up is $\dfrac{1}{4}$.

**Copy and complete the tree diagram. Then use the completed tree diagram for exercises 3–4.**

Heads (H)          Tails (T)

**Toss 2 coins**

| **Event 1** | **Event 2** | **Outcomes** |
| First Coin | Second Coin | |

1. Heads (H) →  Tails (T) ———→ ?
             →  ? ———→ ?

2. Tails (T) →  ? ———→ ?
            →  ? ———→ ?

3. How many possible outcomes are there altogether?

4. What is the probability of each outcome occurring?

*Communicate* ✓

**Use the spinners for exercises 5–7.**

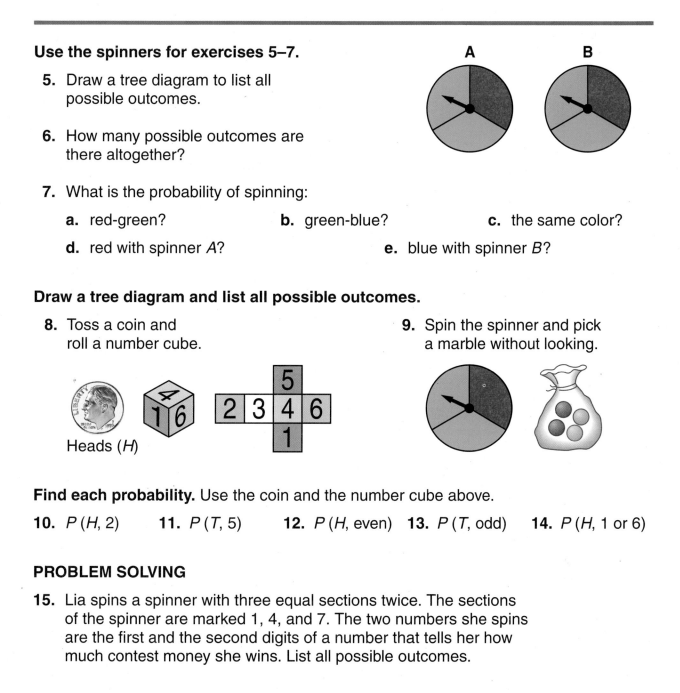

**5.** Draw a tree diagram to list all possible outcomes.

**6.** How many possible outcomes are there altogether?

**7.** What is the probability of spinning:

   **a.** red-green?       **b.** green-blue?       **c.** the same color?

   **d.** red with spinner *A*?       **e.** blue with spinner *B*?

**Draw a tree diagram and list all possible outcomes.**

**8.** Toss a coin and roll a number cube.

Heads (*H*)

**9.** Spin the spinner and pick a marble without looking.

**Find each probability.** Use the coin and the number cube above.

**10.** *P* (*H*, 2)    **11.** *P* (*T*, 5)    **12.** *P* (*H*, even)   **13.** *P* (*T*, odd)    **14.** *P* (*H*, 1 or 6)

**PROBLEM SOLVING**

**15.** Lia spins a spinner with three equal sections twice. The sections of the spinner are marked 1, 4, and 7. The two numbers she spins are the first and the second digits of a number that tells her how much contest money she wins. List all possible outcomes.

**Critical Thinking**

Communicate

**16.** How is a tree diagram like an organized list of possible outcomes?

**17.** Describe how you can use a tree diagram to find the probability of more than one event.

## 7-3 | Independent and Dependent Events

A bag contains 3 cubes: 1 yellow, 1 red, and 1 blue. Pick 2 cubes, one at a time, from the bag without looking. What is the probability of picking a blue and then a red?

You can use tree diagrams to list all possible outcomes for experiments involving more than one event.

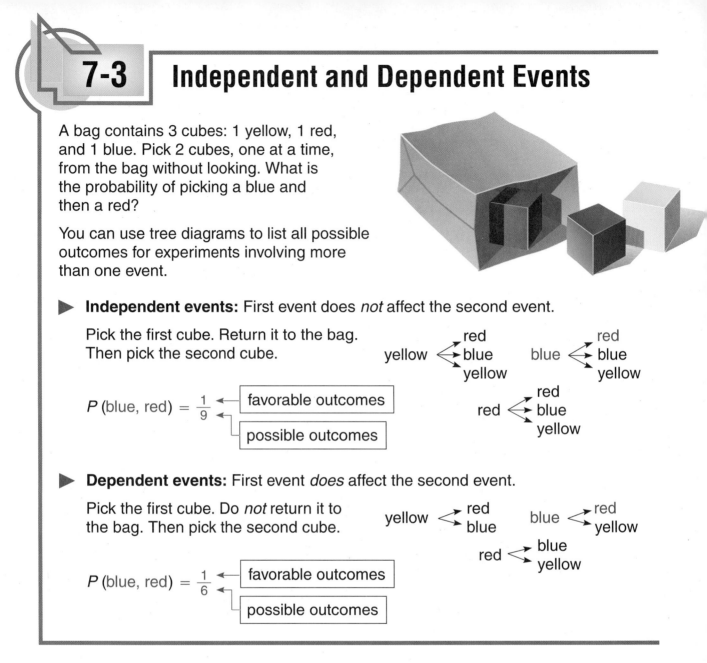

▶ **Independent events:** First event does *not* affect the second event.

Pick the first cube. Return it to the bag. Then pick the second cube.

yellow ⟨ red / blue / yellow

blue ⟨ red / blue / yellow

red ⟨ red / blue / yellow

$P\text{(blue, red)} = \frac{1}{9}$ ← favorable outcomes
← possible outcomes

▶ **Dependent events:** First event *does* affect the second event.

Pick the first cube. Do *not* return it to the bag. Then pick the second cube.

yellow ⟨ red / blue

blue ⟨ red / yellow

red ⟨ blue / yellow

$P\text{(blue, red)} = \frac{1}{6}$ ← favorable outcomes
← possible outcomes

---

### Draw a tree diagram and list all possible outcomes.

1. A bag contains 4 cubes: 2 orange and 2 blue.

   **a.** Pick a cube from the bag at random and put it back. Then pick another cube.

   **b.** Pick a cube from the bag at random and do *not* put it back. Then pick another cube.

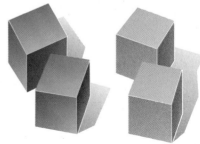

2. A purse contains 5 coins: 2 dimes and 3 nickels. Pick one coin from the purse at random and, without replacing it, pick another coin.

**Draw a tree diagram for the random experiment.
Then find the probability.**

An envelope contains 4 cards marked *A, B, C, D*.

Pick a card and put it back. Then choose another card.

**3.** $P(A, B)$       **4.** $P(C, not\ D)$       **5.** $P(not\ B, D)$

Pick a card and do *not* put it back. Then choose
another card.

**6.** $P(A, B)$       **7.** $P(C, not\ D)$       **8.** $P(not\ B, D)$

## PROBLEM SOLVING

Ben has a bag of fruit: 3 pieces are bananas
and 2 pieces are apples. He ate 2 pieces of
fruit while waiting for the school bus, selecting
them at random one after another.

**9.** Draw a tree diagram and list all possible
outcomes showing which fruit was eaten.

**10.** Find the probability that:

   **a.** both pieces of fruit were bananas.

   **b.** neither piece was a banana.

   **c.** the pieces were the same kind of fruit.

   **d.** at least one piece of fruit was a banana.

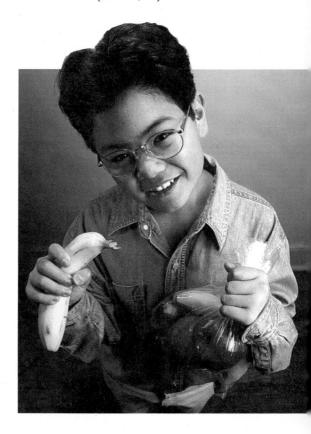

## Challenge

**11.** Laura has 2 quarters and 1 nickel in her pocket.
The pocket has a hole in it and a coin drops out.
She picks up the coin and puts it back into her
pocket. A few minutes later, a coin drops out of
her pocket again. What is the probability that the
two coins that fell have a total value of 30 cents?

# Finding Averages

What was the average number of pounds of newspaper Craig collected each week?

**Craig's Collection of Newspaper for School Drive**

| Week | 1 | 2 | 3 | 4 | 5 | 6 |
|------|---|---|---|---|---|---|
| Number of Pounds | 18 | 23 | 14 | 12 | 27 | 27 |

▶ To find the **average (mean)** of a set of numbers:

- Add the numbers.

$$18 + 23 + 14 + 12 + 27 + 27 = 121$$

- Divide the sum by the number of addends.

$$20 \text{ R1} = 20\tfrac{1}{6} \leftarrow \boxed{\text{average}}$$

$$\text{number of addends} \rightarrow 6\overline{)121} \leftarrow \boxed{\text{sum of addends}}$$

- Check.

$$20\tfrac{1}{6} \times 6 = \frac{121}{\cancel{6}} \times \cancel{6}^{1} = \frac{121}{1} = 121$$

$$\underbrace{20\tfrac{1}{6}}_{\text{average}} \quad \underbrace{\times 6}_{\text{number of addends}} \quad \underbrace{= 121}_{\text{sum of addends}}$$

Craig collected an average of $20\tfrac{1}{6}$ pounds of newspaper each week.

▶ You can use a calculator to find the average of a set of numbers.

Find the average of 87, 83, 90, and 96.

Enter: 87 $\boxed{+}$ 83 $\boxed{+}$ 90 $\boxed{+}$ 96 $\boxed{=}$ $\boxed{\div}$ 4 $\boxed{=}$

Display: $\boxed{\quad 89. \quad}$

The average is 89.

---

**Find the average and check.**

**1.** 82, 85, 85

**2.** 47, 55, 51

**3.** 88, 96, 93, 95

**4.** 36, 39, 32, 45

**5.** 22, 19, 27, 29, 23

**6.** 52, 48, 54, 60, 58

**Use a calculator to find the average for each set of data.**

7.
| Stadium | Capacity |
|---|---|
| Candlestick Park | 66,455 |
| Superdome | 69,065 |
| Georgia Dome | 71,594 |
| L.A. Memorial Coliseum | 92,488 |

8.
| Mountain | Altitude |
|---|---|
| Mt. McKinley | 20,320 ft |
| Mt. Whitney | 14,494 ft |
| Mt. Elbert | 14,433 ft |
| Mt. Rainier | 14,440 ft |

**PROBLEM SOLVING**    Use mental math, paper and pencil, or a calculator. Explain the method you used.

*Computation Method*

9. Jewel scored 82, 81, 85, 92, and 75 in her last five golf games. What was her average score?

10. Tom's exam grades in science were 86, 73, 75, 94, 82, and 70. What was his mean grade?

11. In her first seven basketball games, Angela scored the following number of points: 6, 10, 15, 7, 12, 8, and 5. What was her scoring average?

12. Todd is a running back for his team. In four games he ran 8 yd, 6 yd, 8 yd, and then 10 yd. What was his average yardage per game?

13. Over a period of six months Mark made the following deposits in his savings account: $35, $26, $75, $63, $58, and $25. What was his average monthly deposit?

14. There were eight professional wrestlers in the match. Their weights were: 278 lb, 315 lb, 265 lb, 371 lb, 303 lb, 296 lb, 237 lb, and 359 lb. What was their average weight?

15. In three years the school bus traveled 97,615 miles, 117,341 miles, and 105,198 miles. What was its average mileage per year?

16. In a week-long "Question Contest," 47 contestants answered 14,335 questions. On the average, how many questions did each contestant answer?

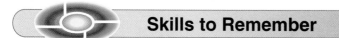

## Skills to Remember

**Order from least to greatest.**

17. 22, 25, 24, 20, 23

18. 16, 19, 20, 14, 18

19. 82, 79, 60, 88, 90

20. 234, 185, 217, 183, 247

## 7-5 Collecting and Organizing Data

Pilar asked the students in her class to name their favorite fish.

► She *collected* the *data*. First she made a list of fish. Then she asked each student in her class to pick his/her favorite fish from the list.

► She *recorded* the *data* in a tally chart or a **frequency table**. She used **tally marks** to record how many students chose each fish. Then she found each total.

► She *organized* the *data* in a table. She arranged the data in order from the greatest number of students to the least number of students.

Which kind of fish was favored by the greatest number of students? the least number?

Organizing data in a table from greatest to least makes it easier to find and compare data.

Angelfish was favored by the greatest number of students. Mollie was favored by the least number of students.

**Favorite Fish**

| Kind of Fish | Tally | Total |
|---|---|---|
| tetra | //// | 4 |
| angelfish | //// //// // | 12 |
| mollie | /// | 3 |
| guppy | //// / | 6 |

/ = 1 and //// = 5

**Favorite Fish**

| Kind | Number |
|---|---|
| angelfish | 12 |
| guppy | 6 |
| tetra | 4 |
| mollie | 3 |

**Copy and complete the frequency table. Then use the completed table for problems 5–7.**

**Fifth Grade Students' Favorite Place To Visit Someday**

| | Place | Tally | Total |
|---|---|---|---|
| 1. | Europe | //// //// /// | ? |
| 2. | Africa | //// // | ? |
| 3. | Caribbean | ? | 10 |
| 4. | Asia | ? | 5 |

5. Organize the data from least to greatest in a table.

6. What was the students' most preferred place? second most preferred place?

7. How many fewer students chose Asia than the Caribbean?

**Make a frequency table.**

8. Mr. O'Donnell asked his students to choose their favorite place to visit from a list of places. The responses are listed below.

| museum | theater | park | zoo | park |
| park | museum | gym | park | gym |
| gym | gym | circus | museum | park |
| theater | park | park | gym | circus |
| park | theater | theater | park | gym |
| zoo | park | zoo | gym | park |

**Use the data in exercise 8.**

9. Organize the data from least to greatest in a table.

10. How many students were asked by Mr. O'Donnell?

11. Which place was favored by the most number of students? by the least number of students?

12. Which place was chosen by 7 students?

13. Explain in your Math Journal:

- What kind of information a frequency table gives you and how this information is useful.

- When you think a frequency table would be more useful than a table that lists only information.

**Finding Together**

*Math Journal*

*Discuss*

**Ask each student in your class how many brothers and/or sisters he/she has.**

14. Record your data in a frequency table.

15. Organize the data from least to greatest in a table.

16. How many students in the class have brothers? How many have sisters?

17. How many students in the class have neither brothers nor sisters?

18. What is the greatest number of brothers or sisters a student has?

19. What other information about each student's brothers and/or sisters could you record?

# 7-6 Working with Data

Romulo has kept a record of the number of points he scored in his first 7 basketball games of the season. Now he is going to *interpret* his scores.

| Game | 1 | 2 | 3 | 4 | 5 | 6 | 7 |
|---|---|---|---|---|---|---|---|
| Score | 6 | 10 | 14 | 8 | 10 | 8 | 7 |

## Hands-On Understanding

**Materials Needed:** connecting cubes

**Step 1** Build stacks with connecting cubes to model Romulo's score in each game.

**Step 2** Compare the tallest and shortest stacks. The **range** of the set is the difference between the number of cubes in the tallest and the number of cubes in the shortest stacks.

How would you calculate the range?
What is the range of Romulo's set of scores?

**Step 3** Arrange the stacks of cubes in order from the shortest to the tallest. The number of cubes in the middle stack is the **median** of the set. If there are two middle stacks, divide the total number of cubes in the two stacks by 2 to find the median.

What is the median of Romulo's set of scores?

**Step 4** Look for stacks that have the same number of cubes. The number that occurs most often is the **mode** of the set. There may be one mode, more than one mode, or none at all.

What are the modes of Romulo's set of scores?

**Step 5** Make 7 even stacks using all the cubes you used to model Romulo's scores. Then count the number of cubes in each stack. The number of cubes in each stack is the **mean** or average of the set.

What is the mean of Romulo's set of scores?
How would you calculate the mean?

**Use connecting cubes to model each set of data. Then find the range, median, mode, and mean for each set of data.**

1. 7, 11, 13, 9
2. 6, 10, 4, 5, 10
3. 15, 11, 13, 6, 9, 6

4. 11, 10, 14, 10, 9, 6
5. 7, 12, 8, 5, 6, 10, 15
6. 5, 9, 11, 8, 10, 7, 6

**Find the range, median, mode, and mean for each set of data.**

7. 39, 31, 39, 27
8. 96, 88, 81, 80, 85

9. 90, 60, 85, 75, 65, 100, 85
10. 31, 59, 73, 96, 30, 96, 118

11. **Rolly's Work Hours**

| Week 1 | Week 2 | Week 3 | Week 4 | Week 5 | Week 6 |
|--------|--------|--------|--------|--------|--------|
| 12 | 17 | 15 | 17 | 12 | 17 |

12. **Cougar Team Scores**

| Game 1 | Game 2 | Game 3 | Game 4 | Game 5 | Game 6 |
|--------|--------|--------|--------|--------|--------|
| 34 | 46 | 54 | 61 | 60 | 67 |

## Communicate

Discuss ✓

13. In what ways are the median and the mode different?

14. In what ways do the median and the mode differ from the mean? from the range?

15. Give a set of data where the mode, median, and mean are all the same.

## Critical Thinking

16. Change one number in the set of data: 6, 4, 8, 9, 6, 5, 8, 7, and 10, so that the range will be 5.

17. Change one number in the set of data: 18, 16, 15, 16, 19, and 21, so that the mode will be 15.

18. Add one number to the set of data: 78, 82, 81, 76, 78, 85, so that the mean will be 80.

19. Add one number to the set of data: 88, 96, 88, 80, and 76, so that the median will be 86.

# Line Plots

Nick's test scores in math are: 100, 90, 70, 80, 75, 80, 95, 90, 90, and 100. He records his scores in a frequency table and then organizes the data in a **line plot**.

### Nick's Math Test Scores

| Score | 100 | 95 | 90 | 85 | 80 | 75 | 70 |
|---|---|---|---|---|---|---|---|
| Tally | // | / | /// |  | // | / | / |
| Frequency | 2 | 1 | 3 | 0 | 2 | 1 | 1 |

 Hands-On Understanding

**Materials Needed:** ruler, paper, colored pencils

**Step 1** Use the data from the table to choose an appropriate scale for the graph.

What is the least score in the data? the greatest score?
What scale would be appropriate for the graph?
What intervals on the scale would you use?

**Step 2** Draw a line and divide it into equal intervals. Label the scale on the line. Start with the least score.

How many equal intervals on the line did you label?

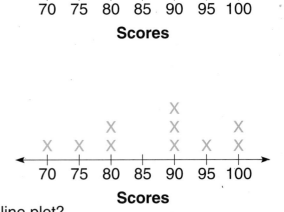

**Step 3** Use an X to represent each score in the data. Mark the corresponding points of all the scores in the data above the line.

How many Xs did you mark on the line plot?

The line plot shows the range and mode of the data.

**1.** What is the range of Nick's test scores? the mode?

The line plot also shows where the data *clusters* (or groups).

2. Around which score do Nick's test scores seem to cluster?

3. How many test scores are 80 or better?

**Use the line plot at the right for problems 4–7.**

4. How many heights are in the data?

5. What is the range of the data? the mode?

6. Around which height do the data seem to cluster?

```
                              X
              X               X
              X           X   X
          X   X   X   X   X   X   X           X
X         X   X   X   X   X   X   X   X
+---+---+---+---+---+---+---+---+---+--->
145 146 147 148 149 150 151 152 153
```

**Heights in cm**

7. How many heights are greater than 150 cm?

**Draw a line plot for each set of data. Then find the range and mode.**

8. Elsa's science test scores:
   100, 93, 93, 96, 89, 89, 89, 96, 94, 95, 88, 92, 91, 89, 88

9. Matt's monthly deposits:
   $25, $30, $27, $30, $29, $28, $29, $26, $29, $32, $27, $29

10. Noontime temperatures:
    65°F, 70°F, 83°F, 78°F, 60°F, 64°F, 72°F

11. Len's bowling scores:
    93, 103, 90, 106, 95, 95, 97, 98, 100, 94

## Communicate

*Discuss* ✓

12. Why is it easy to find the mode and range of a set of data in a line plot?

13. How is a line plot like a bar graph? How is it different?

14. Can you find the median of a set of data in a line plot? Explain your answer.

15. Can you find the mean of a set of data in a line plot? Explain your answer.

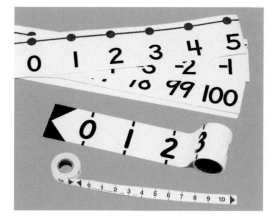

251

# 7-8 Working with Graphs

Graphs are pictures of data. The data are presented in an organized way so that the information being illustrated may be quickly understood.

▶ A **bar graph** presents data so that comparisons of *different* items can be made. It uses bars of different lengths. The length of each bar is proportional to the number the bar represents. The *scale* on the bar graph is divided into equal intervals.

▶ A **line graph** presents data on one item so that changes and trends over time can be identified and comparisons can be made. It uses points and line segments on a grid. The *scale* on the line graph is divided into equal intervals.

▶ A **pictograph** takes the form of a bar graph. It presents data using pictures or symbols. Each picture or symbol represents an assigned amount of data. The *key* for a pictograph tells the number that each picture or symbol represents.

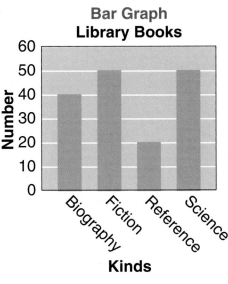

**Bar Graph
Library Books**

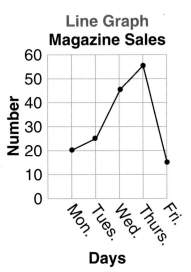

**Line Graph
Magazine Sales**

### Pictograph
**Newspaper Drive**

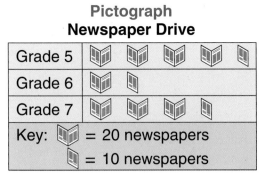

| Grade 5 | |
| --- | --- |
| Grade 6 | |
| Grade 7 | |
| Key: | = 20 newspapers |
| | = 10 newspapers |

▶ A **circle graph** presents the division of a total amount of data. It uses the area of a circle to show how a whole is divided into fractional parts.

**Circle Graph
Vanya's Magazine Collection**

News 6, Fashion 8, Home 3, Sports 5, Hobby 2

**Use the graphs on page 252.**

1. Which graphs use scales divided into equal intervals?

2. Which graphs use vertical and horizontal axes?

3. Which graph shows parts of a whole?

4. Which graph has a key?

5. Which graph compares four items of data?

6. Which graph compares three items of data?

7. How many library books does each unit on the vertical scale represent?

8. Of which kinds of book are there the greatest number? What is the number?

9. Between which two days was the increase in magazine sales the greatest?

10. Between which two days did the magazine sales decrease?

11. How many newspapers does the symbol  represent?

12. What is the number of newspapers collected by each grade?

13. How many magazines are in Vanya's collection?

14. How many home magazines are in Vanya's collection?

15. What fractional part of Vanya's magazine collection is on sports?

16. What fractional part of Vanya's magazine collection is *not* on fashion?

17. Explain in your Math Journal:

    • What advantages a graph has over a table of numerical data.

    • When a bar graph or pictograph is more suitable to use than another type of graph.

*Math Journal*

**Project**

*Communicate*

18. Find examples of bar graphs, pictographs, line graphs, and circle graphs in newspapers or magazines. Choose 2 different examples of each type of graph. Then write 3 questions that can be answered using the information shown in each graph.

# 7-9 | Making Line Graphs

Mr. Moreno organized the ticket sales data for the school play in a **line graph**.

| Monroe School Play Ticket Sales | | | | | | |
|---|---|---|---|---|---|---|
| **Day** | 1 | 2 | 3 | 4 | 5 | 6 |
| **Tickets Sold** | 352 | 453 | 554 | 396 | 503 | 548 |

▶ To make a line graph:

- Use the data from the table to choose an appropriate scale.

- Draw and label the scale on the vertical axis. Start at 0.

- Draw and label the horizontal axis. List the name of each item.

- Locate the points on the graph.

- Connect the points with line segments.

- Write the title of the line graph.

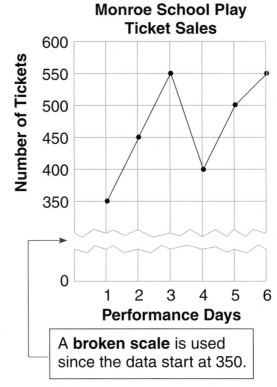

A **broken scale** is used since the data start at 350.

---

**Use the line graph above for problems 1–5.**

1. What data is shown on the horizontal axis?

2. What scale was used for the vertical axis?

3. Which day showed the greatest change in the number of tickets sold? Was the change an increase or decrease?

4. On which day did the play have the least number of tickets sold? the greatest number?

5. About what was the average number of tickets sold each day?

**Copy and complete the line graph.** Use the table.

6.

| Jimenez's Math Test Grades ||
|---|---|
| Test | Grade |
| 1 | 75 |
| 2 | 80 |
| 3 | 100 |
| 4 | 95 |
| 5 | 90 |
| 6 | 95 |

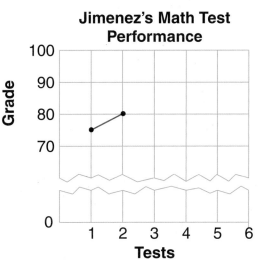

Jimenez's Math Test Performance

**Use the completed line graph.**

7. On which test did Jimenez score the lowest grade? the highest grade? Find the range of his test grades.

8. Between which two tests did Jimenez achieve the largest difference in test grades? What was the difference?

9. On which tests did Jimenez have the same test grade? Find the mode of his test grades.

10. What was the mean of Jimenez's math test grades?

**Make a line graph for each set of data.**

11.

| Booster Club Membership |||||||||
|---|---|---|---|---|---|---|---|---|
| Year | 1989 | 1990 | 1991 | 1992 | 1993 | 1994 | 1995 | 1996 | 1997 |
| Number | 30 | 25 | 40 | 55 | 60 | 70 | 65 | 75 | 80 |

12.

| Soda Machine Profits |||||||
|---|---|---|---|---|---|---|
| Month | Sept. | Oct. | Nov. | Dec. | Jan. | Feb. | Mar. |
| Amount | $16.25 | $17.50 | $15.00 | $10.25 | $12.00 | $14.50 | $15.75 |

 **Connections: Science**

13. Use your own thermometer, listen to weather reports on radio or TV, or read the newspaper. Record the daily high and low temperatures for a week.

14. Use your data to make a line graph that shows the daily high temperatures and low temperatures for the week.

# 7-10 Interpreting Circle Graphs

Mr. Sweeney asked the students in his class to name their favorite kind of videotape.

The **circle graph** at the right shows the data.

How many students are in Mr. Sweeney's class?

▶ To find how many, add the numbers in the sections of the graph.

$$6 + 3 + 5 + 2 + 8 = 24$$

There are 24 students in Mr. Sweeney's class.

**The Favorite Videotapes of Mr. Sweeney's Class**

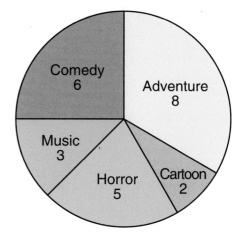

What fractional part of the class chose comedy as its favorite?

▶ To find what fractional part :

- Write the fraction with the number of students who like comedy as the numerator and the number of students in class as the denominator.

- Write the fraction in simplest form.

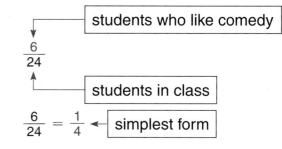

students who like comedy

$\frac{6}{24}$

students in class

$\frac{6}{24} = \frac{1}{4}$ ◀ simplest form

One fourth of the class chose comedy as its favorite.

---

**Use the circle graph above.**

1. What fractional part of Mr. Sweeney's class prefers each kind of videotape:

   **a.** music?     **b.** horror?     **c.** cartoon?     **d.** adventure?

2. How many students chose music or horror videotapes as their favorite? What part of the class do they represent?

3. How many students did *not* choose adventure videotapes as their favorite? What part of the class do they represent?

256

**Use the circle graph at the right.**

**4.** What fractional part of Eva's class chose soccer or baseball as its favorite?

**5.** What fractional part of Eva's class chose football or basketball as its favorite?

**6.** What fractional part of Eva's class did *not* choose tennis as its favorite?

**The Favorite Sports of Eva's Class**

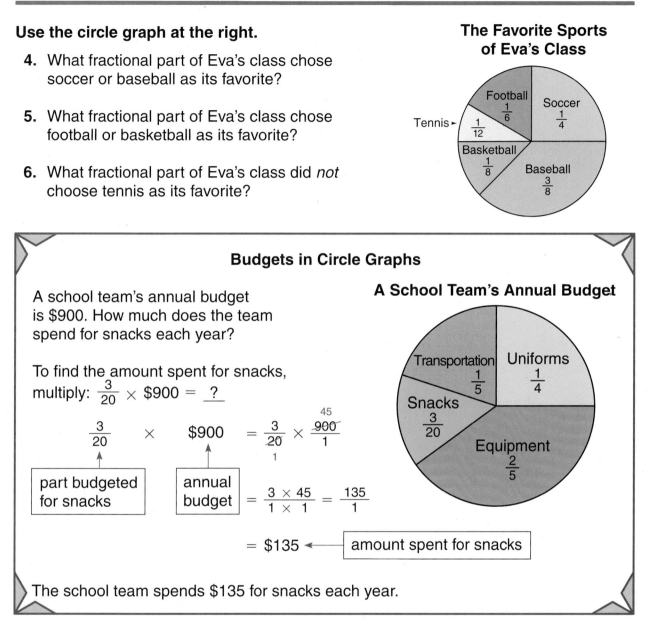

---

**Budgets in Circle Graphs**

A school team's annual budget is $900. How much does the team spend for snacks each year?

To find the amount spent for snacks, multiply: $\frac{3}{20} \times \$900 = \underline{\ ?\ }$

$$\frac{3}{20} \times \$900 = \frac{3}{\underset{1}{\cancel{20}}} \times \frac{\overset{45}{\cancel{900}}}{1}$$

part budgeted for snacks

annual budget

$$= \frac{3 \times 45}{1 \times 1} = \frac{135}{1}$$

$$= \$135 \longleftarrow \boxed{\text{amount spent for snacks}}$$

**A School Team's Annual Budget**

The school team spends $135 for snacks each year.

---

**Use the circle graph above.**

**7.** How much does the school team spend each year for transportation? for equipment?

**8.** How much more money is spent for uniforms than transportation?

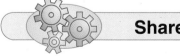

**Share Your Thinking**

Communicate ✓

**9.** Explain what makes the data in a circle graph easy to read and how a circle graph can be useful.

## 7-11  Problem Solving: Use a Model/Diagram

**Problem:**  Half of a class of 24 students have no pets. Four students have only dogs as pets, and five have only cats. The rest of the class have both a cat and a dog. How many students have both a cat and a dog?

**1 IMAGINE**  Picture yourself in the problem.

**2 NAME**

*Facts:*  class of 24 students
Half of the class have no pets.
4 students—dogs
5 students—cats

*Question:*  How many students have both?

**3 THINK**  You can solve this problem by using a Venn diagram. Use circles to represent the groups: no pets, dogs, cats.

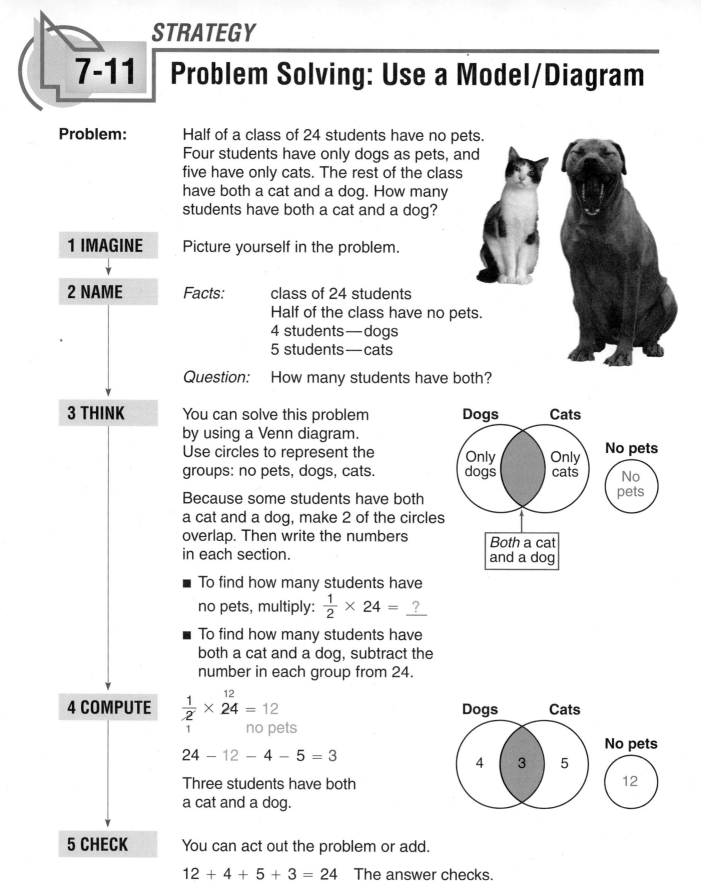

Because some students have both a cat and a dog, make 2 of the circles overlap. Then write the numbers in each section.

- To find how many students have no pets, multiply: $\frac{1}{2} \times 24 = \underline{\ ?\ }$

- To find how many students have both a cat and a dog, subtract the number in each group from 24.

**4 COMPUTE**

$$\frac{1}{\cancel{2}_1} \times \cancel{24}^{12} = 12 \quad \text{no pets}$$

$$24 - 12 - 4 - 5 = 3$$

Three students have both a cat and a dog.

**5 CHECK**  You can act out the problem or add.

$12 + 4 + 5 + 3 = 24$  The answer checks.

**Use a model/diagram to solve each problem.**

1. Dee can join 1 art class from each area: painting and crafts. There are 3 painting classes and 3 crafts classes. What are all the possible combinations of classes she can join?

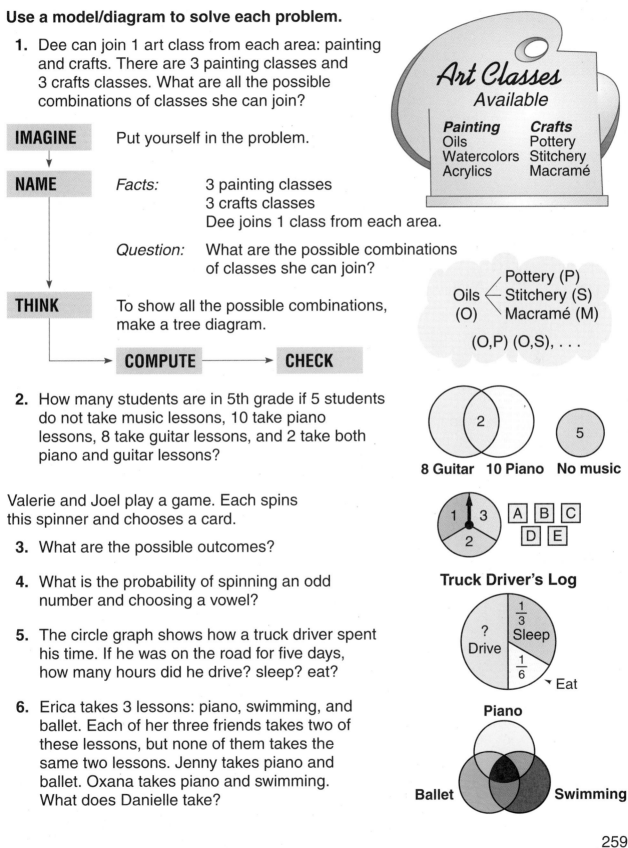

**IMAGINE**    Put yourself in the problem.

**NAME**

*Facts:*    3 painting classes
3 crafts classes
Dee joins 1 class from each area.

*Question:*    What are the possible combinations of classes she can join?

**THINK**    To show all the possible combinations, make a tree diagram.

**COMPUTE** ⟶ **CHECK**

**Art Classes**
*Available*

| Painting | Crafts |
|----------|--------|
| Oils | Pottery |
| Watercolors | Stitchery |
| Acrylics | Macramé |

Oils (O) ← Pottery (P)
Stitchery (S)
Macramé (M)

(O,P) (O,S), . . .

2. How many students are in 5th grade if 5 students do not take music lessons, 10 take piano lessons, 8 take guitar lessons, and 2 take both piano and guitar lessons?

**8 Guitar**    **10 Piano**    **No music**

Valerie and Joel play a game. Each spins this spinner and chooses a card.

3. What are the possible outcomes?

4. What is the probability of spinning an odd number and choosing a vowel?

5. The circle graph shows how a truck driver spent his time. If he was on the road for five days, how many hours did he drive? sleep? eat?

6. Erica takes 3 lessons: piano, swimming, and ballet. Each of her three friends takes two of these lessons, but none of them takes the same two lessons. Jenny takes piano and ballet. Oxana takes piano and swimming. What does Danielle take?

**Truck Driver's Log**

**Piano**

**Ballet**    **Swimming**

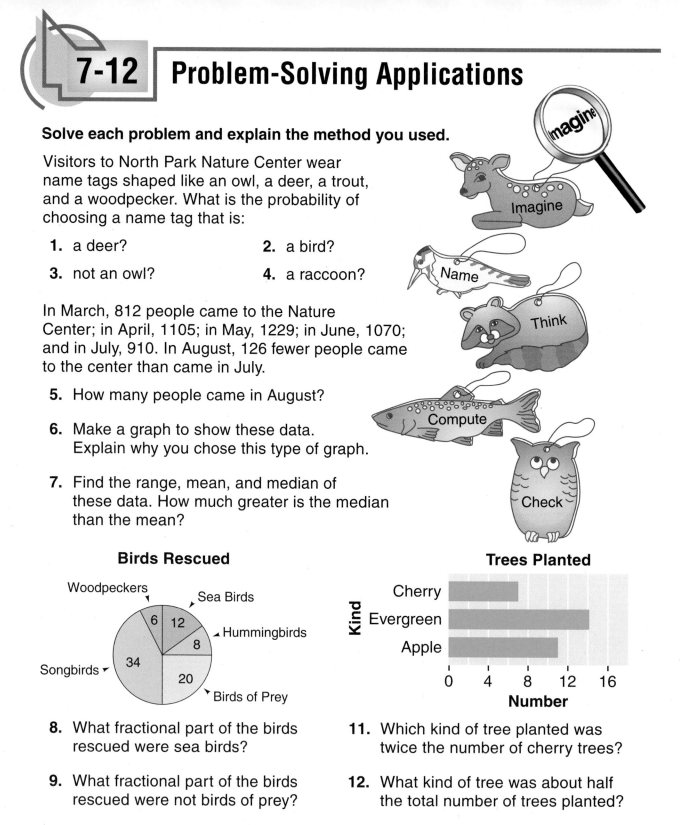

## 7-12 Problem-Solving Applications

**Solve each problem and explain the method you used.**

Visitors to North Park Nature Center wear name tags shaped like an owl, a deer, a trout, and a woodpecker. What is the probability of choosing a name tag that is:

1. a deer?
2. a bird?
3. not an owl?
4. a raccoon?

*Imagine*

*Name*

*Think*

In March, 812 people came to the Nature Center; in April, 1105; in May, 1229; in June, 1070; and in July, 910. In August, 126 fewer people came to the center than came in July.

5. How many people came in August?

6. Make a graph to show these data. Explain why you chose this type of graph.

*Compute*

7. Find the range, mean, and median of these data. How much greater is the median than the mean?

*Check*

### Birds Rescued

Woodpeckers
Sea Birds
6    12
Hummingbirds
8
Songbirds
34
20
Birds of Prey

### Trees Planted

Kind

Cherry
Evergreen
Apple

0   4   8   12   16
**Number**

8. What fractional part of the birds rescued were sea birds?

9. What fractional part of the birds rescued were not birds of prey?

10. Suppose the center budgeted $2400 to rescue birds. How much was spent to rescue hummingbirds?

11. Which kind of tree planted was twice the number of cherry trees?

12. What kind of tree was about half the total number of trees planted?

13. What fractional part of the planted trees produces fruit?

**Use a strategy from the list or use another strategy you know to solve each problem.**

USE THESE STRATEGIES
Use a Model/Diagram
Organized List
Working Backwards
Use Simpler Numbers
Logical Reasoning

14. At the Center there are more squirrels than raccoons and more rabbits than squirrels. Are there more rabbits or raccoons?

15. North Park Center covers $289\frac{1}{4}$ acres. Central Park Center covers $193\frac{1}{8}$ acres. How much smaller is this than North Park Center?

16. The North Park Center sells white, blue, or green shirts in 5 sizes: S, M, L, XL, and XXL. Pictured on each shirt is either an eagle or an owl. How many different kinds of shirts does it sell?

17. In a 5-day period a worker spends $4\frac{1}{2}$ h, $3\frac{1}{4}$ h, $5\frac{1}{8}$ h, $3\frac{3}{8}$ h, and $3\frac{1}{4}$ h pruning trees. What is the average amount of time the worker spends pruning each day?

18. In May, 18 birds' eggs hatched in the Center's incubator. This is $1\frac{1}{2}$ times the number that hatched in April. How many eggs hatched in April?

19. Yesterday 56 people came to the Center. How many people came to the Center to hike if 30 people took classes, 22 went bird-watching, and 12 people did both?

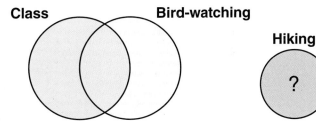

Class        Bird-watching

Hiking

?

20. Two thirds of the visitors on Monday were children. Three fourths of the children came on a school trip. The rest, 21 children, came with their families. How many people visited the Center on Monday?

 **Make Up Your Own**

*Discuss*

21. Invent data about the Nature Center. Then create a graph to show your data. Write a problem that a classmate can solve using your invented data.

261

# Chapter Review and Practice

**Use the number cube to find the probability of each event.** *(See pp. 238–239.)*

**1.** $P(1)$     **2.** $P(2 \text{ or } 3)$     **3.** $P(< 4)$     **4.** $P(5)$

**Draw a tree diagram. List all possible outcomes.** *(See pp. 240–243.)*

**5.** Toss a coin and spin the spinner.

**6.** A bag contains 4 cubes: 3 orange and 1 purple. Pick a cube at random, put it back, and then pick another cube.

**Find each probability.** Use the experiments in exercises 5 and 6.

**7.** $P(T, not \text{ green})$     **8.** $P(H, \text{ red or yellow})$     **9.** $P(\text{orange}, not \text{ purple})$

**Make a frequency table.** *(See pp. 246–247.)*

**10.** Each student in Elsa's class was asked to choose his/her afterschool activity from a list of afterschool activities. The responses are listed below.

| | | | | | | |
|---|---|---|---|---|---|---|
| club | sports | club | club | sports | club | club |
| tutoring | club | sports | sports | club | tutoring | sports |
| sports | club | club | club | sports | club | tutoring |

**Find the range, median, mean, and mode for each set of data.** *(See pp. 244–245, 248–249.)*

**11.  Center School's Enrollment**

| Grade | 4 | 5 | 6 | 7 | 8 |
|---|---|---|---|---|---|
| Number of Students | 68 | 72 | 68 | 54 | 49 |

**12.     Walter's Winning Matches in the Chess Tournament**

| Month | Oct. | Nov. | Dec. | Jan. | Feb. | Mar. |
|---|---|---|---|---|---|---|
| Matches Won | 9 | 8 | 4 | 6 | 8 | 7 |

## PROBLEM SOLVING
**Use the circle graph for problems 15–16.**

*(See pp. 250–257, 258–261.)*

**13.** Draw a line plot for the data in exercise 11.

**14.** Make a line graph for the data in exercise 12.

**15.** What part of the class has fish or cats as pets?

**16.** If there are 60 children in Grade 5, how many in all have pets?

**Kinds of Pets in Grade 5**

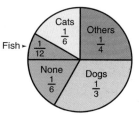

(See *Still More Practice*, p. 483.)

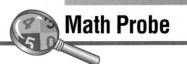

DOUBLE LINE AND DOUBLE BAR GRAPHS

A **double line graph** and a **double bar graph** are used to compare two sets of data. Each set of data is graphed separately, but on the same grid. The *key* identifies the sets of data.

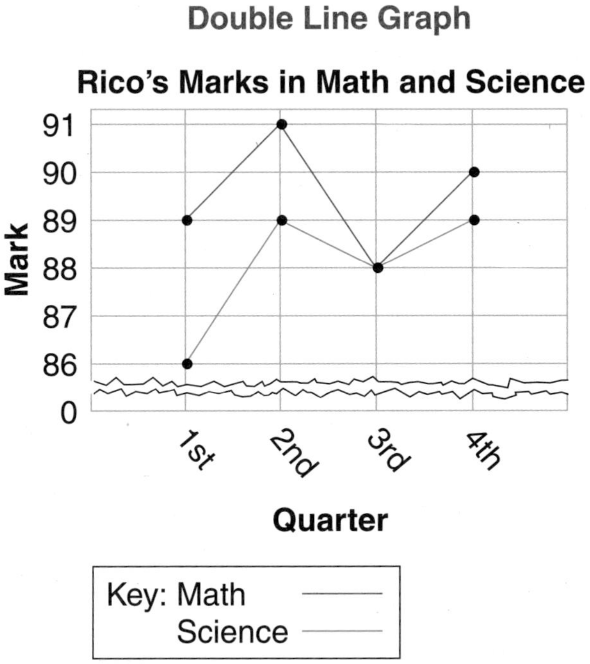

Double Line Graph
Rico's Marks in Math and Science
Mark
91
90
89
88
87
86
0
1st
2nd
3rd
4th
Quarter
Key: Math
Science


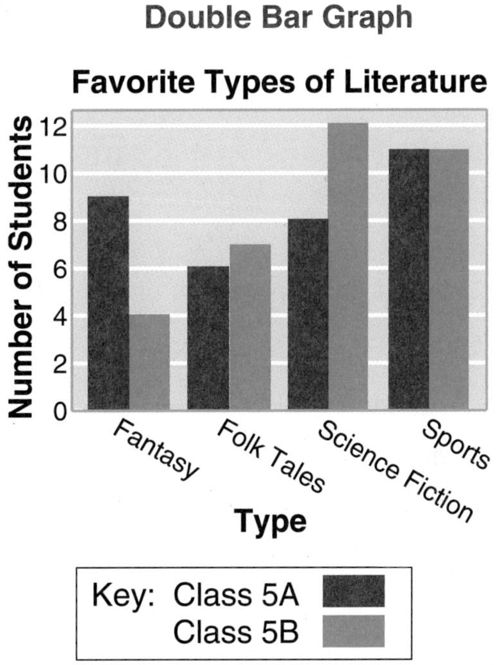

Double Bar Graph
Favorite Types of Literature
Number of Students
12
10
8
6
4
2
0
Fantasy
Folk Tales
Science Fiction
Sports
Type
Key: Class 5A
Class 5B


**PROBLEM SOLVING** Use the graphs above.

1. In which quarter did Rico get the highest math mark?

2. In which quarters did Rico get the same science mark?

3. In which quarter did Rico get the same mark in math and science?

4. In which quarters did Rico's marks increase from the previous quarter?

5. Which type of literature is the least preferred by class 5A? the most preferred?

6. Which type of literature is the most preferred by class 5B? the least preferred?

7. How many more students from class 5B than from class 5A prefer science fiction?

8. Which type of literature is preferred by equal numbers of students in classes 5A and 5B?

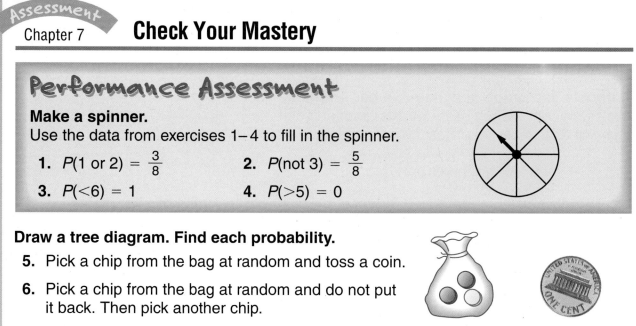

### Performance Assessment

**Make a spinner.**
Use the data from exercises 1–4 to fill in the spinner.

1. $P(1 \text{ or } 2) = \frac{3}{8}$

2. $P(\text{not } 3) = \frac{5}{8}$

3. $P(<6) = 1$

4. $P(>5) = 0$

**Draw a tree diagram. Find each probability.**

5. Pick a chip from the bag at random and toss a coin.

6. Pick a chip from the bag at random and do not put it back. Then pick another chip.

7. $P(\text{yellow}, H)$   8. $P(\text{not red}, T)$   9. $P(\text{red, yellow})$   10. $P(\text{red, not red})$

**Copy and complete the frequency table.**

| | **Number of Members of School Clubs** | | |
|---|---|---|---|
| | **Club** | **Tally** | **Total** |
| **11.** | Drama Club | ⵏ卌 卌 卌 // | ? |
| **12.** | Glee Club | 卌 卌 卌 卌 卌 卌 卌 | ? |
| **13.** | Math and Science Club | ? | 25 |
| **14.** | Debating Club | ? | 15 |

**Find the range, mean, median, and mode for each set of data.**

15.  **School Volunteers**

| Day | Mon. | Tues. | Wed. | Thurs. |
|---|---|---|---|---|
| Number of Students | 15 | 20 | 10 | 15 |

16.  **Bill's Science Test Grades**

| Test | 1 | 2 | 3 | 4 | 5 |
|---|---|---|---|---|---|
| Grade | 85 | 95 | 80 | 100 | 95 |

**PROBLEM SOLVING**   *Use a strategy you have learned.*
Use the circle graph for problems 19–20.

17. Draw a line plot for the data in exercise 16.

18. Make a line graph for the data in exercise 15.

19. What fraction of Verna's allowance is for food?

20. What fraction of Verna's allowance is for books?

**Verna's Allowance**

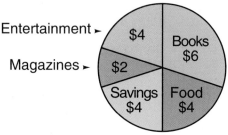

Entertainment ►  $4

Magazines ►  $2

Books $6

Savings $4

Food $4

# Cumulative Test I

**Choose the best answer.**

1. Which shows the standard form of seven billion, ninety-six million?

   **a.** 7,096,000   **b.** 796,000,000
   **c.** 7,096,000,000   **d.** 7,960,000,000

2. Round to the nearest ten cents.

   $4.19

   **a.** $4.00   **b.** $4.09
   **c.** $4.10   **d.** $4.20

3. Estimate.

   86 × $2.98

   **a.** $93.00
   **b.** $100.00
   **c.** $270.00
   **d.** $320.00

4.  2386
   × 453

   **a.** 1,080,858
   **b.** 2,612,118
   **c.** 8,216,014
   **d.** not given

5. Which group shows numbers that are each divisible by 5?

   **a.** 725,840; 1051; 12,750
   **b.** 360,730; 986; 1422
   **c.** 231,620; 814; 2351
   **d.** 2510; 313,155; 21,100

6. Compute. Use the order of operations.

   $47 - 6 + 2 \times 3$

   **a.** 31
   **b.** 47
   **c.** 74
   **d.** 129

7. Which shows the prime factorization of 24?

   **a.** $3 \times 8$
   **b.** $2 \times 6 \times 2$
   **c.** $2 \times 2 \times 2 \times 3$
   **d.** $2 \times 2 \times 3 \times 3$

8. Which is ordered from greatest to least?

   **a.** $\frac{3}{10}, \frac{4}{5}, \frac{7}{10}, \frac{1}{5}$

   **b.** $\frac{1}{24}, \frac{1}{12}, \frac{1}{6}, \frac{1}{2}$

   **c.** $1, \frac{5}{6}, \frac{1}{3}, \frac{1}{2}$

   **d.** none of these

9. $\frac{3}{11} + \frac{5}{11} + \frac{8}{11}$

   **a.** $\frac{16}{33}$   **b.** $1\frac{5}{16}$

   **c.** $1\frac{5}{11}$   **d.** $1\frac{6}{11}$

10. Estimate.

    $14\frac{9}{16} - 9\frac{1}{3}$

    **a.** 4   **b.** 6
    **c.** 15   **d.** 24

11. Choose the improper fraction for $4\frac{3}{5}$.

    **a.** $\frac{12}{5}$   **b.** $\frac{23}{5}$

    **c.** $\frac{20}{3}$   **d.** $\frac{23}{3}$

12. Choose the reciprocal of $2\frac{1}{4}$.

    **a.** $\frac{9}{4}$   **b.** $\frac{8}{9}$

    **c.** $\frac{7}{4}$   **d.** $\frac{4}{9}$

**Compute. Estimate to help you.**

**13.** $27\frac{1}{4}$
$+36\frac{7}{8}$

**14.** $5\frac{7}{8}$
$+12\frac{1}{2}$

**15.** $12\frac{2}{3}$
$-7\frac{1}{6}$

**16.** $47$
$-9\frac{4}{7}$

**17.** $46\frac{1}{5}$
$-19\frac{1}{3}$

**18.** $37 \times 531$

**19.** $48 \times \$81.87$

**20.** $8\overline{)9704}$

**21.** $73\overline{)\$372.30}$

**22.** $\frac{1}{4} \times \frac{2}{3}$

**23.** $4\frac{7}{10} \times 5$

**24.** $\frac{5}{12} \div \frac{1}{6}$

**25.** $\frac{4}{5} \div 4\frac{3}{4}$

**26.** $12 \div \frac{3}{4}$

**27.** $\frac{4}{7} \times 4\frac{2}{3}$

**28.** $7\frac{1}{2} \times 3\frac{3}{5}$

**29.** $2\frac{2}{5} \div 1\frac{1}{7}$

## PROBLEM SOLVING
**Solve each problem and explain the method you used.**
Use the circle graph for problems 33–35.

**30.** Juanita wants to fence in her aunt's triangular garden. The sides measure $4\frac{1}{2}$ feet, $6\frac{2}{3}$ feet, and $3\frac{1}{4}$ feet. About how much fencing must she buy?

**31.** Feng tosses a coin and rolls a 1−6 number cube. List all possible outcomes. What is $P(H, 4)$?  $P(T, 7)$?  $P(H, not\ 3)$?

**32.** Find the range, mean, median, and mode for the data in the table at the right. Then make a line graph.

**Daily Temperature (in °F)**

| Day | S | M | T | W | Th | F | S |
|---|---|---|---|---|---|---|---|
| **Temperature** | 29 | 21 | 31 | 33 | 29 | 30 | 26 |

**33.** How many more hours does Peter spend sleeping than playing? What fractional part of a day is the difference?

**34.** How many hours a day is Peter *not* in school? What fractional part of a day is this?

**35.** What fractional part of a day does Peter spend altogether at school or at play?

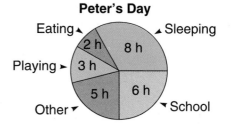

**Peter's Day**

Eating 2 h · Sleeping 8 h · Playing 3 h · Other 5 h · School 6 h

---

## For Rubric Scoring
**Listen for information on how your work will be scored.**

**36.** Each letter in the statements below represents one number in the box. Find out which fraction, whole number, or mixed number to use for each letter.

| $\frac{9}{16}$ | 1 | $1\frac{1}{2}$ | $\frac{3}{4}$ | $2\frac{1}{4}$ |

$$C - A = D \qquad D \times D = E \qquad B - D < E$$

# Geometry 8

One
day, the
triangle began to
feel dissatisfied. "I'm
tired of doing the
same old things," it grumbled.
"There must be more to life." So the
triangle went to see the local shapeshifter.

"How may I help you?" the shapeshifter asked
the triangle.

"I think if I had just one more side and one more
angle," said the triangle, "my life would be more interesting."

"That's easy to do," said the shapeshifter.

From *The Greedy Triangle* by *Marilyn Burns*

**In this chapter you will:**

Classify angles and polygons
Explore congruence, similarity,
  symmetry, and transformations
Use perimeter and circumference
  formulas
Learn about computer procedures in
  coordinate graphs
Solve problems using formulas

**Critical Thinking/Finding Together**

Six equilateral triangles are placed
together to form a hexagon. If the
perimeter of each equilateral triangle
is 20 cm, what is the perimeter of
the hexagon?

## 8-1 | Measuring and Drawing Angles

An **angle** is formed by two rays with a common endpoint. The rays are the **sides** of the angle. The common endpoint is the **vertex** of the angle.

**sides:** $\overrightarrow{DC}$, $\overrightarrow{DE}$     **name:**
**vertex:** $D$        $\angle D$ or $\angle CDE$ or $\angle EDC$

Angles are measured in **degrees (°)**. A **protractor** is used to measure an angle and to draw an angle to a given measure.

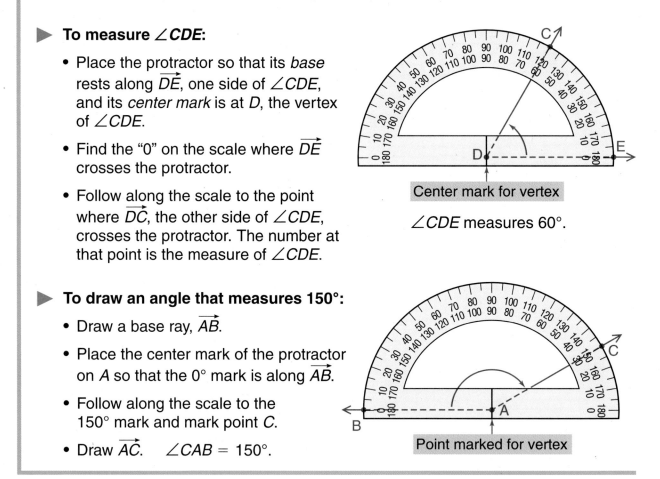

▶ **To measure $\angle CDE$:**

- Place the protractor so that its *base* rests along $\overrightarrow{DE}$, one side of $\angle CDE$, and its *center mark* is at $D$, the vertex of $\angle CDE$.

- Find the "0" on the scale where $\overrightarrow{DE}$ crosses the protractor.

- Follow along the scale to the point where $\overrightarrow{DC}$, the other side of $\angle CDE$, crosses the protractor. The number at that point is the measure of $\angle CDE$.

Center mark for vertex

$\angle CDE$ measures 60°.

▶ **To draw an angle that measures 150°:**

- Draw a base ray, $\overrightarrow{AB}$.

- Place the center mark of the protractor on $A$ so that the 0° mark is along $\overrightarrow{AB}$.

- Follow along the scale to the 150° mark and mark point $C$.

- Draw $\overrightarrow{AC}$.   $\angle CAB = 150°$.

Point marked for vertex

**Name the sides and the vertex of each angle.**

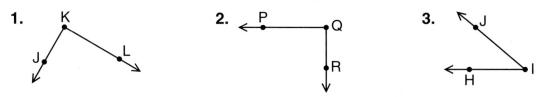

1.  K  J  L

2.  P  Q  R

3.  J  H  I

**Name the angle. Then use a protractor to find the measure.**

**4.**

R

S      T

**5.**

A

B    C

**6.**

X

Y    Z

**7.**

G

F    H

**8.**

X

M    D

**9.** A

C    E

**Use a protractor to draw each angle.**

**10.** 30°      **11.** 45°      **12.** 90°      **13.** 110°      **14.** 135°      **15.** 170°

**Estimate the measure of each angle. Then use a protractor to find the exact measure.**

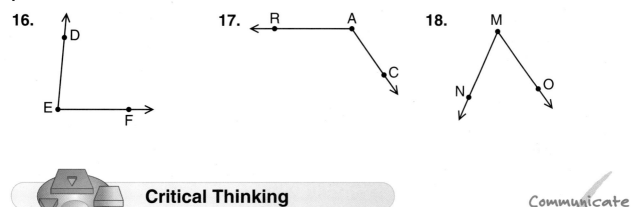

**16.**

D

E    F

**17.** R     A

C

**18.** M

N     O

## Critical Thinking

*Communicate* ✓

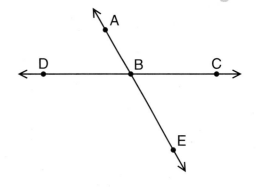

In the figure, $\overleftrightarrow{DC}$ and $\overleftrightarrow{AE}$ intersect at point *B*. Use a protractor to find the measure of each angle.

**19.** ∠*ABC*        **20.** ∠*CBE*

**21.** ∠*ABD*        **22.** ∠*DBE*

**23.** What do your results suggest about the measures of the angles formed by two intersecting lines?

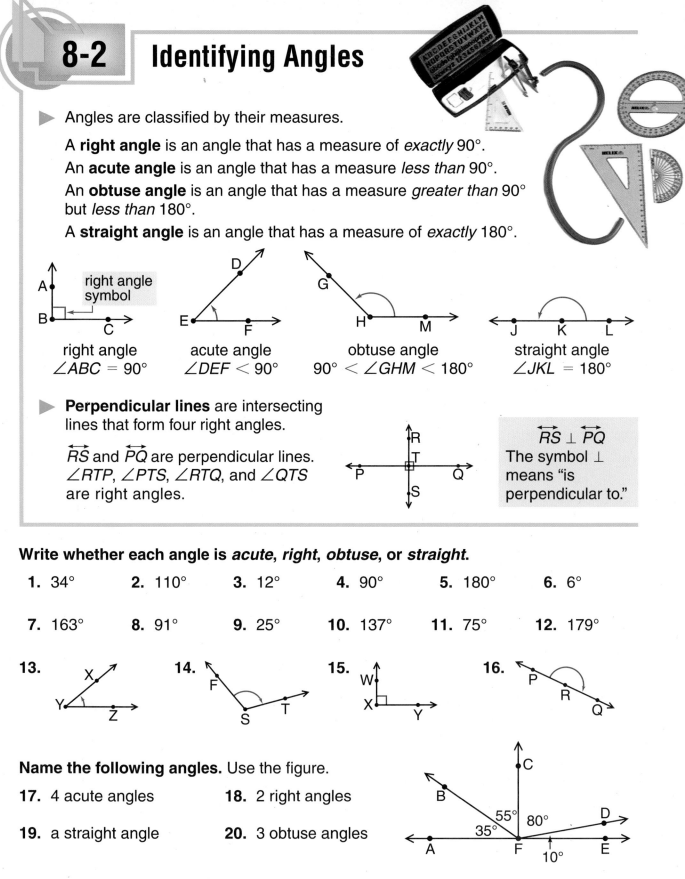

# 8-2 Identifying Angles

▶ Angles are classified by their measures.

A **right angle** is an angle that has a measure of *exactly* 90°.

An **acute angle** is an angle that has a measure *less than* 90°.

An **obtuse angle** is an angle that has a measure *greater than* 90° but *less than* 180°.

A **straight angle** is an angle that has a measure of *exactly* 180°.

| right angle | acute angle | obtuse angle | straight angle |
| --- | --- | --- | --- |
| ∠ABC = 90° | ∠DEF < 90° | 90° < ∠GHM < 180° | ∠JKL = 180° |

▶ **Perpendicular lines** are intersecting lines that form four right angles.

$\overleftrightarrow{RS}$ and $\overleftrightarrow{PQ}$ are perpendicular lines. ∠RTP, ∠PTS, ∠RTQ, and ∠QTS are right angles.

$\overleftrightarrow{RS} \perp \overleftrightarrow{PQ}$
The symbol ⊥ means "is perpendicular to."

**Write whether each angle is *acute, right, obtuse,* or *straight*.**

1. 34°
2. 110°
3. 12°
4. 90°
5. 180°
6. 6°

7. 163°
8. 91°
9. 25°
10. 137°
11. 75°
12. 179°

13.
14.
15.
16.

**Name the following angles.** Use the figure.

17. 4 acute angles

18. 2 right angles

19. a straight angle

20. 3 obtuse angles

270

**Find the measure of each angle.**
Use the figure.

**21.** ∠WED     **22.** ∠DET

**23.** ∠WER     **24.** ∠PEW

**25.** ∠AER     **26.** ∠FEA

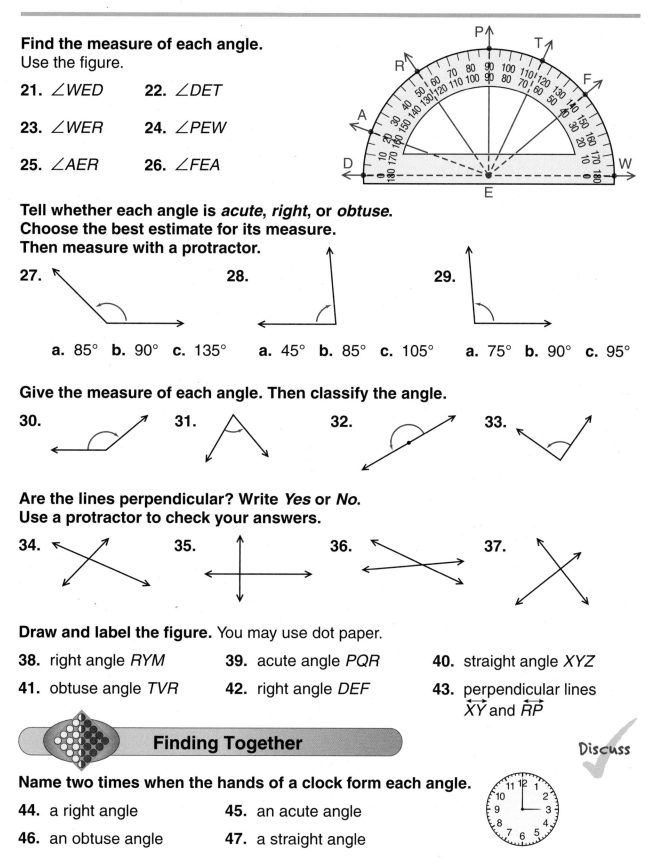

**Tell whether each angle is *acute*, *right*, or *obtuse*.**
**Choose the best estimate for its measure.**
**Then measure with a protractor.**

**27.**

    **a.** 85°  **b.** 90°  **c.** 135°

**28.**

    **a.** 45°  **b.** 85°  **c.** 105°

**29.**

    **a.** 75°  **b.** 90°  **c.** 95°

**Give the measure of each angle. Then classify the angle.**

**30.**     **31.**     **32.**     **33.**

**Are the lines perpendicular? Write *Yes* or *No*.**
**Use a protractor to check your answers.**

**34.**     **35.**     **36.**     **37.**

**Draw and label the figure.** You may use dot paper.

**38.** right angle *RYM*     **39.** acute angle *PQR*     **40.** straight angle *XYZ*

**41.** obtuse angle *TVR*     **42.** right angle *DEF*     **43.** perpendicular lines
                                                      $\overleftrightarrow{XY}$ and $\overleftrightarrow{RP}$

## Finding Together

Discuss

**Name two times when the hands of a clock form each angle.**

**44.** a right angle     **45.** an acute angle

**46.** an obtuse angle     **47.** a straight angle

271

## 8-3 Polygons

**Plane figures** are made up of points that are all in the same plane. They lie on a flat surface.

 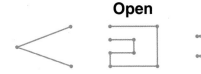 Discover Together

**Materials Needed:** dot paper or geoboard, ruler, protractor, dictionary

A plane figure is either an **open** or a **closed** figure. Look at the figures given below.

**Open**                           **Closed**

1. How are the open figures similar to the closed figures? How are they different?

2. Use a geoboard or dot paper to make several closed figures.

3. Describe each figure you made. How many line segments does each figure have? How many angles?

4. What is the relationship between the number of line segments and the number of angles of each figure?

Some closed plane figures are called polygons. **Polygons** are closed plane figures made up of line segments that meet at vertices but do not cross.

The figures on the geoboards shown below are polygons.

**heptagon**                    **octagon**

5. How many sides does a heptagon have? How many angles?

6. How many sides does an octagon have? How many angles?

7. Use a geoboard or dot paper to model 5 different polygons.

8. What relationship can you find between the number of sides and the number of angles of each of your models?

9. Copy and complete the table of polygons. You may use your polygon models.

| Name of Polygon | Number of Sides | Number of Angles | Model of Polygon |
|---|---|---|---|
| Heptagon | ? | 7 |  |
| Octagon | 8 | ? | ? |
| Nonagon | ? | 9 | ? |
| Decagon | 10 | ? | ? |

Polygons that have all sides of the same length and all angles of the same measure are called **regular polygons**. Look at the figures given below.

10. Trace the figures and make a concept map to classify each as a regular or *not* regular polygon.

## Communicate

Discuss ✓

11. Is a closed plane figure always a polygon? Explain your answer.

12. Can a polygon have sides that are of the same length and angles that are *not* the same measure? Explain your answer.

13. Name some examples of real objects that have polygon shapes.

### Connections: Language Arts

14. Find the meaning of the prefixes *tri, quad, penta, hexa, octa,* and *deca.* Write a story about life in a land where all objects are only shapes beginning with these prefixes.

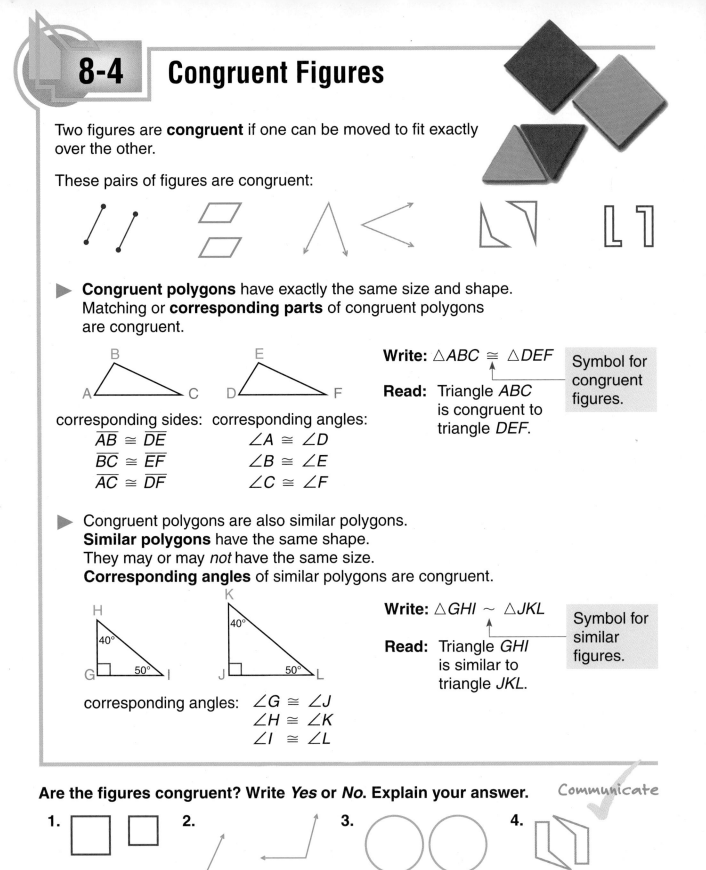

## 8-4 Congruent Figures

Two figures are **congruent** if one can be moved to fit exactly over the other.

These pairs of figures are congruent:

▶ **Congruent polygons** have exactly the same size and shape. Matching or **corresponding parts** of congruent polygons are congruent.

corresponding sides:

$\overline{AB} \cong \overline{DE}$
$\overline{BC} \cong \overline{EF}$
$\overline{AC} \cong \overline{DF}$

corresponding angles:

$\angle A \cong \angle D$
$\angle B \cong \angle E$
$\angle C \cong \angle F$

**Write:** $\triangle ABC \cong \triangle DEF$

Symbol for congruent figures.

**Read:** Triangle *ABC* is congruent to triangle *DEF*.

▶ Congruent polygons are also similar polygons.
**Similar polygons** have the same shape.
They may or may *not* have the same size.
**Corresponding angles** of similar polygons are congruent.

corresponding angles:   $\angle G \cong \angle J$
$\angle H \cong \angle K$
$\angle I \cong \angle L$

**Write:** $\triangle GHI \sim \triangle JKL$

Symbol for similar figures.

**Read:** Triangle *GHI* is similar to triangle *JKL*.

**Are the figures congruent? Write *Yes* or *No*. Explain your answer.**   Communicate

1.   2.   3.   4.

**Copy and complete.**

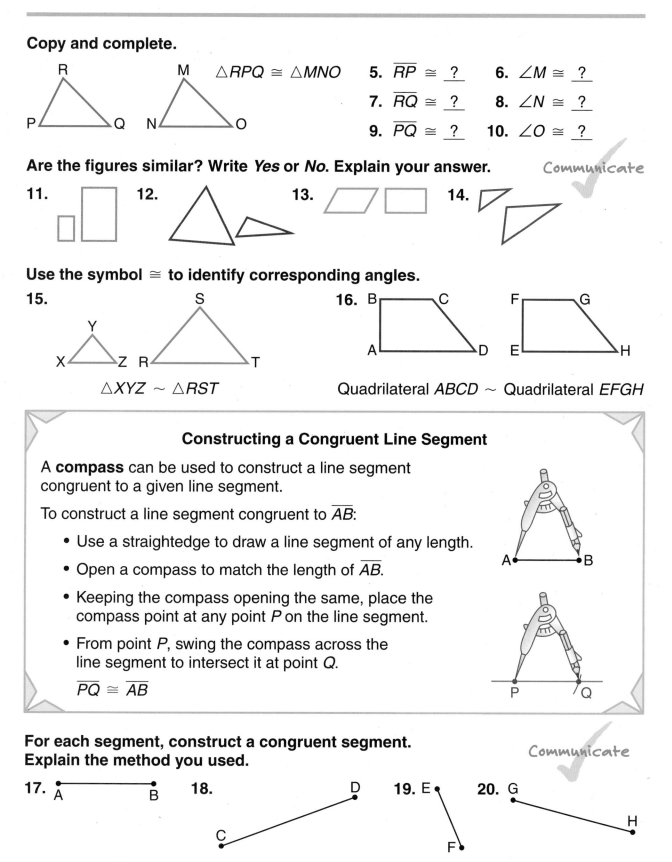

△RPQ ≅ △MNO

**5.** $\overline{RP}$ ≅ __?__    **6.** ∠M ≅ __?__

**7.** $\overline{RQ}$ ≅ __?__    **8.** ∠N ≅ __?__

**9.** $\overline{PQ}$ ≅ __?__    **10.** ∠O ≅ __?__

**Are the figures similar? Write *Yes* or *No*. Explain your answer.**    *Communicate* ✓

**11.**

**12.**

**13.**

**14.**

**Use the symbol ≅ to identify corresponding angles.**

**15.**

△XYZ ∼ △RST

**16.** B — C    F — G

Quadrilateral *ABCD* ∼ Quadrilateral *EFGH*

### Constructing a Congruent Line Segment

A **compass** can be used to construct a line segment congruent to a given line segment.

To construct a line segment congruent to $\overline{AB}$:

- Use a straightedge to draw a line segment of any length.

- Open a compass to match the length of $\overline{AB}$.

- Keeping the compass opening the same, place the compass point at any point *P* on the line segment.

- From point *P*, swing the compass across the line segment to intersect it at point *Q*.

$\overline{PQ}$ ≅ $\overline{AB}$

A •———• B

P      Q

**For each segment, construct a congruent segment. Explain the method you used.**    *Communicate* ✓

**17.** A •———• B

**18.** C •——————• D

**19.** E •, F •

**20.** G •————• H

275

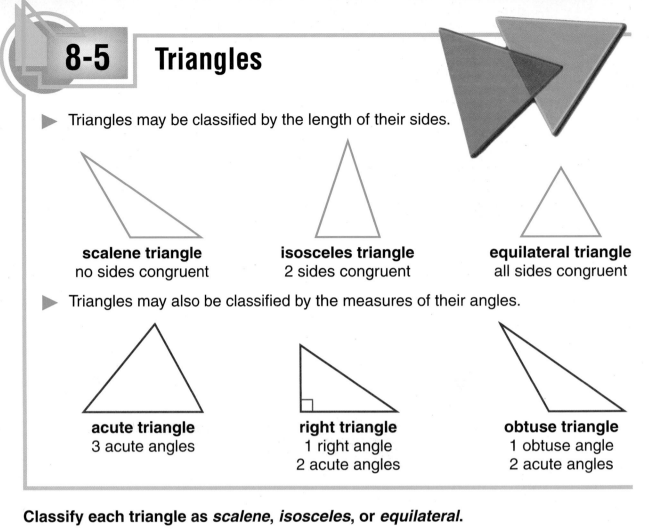

# 8-5 Triangles

▶ Triangles may be classified by the length of their sides.

**scalene triangle**
no sides congruent

**isosceles triangle**
2 sides congruent

**equilateral triangle**
all sides congruent

▶ Triangles may also be classified by the measures of their angles.

**acute triangle**
3 acute angles

**right triangle**
1 right angle
2 acute angles

**obtuse triangle**
1 obtuse angle
2 acute angles

**Classify each triangle as *scalene*, *isosceles*, or *equilateral*.**

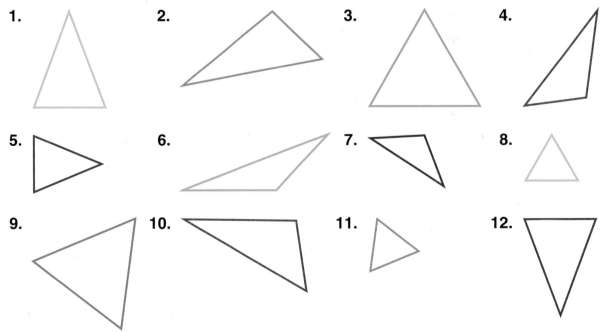

1.

2.

3.

4.

5.

6.

7.

8.

9.

10.

11.

12.

**Classify each triangle as *acute*, *right*, or *obtuse*.**

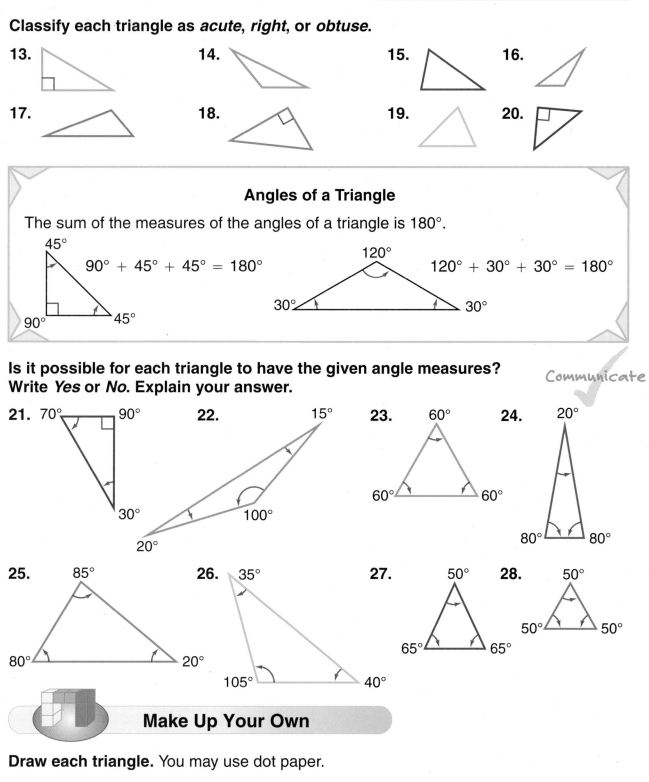

13.

14.

15.

16.

17.

18.

19.

20.

**Angles of a Triangle**

The sum of the measures of the angles of a triangle is 180°.

$90° + 45° + 45° = 180°$

$120° + 30° + 30° = 180°$

**Is it possible for each triangle to have the given angle measures?
Write *Yes* or *No*. Explain your answer.**

*Communicate* ✓

21. 70° 90° 30° 20°

22. 15° 100°

23. 60° 60° 60°

24. 20° 80° 80°

25. 85° 80° 20°

26. 35° 105° 40°

27. 50° 65° 65°

28. 50° 50° 50°

**Make Up Your Own**

**Draw each triangle.** You may use dot paper.

29. an acute isosceles triangle

30. a right scalene triangle

31. a right isosceles triangle

32. an obtuse scalene triangle

277

# 8-6 Quadrilaterals

Some quadrilaterals have special names.

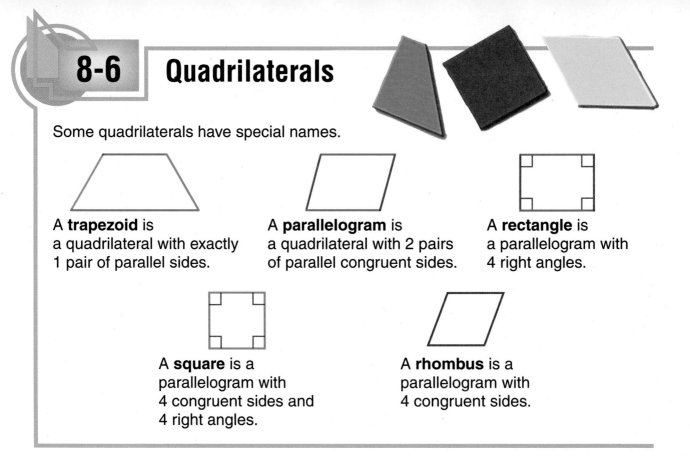

A **trapezoid** is a quadrilateral with exactly 1 pair of parallel sides.

A **parallelogram** is a quadrilateral with 2 pairs of parallel congruent sides.

A **rectangle** is a parallelogram with 4 right angles.

A **square** is a parallelogram with 4 congruent sides and 4 right angles.

A **rhombus** is a parallelogram with 4 congruent sides.

**Classify each quadrilateral as a *parallelogram*, a *rectangle*, a *square*, a *rhombus*, or a *trapezoid*.**

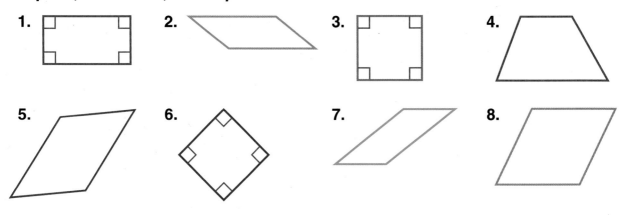

1.  2.  3.  4.

5.  6.  7.  8.

**Use the quadrilaterals in exercises 1–8.**

9. Which have 4 right angles?

10. Which have 4 congruent sides?

11. Which have no right angles?

12. Which 2 figures are congruent?

13. Which have 2 pairs of parallel sides?

14. Which have 4 congruent sides, but no right angles?

278

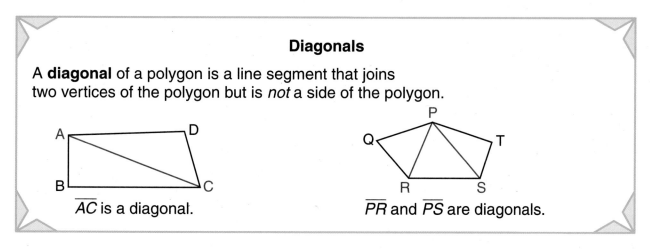

## Diagonals

A **diagonal** of a polygon is a line segment that joins two vertices of the polygon but is *not* a side of the polygon.

$\overline{AC}$ is a diagonal.

$\overline{PR}$ and $\overline{PS}$ are diagonals.

**Use the above polygons for exercises 15–16.**

**15.** Name another diagonal in quadrilateral *ABCD*.

**16.** Name another 3 diagonals in pentagon *PQRST*.

**Trace each figure. Then draw all its diagonals and count how many you have drawn.**

**17.**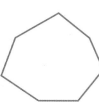

**18.**

**19.**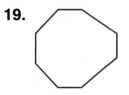

**Write *always*, *sometimes*, or *never*. Explain your answers.**   *Communicate* ✓

**20.** A square is a regular polygon.

**21.** A rhombus is a square.

**22.** A triangle has no diagonals.

**23.** A square is a rectangle.

**24.** A quadrilateral has no parallel sides.

**25.** A trapezoid is a parallelogram.

 ### Connections: Social Studies

A **tangram** is a geometric puzzle that originated in China over 4000 years ago. It starts out as a square, and is then cut into 7 prescribed pieces.

**26.** Trace the 7 pieces of the tangram shown at the right. Cut out the pieces. Then rearrange them to form different quadrilaterals. How many quadrilaterals can you make? Name them.

279

# Perimeter of a Polygon

Mr. Seymour makes a footpath around his garden. The garden is shaped like a polygon. How many yards long is the footpath?

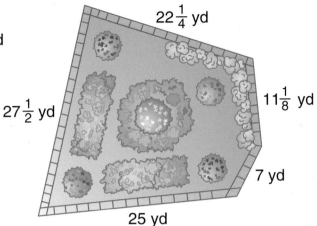

To find the length of the footpath, find the perimeter of the garden.

The **perimeter of a polygon** is the distance around the polygon. To find the perimeter, add the measures of all its sides.

Let $P$ stand for *perimeter*.

$$P = 27\tfrac{1}{2} \text{ yd} + 22\tfrac{1}{4} \text{ yd} + 11\tfrac{1}{8} \text{ yd} + 7 \text{ yd} + 25 \text{ yd}$$

$$P = 27\tfrac{4}{8} \text{ yd} + 22\tfrac{2}{8} \text{ yd} + 11\tfrac{1}{8} \text{ yd} + 7 \text{ yd} + 25 \text{ yd}$$

$$P = 92\tfrac{7}{8} \text{ yd}$$

The footpath around the garden is $92\tfrac{7}{8}$ yd.

**Find the perimeter of each polygon.**

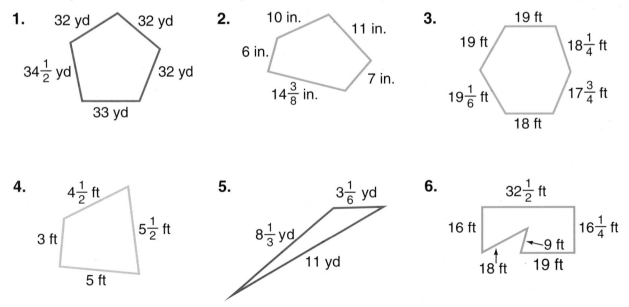

1. 32 yd, 32 yd, $34\tfrac{1}{2}$ yd, 32 yd, 33 yd

2. 10 in., 11 in., 6 in., 7 in., $14\tfrac{3}{8}$ in.

3. 19 ft, 19 ft, $18\tfrac{1}{4}$ ft, $19\tfrac{1}{6}$ ft, $17\tfrac{3}{4}$ ft, 18 ft

4. $4\tfrac{1}{2}$ ft, 3 ft, $5\tfrac{1}{2}$ ft, 5 ft

5. $3\tfrac{1}{6}$ yd, $8\tfrac{1}{3}$ yd, 11 yd

6. $32\tfrac{1}{2}$ ft, 16 ft, $16\tfrac{1}{4}$ ft, 9 ft, 18 ft, 19 ft

280

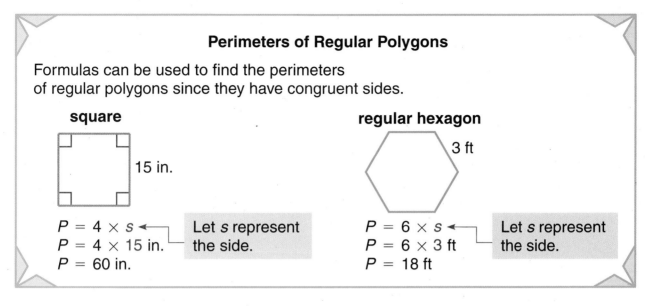

## Perimeters of Regular Polygons

Formulas can be used to find the perimeters
of regular polygons since they have congruent sides.

**square**

15 in.

$P = 4 \times s$ ← Let $s$ represent the side.
$P = 4 \times 15$ in.
$P = 60$ in.

**regular hexagon**

3 ft

$P = 6 \times s$ ← Let $s$ represent the side.
$P = 6 \times 3$ ft
$P = 18$ ft

**Find the perimeter of each regular polygon.**

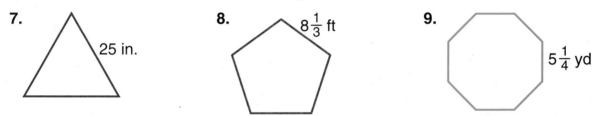

**7.** 25 in.

**8.** $8\frac{1}{3}$ ft

**9.** $5\frac{1}{4}$ yd

### PROBLEM SOLVING

**10.** The Great Pyramid in Egypt
has a square base. The length
of a side of the base is 230 meters.
What is its perimeter?

**11.** A park is shaped like a
regular hexagon. Each of its
sides is 26 yd long. Find the
perimeter of the park.

**12.** Mr. Jones left Dorado and traveled to
Trinidad, La Jolla, Springs, Junction,
and then back to Dorado. His route is in
the shape of a polygon. How many
miles did he travel?

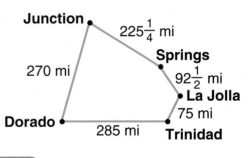

Junction
$225\frac{1}{4}$ mi
Springs
270 mi
$92\frac{1}{2}$ mi
La Jolla
75 mi
Dorado
285 mi
Trinidad

### Skills to Remember

**Compute.**

**13.** $2 \times 9 + 2 \times 18$

**14.** $2 \times 15 + 2 \times 25$

**15.** $2 \times (13 + 20)$

**16.** $2 \times (18 + 42)$

**17.** $2 \times 2\frac{1}{2} + 2 \times 3\frac{1}{2}$

**18.** $2 \times 4\frac{1}{3} + 2 \times 8\frac{1}{3}$

**19.** $2 \times (5\frac{1}{2} + 8\frac{1}{3})$

**20.** $2 \times (7\frac{1}{4} + 8\frac{1}{6})$

**21.** $2 \times (3\frac{2}{3} + 4\frac{1}{6})$

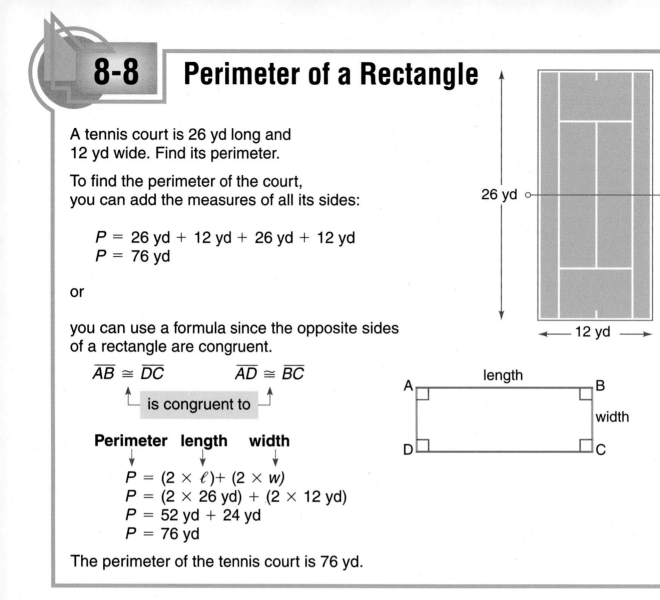

## 8-8 Perimeter of a Rectangle

A tennis court is 26 yd long and 12 yd wide. Find its perimeter.

To find the perimeter of the court, you can add the measures of all its sides:

$P$ = 26 yd + 12 yd + 26 yd + 12 yd
$P$ = 76 yd

or

you can use a formula since the opposite sides of a rectangle are congruent.

$\overline{AB} \cong \overline{DC}$ $\overline{AD} \cong \overline{BC}$

is congruent to

**Perimeter length width**

$P = (2 \times \ell) + (2 \times w)$
$P = (2 \times 26$ yd$) + (2 \times 12$ yd$)$
$P = 52$ yd $+ 24$ yd
$P = 76$ yd

The perimeter of the tennis court is 76 yd.

**Find the perimeter of each rectangle.**

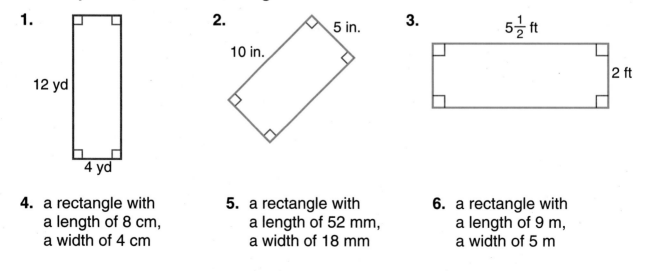

**1.** 12 yd, 4 yd

**2.** 5 in., 10 in.

**3.** $5\frac{1}{2}$ ft, 2 ft

**4.** a rectangle with a length of 8 cm, a width of 4 cm

**5.** a rectangle with a length of 52 mm, a width of 18 mm

**6.** a rectangle with a length of 9 m, a width of 5 m

**Find the perimeter of each rectangle.
Write the letter of the correct answer.**

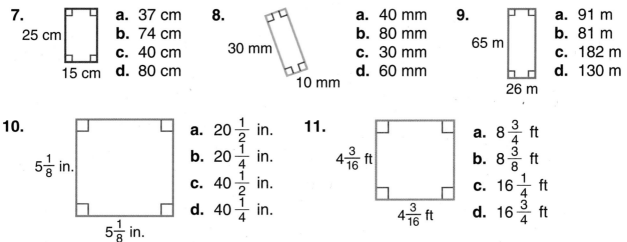

**7.**   25 cm   15 cm
a. 37 cm
b. 74 cm
c. 40 cm
d. 80 cm

**8.**   30 mm   10 mm
a. 40 mm
b. 80 mm
c. 30 mm
d. 60 mm

**9.**   65 m   26 m
a. 91 m
b. 81 m
c. 182 m
d. 130 m

**10.**   $5\frac{1}{8}$ in.   $5\frac{1}{8}$ in.
a. $20\frac{1}{2}$ in.
b. $20\frac{1}{4}$ in.
c. $40\frac{1}{2}$ in.
d. $40\frac{1}{4}$ in.

**11.**   $4\frac{3}{16}$ ft   $4\frac{3}{16}$ ft
a. $8\frac{3}{4}$ ft
b. $8\frac{3}{8}$ ft
c. $16\frac{1}{4}$ ft
d. $16\frac{3}{4}$ ft

## PROBLEM SOLVING
**Draw a diagram for each problem. Then use a formula to solve.**

**12.** Stephanie is making a quilt cover 72 in. wide and 80 in. long. Find the perimeter of the cover.

**13.** A room is 9 yd long and $6\frac{1}{4}$ yd wide. How much molding is needed to go around the ceiling?

**14.** Mr. Williams's desk is 8 ft long and 4 ft wide. Find the perimeter of the desk.

**15.** How many feet of ribbon will be needed to put a ribbon edge around a scarf $1\frac{1}{2}$ ft wide and 6 ft long?

**16.** How many feet of trim will it take to border a rug that is $5\frac{2}{3}$ ft wide and $8\frac{1}{6}$ ft long?

**17.** How many feet of satin will be needed to trim the edges of a blanket that is $6\frac{1}{4}$ ft wide and $12\frac{1}{8}$ ft long?

**18.** A field in the shape of a rectangle is 550 yd wide and 880 yd long. If Karen jogs around the field twice, how many yards does she jog?

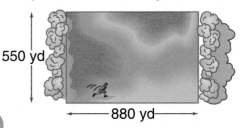

550 yd

880 yd

### Challenge

**19.** A painting is shaped like a rectangle. Its length is 16 in. and its perimeter is 56 in. Find the width of the painting.

**20.** The perimeter of a regular quadrilateral is 192 in. Find the length of one side.

Algebra

# 8-9 | Circles

> A **circle** is a plane figure. All points of the circle are the same distance from a given point, called the **center**.

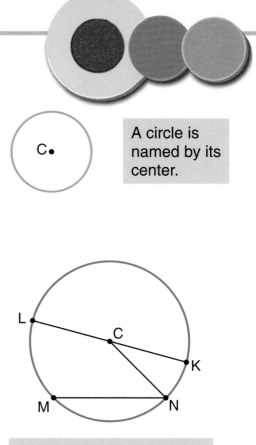

A circle is named by its center.

Point $C$ is the center of circle $C$.

> The parts of a circle have special names.

A **chord** is a line segment with its endpoints on the circle. $\overline{LK}$ and $\overline{MN}$ are chords of circle $C$.

A **diameter** is a chord that passes through the center. $\overline{LK}$ is a diameter of circle $C$.

A **radius** is a line segment with one endpoint at the center of the circle and the other endpoint on the circle. $\overline{CL}$, $\overline{CK}$, $\overline{CN}$ are radii (plural of radius) of circle $C$.

The length of the diameter is twice the length of the radius.

$$d = 2 \times r \qquad r = d \div 2$$

---

**Use the circle at the right.**

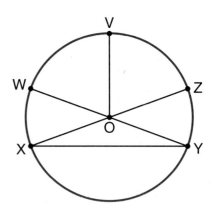

1. Name the circle and its center.

2. Name 5 points of the circle.

3. How many chords of the circle are shown? Name them.

4. Is $\overline{XY}$ a diameter of the circle? Explain why or why not.

5. How many diameters of the circle are shown? Name them.

6. How many radii of the circle are shown? Name them.

**Match. Write the correct letter.**

7. chord  _?_

8. radius  _?_

9. diameter  _?_

10. center  _?_

  **a.** a chord that passes through the center of the circle

  **b.** a point that names the circle

  **c.** a line segment joining any two points on the circle

  **d.** a line segment drawn from the center of the circle to any point on the circle

### Constructing a Circle

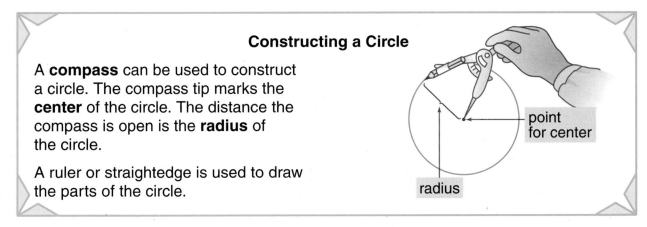

A **compass** can be used to construct a circle. The compass tip marks the **center** of the circle. The distance the compass is open is the **radius** of the circle.

A ruler or straightedge is used to draw the parts of the circle.

**Use a compass to construct a circle. Then do the following:**

11. Label the center point, *X*.

12. Draw chord $\overline{MN}$.

13. Draw diameter $\overline{VT}$.

14. Draw radius $\overline{XR}$.

**Use a compass and a ruler to construct a circle with a:**

15. radius of 2 cm.

16. radius of $1\frac{1}{2}$ in.

17. diameter of 6 cm.

18. diameter of 5 m.

### PROBLEM SOLVING

19. The diameter of a circular track is 140 yd. What is the radius?

20. The radius of a circular flower bed is $8\frac{5}{6}$ ft. What is the diameter?

 **Connections: Art**

21. Use a compass or circular templates, a ruler, and a protractor to draw a picture of any familiar object using only circles and polygons.

# Estimating Circumference

The distance around a circle is called
the **circumference** (C) of the circle.

Do this experiment:

Mark a point A on a circle and on a sheet of paper.
Roll the circle along the paper until A returns
to its original position. Mark this point B on the paper.
The line segment $\overline{AB}$ is the same length as the circumference.

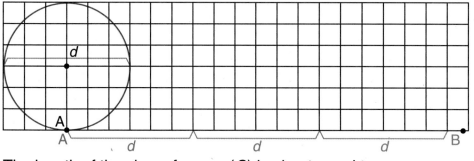

The length of the circumference (C) is about equal to
*three times* the length of the diameter (d).

$$C \approx 3 \times d$$

The symbol ≈ means
"is approximately equal to."

**Study this problem.**

A fountain has a diameter of 16 ft.
Estimate the circumference of the fountain.

$$C \approx 3 \times d$$
$$\approx 3 \times 16 \text{ ft}$$
$$\approx 48 \text{ ft}$$

The circumference of the fountain is about 48 ft.

**Estimate the circumference of each circle.**

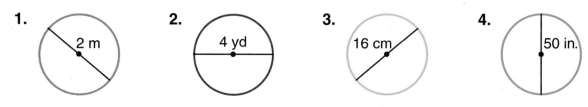

1. 2 m

2. 4 yd

3. 16 cm

4. 50 in.

**Use the formula to find the approximate circumference of each circle. Write the letter of the correct answer.**

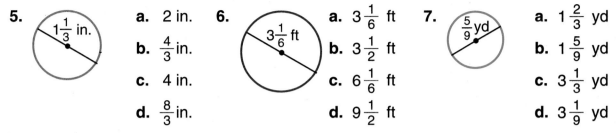

**5.** $1\frac{1}{3}$ in.

  **a.** 2 in.
  **b.** $\frac{4}{3}$ in.
  **c.** 4 in.
  **d.** $\frac{8}{3}$ in.

**6.** $3\frac{1}{6}$ ft

  **a.** $3\frac{1}{6}$ ft
  **b.** $3\frac{1}{2}$ ft
  **c.** $6\frac{1}{6}$ ft
  **d.** $9\frac{1}{2}$ ft

**7.** $\frac{5}{9}$ yd

  **a.** $1\frac{2}{3}$ yd
  **b.** $1\frac{5}{9}$ yd
  **c.** $3\frac{1}{3}$ yd
  **d.** $3\frac{1}{9}$ yd

---

### Estimating the Circumference Using the Radius

A circular rug has a radius of 2 meters.
Estimate the circumference of the rug.

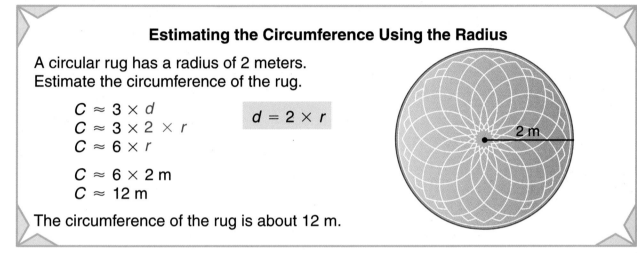

$C \approx 3 \times d$
$C \approx 3 \times 2 \times r$
$C \approx 6 \times r$

$\boxed{d = 2 \times r}$

$C \approx 6 \times 2$ m
$C \approx 12$ m

2 m

The circumference of the rug is about 12 m.

---

**Estimate the circumference of each circle.**

**8.** 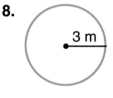 3 m

**9.** 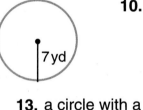 7 yd

**10.** 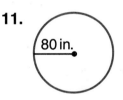 20 ft

**11.** 80 in.

**12.** a circle with a diameter of 150 cm

**13.** a circle with a radius of 65 mm

**14.** a circle with a diameter of $12\frac{1}{3}$ in.

### PROBLEM SOLVING

**15.** A circular table has a radius of $2\frac{1}{2}$ ft. Estimate the circumference.

**16.** A Ferris wheel has a diameter of 80 ft. Estimate the circumference.

 **Connections: Science**

**17.** Earth's equator is a circle with a radius of 6378 km. Estimate Earth's circumference.

# Lines of Symmetry

If a figure can be folded along a line so that the two halves are congruent, the figure has **line symmetry**.
The fold line is called the **line of symmetry**.

Some figures have *more than one* line of symmetry.

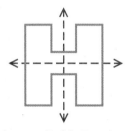

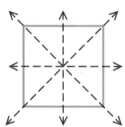

This capital letter has *two* lines of symmetry.

This square has *four* lines of symmetry.

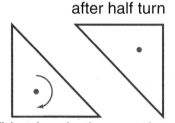

line of symmetry

▶ If a figure can be turned halfway around a point so that it looks exactly the same, the figure has **half-turn symmetry**.

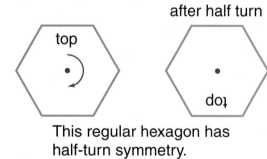

after half turn

top

dot

This regular hexagon has half-turn symmetry.

after half turn

This triangle does *not* have half-turn symmetry.

**Is the dashed line in each figure a line of symmetry? Write *Yes* or *No*.**

1.    2.    3.    4.

**Trace each figure. Then draw all lines of symmetry for each.**

5.    6.    7.    8.

**Find how many lines of symmetry each figure has.**

9.

10.

11.

12.

**Does the figure have half-turn symmetry? Write *Yes* or *No*.**

13.

14.

15.

16.

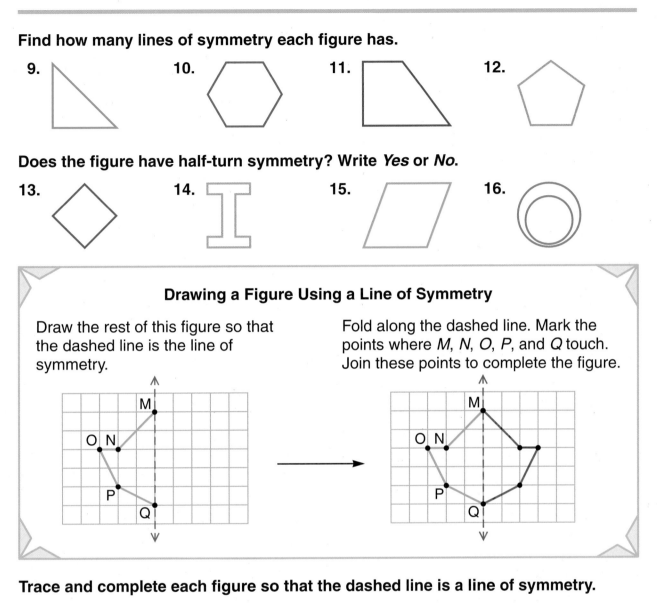

**Drawing a Figure Using a Line of Symmetry**

Draw the rest of this figure so that the dashed line is the line of symmetry.

Fold along the dashed line. Mark the points where *M, N, O, P,* and *Q* touch. Join these points to complete the figure.

**Trace and complete each figure so that the dashed line is a line of symmetry.**

17.

18.

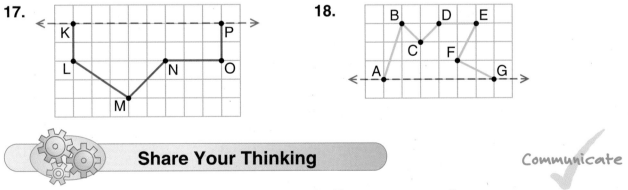

**Share Your Thinking**

Communicate ✓

19. Can a figure have line symmetry but no half-turn symmetry? line symmetry and half-turn symmetry? half-turn symmetry but no line symmetry? Explain your answers.

# 8-12 Transformations

There are three basic types of movements
of geometric figures in a plane:

- **slide** (or translation) — Every point of a figure moves the same
  distance and in the same direction.

- **flip** (or reflection) — A figure is flipped over a line so that
  its mirror image is formed.

- **turn** (or rotation) — A figure is turned around a center point.

## Hands-On Understanding

**Materials Needed:** pattern blocks, grid paper, ruler

| Step 1 | Fold a sheet of grid paper in thirds. Open it up and label each section *A*, *B*, or *C*. |

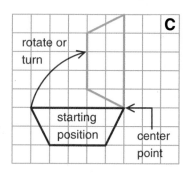

| Step 2 | Place a trapezoid in section *A* and trace around it to record a starting position. **Slide**, or move, the trapezoid along the lines going up 3 units and right 2 units as shown. |

Did the trapezoid change its size or shape?
What changed when you slid the trapezoid?

| Step 3 | Now trace the trapezoid in section *B* to record a starting position. **Flip**, or turn over, the trapezoid across a line of symmetry as shown. |

Are the two trapezoids in section *B* congruent?
How did the position of the trapezoid change?

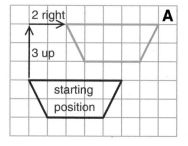

| Step 4 | Trace the trapezoid in section *C* to record a starting position. **Turn**, or rotate, the trapezoid around a vertex as shown. |

What changed when you turned the trapezoid?
What did *not* change?

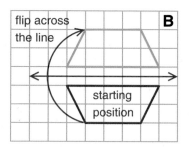

**Decide whether figure _B_ is a result of moving figure _A_.**
**Write _Yes_ or _No_. Explain your answers.**

1.
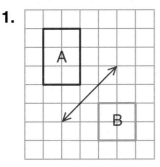

2.
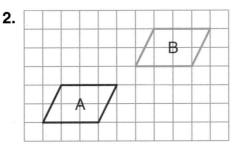

3.
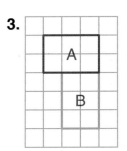

**Copy each figure on grid paper. Then draw a second figure to show the result of a slide, flip, or turn. Use pattern blocks to model.**

4. Slide parallelogram _RSTP_ down 4 units and left 5 units.

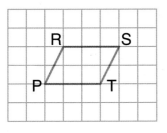

5. Flip square _NOLM_ across the line.

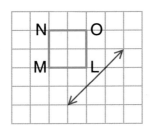

6. Turn triangle _ABC_ around vertex _C_.

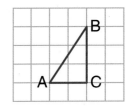

## Communicate

Discuss ✓

7. Does sliding, flipping, or turning a figure change either its size or shape?

8. How can you tell if the second figure is a result of moving the first figure?

9. In the figures at the right, name the motion(s) when _A_ is moved to _B_; _A_ is moved to _C_; _A_ is moved to _D_. Explain your answers.

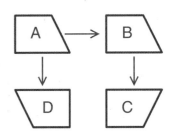

## Project

10. To create quilting patterns, transformations are used. Find examples of Early American quilt patterns and then create your own design using transformations.

# TECHNOLOGY

## Procedures and Coordinate Graphs

Algebra ✓

Procedures can be used to graph ordered pairs. LOGO is one computer language that can be programmed to do this.

> To create a LOGO procedure:
>
> - Choose a procedure name (using no spaces).
> - Type the word TO followed by the procedure name. Then press ENTER↵.
> - Enter the commands of the procedure, pressing ENTER↵ after each command line.
> - Type the word END after all the commands have been entered. Then press ENTER↵.

A LOGO procedure contains commands that tell the turtle how to move.

To reuse a procedure, just enter the procedure name.

▶ The procedure below tells the turtle to draw a square.

TO SQUARE
REPEAT 4 [FD 100 RT 90]
END

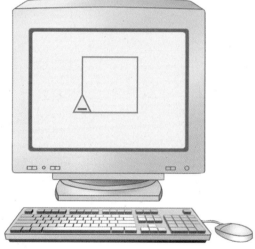

Think of a computer screen as a coordinate grid divided into four sections. The origin, the center of the screen, is referred to as the HOME position and is at point (0,0).

▶ Use the LOGO command SETXY to move the turtle and draw a line to a specific point on the screen.

TO SQUAREAGAIN
SQUARE
PU
SETXY [10 10]
PD
SQUARE
END

space

moves the turtle to the point (10, 10)

2nd coordinate

1st coordinate

A procedure may be used within another procedure.

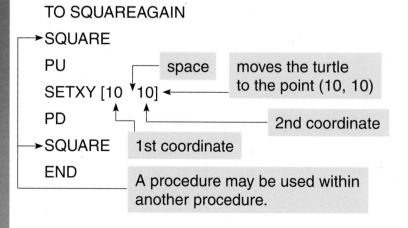

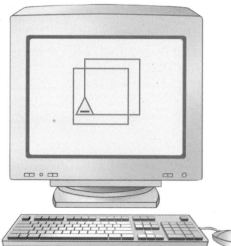

**Use each procedure to draw each figure.**

Algebra ✓

1. TO TRIANGLE
   RT 30
   FD 90
   REPEAT 2 [RT 120 FD 90]
   END

2. TO SQUARE
   REPEAT 4 [FD 90 RT 90]
   END

3. TO ENCLOSE
   SQUARE
   TRIANGLE
   END

4. TO PINWHEEL
   REPEAT 4 [RT 90 SQUARE]
   REPEAT 4 [TRIANGLE]
   END

5. TO RTTRIANGLE
   SETXY [100 100]
   SETXY [100 0]
   SETXY [0 0]
   END

6. TO RECTANGLE
   SETXY [0 100]
   SETXY [50 100]
   SETXY [50 0]
   SETXY [0 0]
   END

**PROBLEM SOLVING**
**Use the coordinate grid.**

7. Name the coordinates of the vertices of the figure, in order.

8. Write a LOGO procedure to draw this figure. Use the SETXY command.

9. Write a LOGO procedure to show a slide of the figure 2 units to the right. Use the SETXY command.

10. Write a LOGO procedure to slide the figure along the slide arrow so point C is at (9, 2).

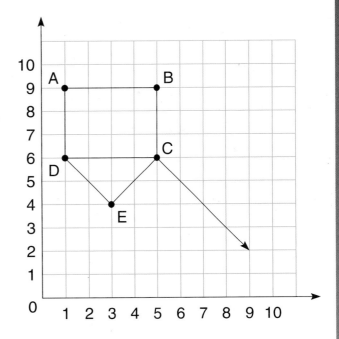

11. Write a LOGO procedure to draw all the lines of symmetry of rectangle ABCD. Use the SETXY command.

12. Write a LOGO procedure to draw a figure on a coordinate grid. Use the SETXY command.

293

# 8-14 Problem Solving: Use Formulas

*Algebra* ✓

**Problem:** A steeple shaped like a square pyramid is on top of a building. The dimensions of the pyramid are shown in the diagram. Find the perimeter of *one* triangular face of the steeple.

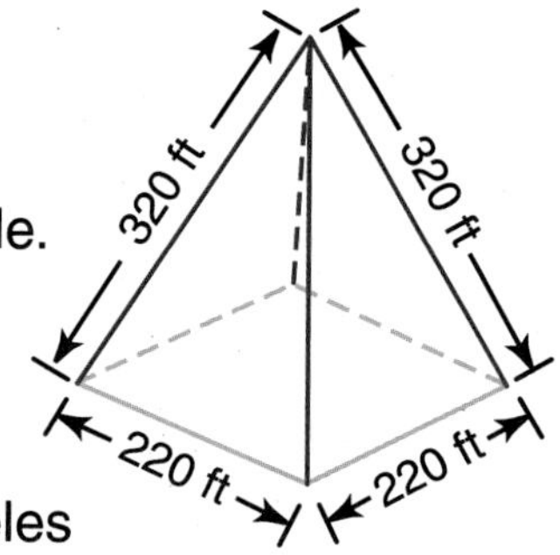

**1 IMAGINE** Draw a picture of the face of the steeple.

**2 NAME** *Facts:* Face of the steeple is an isosceles triangle (two congruent sides).

*Question:* What is the perimeter of *one* triangular face of the steeple?

**3 THINK** The distance around an object is its perimeter.

Write a *formula* for the perimeter of an isosceles triangle.
Let *a* represent the length of each congruent side.
Let *b* represent the length of the third side.

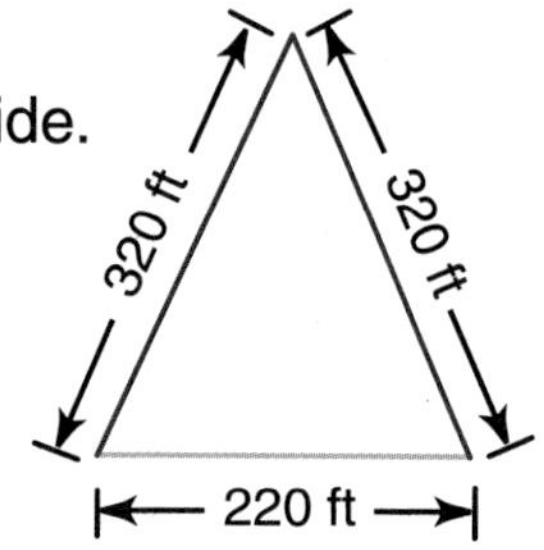

$$P = a + a + b$$

two congruent sides

$$P = (2 \times a) + b$$

**4 COMPUTE** $P = (2 \times 320 \text{ ft}) + 220 \text{ ft}$

$P = 640 \text{ ft} + 220 \text{ ft}$

$P = 860 \text{ ft}$

The perimeter of one triangular face of the steeple is 860 ft.

**5 CHECK** Use the formula $P = a + a + b$ to check your computation.

$$P = a + a + b$$

$$860 \text{ ft} \stackrel{?}{=} 320 \text{ ft} + 320 \text{ ft} + 220 \text{ ft}$$

$$860 \text{ ft} = 860 \text{ ft} \qquad \text{The answer checks.}$$

**Use formulas to solve each problem.**

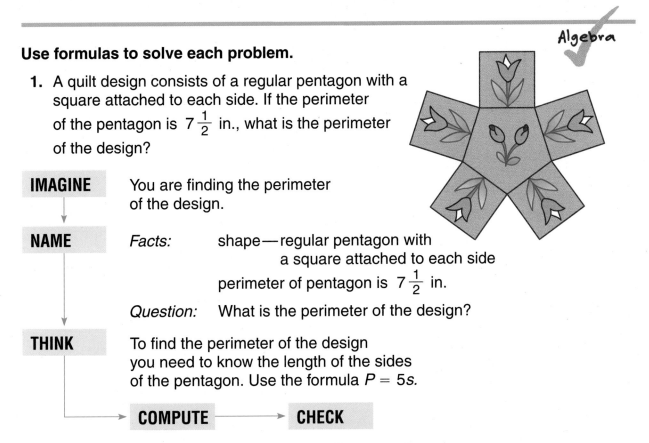

1. A quilt design consists of a regular pentagon with a square attached to each side. If the perimeter of the pentagon is $7\frac{1}{2}$ in., what is the perimeter of the design?

**IMAGINE**    You are finding the perimeter of the design.

**NAME**    *Facts:*    shape—regular pentagon with a square attached to each side

                     perimeter of pentagon is $7\frac{1}{2}$ in.

         *Question:*    What is the perimeter of the design?

**THINK**    To find the perimeter of the design you need to know the length of the sides of the pentagon. Use the formula $P = 5s$.

           **COMPUTE**  ⟶  **CHECK**

2. The perimeter of a gazebo floor shaped like a regular octagon is 96 ft. What is the length of one side of the floor?

3. A merry-go-round at the children's playground has a radius of 5 ft. Estimate the circumference of the merry-go-round.

4. A parallelogram has two pairs of congruent sides. Find the perimeter of a parallelogram whose parallel sides have lengths of 21 ft and 29 ft.

5. The sum of the angles of a quadrilateral equals 360°. Figure *MGRL* is a rhombus. If $\angle M \cong \angle R$ and $\angle G \cong \angle L$, and $\angle G$ equals 50°, what is the measure of $\angle R$?

6. The parallel sides of this window measure 3 ft and 5 ft. The other sides of the window are congruent. If the perimeter is 16 ft, what is the length of each congruent side?

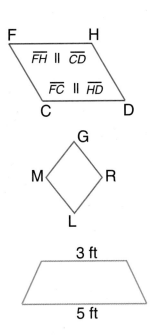

$\overline{FH} \parallel \overline{CD}$

$\overline{FC} \parallel \overline{HD}$

# Problem-Solving Applications

**Solve each problem and explain the method you used.**

1. Darlene drew right angle *DAR*. Name the two rays she drew.

2. Arnie's polygon has 2 pairs of parallel sides but no right angles. What might his polygon be?

3. Josh drew a square and a rhombus. Then he drew the diagonals of each. Name the types of triangles he formed.

4. In a quilt 2 congruent isosceles right triangles were joined at one side. What possible figures were formed?

5. Cleo put new trim around the edge of a circular rug that had a radius of 2 ft. About how much trim did she use?

6. Helen drew figure *FHMSVR*.
   a. Name the two rays.
   b. Name an acute angle and an obtuse angle.
   c. Name the polygon.
   d. $\overline{HF}$ is 2 cm long. Find the perimeter.

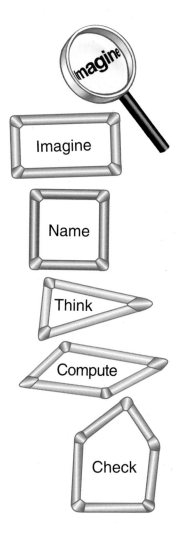

**Use the drawing for problems 7–10.**

7. A rectangular rug has this pattern. What polygon is congruent to △*AGC*?

8. Classify quadrilateral *FBCG*.

9. How many trapezoids are in this pattern?

10. How many lines of symmetry does this figure have? Does it have half-turn symmetry?

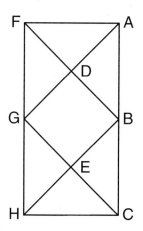

**Choose a strategy from the list or use another strategy you know to solve each problem.**

USE THESE STRATEGIES
Use a Model/Drawing
Logical Reasoning
Working Backwards
Use Formulas
Guess and Test

**11.** Judy makes a quilt square that has a perimeter of 18 in. How long is each side?

**12.** Ed drew 2 rays from the vertex of straight angle *FHM*. $\angle FHC$ equals 35°, and $\angle MHD$ equals twice that. How many degrees is $\angle CHD$?

**13.** A rectangular rug has a length of $7\frac{1}{2}$ ft. Its width is $2\frac{1}{4}$ ft less than its length. What is its perimeter?

**14.** What is the total number of diagonals that can be drawn in a square? in a regular pentagon? in a regular hexagon?

**15.** In this design, Rita painted the equilateral triangle blue, the rectangles yellow, and the obtuse triangles green. She painted the isosceles triangle orange. Finally she painted the remaining figures red. Name them.

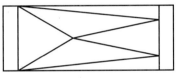

**Write *all, some,* or *none* to make true statements.**

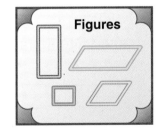

Figures

**16.** _?_ of these figures are polygons.

**17.** _?_ of these figures have acute angles.

**18.** _?_ of these figures are parallelograms.

**19.** Each of four friends made 1 of these quilt squares. Tina's has 4 triangles and 2 trapezoids. Frank's has the most rectangles. Julio's has 7 right triangles. Hope's has triangles, squares, and trapezoids. Which friend drew each square?

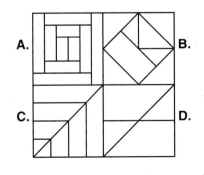

A.  B.

C.  D.

**Make Up Your Own**

**20.** Use the drawing of the rug on page 296 and these data to write two problems. Then solve them. Share your work with a classmate.

$\overline{FA} = 4$ ft
$\overline{FH} = 8$ ft

Discuss

297

# Chapter Review and Practice

**Use your protractor. Write the measure of each angle.** *(See pp. 268–283, 286–287.)*
**Then classify the angle.**

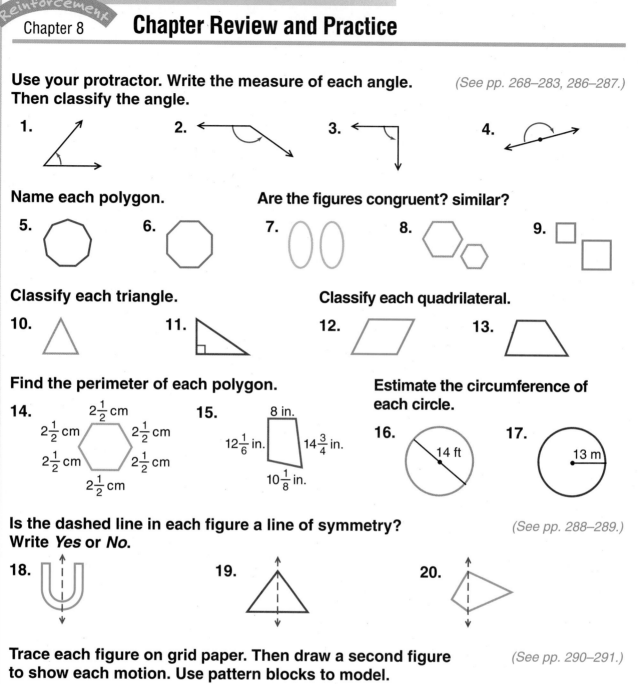

**1.**

**2.**

**3.**

**4.**

**Name each polygon.**

**Are the figures congruent? similar?**

**5.**

**6.**

**7.**

**8.**

**9.**

**Classify each triangle.**

**Classify each quadrilateral.**

**10.**

**11.**

**12.**

**13.**

**Find the perimeter of each polygon.**

**Estimate the circumference of each circle.**

**14.** $2\frac{1}{2}$ cm  $2\frac{1}{2}$ cm  $2\frac{1}{2}$ cm  $2\frac{1}{2}$ cm  $2\frac{1}{2}$ cm  $2\frac{1}{2}$ cm

**15.** 8 in.  $12\frac{1}{6}$ in.  $14\frac{3}{4}$ in.  $10\frac{1}{8}$ in.

**16.** 14 ft

**17.** 13 m

**Is the dashed line in each figure a line of symmetry?** *(See pp. 288–289.)*
**Write *Yes* or *No*.**

**18.**

**19.**

**20.**

**Trace each figure on grid paper. Then draw a second figure** *(See pp. 290–291.)*
**to show each motion. Use pattern blocks to model.**

**21.** Slide triangle *ABC* up 4 units and left 3 units.

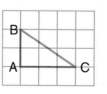

**22.** Turn rhombus *DGFE* around vertex *G*.

## PROBLEM SOLVING

*(See pp. 282–285, 294–297.)*

**23.** A tennis court is 78 ft long and 36 ft wide. Find its perimeter.

**24.** The diameter of a bicycle wheel is 28 in. Find its radius.

*(See Still More Practice, p. 484.)*

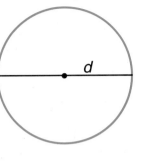

PI ($\pi$)

Mathematicians have shown that for every circle, the circumference ($C$) divided by the diameter ($d$) is always the same value. This value is represented by the Greek letter $\pi$ (read: "**pi**").

$$\frac{\text{Circumference}}{\text{diameter}} = \frac{C}{d} = \pi$$

Computers have calculated the value of $\pi$ to millions of decimal places. Approximate values of $\pi$ that are commonly used are $3.14$ and $\frac{22}{7}$.

▶ The *circumference (C) of a circle* can be found by:

- multiplying $\pi$ by the length of the diameter ($d$), or

- multiplying $\pi$ by twice the length of the radius ($r$).

circumference   diameter
$$C = \pi \times d$$

circumference   radius
$$C = \pi \times 2 \times r$$

Using $\pi \approx \frac{22}{7}$, find the circumference of a circle when:

diameter = 7 mm

$C = \pi \times d$

$C \approx \frac{22}{7} \times 7 \text{ mm} \approx \frac{22}{\overset{1}{\cancel{7}}} \times \frac{\overset{1}{\cancel{7}}}{1} \text{ mm}$

$C \approx \frac{22}{1} \text{ mm} \approx 22 \text{ mm}$

radius = 28 in.

$C = \pi \times 2 \times r$

$C \approx \frac{22}{7} \times 2 \times 28 \text{ in.}$

$C \approx \frac{22}{\underset{1}{\cancel{7}}} \times \frac{2}{1} \times \frac{\overset{4}{\cancel{28}}}{1} \text{ in.}$

$C \approx \frac{176}{1} \text{ in.} \approx 176 \text{ in.}$

**Find the circumference of each circle.** Use $\pi \approx \frac{22}{7}$.

**1.** diameter = 35 in.

**2.** radius = 14 cm

**3.** diameter = $10\frac{1}{2}$ ft

**4.** radius = $3\frac{1}{2}$ m

**5.** diameter = 12 yd

**6.** radius = 7 mi

**PROBLEM SOLVING**

**7.** If the diameter of a circle is doubled, what happens to its circumference?

# Check Your Mastery

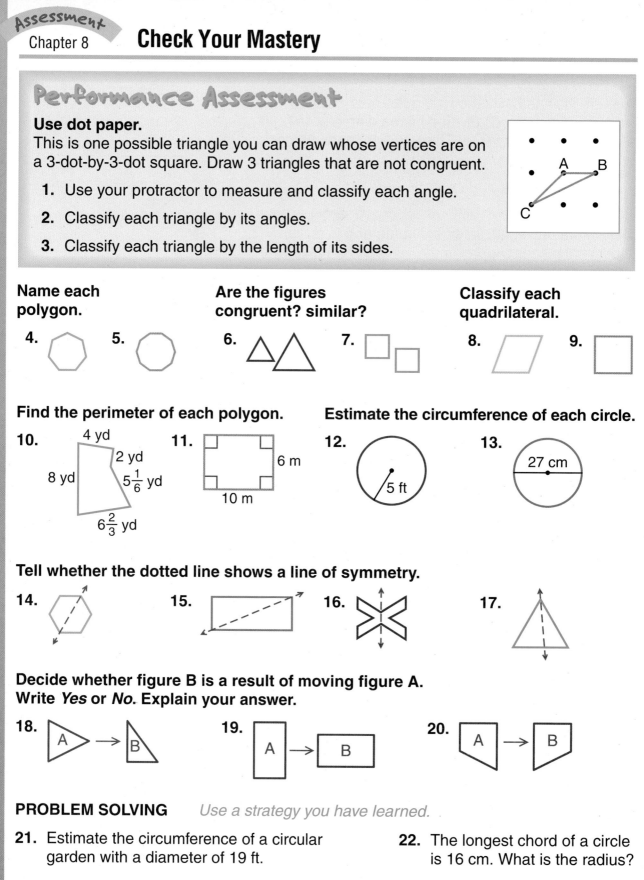

*Performance Assessment*

**Use dot paper.**
This is one possible triangle you can draw whose vertices are on a 3-dot-by-3-dot square. Draw 3 triangles that are not congruent.

1. Use your protractor to measure and classify each angle.

2. Classify each triangle by its angles.

3. Classify each triangle by the length of its sides.

**Name each polygon.**

4.

5.

**Are the figures congruent? similar?**

6.

7.

**Classify each quadrilateral.**

8.

9.

**Find the perimeter of each polygon.**

10. 4 yd
2 yd
8 yd
$5\frac{1}{6}$ yd
$6\frac{2}{3}$ yd

11. 6 m
10 m

**Estimate the circumference of each circle.**

12. 5 ft

13. 27 cm

**Tell whether the dotted line shows a line of symmetry.**

14.

15.

16.

17.

**Decide whether figure B is a result of moving figure A. Write *Yes* or *No*. Explain your answer.**

18. A → B

19. A → B

20. A → B

**PROBLEM SOLVING**   *Use a strategy you have learned.*

21. Estimate the circumference of a circular garden with a diameter of 19 ft.

22. The longest chord of a circle is 16 cm. What is the radius?

# Measurement Topics

# 9

**In this chapter you will:**

Investigate customary units of length, capacity,
 and weight
Read Fahrenheit and Celsius temperature scales
Learn about time zones
Compute customary units with regrouping
Solve multi-step problems

**Critical Thinking/Finding Together**

A ship's watch began at midnight on the mid-Atlantic
with one bell rung to signal the time. If one additional
bell is rung as each half hour passes and 9 bells were
just rung, what time is it?

## Midnight on the mid-Atlantic

Nothing blacker than the water,
nothing wider than the sky.
Pitch and toss, pitch and toss.
The Big Dipper might just ladle
a drink out of the sea.
Midnight on the mid-Atlantic is...

From *Nine O'Clock Lullaby* by *Marilyn Singer*

# 9-1 | Relating Customary Units of Length

## Discover Together

**Materials Needed:** inch ruler or measuring tape, paper, pencil

The **inch (in.)**, **foot (ft)**, **yard (yd)**, and **mile (mi)** are customary units of length.

| |
|---|
| 12 inches (in.) = 1 foot (ft) |
| 3 feet = 1 yard (yd) |
| 5280 ft or 1760 yd = 1 mile (mi) |

1. Choose the following objects to measure:
   • two objects that are longer than 1 inch but less than 1 foot,
   • two objects that are between 1 foot and 1 yard long,
   • two objects that are longer than 1 yard.

2. Estimate the length of each object. Then use a ruler or a measuring tape to measure each of them. Record your answers in a table like the one shown.

| Object | Estimate | Measure |
|---|---|---|
| | | |

3. What unit of measure did you use for lengths between 1 inch and 1 foot? between 1 foot and 1 yard? longer than 1 yard?

4. How does each estimate in your table compare with the actual measurement?

5. What unit would you use to measure the width of your math book? Why?

6. What unit would you use to measure the height of a table? Why?

7. What unit would you use to measure the distance of a race? Why?

8. What unit would you use to measure the distance between New York City and Washington, DC? Why?

Sometimes we use two units instead of one to give a measurement. It is usually easier to think about a person's height as 5 ft 4 in. rather than 64 in. To rename larger units as smaller units, *multiply;* to rename smaller units as larger units, *divide.*

9. Describe how you would rename 5 ft 4 in. as 64 in.

**10.** Rename 58 inches as feet and inches. Explain the method you used.

You can also use a ruler to measure the length of an object to the nearest inch, nearest $\frac{1}{2}$ inch, nearest $\frac{1}{4}$ inch, and nearest $\frac{1}{8}$ inch.

**11.** Lay your ruler along the crayon at the right. Is the length closer to 3 inches or to 4 inches?

**12.** What is the length of the crayon to the nearest inch?

**13.** Is the length of the crayon closer to 3 inches or to $3\frac{1}{2}$ inches? What is the length to the nearest $\frac{1}{2}$ inch?

**14.** Is the length of the crayon closer to 3 inches or to $3\frac{1}{4}$ inches? What is the length to the nearest $\frac{1}{4}$ inch?

**15.** Is the length of the crayon closer to 3 inches or to $3\frac{1}{8}$ inches? What is the length to the nearest $\frac{1}{8}$ inch?

**16.** Why is measuring to the nearest $\frac{1}{8}$ inch more precise than measuring to the nearest $\frac{1}{4}$ or $\frac{1}{2}$ inch?

**17.** Measure each to the nearest inch, nearest $\frac{1}{2}$ inch, nearest $\frac{1}{4}$ inch, and nearest $\frac{1}{8}$ inch.

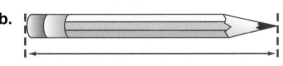

**a.**   **b.**

# Communicate

Discuss

**18.** Why are there different units of measurement?

**19.** How do you decide which customary unit of length to use in a particular situation?

**20.** Give examples of when an estimate of length is needed and when a precise measurement is essential.

**21.** You needed 85 in. of ribbon. You bought 8 ft of ribbon. Did you have enough ribbon? If so, will you have any left over? How much? Explain your answer.

*Update your skills. See page 15.*

# 9-2 Relating Customary Units of Capacity

Raul needs to put 6 gallons of water in his aquarium. He is using a quart jar to fill it. How many times will he need to fill the jar to get 6 gallons of water into the aquarium?

To find how many times Raul will need to fill the jar, find how many quarts are in 6 gallons or rename 6 gallons as quarts.

| Customary Units of Capacity |
|---|
| 8 fluid ounces (fl oz) = 1 cup (c) |
| 2 cups = 1 pint (pt) |
| 2 pints = 1 quart (qt) |
| 2 quarts = 1 half gallon |
| 4 quarts = 1 gallon (gal) |

▶ To rename customary units of capacity:
  • *Multiply* to rename larger units as smaller units.
  • *Divide* to rename smaller units as larger units.

6 gal = _?_ qt
6 gal = $(6 \times 4)$ qt
6 gal = 24 qt

Think:
1 gal = 4 qt

Raul will need to fill the quart jar 24 times to get 6 gallons of water.

## Study these examples.

13 pt = _?_ qt

13 pt = $(13 \div 2)$ qt

13 pt = $6\frac{1}{2}$ qt

Think:
2 pt = 1 qt

23 qt = _?_ gal _?_ qt

23 qt = 5 gal 3 qt

$$\begin{array}{r} 5 \text{ R3} \\ 4\overline{)23} \\ -20 \\ \hline 3 \end{array}$$ ⌐remaining
                                              ⌐ quarts

## Copy and complete.

**1.** 6 pt = _?_ qt

**2.** 22 qt = _?_ gal

**3.** 4 qt = _?_ pt

**4.** 4 c = _?_ fl oz

**5.** 16 pt = _?_ gal

**6.** 28 fl oz = _?_ c

**7.** 22 fl oz = _?_ c _?_ fl oz

**8.** 23 c = _?_ pt _?_ c

## Compare. Use <, =, or >.

**9.** 42 fl oz _?_ 5 c 2 fl oz

**10.** 22 qt _?_ 5 gal 3 qt

**11.** 2 qt _?_ 5 pt

**12.** 25 c _?_ 6 qt

**Find the picture that matches each measure. Then complete.**

**13.** ? cups of vinegar

**14.** ? pints of bottled water

**15.** ? fluid ounces of lemonade

**16.** ? quarts of milk

**Do the pictures show the correct amount for exercises 17–20? Explain why or why not.**

Communicate ✓

**17.** 8 fl oz honey

**18.** 1 c ketchup

**19.** 8 qt paint

**20.** 1 pt salad dressing

## PROBLEM SOLVING

**21.** Dale bought 4 gal of milk. The milk came in half gallons. How many half gallons of milk did she buy?

**22.** If Sally mixes $\frac{1}{2}$ c of poster paint with $\frac{1}{4}$ c of water, how many fluid ounces will she have?

**23.** Harvey wanted to buy 3 gal of honey. The beekeeper had 10 qt of honey on hand. Was Harvey able to purchase the amount of honey he wanted? Why or why not?

**24.** Philip needs to buy 1 gal of paint. The store sells 1 gal of paint for $18.49 or 1 qt of paint for $4.85. Which is the less expensive way for him to buy 1 gal of paint? Why?

 **Calculator Activity**

Discuss ✓

**25.** A leaky faucet drips 2 fl oz of water each hour. About how many gallons of water are lost from the faucet in a week? in a month? in a year? Share your results with a classmate.

*Update your skills. See page 15.*

## 9-3 **Relating Customary Units of Weight**

Estimate the weights of some
classroom objects.

**Materials Needed:** balance scale, pencils, almanac

The **ounce (oz)**, **pound (lb)**, and **ton (T)**
are customary units of weight.

| 16 ounces (oz) = 1 pound (lb) |
| 2000 pounds = 1 ton (T) |

A pencil weighs about one ounce (oz).

1. Hold a pencil in your hand and feel its weight.
   Find 3 classroom objects each of which would
   weigh about 1 oz.

2. Place the pencil on one side of a balance scale.
   Then place each object that you found, one at
   a time, on the other side of the scale.

3. Is the weight of the object less than, equal to,
   or greater than 1 oz? Record your findings
   in a table like the one shown.

| Object | <, =, or > 1 oz |
|--------|-----------------|
|        |                 |
|        |                 |

Sixteen ounces equal one pound (lb). Combine your
pencils so you have enough to weigh about 1 lb.

4. About how many pencils are in 1 lb?

5. Find 3 classroom objects each of which seems
   to weigh about 1 lb. Weigh each object on the
   balance scale using the pencils on the other
   side of the scale.

6. Is the weight of each object less than, equal to,
   or greater than 1 lb? Record your findings
   in a table like the one shown.

| Object | <, =, or > 1 lb |
|--------|-----------------|
|        |                 |
|        |                 |

7. Compare your findings with those of other groups'
   findings. Make a class list of objects that weigh
   about 1 oz and about 1 lb.

Two thousand pounds equal one ton (T).

8. About how many pencils are in 1 T.

9. Name some objects that would weigh about 1 T or more than 1 T.

10. Why are you less likely to use the ton than the pound or the ounce as a unit of weight in your everyday life?

Customary units of weight can also be renamed by multiplying or dividing.

11. Describe how you would rename 3 tons as 6000 pounds.

12. Rename 4 T 105 lb as pounds. Explain the method you used.

13. Why do you multiply to rename tons as pounds? divide to rename ounces as pounds?

## Communicate

Discuss ✓

14. What unit would you use to measure the weight of an elephant? a bag of flour? a slice of cheese? Explain your answers.

15 When might you need to know the weight of an object?

16. A sign on a bridge lists a load limit of 4 tons. Can a truck with a loaded weight of 12,000 lb safely cross the bridge? Why or why not?

17. Estimate the weight of the items, then write them in order from heaviest to lightest.
    a. a book, a pen, a letter, a calculator
    b. a bicycle, a motorcycle, a shopping cart, a truck
    c. a bowling ball, a golf ball, a basketball, a Ping-Pong ball

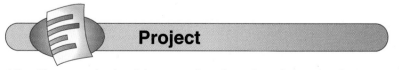
**Project**

Communicate ✓

18. Research the history of units of weights in the customary (English) system. Include an explanation of Troy units and avoirdupois units.

## 9-4 Temperature

A **thermometer** is used to measure temperature.

Temperature can be measured in **degrees Fahrenheit (°F)**, or in **degrees Celsius (°C)**.

▶ The thermometer shows some common temperatures as measured in degrees Fahrenheit and in degrees Celsius.

Water freezes at 32°F or 0°C, and boils at 212°F or 100°C.

▶ Use (⁻) sign to write temperatures below zero.

Write: ⁻10°F
Read: 10 degrees Fahrenheit below zero

Write: ⁻5°C
Read: 5 degrees Celsius below zero

▶ If you know the starting temperature and how many degrees the temperature rises or falls, you can find the final temperature.

Starting Temperature: 37°F
Change: rises 8°
Final Temperature: 45°F

Think:
37° + 8° = 45°

Starting Temperature: 12°C
Change: falls 16°
Final Temperature: ⁻4°C

Think: from 12° to 0° ⟶ 12°
12° + _?_ = 16° ⟶ 16° − 12° = 4°
4° below zero = ⁻4°

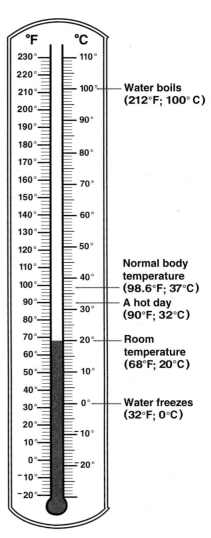

°F    °C

230°
220°
210°    100° — Water boils
200°              (212°F; 100° C)
190°    90°
180°
170°    80°
160°    70°
150°
140°    60°
130°
120°    50°
110°
100°    40°    Normal body
90°                temperature
80°    30°    (98.6°F; 37°C)
70°              A hot day
60°    20°    (90°F; 32°C)
50°    10°    Room
40°                temperature
30°    0°    (68°F; 20°C)
20°
10°    10°    Water freezes
0°                (32°F; 0°C)
⁻10°    20°
⁻20°

**Choose the most reasonable temperature for each.**

1. ice skating outdoors    **a.** ⁻40°F    **b.** 10°F    **c.** 60°F

2. oven temperature to bake a cake    **a.** 60°F    **b.** 120°F    **c.** 350°F

3. a summer day in Miami    **a.** 20°C    **b.** 75°C    **c.** 35°C

4. snow skiing    **a.** ⁻10°C    **b.** 20°C    **c.** 40°C

**Write each temperature.**

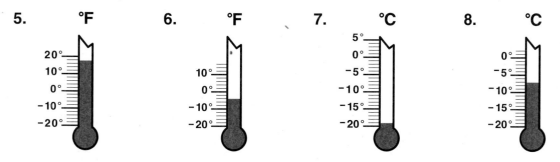

5. °F      6. °F      7. °C      8. °C

**Copy and complete the table.**

|      | Starting Temperature | Change | Final Temperature |
|------|---------------------|--------|-------------------|
| 9.   | 26°F                | rises 6° | ?               |
| 10.  | 3°F                 | falls 10° | ?              |
| 11.  | 19°C                | rises 4° | ?               |
| 12.  | 11°C                | falls 20° | ?              |

## PROBLEM SOLVING

13. The temperature yesterday was 13°F in the morning and ⁻4°F in the evening. How many degrees did the temperature drop?

14. A snowstorm drove the temperature down 3°C each hour. The thermometer read 8°C before the storm began. What did it read 4 hours later?

15. During the week the temperature each day at noon was 25°C, 23°C, 20°C, 22°C, 22°C, 18°C, and 17°C. What was the average daily noon temperature?

16. The morning temperatures during the school week were 37°F, 45°F, 41°F, 21°F, and 26°F. What was the average daily morning temperature?

## Skills to Remember

**Compute.**

17. 5 × 60

18. 13 × 7

19. 6 × 12

20. 8 × 100

21. 3 × 60 + 8

22. 5 × 7 + 4

23. 2 × 12 + 5

24. 3 × 100 + 2

25. 480 ÷ 60

26. 242 ÷ 7

27. 138 ÷ 12

28. 4000 ÷ 100

**Skip-count to make each pattern.**

29. by 5 from 0 to 60

30. by 10 from 0 to 360

## 9-5 Units of Time

The **second (s)**, **minute (min)**, **hour (h)**, **day (d)**, **week (wk)**, **month (mo)**, **year (y)**, and **century (cent.)** are units of time.

> To rename units of time:
> • *Multiply* to rename larger units as smaller units.
> • *Divide* to rename smaller units as larger units.

60 seconds (s) = 1 minute (min)
60 minutes = 1 hour (h)
24 hours = 1 day (d)
7 days = 1 week (wk)
12 months (mo) = 1 year (y)
365 days = 1 year
100 years = 1 century (cent.)

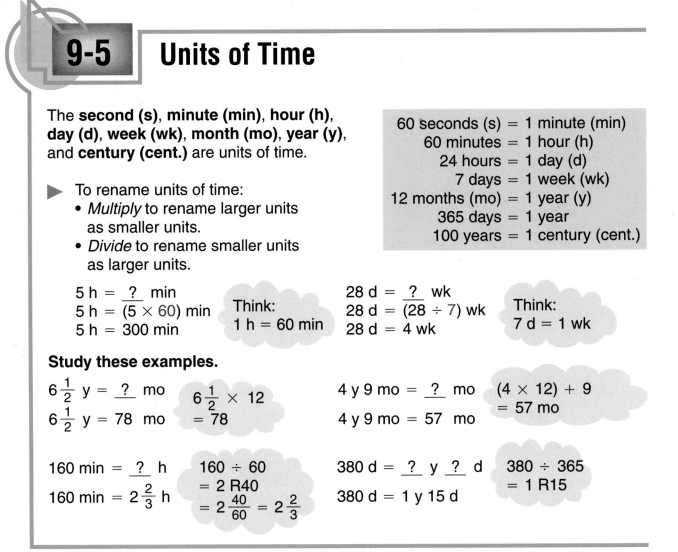

5 h = _?_ min
5 h = (5 × 60) min
5 h = 300 min

Think: 1 h = 60 min

28 d = _?_ wk
28 d = (28 ÷ 7) wk
28 d = 4 wk

Think: 7 d = 1 wk

**Study these examples.**

$6\frac{1}{2}$ y = _?_ mo
$6\frac{1}{2}$ y = 78 mo

$6\frac{1}{2}$ × 12 = 78

4 y 9 mo = _?_ mo
4 y 9 mo = 57 mo

(4 × 12) + 9 = 57 mo

160 min = _?_ h
160 min = $2\frac{2}{3}$ h

160 ÷ 60 = 2 R40 = $2\frac{40}{60}$ = $2\frac{2}{3}$

380 d = _?_ y _?_ d
380 d = 1 y 15 d

380 ÷ 365 = 1 R15

---

**Write *s*, *min*, *h*, *d*, *wk*, or *mo* to complete.**

1. Baseball season lasts about 7 _?_ .

2. Jane exercised for 15 _?_ .

3. The lightning flashed for about 3 _?_ .

4. Leo's cold lasted 1 _?_ .

5. The circus performed 263 _?_ last year.

6. The movie was about 2 _?_ long.

**Copy and complete.** Explain the method you used.    *Communicate* ✓

7. 9 min = _?_ s

8. 4 d = _?_ h

9. $2\frac{1}{2}$ y = _?_ mo

10. 400 y = _?_ cent.

11. 42 d = _?_ wk

12. 260 min = _?_ h

13. 192 min = _?_ h _?_ min

14. 300 wk = _?_ y _?_ wk

15. 7 y 5 mo = _?_ mo

16. 220 s = _?_ min _?_ s

## Computing Elapsed Time

School begins at 8:30 A.M. and ends at 2:45 P.M. How much time does Anna spend in school?

To find how much time, find the **elapsed time** from 8:30 A.M. to 2:45 P.M. Count the number of hours and then the number of minutes.

> From 8:30 A.M. to 2:30 P.M. is 6 h.
> From 2:30 P.M. to 2:45 P.M. is 15 min.

Anna spends 6 h 15 min in school.

**Find the elapsed time.**

**17.** from 2:15 P.M. to 5:30 P.M.

**18.** from 6:55 A.M. to 8:30 A.M.

**19.** from 9:30 A.M. to 4:15 P.M.

**20.** from 8:20 A.M. to 5:30 P.M.

**21.** from 10:25 P.M. to 6:38 A.M.

**22.** from 3:10 P.M. to 7:23 A.M.

**23.** Explain in your Math Journal why we need A.M. and P.M. when referring to time.

Math Journal

## PROBLEM SOLVING

**24.** Tim ran the marathon in 4 h 13 min. Neil ran the marathon in 310 min. Who ran the marathon in less time?

**25.** Elsa practiced the piano for 2 h 20 min. If she began at 2:50 P.M., at what time did she finish?

**26.** Melissa has to be at school at 8:10 A.M. She takes 25 minutes to shower and get dressed, 20 minutes to eat breakfast, and 18 minutes to walk to school. What is the latest time she should get up?

 **Connections: Science**

**27.** The Earth takes $365\frac{1}{4}$ days or 1 year to complete its orbit of the Sun. To account for the $\frac{1}{4}$ day, a leap year of 366 days occurs every 4 years. How many days are there in 4 consecutive years?

## 9-6  Time Zones

The United States is divided into six time zones.
This map shows **four time zones** of the United States:
**Pacific**, **Mountain**, **Central**, and **Eastern**.

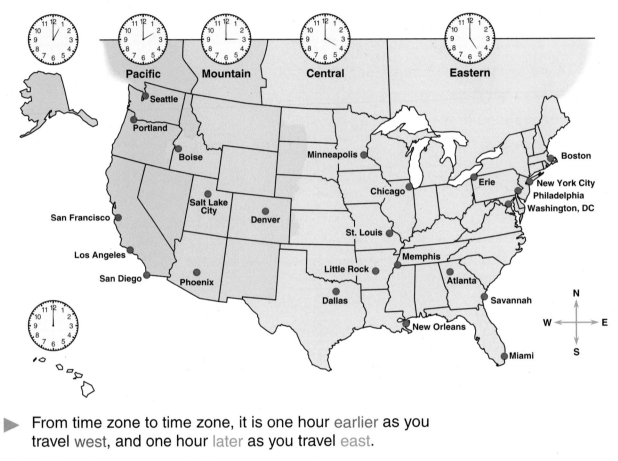

▶ From time zone to time zone, it is one hour earlier as you travel west, and one hour later as you travel east.

When it is 3:00 A.M. in Phoenix,
it is 2:00 A.M. in San Francisco.

When it is 4:00 P.M. in Chicago,
it is 5:00 P.M. in New York.

**Write the time zone where each is located.** Use the map above.

1. Boise

2. Portland

3. Dallas

4. St. Louis

5. San Diego

6. Washington, DC

7. Denver

8. Miami

9. Why do you think that time gets earlier as you move from east to west?

**Copy and complete each table.** Use the map on page 312.

| | Time Zone | Time | | | |
|---|---|---|---|---|---|
| **10.** | Pacific | 7:00 A.M. | ? | ? | ? |
| **11.** | Mountain | ? | ? | 11:30 A.M. | ? |
| **12.** | Central | ? | 1:15 P.M. | ? | ? |
| **13.** | Eastern | ? | ? | ? | 10:45 P.M. |

| | Cities | Time | | | |
|---|---|---|---|---|---|
| **14.** | Philadelphia | 9:30 P.M. | ? | ? | ? |
| **15.** | Memphis | ? | 3:15 P.M. | ? | ? |
| **16.** | Salt Lake City | ? | ? | 7:20 A.M. | ? |
| **17.** | Los Angeles | ? | ? | ? | 8:45 A.M. |

**PROBLEM SOLVING** Use the map on page 312.

18. Sandra wants to call a friend in St. Louis at 4:00 P.M. At what time should she call from Seattle?

19. Darin called his aunt in Boise from Savannah at 8:00 P.M. What time was it in Boise when he called?

20. A plane bound for Minneapolis leaves Philadelphia at 9:00 A.M. If the flight takes 2 hours, what time does the plane arrive in Minneapolis?

21. Mr. Kenney took a nonstop flight from New York to San Francisco. His plane left New York at 11:00 A.M. Eastern time and arrived in San Francisco at 1:30 P.M. Pacific time. How long was his flight?

**Finding Together**

*Discuss* ✓

22. **a.** Use your social studies textbook or an almanac to look up the other two time zones of the United States.

   **b.** List 4 cities in each time zone.

   **c.** At midnight in Houston, what time is it in each city on your list?

# Computing Customary Units

▶ To add customary units:

- Add like units. Start with smaller units.
- Rename units as needed. Regroup.

$$
\begin{array}{r}
5 \text{ ft } 10 \text{ in.} \\
+\ 8 \text{ ft }\ \ 6 \text{ in.} \\
\hline
13 \text{ ft } 16 \text{ in.}
\end{array}
= 13 \text{ ft} + 1 \text{ ft} + 4 \text{ in.} = 14 \text{ ft } 4 \text{ in.}
$$

16 in. = 12 in. + 4 in.

= 1 ft + 4 in.

▶ To subtract customary units:

- Rename units as needed. Regroup.
- Subtract like units. Start with smaller units.

$$
\begin{array}{r}
\overset{3}{\cancel{4}} \text{ gal } \overset{6}{\cancel{2}} \text{ qt} \\
-\ 2 \text{ gal } 3 \text{ qt} \\
\hline
1 \text{ gal } 3 \text{ qt}
\end{array}
$$

2 qt < 3 qt. Rename 4 gal 2 qt.

4 gal 2 qt = 3 gal + 1 gal + 2 qt

= 3 gal + 4 qt + 2 qt

= 3 gal + 6 qt

**Study these examples.**

$$
\begin{array}{r}
6 \text{ yd } 1 \text{ ft} \\
+\ 5 \text{ yd } 1 \text{ ft} \\
\hline
11 \text{ yd } 2 \text{ ft}
\end{array}
$$

$$
\begin{array}{r}
8 \text{ lb } 17 \text{ oz} \\
+\ \ \ \ \ 15 \text{ oz} \\
\hline
8 \text{ lb } 32 \text{ oz}
\end{array}
= 8 \text{ lb} + 2 \text{ lb}
$$
$$
= 10 \text{ lb}
$$

$$
\begin{array}{r}
\overset{8}{\cancel{9}} \text{ h} \overset{60 \text{ min}}{\ } \\
-\ \ \ \ \ 50 \text{ min} \\
\hline
8 \text{ h } 10 \text{ min}
\end{array}
$$

**Add.**

1.  $\begin{array}{r} 8 \text{ yd } 5 \text{ in.} \\ +3 \text{ yd } 4 \text{ in.} \\ \hline \end{array}$

2.  $\begin{array}{r} 17 \text{ ft } 2 \text{ in.} \\ +8 \text{ ft } 9 \text{ in.} \\ \hline \end{array}$

3.  $\begin{array}{r} 2 \text{ mi } 450 \text{ yd} \\ +1 \text{ mi } 330 \text{ yd} \\ \hline \end{array}$

4.  $\begin{array}{r} 6 \text{ c } 5 \text{ fl oz} \\ +3 \text{ c } 2 \text{ fl oz} \\ \hline \end{array}$

5.  $\begin{array}{r} 2 \text{ qt } 1 \text{ pt} \\ +3 \text{ qt } 1 \text{ pt} \\ \hline \end{array}$

6.  $\begin{array}{r} 2 \text{ gal } 2 \text{ qt} \\ +5 \text{ gal } 3 \text{ qt} \\ \hline \end{array}$

7.  $\begin{array}{r} 2 \text{ lb } 12 \text{ oz} \\ +4 \text{ lb } 12 \text{ oz} \\ \hline \end{array}$

8.  $\begin{array}{r} 2 \text{ h } 51 \text{ min} \\ +4 \text{ h } 29 \text{ min} \\ \hline \end{array}$

9.  $\begin{array}{r} 4 \text{ wk } 5 \text{ d} \\ +7 \text{ wk } 6 \text{ d} \\ \hline \end{array}$

10. $\begin{array}{r} 13 \text{ ft } 10 \text{ in.} \\ +\ 5 \text{ ft }\ \ 9 \text{ in.} \\ \hline \end{array}$

11. $\begin{array}{r} 4 \text{ mi }\ \ 870 \text{ yd} \\ +3 \text{ mi } 1085 \text{ yd} \\ \hline \end{array}$

12. $\begin{array}{r} 7 \text{ pt } 3 \text{ c} \\ +2 \text{ pt } 1 \text{ c} \\ \hline \end{array}$

**Subtract.**

**13.**  10 yd 2 ft
  − 4 yd 1 ft

**14.**  3 ft 10 in.
  − 1 ft 10 in.

**15.**  10 gal 1 qt
  − 7 gal 2 qt

**16.**  9 lb 3 oz
  − 3 lb 5 oz

**17.**  8 pt 1 c
  − 2 pt

**18.**  6 T 100 lb
  − 2 T 800 lb

**19.**  5 h 10 min
  − 3 h 40 min

**20.**  6 y  8 mo
  − 2 y 10 mo

**21.**  6 qt
  − 2 qt 1 pt

**Find the sum or difference.**

**22.** 7 pt + 2 pt 1 c

**23.** 6 ft 10 in. − 11 in.

**24.** 12 yd 1 ft + 2 ft

**25.** 5 d 10 h − 16 h

**26.** 10 lb 5 oz + 16 lb 12 oz

**27.** 18 c 5 fl oz − 13 c 7 fl oz

## PROBLEM SOLVING

**28.** Alfonso needs 1 ft 3 in. of ribbon to wrap one present and 1 ft 11 in. of ribbon to wrap another one. How much ribbon does he need in all?

**29.** Three packages weigh a total of 19 lb 4 oz. Two of these packages weigh 12 lb 7 oz. What is the weight of the third package?

**30.** Nestor worked for 6 h 45 min. Carla worked 4 h 20 min more than Nestor. How much time did Carla work?

**31.** A barrel holds 14 gal 1 qt of liquid. After removing 10 gal 3 qt, how much liquid is in the barrel?

**32.** Max weighs 82 lb 6 oz. He stands on a scale with his cat and the scale reads 95 lb 2 oz. How much does his cat weigh?

 **Challenge**

**33.** Jean bought 3 gal 2 qt of paint. She used 1 gal 3 qt to paint the walls of her room and some more to paint the kitchen. She had 2 qt of paint left over. How much paint did she use to paint the kitchen?

## 9-8 Problem Solving: Multi-Step Problem

**Problem:** Marina Petro worked from 8:15 A.M. to 5:30 P.M. on Monday. She spent 45 min for lunch. She was told she had worked only 7 hours. Marina disagreed and asked her employer to check her time card. Who was correct?

**1 IMAGINE**

Put yourself in the problem.
Draw a clock.

**2 NAME**

*Facts:* worked from 8:15 A.M. to 5:30 P.M. lunch for 45 min

*Question:* How long did Marina work on Monday?

**3 THINK**

*Step 1* To find the time difference between 8:15 A.M. and 5:30 P.M.:

Find the difference.
8:15 A.M. − 5:15 P.M. = _?_ h
5:30 P.M. − 5:15 P.M. = _?_ min
Find the difference.
Add the two differences.

*Step 2* Subtract 45 min from the time difference between 8:15 A.M. and 5:30 P.M.

**4 COMPUTE**

8:15 A.M. − 5:15 P.M. = 9 h ⎤
5:30 P.M. − 5:15 P.M. = 15 min ⎦ 9 h + 15 min

(9 h + 15 min) − 45 min = _?_

$$\begin{array}{r} \overset{8}{\cancel{9}} \text{ h } \overset{75}{\cancel{15}} \text{ min} = 8 \text{ h } 75 \text{ min} \\ - \qquad 45 \text{ min} = \qquad 45 \text{ min} \\ \hline 8 \text{ h } 30 \text{ min} \end{array}$$

Marina was correct. 8 h 30 min > 7 h

**5 CHECK**

Count on using the clock above to check that Marina worked 8 h 30 min.

316

**Use the Multi-Step Problem strategy to solve each problem.**

1. Last week the average temperature was 18°C.
   The daily temperatures this week were: 21°C, 17°C,
   25°C, 22°C, 23°C, 18°C, 21°C. How many
   degrees did the average temperature increase?

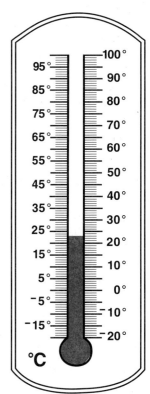

**IMAGINE**  Put yourself in the problem.

**NAME**  *Facts:*  Last week's average—18°C
   This week's temperatures—21°C,
   17°C, 25°C, 22°C, 23°C, 18°C, 21°C

   *Question:*  How many degrees did the
   average temperature increase?

**THINK**  First find the average temperature for this week.
   Then subtract to find the increase.

**COMPUTE** ⟶ **CHECK**

2. It is 10:45 A.M. in Savannah, Georgia. Sharon wants
   to phone her friend when it is 10:00 A.M. in Denver, Colorado.
   How much longer must she wait before phoning her friend?

3. The Morse family attended the school concert at 7:30 P.M.
   The concert lasted 1 hour 50 min. If it took them 15 minutes
   to drive home, what time did they arrive home?

4. A barrel holds 14 gal 1 qt of water. A gardener used 8 gal 2 qt
   to water the flowers and 2 gal 1 qt to fill the birdbath. How much
   water was left?

5. Jan bought 4 bags of oatmeal cookies and 3 bags of raisin
   cookies. Each bag of oatmeal cookies weighed 2 lb 7 oz,
   and each bag of raisin cookies weighed 1 lb 12 oz. What
   was the total weight of the cookies Jan bought?

6. Margaret leaves Newark, New Jersey, at 2:30 P.M. on Monday.
   She arrives in Honolulu, Hawaii, 13 hours later. Time in
   Hawaii is 2 hours earlier than in the Pacific time zone.
   What day and time will it be when she arrives?

# Problem-Solving Applications

**Solve each problem and explain the method you used.**

1. A frozen yogurt cart at the Midwood Mall weighs about half a ton. About how many pounds does the cart weigh?

2. The yogurt cart's awning is 50 in. high. The awning on a nearby jewelry cart is 4 ft 9 in. high. Which awning is higher?

3. The jewelry cart owner opens it at 11:25 A.M. and closes it at 10:00 P.M. How long is the cart open each day?

4. Each side of a square sign is 2 ft 3 in. How long is the trim that goes around it?

5. How many pints of yogurt are there in a 2-gallon container?

6. The temperature outdoors was 48°F. Inside the mall the temperature was 70°F. How much colder was it outdoors?

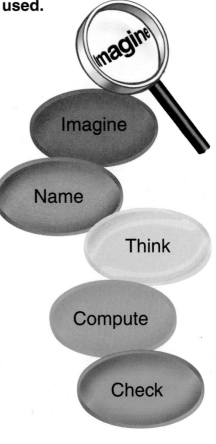

**Use the chart for problems 7 and 8.**

7. What is the cost per ounce of each special?

8. Which special is the best buy?

9. The jewelry cart has a rectangular sign 3 ft long and 2 ft 5 in. wide. What is its perimeter?

10. This pictograph shows the number of yogurt cones sold last Friday. How many more peach than melon cones were sold?

11. What symbol would be used to represent 3 cones? 9 cones?

| Today's Specials | |
|---|---|
| 5 | 6-oz cups for $3.00 |
| 4 | 8-oz cups for $3.00 |
| 2 | pints        for $2.50 |

**Yogurt Sales**

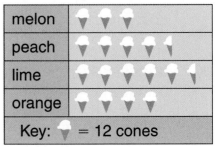

| melon | ▼ ▼ ▼ |
|---|---|
| peach | ▼ ▼ ▼ ▼ ▼ |
| lime | ▼ ▼ ▼ ▼ ▼ ▼ |
| orange | ▼ ▼ ▼ ▼ |
| Key: ▼ = 12 cones | |

**Choose a strategy from the list or use another strategy you know to solve each problem.**

USE THESE STRATEGIES
Multi-Step Problem
Use Formulas
Hidden Information
Use a Model/Drawing
Organized List
Find a Pattern

12. Amelia opened the jewelry cart at 9:15 A.M. and worked for $5\frac{1}{4}$ h. Then Marie took over until 9:45 P.M. How long did Marie work?

13. There are 4 carts at the mall. Each cart is 4 in. taller than the next. The tallest cart is 6 ft 2 in. What are the heights of the two shortest carts?

14. Five flavors of yogurt are sold. How many possible combinations of 3 different flavors can Jules order?

15. The diagonal of the square in the mall sign is 3 ft 9 in. What is the approximate circumference of the sign?

16. For every quart of yogurt you buy, you will get a $2\frac{1}{4}$-oz package of topping. Roy bought 2 gallons of yogurt. How many ounces of topping did Roy get?

17. David won a charm at the jewelry cart by naming the tenth number in the sequence 1, 3, 7, 15, . . . What number did he name?

18. Cheryl gives this business card to each new customer. Measure each side to the nearest $\frac{1}{8}$ inch. Then find the perimeter.

Gems 'n' Jewels

With every $25 purchase receive $2 discount.

**Midwood Mall**

19. The yogurt cart features a special on peach and lime. Fifteen people bought a pint of each. If 2 dozen pints of peach and 20 pints of lime were sold, how many people bought a pint of yogurt on sale?

**Make Up Your Own**

Communicate ✓

20. Write a multi-step problem. Then have a classmate solve it.

**Write the letter of the best estimate.**                    (See pp. 302–309.)

1. width of a camera            a. 8 yd      b. 8 ft       c. 8 in.
2. capacity of a blender        a. 2 gal     b. 2 c        c. 2 qt
3. weight of a whale            a. $1\frac{2}{3}$ oz   b. $1\frac{2}{3}$ T   c. $1\frac{2}{3}$ lb
4. temperature on a beach day   a. 40°F      b. 50°F       c. 90°F
5. temperature for water to freeze   a. 0°C   b. 32°C       c. 10°C

**Copy and complete.**                         (See pp. 302–307, 310–311.)

6. 72 in. = _?_ yd          7. 490 min = _?_ h        8. 112 oz = _?_ lb
9. 5 qt = _?_ pt            10. 4 yd = _?_ ft         11. 12 min = _?_ s
12. 3 c = _?_ fl oz         13. 6 gal = _?_ pt        14. 2 T = _?_ lb

**Copy and complete the table.** You may use the map on page 312.    (See pp. 312–313.)

| | City | Time | | | |
|---|---|---|---|---|---|
| 15. | Washington, DC | 6:00 A.M. | ? | ? | ? |
| 16. | Chicago, Illinois | ? | 7:30 A.M. | ? | ? |
| 17. | Denver, Colorado | ? | ? | 8:15 P.M. | ? |
| 18. | Los Angeles, California | ? | ? | ? | 9:45 P.M. |

**Add or subtract.**                              (See pp. 314–315.)

19.   4 gal 2 qt        20.    5 ft  2 in.        21.   4 wk 1 d
    − 1 gal 3 qt            + 11 ft 11 in.            − 2 wk 5 d

22.   2 lb 10 oz        23.   4 yd 18 in.         24.   9 pt 1 c
    + 5 lb  9 oz             − 2 yd 26 in.             + 2 pt 1 c

**PROBLEM SOLVING**                        (See pp. 308–311, 316–319.)

25. The temperature last Monday was 12°F in the morning and ⁻8°F in the evening. How many degrees did the temperature drop?

26. David jogs around the $2\frac{1}{2}$-mile perimeter of the lake 6 days a week. On Sundays he jogs around the lake twice. How many miles around the lake does David jog in one week?

(See *Still More Practice*, p. 485.)

## TESSELLATIONS

A **tessellation** is a pattern formed by covering a plane surface with a set of polygons such that no polygons overlap and no gaps exist between the polygons.

You can use a slide, a flip, or a turn to tessellate.

Study the following tessellations.

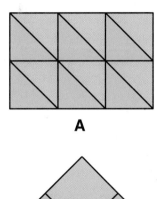

**A**

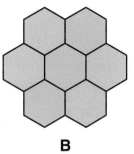

**B**

**C**

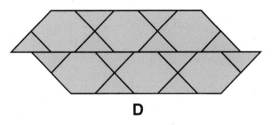

**D**

## PROBLEM SOLVING

What polygons are used in each tessellation?

**1.** *A*          **2.** *B*          **3.** *C*          **4.** *D*

**5.** What room in your house has a tessellating pattern? What polygons are in the pattern?

**6.** Do all regular polygons form a tessellation?

**7.** Create your own tessellation by using a combination of regular polygons.

**8.** Research the Dutch artist *M. C. Escher*. Report to the class how he used tessellations.

*Communicate* ✓

**Performance Assessment**

**Use the table.**
Mia recorded data about her pets in this table.

| Pet | Age | Weight |
|-----|-----|--------|
| Rex | 4 y 2 mo | 42 lb 10 oz |
| Tiny | 1 y 10 mo | 1 lb 13 oz |
| Goldie | 2 y 6 mo | 9 lb 8 oz |

1. How much older is Goldie than Tiny?

2. What is the combined weight of her pets?

3. Mia weighs 70 lb. How much less does Rex weigh?

4. Write and solve a problem using the data.

**Write the letter of the best estimate.**

5. length of a bed    **a.** 6 in.    **b.** 6 yd    **c.** 6 ft

6. weight of a bag of flour    **a.** 6 lb    **b.** 6 oz    **c.** 6 T

7. capacity of a large bowl    **a.** 4 gal    **b.** 4 pt    **c.** 4 qt

8. temperature on a cold, snowy day    **a.** 0°C    **b.** −10°C    **c.** 10°C

9. temperature on a good day to swim    **a.** 5°F    **b.** 45°F    **c.** 90°F

**Compare. Write <, =, or >.**

10. 42 ft __?__ 14 yd     11. 3 qt __?__ 7 pt     12. 1 gal 5 qt __?__ 2 gal 2 pt

13. 15 c __?__ 4 qt     14. 120 in. __?__ 10 ft     15. 350 min __?__ 3 h

**Copy and complete the table.**

| | Time Zone | Time | | | |
|-----|-----------|------|------|------|------|
| 16. | Pacific | 11:30 P.M. | ? | ? | ? |
| 17. | Mountain | ? | 9:15 P.M. | ? | ? |
| 18. | Central | ? | ? | 8:45 A.M. | ? |
| 19. | Eastern | ? | ? | ? | 6:00 A.M. |

**PROBLEM SOLVING**    *Use a strategy you have learned.*

20. Chris left her house at 7:45 A.M. and came home at 5:30 P.M. How long was she away from home?

21. The temperature at midnight was −6°C. It rose to 3°C by 8:00 A.M. How many degrees did it rise?

# Cumulative Review III

**Choose the best answer.**

**1.** Which is ordered from greatest to least?

   **a.** 2.3, 2.4, 2.0, 2.9
   **b.** 0.14, 0.16, 0.18, 0.2
   **c.** 7.43, 7.42, 7.41, 7.4
   **d.** none of these

**2.** How much more than

$658 - 309$ is

$658 \times 309$?

   **a.** 22,208
   **b.** 202,971
   **c.** 202,973
   **d.** 203,671

**3** $16)\overline{\$138.88}$

   **a.** $8.68
   **b.** $9.38
   **c.** $18.68
   **d.** $19.38

**4.** Choose the simplest form of the mixed number.

$27\frac{20}{15}$

   **a.** $27\frac{1}{3}$   **b.** $27\frac{3}{4}$
   **c.** $28\frac{1}{3}$   **d.** $28\frac{1}{2}$

**5.** $\begin{array}{r} 23\frac{3}{8} \\ -17\frac{3}{4} \\ \hline \end{array}$

   **a.** $5\frac{5}{8}$   **b.** $6\frac{5}{8}$
   **c.** $6\frac{3}{4}$   **d.** $5\frac{3}{4}$

**6.** $5\frac{1}{5} \div 5$

   **a.** $\frac{1}{5}$   **b.** $1\frac{1}{26}$
   **c.** 26   **d.** not given

**7.** Which is a true statement about the data?

**Average Weekly Temperature (in °F)**

| Week | 1 | 2 | 3 | 4 | 5 | 6 | 7 |
|---|---|---|---|---|---|---|---|
| Temperature | 28 | 20 | 33 | 34 | 28 | 30 | 21 |

   **a.** median = 33   **b.** median = mode
   **c.** mean > median   **d.** range = 13

**8.** Use the spinner. Which is a true probability statement?

   **a.** $P(3) = \frac{1}{8}$
   **b.** $P(not\ 3) = \frac{1}{3}$
   **c.** $P(3) = \frac{3}{8}$
   **d.** $P(not\ 3) = \frac{1}{5}$

**9.** Which type of angle is shown?

   **a.** acute   **b.** obtuse
   **c.** scalene   **d.** right

**10.** Which is true about the polygons?

   **a.** congruent, *not* similar
   **b.** congruent and similar
   **c.** similar, *not* congruent
   **d.** none of these

# Ongoing Assessment III

## For Your Portfolio

**Solve each problem. Explain the steps and the strategy
or strategies you used for each. Then choose one from
problems 1–3 for your Portfolio.**

1. Tina wants to put a fence around
   the enclosed area shown at the
   right. To decide how much fencing
   to use, should she find its area or its
   perimeter? Explain. Then compute
   the correct measurement.

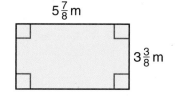

$5\frac{7}{8}$ m

$3\frac{3}{8}$ m

2. Dawind has a circular rug in his room. The diameter is
   $3\frac{1}{3}$ meters. Estimate the circumference.

3. A forward on the Lansing varsity basketball team is 6 ft 4 in. tall.
   A guard is 5 ft 11 in. The center is 6 ft 9 in. What is the average
   (mean) height of the three players?

### Tell about it.

4. Explain how you found your estimate in problem 2.

*Communicate*

5. To solve problem 3, did you change feet to inches or inches
   to feet? Could you have done it the other way? Try it.

---

## For Rubric Scoring

**Listen for information on how your work will be scored.**

Feng's grandfather gave him $120 for his
birthday. The circle graph shows how he
spent the money.

**Feng's Expenses**

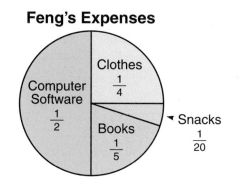

6. How much money did Feng spend on
   each item?

7. Use graph paper. Make and label a
   different kind of graph that shows the
   money (dollars) that Feng spent on
   each item.

8. How would the circle graph look different if Feng spent the same
   amount on clothes and snacks? What amount of money is that?

# Decimals: Addition and Subtraction

# 10

**Speed!**

no hands
down the hill
no hands
just the wheel

brisk breeze
in my hair
such ease
not a care

my feet
steer the bike
my seat
sitting tight

wheels spin
this is speed!
wheels spin
all **I** need

*Monica Kulling*

**In this chapter you will:**

Estimate, add, and subtract decimals
Learn about computer databases
Solve problems with extra information

**Critical Thinking/Finding Together**

A cyclist biked one tenth of a mile less
on Tuesday than on Monday. He biked
five miles farther on Wednesday than on
Tuesday. He biked 13.8 miles on Monday.
Which day did he bike the farthest?

# **Decimal Sense**

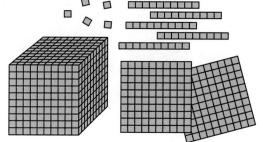

Decimals may be represented on a number line. As with whole numbers, a greater decimal is located to the right of a lesser decimal.

Study these number lines:

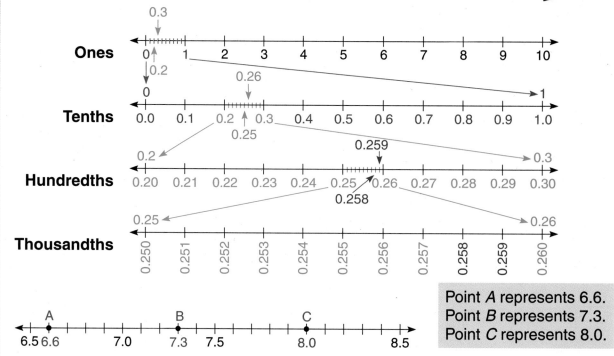

Point *A* represents 6.6.
Point *B* represents 7.3.
Point *C* represents 8.0.

**Name the decimal represented by *A*, *B*, and *C* on each number line.**

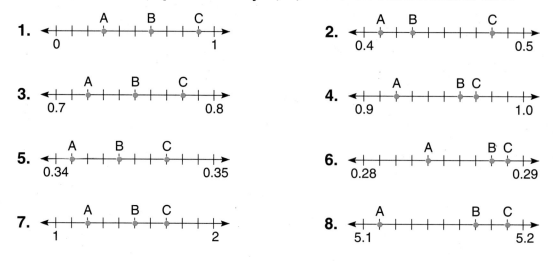

1.
   0 ... 1

2.
   0.4 ... 0.5

3.
   0.7 ... 0.8

4.
   0.9 ... 1.0

5.
   0.34 ... 0.35

6.
   0.28 ... 0.29

7.
   1 ... 2

8.
   5.1 ... 5.2

**Name the decimal for each point on the number line.**

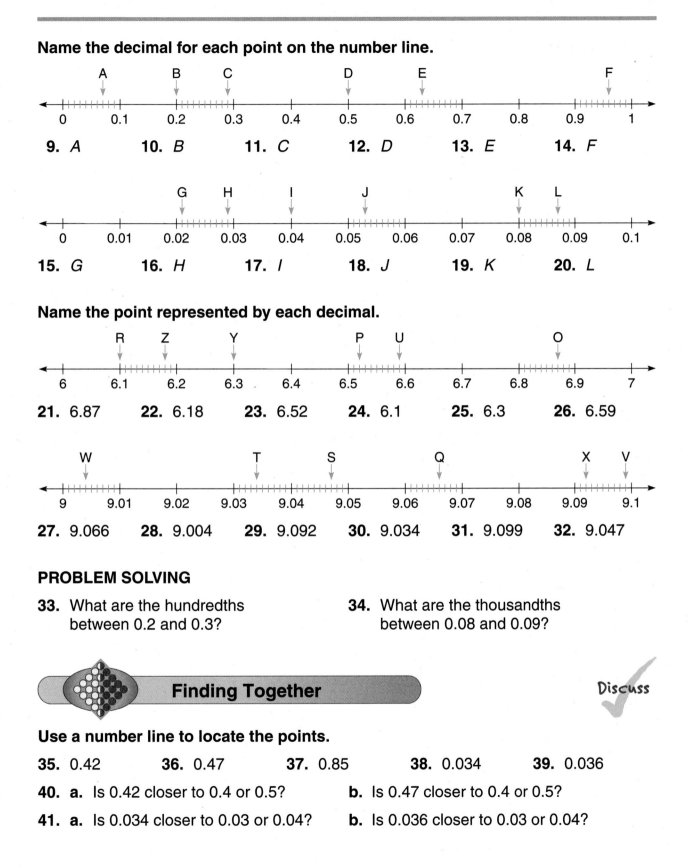

**9.** A      **10.** B      **11.** C      **12.** D      **13.** E      **14.** F

**15.** G      **16.** H      **17.** I      **18.** J      **19.** K      **20.** L

**Name the point represented by each decimal.**

**21.** 6.87      **22.** 6.18      **23.** 6.52      **24.** 6.1      **25.** 6.3      **26.** 6.59

**27.** 9.066      **28.** 9.004      **29.** 9.092      **30.** 9.034      **31.** 9.099      **32.** 9.047

## PROBLEM SOLVING

**33.** What are the hundredths between 0.2 and 0.3?

**34.** What are the thousandths between 0.08 and 0.09?

### Finding Together

*Discuss*

**Use a number line to locate the points.**

**35.** 0.42      **36.** 0.47      **37.** 0.85      **38.** 0.034      **39.** 0.036

**40.** **a.** Is 0.42 closer to 0.4 or 0.5?      **b.** Is 0.47 closer to 0.4 or 0.5?

**41.** **a.** Is 0.034 closer to 0.03 or 0.04?      **b.** Is 0.036 closer to 0.03 or 0.04?

# Decimals and Place Value

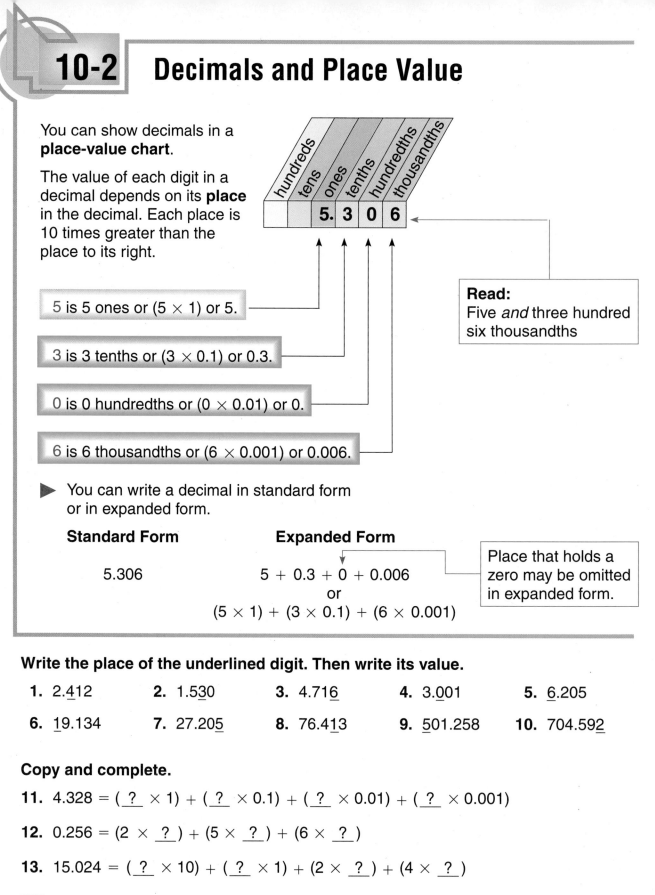

You can show decimals in a **place-value chart**.

The value of each digit in a decimal depends on its **place** in the decimal. Each place is 10 times greater than the place to its right.

hundreds tens ones tenths hundredths thousandths

**5. 3 0 6**

5 is 5 ones or (5 × 1) or 5.

3 is 3 tenths or (3 × 0.1) or 0.3.

0 is 0 hundredths or (0 × 0.01) or 0.

6 is 6 thousandths or (6 × 0.001) or 0.006.

**Read:**
Five *and* three hundred six thousandths

► You can write a decimal in standard form or in expanded form.

| **Standard Form** | **Expanded Form** |
|---|---|
| 5.306 | 5 + 0.3 + 0 + 0.006 |
| | or |
| | (5 × 1) + (3 × 0.1) + (6 × 0.001) |

Place that holds a zero may be omitted in expanded form.

**Write the place of the underlined digit. Then write its value.**

**1.** 2.4̲12

**2.** 1.5̲30

**3.** 4.716̲

**4.** 3.0̲01

**5.** 6̲.205

**6.** 1̲9.134

**7.** 27.20̲5

**8.** 76.41̲3

**9.** 5̲01.258

**10.** 704.59̲2

**Copy and complete.**

**11.** 4.328 = ( ?̲ × 1) + ( ?̲ × 0.1) + ( ?̲ × 0.01) + ( ?̲ × 0.001)

**12.** 0.256 = (2 × ?̲ ) + (5 × ?̲ ) + (6 × ?̲ )

**13.** 15.024 = ( ?̲ × 10) + ( ?̲ × 1) + (2 × ?̲ ) + (4 × ?̲ )

**Write each number in standard form.**

**14.** two and nine thousandths

**15.** fifty-four and eight tenths

**16.** six and five hundredths

**17.** eleven and one thousandth

**Write each in expanded form.**

**18.** 4.512     **19.** 3.014     **20.** 5.025     **21.** 2.107     **22.** 6.51

**23.** 13.15     **24.** 131.5     **25.** 1.315     **26.** 0.315     **27.** 13.152

**Write each in standard form.**

**28.** 8 + 0.1 + 0.05 + 0.003

**29.** 4 + 0.4 + 0.08 + 0.006

**30.** 0.5 + 0.05 + 0.004

**31.** 0.7 + 0.09 + 0.009

**32.** 9 + 0.04 + 0.006

**33.** 30 + 0.5 + 0.007

**34.** 200 + 0.7 + 0.001

**35.** 400 + 10 + 0.05 + 0.009

**For which numbers in the box:**

**36.** does the 5 have a value of 0.005?

**37.** does the 3 have a value of 0.03?

**38.** does the 6 have a value of 0.6?

| 35.337 | 8.615 |
| 98.545 | 0.516 |
| 7.653 | 0.238 |

**PROBLEM SOLVING**

**39.** Write a decimal in tenths that is less than 2.1 and greater than 1.8.

**40.** A car travels at a speed of 0.915 miles per minute. Write the speed in expanded form.

## Skills to Remember

**Align and add.**

**41.** 478 + 96

**42.** 5509 + 693

**43.** 857 + 9278

**44.** 507 + 38 + 4

**45.** 45 + 317 + 6

**46.** 312 + 9 + 63

## 10-3 Adding Decimals

David has 3 strips of wood measuring 0.28 m, 0.6 m, and 0.09 m, respectively. How many meters of wood does he have?

To find how many meters of wood, add: 0.28 + 0.6 + 0.09 = ?

▶ You can use base ten blocks to model 0.28 + 0.6 + 0.09.

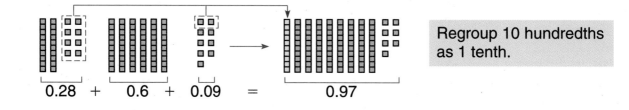

| 0.28 | + | 0.6 | + | 0.09 | = | 0.97 |

Regroup 10 hundredths as 1 tenth.

▶ To add decimals, add the same way as you add whole numbers.

| Line up the decimal points. | Add the hundredths. Regroup. | Add the tenths. | Write the decimal point in the sum. |
|---|---|---|---|
| 0.28<br>0.60 ← ⎡0.6 = 0.60⎤<br>+0.09 | ¹<br>0.28<br>0.60<br>+0.09<br>7 | ¹<br>0.28<br>0.60<br>+0.09<br>97 | 0.28<br>0.60<br>+0.09<br>0.97 |

David has 0.97 m of wood.

**Study these examples.**

| ¹<br>0.9<br>+0.3<br>1.2 | ¹<br>0.39<br>+0.23<br>0.62 | 0.53<br>+0.40<br>0.93 | ¹ ¹<br>0.34<br>0.72<br>+0.54<br>1.60 = 1.6 |

---

**Use base ten blocks to model each sum. Then write the sum.**

**1.**   0.2
      +0.5

**2.**   0.63
      +0.03

**3.**   0.42
      +0.54

**4.**   0.3
      0.4
      +0.2

**5.**   0.05
      0.82
      +0.12

**Find the sum.**

| 6. | 7. | 8. | 9. | 10. |
|---|---|---|---|---|
| 0.39 <br> + 0.05 | 0.49 <br> + 0.38 | 0.8 <br> + 0.39 | 0.98 <br> + 0.32 | 0.87 <br> + 0.48 |

| 11. | 12. | 13. | 14. | 15. |
|---|---|---|---|---|
| 0.6 <br> 0.5 <br> + 0.8 | 0.09 <br> 0.75 <br> + 0.24 | 0.7 <br> 0.29 <br> + 0.43 | 0.4 <br> 0.75 <br> + 0.6 | 0.07 <br> 0.3 <br> + 0.9 |

**Align and add.**

**16.** 0.2 + 0.79

**17.** 0.03 + 0.9

**18.** 0.54 + 0.05

**19.** 0.38 + 0.06

**20.** 0.72 + 0.3

**21.** 0.7 + 0.97

**22.** 0.6 + 0.54 + 0.05

**23.** 0.82 + 0.6 + 0.05

**24.** 0.2 + 0.08 + 0.32

**25.** 0.9 + 0.01 + 0.65

**Find the perimeter.**

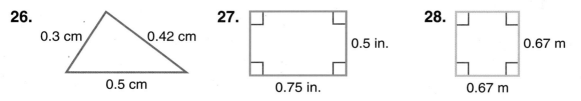

**26.** 0.3 cm, 0.42 cm, 0.5 cm

**27.** 0.5 in., 0.75 in.

**28.** 0.67 m, 0.67 m

## PROBLEM SOLVING

**29.** Rainfall for two days was measured as 0.24 in. and 0.39 in. at the city airport. What was the total rainfall measured over the two days?

**30.** Chana has 3 packages of cheese weighing 0.24 lb, 0.69 lb, and 0.8 lb, respectively. How many pounds of cheese does she have?

**True or false? Explain your answer.**

*Communicate* ✓

**31.** The sum of two decimals less than 1 is always less than 1.

**32.** The sum of two decimals greater than 0.5 is always greater than 1.

**Project**

**33.** Find newspaper or magazine articles demonstrating the use of decimals in real-life situations. Make a poster from these clippings. Display your work in the classroom.

## 10-4 Estimate Decimal Sums

A bicycle trail has three sections measuring 5.5 mi, 6.45 mi, and 7.62 mi. About how long is the bicycle trail?

To find about how long, estimate the sum: 5.5 + 6.45 + 7.62

You can use rounding or front-end estimation to estimate a decimal sum.

▶ To **estimate** a *decimal sum* by *rounding*:

- Round the decimals to the greatest *nonzero* place of the smallest number.

- Add the rounded numbers.

$$
\begin{array}{rcl}
5.5 & \longrightarrow & 6 \\
6.45 & \longrightarrow & 6 \\
+\,7.62 & \longrightarrow & +8 \\
\hline
& \text{about} & 20
\end{array}
$$

▶ To **estimate** a *decimal sum* by *front-end estimation*:

- Add the *nonzero* front digits.

- Write zeros for the other digits.

$$
\begin{array}{r}
5.5 \\
6.45 \\
+\,7.62 \\
\hline
\text{about } 18.00
\end{array}
$$

The bicycle trail is about 18 to 20 mi long.

### Study these examples.

$$
\begin{array}{rcl}
0.591 & \longrightarrow & 0.6 \\
+\,0.305 & \longrightarrow & +0.3 \\
\hline
& \text{about} & 0.9
\end{array}
\qquad
\begin{array}{r}
0.591 \\
+\,0.305 \\
\hline
\text{about } 0.800
\end{array}
$$

$$
\begin{array}{rcl}
23.31 & \longrightarrow & 20 \\
+\,46.672 & \longrightarrow & +50 \\
\hline
& \text{about} & 70
\end{array}
\qquad
\begin{array}{r}
23.31 \\
+\,46.672 \\
\hline
\text{about } 60.000
\end{array}
$$

So the exact sum is between 0.8 and 0.9.

So the exact sum is between 60 and 70.

**Write the letter of the best estimated sum.**

1. 10.93 + 6.1      **a.** 17      **b.** 15      **c.** 11      **d.** 18

2. 0.872 + 0.141 + 0.56      **a.** 1.3      **b.** 1.2      **c.** 1.4      **d.** 1.1

**Estimate the sum by rounding.**

| 3. | 4. | 5. | 6. | 7. |
|---|---|---|---|---|
| 0.57 | 6.6 | 8.57 | 0.771 | 5.412 |
| 0.91 | 1.8 | 0.73 | 0.567 | 2.793 |
| + 0.3 | + 4.2 | + 0.59 | + 0.48 | + 0.137 |

**8.** 7.39 + 5.3    **9.** 0.554 + 0.94    **10.** 3.07 + 7.5 + 4.273

**Estimate the sum. Use front-end estimation.**

| 11. | 12. | 13. | 14. | 15. |
|---|---|---|---|---|
| 0.19 | 7.8 | 2.65 | 0.228 | 3.791 |
| 0.74 | 5.2 | 6.2 | 0.376 | 4.38 |
| + 0.8 | + 4.4 | + 5.93 | + 0.59 | + 7.332 |

**16.** 3.2 + 6.43    **17.** 0.257 + 0.65    **18.** 1.708 + 6.391 + 3.94

**Estimate by both rounding and front-end estimation.
Between what two numbers will the exact sum be?**

| 19. | 20. | 21. | 22. | 23. |
|---|---|---|---|---|
| 0.93 | 3.283 | 50.78 | 35.472 | 68.24 |
| + 0.564 | + 8.59 | + 18.9 | + 25.29 | + 40.168 |

| 24. | 25. | 26. | 27. | 28. |
|---|---|---|---|---|
| 5.23 | 8.61 | 45.31 | 2.653 | 19.134 |
| 4.7 | 2.315 | 88.2 | 3.91 | 23.14 |
| + 6.5 | + 7.83 | + 92.7 | + 4.32 | + 37.421 |

**29.** 17.08 + 25.9    **30.** 3.07 + 2.54 + 4.654    **31.** 37.91 + 59.6 + 27.732

## PROBLEM SOLVING

**32.** Elaine rode her bike 3.45 mi on Friday, 5.38 mi
on Saturday, and 6.35 mi on Sunday. About how
many miles did she ride her bike in these three days?

 **Share Your Thinking**    Communicate

**Complete the statement to make it true. Write *less than* or
*greater than*. Explain your answer.**

**33.** When rounding down the addends,
the estimated sum is __?__ the
actual sum.

**34.** When rounding up the addends,
the estimated sum is __?__ the
actual sum.

**35.** The estimated sum by front-end estimation is __?__ the actual sum.

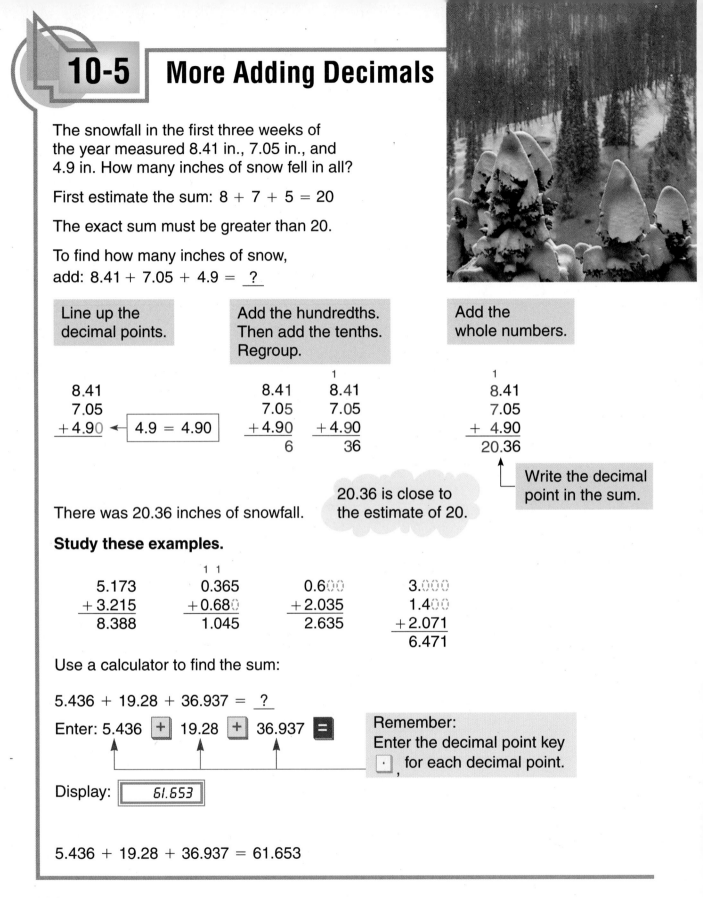

## 10-5 More Adding Decimals

The snowfall in the first three weeks of the year measured 8.41 in., 7.05 in., and 4.9 in. How many inches of snow fell in all?

First estimate the sum: $8 + 7 + 5 = 20$

The exact sum must be greater than 20.

To find how many inches of snow, add: $8.41 + 7.05 + 4.9 = $ ?

| Line up the decimal points. | Add the hundredths. Then add the tenths. Regroup. | Add the whole numbers. |
|---|---|---|

Line up the decimal points.

```
  8.41
  7.05
+ 4.90   ← 4.9 = 4.90
```

Add the hundredths. Then add the tenths. Regroup.

```
                1
  8.41        8.41
  7.05        7.05
+ 4.90      + 4.90
     6         36
```

Add the whole numbers.

```
     1
  8.41
  7.05
+  4.90
 20.36
```

Write the decimal point in the sum.

There was 20.36 inches of snowfall.

20.36 is close to the estimate of 20.

### Study these examples.

```
              1 1
   5.173     0.365       0.600      3.000
 + 3.215   + 0.680     + 2.035      1.400
   8.388     1.045       2.635    + 2.071
                                    6.471
```

Use a calculator to find the sum:

$5.436 + 19.28 + 36.937 = $ ?

Enter: 5.436 [+] 19.28 [+] 36.937 [=]

Remember: Enter the decimal point key [ · ], for each decimal point.

Display: | 61.653 |

$5.436 + 19.28 + 36.937 = 61.653$

**Estimate. Then find the sum.**

| 1. | 3.6<br>+2.8 | 2. | 3.02<br>+4.06 | 3. | 4.12<br>+5.63 | 4. | 0.597<br>+0.802 | 5. | 3.125<br>+7.431 |
|---|---|---|---|---|---|---|---|---|---|

| 6. | 37.01<br>+ 2.69 | 7. | 29.6<br>+ 3.49 | 8. | 42.75<br>+50.8 | 9. | 4.071<br>+15.32 | 10. | 56.021<br>+ 3.123 |
|---|---|---|---|---|---|---|---|---|---|

| 11. | 5.4<br>3.2<br>+7.6 | 12. | 7.36<br>9.43<br>+5.72 | 13. | 0.825<br>0.914<br>+0.203 | 14. | 16.3<br>25.7<br>+32.4 | 15. | 9.435<br>7.362<br>+8.417 |
|---|---|---|---|---|---|---|---|---|---|

**Align and add.**

**16.** 7.05 + 9.5     **17.** 17 + 4.5 + 1.15     **18.** 2.114 + 4 + 1.07

**19.** 28.72 + 6.8     **20.** 7.424 + 3.005 + 10.1     **21.** 6.9 + 3.08 + 1.247

**22.** 7.602 + 0.98     **23.** 6.004 + 27.31 + 9.5     **24.** 0.63 + 7.819 + 24.8

**Compare. Write <, =, or >. You may use a calculator.**

**25.** 5.6 + 7.82 _?_ 13.52     **26.** 35.5 + 19.8 + 0.63 _?_ 55.73

**27.** 7.15 _?_ 2.079 + 5.08     **28.** 35.195 _?_ 24.08 + 5 + 6.115

**29.** 0.668 + 6.584 _?_ 7.052     **30.** 7.19 + 0.583 + 2.745 _?_ 9.518

## PROBLEM SOLVING

**31.** Tara biked 13.8 laps in the morning and 14.75 laps in the afternoon. How many laps did she bike in all?

**32.** Aldo ran 9.8 mi, Greg ran 13.7 mi, and Victor ran 12.5 mi. What was the total distance for the three?

**33.** The leading team's score in the Decimal Olympics was 40.816 points. The final team's three players scored 14.21, 12.924, and 13.689 points. Did they have enough points to take the lead?

## Calculator Activity

*Algebra* ✓

**Find the missing digits.** Use Guess and Test.

| 34. | ☐8.6 7<br>+3 5.☐9<br>7 4.4 6 | 35. | 2☐.5 6<br>+ 3.☐5<br>3 2.2☐ | 36. | 3 9.☐☐2<br>+☐6.3 4☐<br>9☐.7 2 9 | 37. | 5☐.5 ☐4<br>+ 6.☐7☐<br>☐2.3 0 1 |
|---|---|---|---|---|---|---|---|

**Subtracting Decimals**

Aileen jumped 0.9 m on her first jump and 0.78 m on her second jump. How much farther was her first jump than her second jump?

To find how much farther her first jump was, subtract: 0.9 − 0.78 = __?__

▶ You can use base ten blocks to model 0.9 − 0.78.

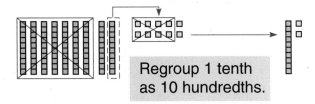

Regroup 1 tenth as 10 hundredths.

0.9 − 0.78 = 0.12

▶ To subtract decimals, subtract the same way as you subtract whole numbers.

| Line up the decimal points. | Regroup. Subtract the hundredths. | Subtract the tenths. | Write the decimal point in the difference. |
|---|---|---|---|
| $\begin{array}{r} 0.90 \\ -0.78 \end{array}$  0.9 = 0.90 | $\begin{array}{r} \overset{8\ 10}{0.9\cancel{0}} \\ -0.7\ 8 \\ \hline 2 \end{array}$ | $\begin{array}{r} \overset{8\ 10}{0.9\cancel{0}} \\ -0.7\ 8 \\ \hline 1\ 2 \end{array}$ | $\begin{array}{r} \overset{8\ 10}{0.9\cancel{0}} \\ -0.7\ 8 \\ \hline 0.1\ 2 \end{array}$ |

Aileen jumped 0.12 m farther on her first jump.

**Study these examples.**

$\begin{array}{r} 0.8 \\ -0.3 \\ \hline 0.5 \end{array}$
$\qquad$
$\begin{array}{r} 0.69 \\ -0.52 \\ \hline 0.17 \end{array}$
$\qquad$
$\begin{array}{r} 0.73 \\ -0.40 \\ \hline 0.33 \end{array}$
$\qquad$
$\begin{array}{r} \overset{4\ 18}{0.\cancel{5}\cancel{8}} \\ -0.3\ 9 \\ \hline 0.1\ 9 \end{array}$

**Use base ten blocks to model each difference. Then write the difference.**

**1.** $\begin{array}{r} 0.7 \\ -0.2 \end{array}$
**2.** $\begin{array}{r} 0.75 \\ -0.2 \end{array}$
**3.** $\begin{array}{r} 0.95 \\ -0.54 \end{array}$
**4.** $\begin{array}{r} 0.7 \\ -0.25 \end{array}$
**5.** $\begin{array}{r} 0.76 \\ -0.08 \end{array}$

**Find the difference.**

| | | | | | | | | | |
|---|---|---|---|---|---|---|---|---|---|
| **6.** | 0.08 <br> − 0.04 | **7.** | 0.67 <br> − 0.36 | **8.** | 0.63 <br> − 0.38 | **9.** | 0.84 <br> − 0.46 | **10.** | 0.51 <br> − 0.29 |
| **11.** | 0.9 <br> − 0.2 | **12.** | 0.78 <br> − 0.3 | **13.** | 0.4 <br> − 0.06 | **14.** | 0.9 <br> − 0.37 | **15.** | 0.7 <br> − 0.54 |

**Align and subtract.**

**16.** 0.97 − 0.6     **17.** 0.39 − 0.2     **18.** 0.8 − 0.17

**19.** 0.49 − 0.24     **20.** 0.97 − 0.5     **21.** 0.5 − 0.09

**22.** 0.89 − 0.7     **23.** 0.6 − 0.16     **24.** 0.61 − 0.3

**25.** 0.92 − 0.3     **26.** 0.8 − 0.51     **27.** 0.47 − 0.06

**Add or subtract to find the next 2 terms in each set.** *Algebra* ✓

**28.** 0.1, 0.5, 0.9, 1.3, ___ , ___

**29.** 0.28, 0.31, 0.34, 0.37, ___ , ___

**30.** 0.9, 0.85, 0.8, 0.75, ___ , ___

**31.** 0.85, 0.7, 0.55, 0.4, ___ , ___

## PROBLEM SOLVING

**32.** What is the difference between 0.9 and 0.09?

**33.** How much less than 0.91 is 0.4?

**34.** Max had 0.85 m of ribbon. He used 0.5 m for a gift. How much of the ribbon was *not* used for the gift?

**35.** Elma walked 0.9 mi on Thursday. She walked 0.25 mi less on Friday. How far did she walk on Friday?

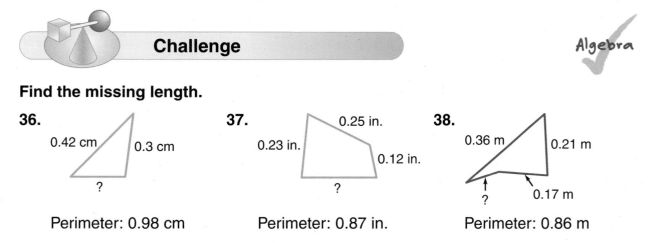

**Challenge**     *Algebra* ✓

**Find the missing length.**

**36.** 0.42 cm, 0.3 cm, ? — Perimeter: 0.98 cm

**37.** 0.25 in., 0.23 in., 0.12 in., ? — Perimeter: 0.87 in.

**38.** 0.36 m, 0.21 m, 0.17 m, ? — Perimeter: 0.86 m

**Estimate Decimal Differences**

The horseback riding trail is 34.35 km. Jesse has ridden 17.78 km. About how much farther must he ride to finish the trail?

To find how much farther, estimate the difference: 34.35 − 17.78

You can use rounding or front-end estimation to estimate a decimal difference.

▶ To **estimate** a *decimal difference* by *rounding*:

- Round the decimals to the greatest *nonzero* place of the smaller number.

$$
\begin{array}{r}
34.35 \longrightarrow 30 \\
- 17.78 \longrightarrow - 20 \\
\hline
\text{about } 10
\end{array}
$$

- Subtract the rounded numbers.

▶ To **estimate** a *decimal difference* by *front-end estimation*:

- Subtract the *nonzero* front digits.

$$
\begin{array}{r}
34.35 \\
- 17.78 \\
\hline
\text{about } 20.00
\end{array}
$$

- Write zeros for the other digits.

Jesse needs to ride about 10 to 20 km farther.

**Study these examples.**

$$
\begin{array}{r}
0.86 \longrightarrow 0.9 \\
- 0.3 \longrightarrow - 0.3 \\
\hline
\text{about } 0.6
\end{array}
\qquad
\begin{array}{r}
0.86 \\
- 0.3 \\
\hline
\text{about } 0.50
\end{array}
\qquad
\begin{array}{r}
0.93 \longrightarrow 0.9 \\
- 0.451 \longrightarrow - 0.5 \\
\hline
\text{about } 0.4
\end{array}
\qquad
\begin{array}{r}
0.93 \\
- 0.451 \\
\hline
\text{about } 0.500
\end{array}
$$

So the exact difference is between 0.5 and 0.6.

So the exact difference is between 0.4 and 0.5.

---

**Write the letter of the best estimated difference.**

**1.** 0.89 − 0.22     **a.** 0.7     **b.** 0.8     **c.** 0.5     **d.** 0.9

**2.** 18.19 − 7.23     **a.** 12     **b.** 9     **c.** 11     **d.** 8

**3.** 0.506 − 0.38     **a.** 0.1     **b.** 0.3     **c.** 0.4     **d.** 0.5

**Estimate the difference by rounding.**

| 4. | 0.73<br>− 0.4 | 5. | 7.3<br>− 2.16 | 6. | 0.582<br>− 0.43 | 7. | 5.879<br>− 3.71 | 8. | 26.259<br>− 13.4 |
|----|----|----|----|----|----|----|----|----|----|

**9.** 0.476 − 0.32          **10.** 14.8 − 9.223          **11.** 50.78 − 9.6

**Estimate the difference. Use front-end estimation.**

| 12. | 0.87<br>− 0.4 | 13. | 0.695<br>− 0.26 | 14. | 9.347<br>− 8.12 | 15. | 23.754<br>− 12.412 | 16. | 35.471<br>− 11.53 |
|----|----|----|----|----|----|----|----|----|----|

**17.** 0.735 − 0.54          **18.** 26.73 − 14.52          **19.** 95.143 − 23.21

**Estimate by both rounding and front-end estimation.
Between what two numbers will the exact difference be?**

| 20. | 0.986<br>− 0.21 | 21. | 52.49<br>− 19.6 | 22. | 63.231<br>− 49.16 | 23. | 35.47<br>− 12.529 | 24. | 69.3<br>− 12.135 |
|----|----|----|----|----|----|----|----|----|----|
| 25. | 3.89<br>− 1.158 | 26. | 78.5<br>− 14.371 | 27. | 84.53<br>− 28.165 | 28. | 69.451<br>− 12.3 | 29. | 92.473<br>− 27.51 |

**30.** 30.64 − 19.3          **31.** 49.72 − 21.514          **32.** 94.713 − 78.4

## Choose a Computation Method

**Solve and explain the method you used. Write whether
you estimated or found an exact answer.**

**33.** Lani needs 9.5 m of ribbon. She has 2.8 m. About how many more meters of ribbon does she need?

**34.** From a 5.3 ft piece of rope, Omar cut off a piece and had 2.95 ft left. How much rope did he cut off?

**35.** Jason is 136.5 cm tall. He marked this length on the ground, then did a running jump. He jumped a distance of 152.3 cm. How much longer was his jump than his height?

**36.** Ruth tries to run on the treadmill at least 8 mi a week. Last week, she ran 1.45 mi on Tuesday, 1.7 mi on Thursday, and 2.25 mi on Saturday. Did she meet her goal of 8 mi last week?

**More Subtracting Decimals**

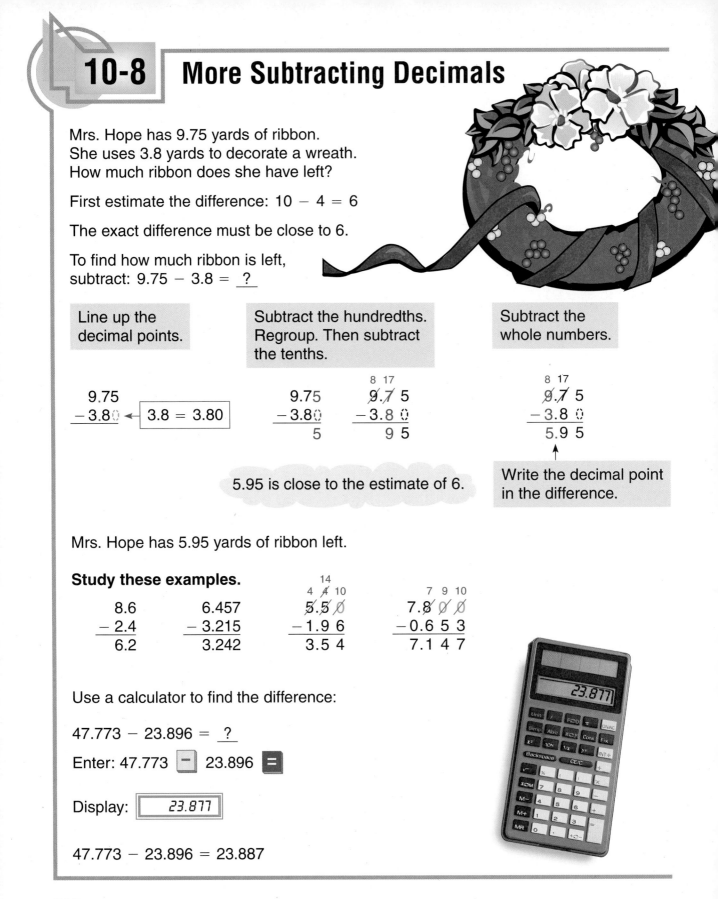

Mrs. Hope has 9.75 yards of ribbon.
She uses 3.8 yards to decorate a wreath.
How much ribbon does she have left?

First estimate the difference: $10 - 4 = 6$

The exact difference must be close to 6.

To find how much ribbon is left,
subtract: $9.75 - 3.8 = \underline{?}$

| Line up the decimal points. | Subtract the hundredths. Regroup. Then subtract the tenths. | Subtract the whole numbers. |
|---|---|---|
| 9.75<br>− 3.80 ← 3.8 = 3.80 | $\begin{array}{r}9.75\\-3.80\\\hline 5\end{array}$  $\begin{array}{r}{}^{8\ 17}\\ \cancel{9}.\cancel{7}\,5\\-3.8\,0\\\hline 9\,5\end{array}$ | $\begin{array}{r}{}^{8\ 17}\\ \cancel{9}.\cancel{7}\,5\\-3.8\,0\\\hline 5.9\,5\end{array}$ |

5.95 is close to the estimate of 6.

Write the decimal point in the difference.

Mrs. Hope has 5.95 yards of ribbon left.

**Study these examples.**

$\begin{array}{r}8.6\\-2.4\\\hline 6.2\end{array}$
$\begin{array}{r}6.457\\-3.215\\\hline 3.242\end{array}$
$\begin{array}{r}{}^{\quad\ 14}\\{}^{4\ \cancel{4}\ 10}\\ \cancel{5}.\cancel{5}\,\cancel{0}\\-1.9\,6\\\hline 3.5\,4\end{array}$
$\begin{array}{r}{}^{7\ 9\ 10}\\ 7.\cancel{8}\,\cancel{0}\,\cancel{0}\\-0.6\,5\,3\\\hline 7.1\,4\,7\end{array}$

Use a calculator to find the difference:

$47.773 - 23.896 = \underline{?}$

Enter: 47.773 [−] 23.896 [=]

Display: | 23.877 |

$47.773 - 23.896 = 23.887$

**Estimate. Then find the difference.**

1.  5.6
    − 2.4

2.  7.03
    − 2.01

3.  9.37
    − 4.26

4.  0.646
    − 0.523

5.  4.549
    − 1.317

6.  8.515
    − 7.6

7.  17.51
    − 8.4

8.  17.34
    − 3.545

9.  9.763
    − 7.52

10. 13.719
    − 1.9

**Align and subtract.**

11. 7.006 − 3.489

12. 6.034 − 2.05

13. 21.7 − 8.34

14. 7.22 − 3.405

15. 9.459 − 6.48

16. 19.42 − 2.579

17. 40.16 − 25.714

18. 29.7 − 14.634

19. 38.1 − 9.134

**Compare. Write $<$, $=$, or $>$.** You may use a calculator.

20. 9.32 − 5.171 __?__ 4.149

21. 8.5 − 4.062 __?__ 5.438

22. 4.549 __?__ 12.6 − 7.051

23. 5.72 __?__ 7.73 − 2.104

24. 40.16 − 25.714 __?__ 14.5

25. 24.714 − 9.3 __?__ 15.414

**Find the missing minuend.**

Algebra ✓

26. ? − 3.6
    4.5

27. ? − 4.59
    3.36

28. ? − 0.532
    0.284

29. ? − 2.109
    5.145

30. ? − 4.062
    3.149

**PROBLEM SOLVING**

31. Cesar is 1.52 m tall. Cheryl is 1.176 m tall. How much taller is Cesar than Cheryl?

32. Dean had 2.75 qt of paint. He used some and had 0.6 qt left. How much paint did he use?

## Mental Math

**Compute.**

33. 6.145
    − 2

34. 5
    + 2.143

35. 9.53
    + 7

36. 8.57
    − 4

37. 17.539
    − 9

# TECHNOLOGY

## Databases

A **database** is a set of related information organized into files. Each file is a collection of data relating to a particular subject.

The file below displays data related to Customer Sales.

Field Name

Customer Sales ← File Name

| Last Name | First Name | Address | Zip Code | Age | Items Sold | Amount |
|-----------|-----------|---------|----------|-----|-----------|--------|
| Weiss | Carl | 44 Main St. | 48019 | 24 | 3 | $224 |
| Ruisa | Ellen | 121 4th Ave. | 46193 | 25 | 5 | $1290 |
| Ryan | John | 3 Huron Blvd. | 48109 | 29 | 3 | $15 |
| Zake | Bob | 980 Bedford Dr. | 47104 | 31 | 7 | $2297 |
| Tucci | Wilma | 88 4th Ave. | 48114 | 31 | 12 | $410 |
| Hernandez | Ruis | 76 Hill St. | 48121 | 52 | 6 | $3900 |

record

field

Most databases display a file in columns and rows. Each file contains **records** and each record displayed as one row contains data items. Each data item is placed in a separate **field**.

The file above is arranged by age from youngest to oldest. Ordering data by a particular field makes it easy to obtain information.

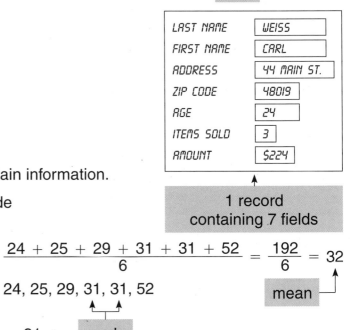

| LAST NAME | WEISS |
| FIRST NAME | CARL |
| ADDRESS | 44 MAIN ST. |
| ZIP CODE | 48019 |
| AGE | 24 |
| ITEMS SOLD | 3 |
| AMOUNT | $224 |

1 record containing 7 fields

Find the range, mean, median, and mode of the customers' ages.

$52 - 24 = 28$ ← range

$$\frac{24 + 25 + 29 + 31 + 31 + 52}{6} = \frac{192}{6} = 32$$

24, 25, 29, 31, 31, 52

24, 25, 29, 31, 31, 52          mean

$$\frac{29 + 31}{2} = 30$$ ← median

$31$ ← mode

342

**Use the database on page 342.**

1. How many field names does the Customer Sales file contain? List them.

2. How many records does the Customer Sales file contain?

3. List all the data in the second record.

4. What information is in the third field in Ellen Ruisa's record?

5. Find the range of the number of items sold.

6. Find the mean of the number of items sold.

7. Sort the file by Amount, from least to greatest. List the new order of the records.

8. Sort the file by Zip Code, from greatest to least. List the new order of the records.

**Use the database file below.**

| Last Name | First Name | Event (run) | Time (s) | Won | Lost |
|-----------|-----------|-------------|----------|-----|------|
| Smith | Gabe | 100-m | 10.03 | 6 | 9 |
| Gibbons | John | 200-m | 19.75 | 9 | 6 |
| Walker | Theresa | 100-m | 10.11 | 4 | 11 |
| Connors | Jackie | 100-m | 9.92 | 8 | 7 |
| Zamora | Ernesto | 200-m | 19.80 | 7 | 8 |
| Ferrara | Frances | 200-m | 20.09 | 1 | 14 |
| Innes | Sheryl | 200-m | 19.99 | 5 | 10 |
| Avillo | Grace | 100-m | 10.05 | 2 | 13 |
| Tran | Hui | 100-m | 9.98 | 8 | 7 |

9. List the field names of the file.

10. How many records does the file contain?

11. Sort the file by Time from least to greatest. List the new order of the records.

12. Find the range of the Time for the 100-m run; the 200-m run.

13. Find the median Time for the 100-m run; the 200-m run.

14. Find the mean number of wins for the 100-m run; the 200-m run.

15. Find the mean number of losses for the 100-m run; the 200-m run.

343

# 10-10 Problem Solving: Extra Information

**Problem:**   The Blackstones drove 145.2 mi the first day and 203.9 mi the next day of their vacation. They spent $15 for gas each day. How many miles did they travel?

**1 IMAGINE**   Picture the Blackstones checking their mileage.

**2 NAME**   *Facts:*   drove 145.2 mi one day
drove 203.9 mi the next day
spent $15 each day for gas

*Question:*   How many miles did they travel?

**3 THINK**   This problem contains extra information that is not needed to solve the problem.

You need to find the total mileage.

To do that, you do not need to know how much money they spent for gas.

To find the total mileage, add.

**4 COMPUTE**   145.2 + 203.9 = __?__
      mi       mi    total mileage

First estimate. 100 + 200 = 300

$$\begin{array}{r} \overset{1}{\phantom{0}}145.2 \\ +\,203.9 \\ \hline 349.1 \end{array}$$  total mileage

349.1 is close to the estimate of 300.

The total mileage is 349.1 mi.

**5 CHECK**   Use the commutative property and a calculator to check.

$$\begin{array}{r} \overset{1}{\phantom{0}}203.9 \\ +\,145.2 \\ \hline 349.1 \end{array}$$  The answer checks.

**Identify the extra information. Then solve each problem.**

1. Tony is saving to buy a CD player that costs $68.95. He won 3 CDs at the carnival. If he has already saved $43.50, how much more money does he need?

| IMAGINE | Put yourself in the problem. |
|---------|------------------------------|

| NAME | *Facts:* | CD player costs $68.95 |
|------|----------|------------------------|
| | | Tony has 3 CDs. |
| | | He has saved $43.50. |

*Question:* How much more money does he need?

THINK
This problem has extra information.
You only need to know the cost of the CD player and how much Tony has already saved.

Subtract to find the difference.

$68.95 − $43.50 = ?

COMPUTE ⟶ CHECK

2. Paul has three wood planks, one measuring 0.5 m, another 0.8 m, and the third 1.6 m. He also has a circular piece of wood measuring 0.4 m in diameter. What is the total length of the three wood planks?

3. Chen rides his bicycle for 30 minutes four days a week. One week he clocked the mileage at 14.2 km, 12.6 km, 10.9 km, and 13.3 km. What was the total mileage?

4. Cathy bought 5 lb of tomatoes at $1.08 a pound and 2 heads of lettuce at $0.89 each. How much did the tomatoes cost?

5. Ken's math scores for the month were 92, 93, 90, and 81. His creative writing score was 91. If the score of Ken's next math test is 99, by how many points will his math average increase?

## 10-11 | Problem-Solving Applications

**Solve each problem and explain the method you used.**

1. An organic string bean is 4.6 cm long. A nonorganic bean is 6.42 cm long. How much longer is the nonorganic bean?

2. Andy buys 1.05 kg of organic oranges and 0.96 kg of organic grapefruit. What is the total mass of the fruit Andy bought?

3. Missy measured an organic carrot's length in tenths of centimeters. Then she rounded its length to 11 cm. What is the longest length she could have measured?

4. Juan has 1.243 kg of organic flour. His recipe calls for 2 kg of flour. How much more flour does he need?

5. Organic strawberries cost $1.45 for a pint and $2.78 for a quart. Jen buys 1 pint and 2 quarts of strawberries. About how much does she spend?

6. Alma buys four organic apples. They have masses of 154.5 g, 120 g, 127.72 g, and 151.19 g. What is the total mass of the apples?

7. An organic peach weighed 142.3 g. Its pit weighed 18.48 g. How much did its skin and flesh weigh?

8. An apricot weighed 4.5 oz before drying. After drying, it weighed 1.375 oz. How many ounces of water did it lose while drying?

9. The line graph shows the amount of produce sold each month. In which months did Pélé sell about 2.5 metric tons of produce?

10. How much more did Pélé's Produce sell in June than in April?

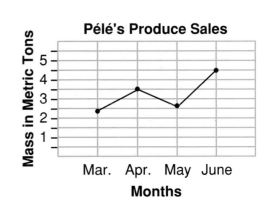

346

**Choose a strategy from the list or use another strategy you know to solve each problem.**

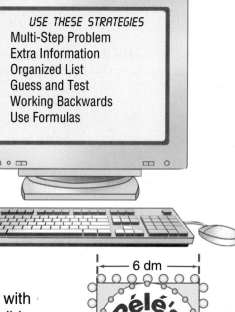

USE THESE STRATEGIES
Multi-Step Problem
Extra Information
Organized List
Guess and Test
Working Backwards
Use Formulas

11. A bag of 12 organic onions costs $3.49. How much would 2 bags cost?

12. Mary Ann bought some fruit. She gave 1.4 kg of pears to Jill, who gave her 1.15 kg of melon. Then she had 3 kg of fruit altogether. How much fruit had she bought?

13. Thea's organic tomato weighs 0.145 kg more than Fran's. Together their tomatoes weigh 3.945 kg. How much does Thea's tomato weigh?

14. Pélé trimmed the circle and the square of his sign with lights. About how many more decimeters of lights did he use around the square than around the circle?

6 dm

**Use this chart for problems 15–17.**

15. Belinda bought 1 of each fruit. How much change did she receive from $10?

16. Rich spent exactly $3.83. What fruits did he purchase?

17. Ms. Fermat buys 3 different fruits. What combinations of fruits can she purchase? What is the most expensive combination?

### Organic Fruit Prices

| | |
|---|---|
| Apples | $0.49 each |
| Pears | $0.39 each |
| Kiwis | $0.75 each |
| Melons | $1.89 each |
| Mangos | $2.95 each |

**Use this chart for problems 18–20.**

18. How much more expensive is it to buy 1 lb of each organic vegetable than nonorganic vegetable?

19. Bill buys 5 pounds of spinach, some organic and some nonorganic. He spends $8.02. How many pounds of organic spinach does he buy?

### Vegetable Prices (per lb)

| Food | Organic | Nonorganic |
|---|---|---|
| Beets | $1.19 | $0.89 |
| Carrots | $0.98 | $0.45 |
| Onions | $1.25 | $0.99 |
| Spinach | $2.09 | $1.28 |

**Make Up Your Own**

Discuss ✓

20. Write a problem using the data in a graph or chart in this lesson. Then solve it. Share your work with a classmate.

# Chapter Review and Practice

**Name the decimal for each point on the number line.** *(See pp. 326–327.)*

```
        G    H        I          J              K    L
        ↓    ↓        ↓          ↓              ↓    ↓
←──┼────────┼────────┼┼┼┼┼┼┼┼┼┼┼─┼──────┼┼┼┼┼┼┼┼┼┼┼┼──────┼──────┼┼┼┼┼┼┼┼┼┼┼┼──────┼──→
  10    10.01    10.02    10.03    10.04    10.05    10.06    10.07    10.08    10.09    10.1
```

**1.** G **2.** H **3.** I **4.** J **5.** K **6.** L

**Write the place of the underlined digit. Then write its value.** *(See pp. 328–329.)*

**7.** 3<u>6</u>.02  **8.** 2.7<u>5</u>  **9.** 0.96<u>3</u>  **10.** 47.<u>9</u>12

**Write each in expanded form.**

**11.** 470.47  **12.** 39.62  **13.** 50.2  **14.** 49.308

**Estimate by both rounding and front-end estimation. Between what two numbers will the exact sum or exact difference be?** *(See pp. 332–333, 338–339.)*

**15.**  $\begin{array}{r} 0.97 \\ + 0.465 \\ \hline \end{array}$  **16.**  $\begin{array}{r} 5.575 \\ + 6.81 \\ \hline \end{array}$  **17.**  $\begin{array}{r} 0.753 \\ - 0.52 \\ \hline \end{array}$  **18.**  $\begin{array}{r} 4.76 \\ - 2.135 \\ \hline \end{array}$  **19.**  $\begin{array}{r} 7.52 \\ 3.153 \\ + 8.64 \\ \hline \end{array}$

**Add.** *(See pp. 330–331, 334–335.)*

**20.**  $\begin{array}{r} 0.58 \\ + 0.69 \\ \hline \end{array}$  **21.**  $\begin{array}{r} 3.142 \\ + 13.236 \\ \hline \end{array}$  **22.**  $\begin{array}{r} 0.4 \\ + 0.63 \\ \hline \end{array}$  **23.**  $\begin{array}{r} 3.25 \\ + 1.7 \\ \hline \end{array}$  **24.**  $\begin{array}{r} 17.154 \\ + 5.24 \\ \hline \end{array}$

**25.** 5.2 + 8.13 + 9.152  **26.** 13.21 + 5.358 + 0.259

**Subtract.** *(See pp. 336–337, 340–341.)*

**27.**  $\begin{array}{r} 6.85 \\ - 0.72 \\ \hline \end{array}$  **28.**  $\begin{array}{r} 20.84 \\ - 9.18 \\ \hline \end{array}$  **29.**  $\begin{array}{r} 0.9 \\ - 0.254 \\ \hline \end{array}$  **30.**  $\begin{array}{r} 72.35 \\ - 8.513 \\ \hline \end{array}$  **31.**  $\begin{array}{r} 17.9 \\ - 6.129 \\ \hline \end{array}$

**32.** 5.2 − 3.75  **33.** 15.67 − 3.4  **34.** 19.1 − 4.853

## PROBLEM SOLVING

*(See pp. 334–335, 340–341, 344–347.)*

**35.** Find the perimeter of a rectangle with a diagonal that measures 45.5 cm and with sides of 17.5 cm and 42 cm.

**36.** Ana had 3.75 pt of milk. She used some for a recipe and had 1.5 pt left. How much did she use?

*(See Still More Practice, p. 486.)*

## SHORT WORD NAMES

Scientists often use *short word names* when writing large numbers.

**Study these examples.**

$4,000,000 = 4 \times 1,000,000 = 4 \times 1$ million $= 4$ million

**Standard form:** 4,000,000          **Short word name:** 4 million

$1,200,000 = 1\frac{200,000}{1,000,000}$ million $= 1\frac{200,000}{1,000,000}$ million

$\quad\quad\quad\quad = 1\frac{2}{10}$ million $= 1.2$ million

**Standard form:** 1,200,000          **Short word name:** 1.2 million

$3,580,000,000 = 3\frac{580,000,000}{1,000,000,000}$ billion $= 3\frac{580,000,000}{1,000,000,000}$ billion

$\quad\quad\quad\quad = 3\frac{58}{100}$ billion $= 3.58$ billion

**Standard form:** 3,580,000,000          **Short word name:** 3.58 billion

**Write the short word name.**

| | | | |
|---|---|---|---|
| **1.** 6,000,000 | **2.** 8,000,000 | **3.** 5,000,000,000 | **4.** 9,000,000,000 |
| **5.** 7,800,000 | **6.** 6,500,000 | **7.** 8,300,000,000 | **8.** 5,600,000,000 |
| **9.** 5,760,000 | **10.** 3,540,000 | **11.** 9,214,000,000 | **12.** 3,469,000,000 |

**Write in standard form.**

| | | | |
|---|---|---|---|
| **13.** 7.9 million | **14.** 8.03 million | **15.** 9.2 billion | **16.** 4.69 billion |
| **17.** 6.12 million | **18.** 7.08 billion | **19.** 4.14 million | **20.** 5.05 billion |
| **21.** 3.215 million | **22.** 2.109 billion | **23.** 7.062 million | **24.** 9.008 billion |

## PROBLEM SOLVING

**25.** Look for examples of short word names in textbooks, almanacs, and newspapers. Summarize the ways in which they are used.

## Performance Assessment

**Use the number line.**

1. Name the decimal for points *X* and *V*.

X    V

19.09    19.1

**Draw a number line and locate each point.**

2. *S* = 19.047          3. *T* = 19.034          4. *D* = 19.04

5. Name the thousandths between 19.0 and 19.010.

**Write the place of the underlined digit.
Then write its value.**

6. 84.2<u>6</u>8          7. 5.23<u>9</u>          8. 873.<u>1</u>

**Write each in expanded form.**

9. 47.04          10. 5.902          11. 0.593

**Estimate by both rounding and front-end estimation.
Between what two numbers will the exact sum
or exact difference be?**

12.    0.86
   + 0.683

13.    7.53
   2.752
  + 4.385

14.    0.853
  − 0.52

15.    58.457
  − 13.3

**Add or subtract.**

16.    0.516
  + 0.47

17.   6.8
  + 0.72

18.    0.595
  − 0.41

19.    12.79
  −  3.581

20. 1.23 + 3.517 + 12.3          21. 13.236 + 8.2 + 5.34

22. 6.85 − 2.4          23. 34.9 − 8.183

**PROBLEM SOLVING**    *Use a strategy you have learned.*

24. The sum of 0.497 and another
number is 0.91. Find the
other number.

25. Betty has 3 pieces of fabric
measuring 0.45 m, 0.24 m, and
0.3 m. Is the total length
more or less than one meter?

# Decimals: Multiplication and Division 11

## Sand Dollar

What can we buy
with this loose
money?

It spilled
from the green silk
pocket
of the sea
a white coin tossed up
a careless gift  wet
shining
at the water's edge

Who can break a dollar?

What a bargain! Five
white doves
ready to fly to your hand

Sea change!

*Barbara Juster Esbensen*

### In this chapter you will:

Multiply and divide by powers of ten
Estimate decimal products and
   quotients
Multiply and divide decimals and
   money
Write a number sentence to solve
   problems

### Critical Thinking/Finding Together

You bought some supplies that cost
$2.59 and paid with $10. What is the
least possible combination of bills and
coins you could receive as change?

351

# Multiplying by 10, 100, and 1000  *Algebra* ✓

## Discover Together

**Materials Needed:** calculator, paper, pencil, almanac

Copy the given table. Use a calculator to complete it.
Look for patterns.

|     | *n*   | 10 × *n* | 100 × *n* | 1000 × *n* |
|-----|-------|----------|-----------|------------|
| **1.** | 0.352 | ?        | ?         | ?          |
| **2.** | 0.74  | ?        | ?         | ?          |
| **3.** | 0.6   | ?        | ?         | ?          |

Compare the position of the decimal point in *n* with the position
of the decimal point in 10 × *n*;  in 100 × *n*;  in 1000 × *n*.

**4.** What patterns do you notice in your complete table?
   What happens to the decimal point when you multiply
   a decimal by 10? by 100? by 1000?

**5.** Examine the products in exercise 3. What happens when
   there are not enough places to move the decimal point
   as far to the right as needed?

Predict the products. Use the patterns to help you.

| **6.** | **7.** | **8.** | **9.** |
|--------|--------|--------|--------|
| 10 × 3.628 | 10 × 9.65 | 10 × 0.5 | 10 × 4.8 |
| 100 × 3.628 | 100 × 9.65 | 100 × 0.5 | 100 × 4.8 |
| 1000 × 3.628 | 1000 × 9.65 | 1000 × 0.5 | 1000 × 4.8 |

Use a calculator to check your predictions.

**10.** Write a rule that you can use to multiply a decimal
   by 10, 100, and 1000.

Use your rule to find the product. Check by using a calculator.

**11.** 10 × 0.02   **12.** 1000 × 0.691   **13.** 100 × 0.03   **14.** 1000 × 0.007

**15.** 10 × 37.9   **16.** 100 × 1.7   **17.** 1000 × 2.63   **18.** 10 × 0.296

Now use a calculator to complete each multiplication sentence.

**19.** $20 \times 0.4 = \underline{?}$  **20.** $70 \times 0.9 = \underline{?}$  **21.** $40 \times 0.6 = \underline{?}$

$200 \times 0.4 = \underline{?}$  $700 \times 0.9 = \underline{?}$  $400 \times 0.6 = \underline{?}$

**22.** What pattern do you notice?

**23.** Write a rule that you can use to multiply a decimal by a multiple of 10 or 100. Use the rule for multiplying a decimal by 10, 100, and 1000 as a model.

Use your rule to find the products. Check by using a calculator.

**24.** $500 \times 0.9$   **25.** $900 \times 0.7$   **26.** $40 \times 0.8$   **27.** $30 \times 0.2$

**28.** $80 \times 0.8$   **29.** $90 \times 0.5$   **30.** $300 \times 0.9$   **31.** $600 \times 0.7$

## Communicate

Discuss

**32.** Describe in your Math Journal the pattern formed by the number of zeros in 10, 100, and 1000 and the number of places the decimal point "moves" when you multiply by these numbers.

**33.** When you multiply a decimal by a multiple of 10 or 100, why does the decimal point move to the right rather than to the left?

**34.** Find the missing factor. Explain your answers.

**a.** $\underline{?} \times 0.309 = 309$   **b.** $\underline{?} \times 0.028 = 0.28$   **c.** $\underline{?} \times 0.054 = 5.4$

**d.** $10 \times \underline{?} = 32.13$   **e.** $1000 \times \underline{?} = 1580$   **f.** $100 \times \underline{?} = 350$

**35.** Multiply each of the factors in box *B* by one of the factors in box *A*. Write each multiplication sentence.

| 50 | 800 |
| 60 | 300 |
| 40 | 700 |
| **A** | |

| 0.1 | 0.4 | 0.7 |
| 0.2 | 0.5 | 0.8 |
| 0.3 | 0.6 | 0.9 |
| **B** | | |

**Project**

**Make a *Math in History* poster.**

Communicate

**36.** Research one ancient Greek mathematician and his contributions to the development of mathematics. Create a poster on this mathematician. Display your work in the classroom.

353

# Estimating Decimal Products

Ms. Millar drove for 3.8 hours at a speed of 48.95 miles an hour. About how far did she drive?

To find about how far, estimate: 3.8 × 48.95

▶ To **estimate** a *decimal product*:

- Round each factor to its greatest place.

- Multiply the rounded factors.

$$
\begin{array}{r}
48.95 \longrightarrow 50 \\
\times\ \ 3.8 \longrightarrow \times\ \ 4 \\
\hline
\text{about}\ \ 200
\end{array}
$$

Ms. Millar drove about 200 miles.

Both factors are rounded *up*. The actual product is *less than* 200.

**Study these examples.**

$$
\begin{array}{r}
0.734 \longrightarrow 0.7 \\
\times 22.86 \longrightarrow \times\ 20 \\
\hline
\text{about}\ \ 14
\end{array}
$$

Both factors are rounded *down*.

$$
\begin{array}{r}
0.56 \longrightarrow 0.6 \\
\times\ 9.7 \longrightarrow \times\ 10 \\
\hline
\text{about}\ \ 6
\end{array}
$$

Both factors are rounded *up*.

The actual product is *greater than* 14.

The actual product is *less than* 6.

**Estimate each product. Then tell whether the actual product is *greater than* or *less than* the estimated product.**

| | | | | |
|---|---|---|---|---|
| **1.** 4.81 × 2.6 | **2.** 3.45 × 4.3 | **3.** 5.56 × 9.7 | **4.** 0.75 × 9.5 | **5.** 0.88 × 9.8 |
| **6.** 4.376 × 8.2 | **7.** 9.135 × 4.2 | **8.** 4.836 × 6.7 | **9.** 7.036 × 2.31 | **10.** 5.645 × 3.84 |
| **11.** 13.96 × 0.84 | **12.** 24.69 × 0.23 | **13.** 17.68 × 0.55 | **14.** 0.146 × 29.34 | **15.** 0.341 × 32.49 |
| **16.** 15.435 × 0.48 | **17.** 28.776 × 0.76 | **18.** 45.186 × 0.35 | **19.** 83.607 × 0.64 | **20.** 92.487 × 0.92 |

## Estimation by Clustering

When a number of addends "clusters" around a certain number, an estimate for the sum may be obtained by multiplying that number by the number of addends.

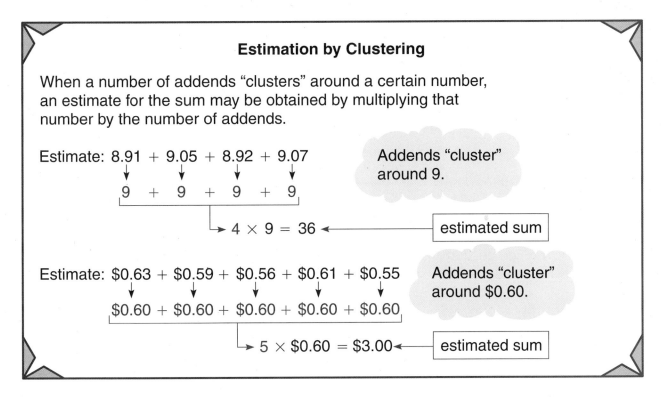

Estimate: 8.91 + 9.05 + 8.92 + 9.07

9 + 9 + 9 + 9

↳ 4 × 9 = 36 ◄—— estimated sum

Addends "cluster" around 9.

Estimate: $0.63 + $0.59 + $0.56 + $0.61 + $0.55

$0.60 + $0.60 + $0.60 + $0.60 + $0.60

↳ 5 × $0.60 = $3.00 ◄—— estimated sum

Addends "cluster" around $0.60.

**Estimate the sum.** Use clustering.

**21.** 9.8 + 10.15 + 11.2

**22.** 71.3 + 67.89 + 70.02 + 73 + 69.18

**23.** 0.93 + 1.1 + 1.08 + 0.9

**24.** 2.05 + 1.986 + 2.014 + 1.895 + 2.1

**25.** $0.84 + $0.77 + $0.81 + $0.79

**26.** $.35 + $.41 + $.39 + $.44 + $.36

**27.** $.53 + $.48 + $.54 + $.46 + $.51

**28.** $.99 + $1.01 + $.96 + $1.10 + $.95

### PROBLEM SOLVING

**29.** Jon runs 5.3 miles in one hour. At this rate, about how far could he run in 1.7 hours?

**30.** Mila can swim 18.55 meters in one minute. About how far can she swim in 4.75 minutes?

**31.** If one sample of ore weighs 23.8 g, about how many grams will 87 equal samples weigh?

### Skills to Remember

**Estimate. Then find the product.**

**32.** 17 × 69

**33.** 540 × 7

**34.** 65 × 158

**35.** 150 × 700

**36.** 157 × 309

**37.** 104 × 503

**38.** 407 × 873

**39.** 1809 × 480

# Multiplying Decimals by Whole Numbers

If one cup of skim milk contains 0.31 grams of calcium, how much calcium is in 11 cups of skim milk?

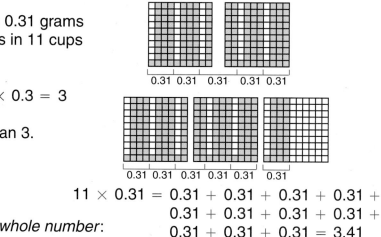

0.31  0.31   0.31   0.31 0.31

0.31  0.31   0.31   0.31 0.31   0.31

First estimate the product: $10 \times 0.3 = 3$

The actual product is greater than 3.

To find how much calcium, multiply: $11 \times 0.31 = \underline{\ ?\ }$

$11 \times 0.31 = 0.31 + 0.31 + 0.31 + 0.31 +$
$0.31 + 0.31 + 0.31 + 0.31 +$
$0.31 + 0.31 + 0.31 = 3.41$

▶ To **multiply** a *decimal* by a *whole number*:

- Multiply as you would with whole numbers.
- Count the number of decimal places in the decimal factor.
- Mark off the *same number* of decimal places in the product.

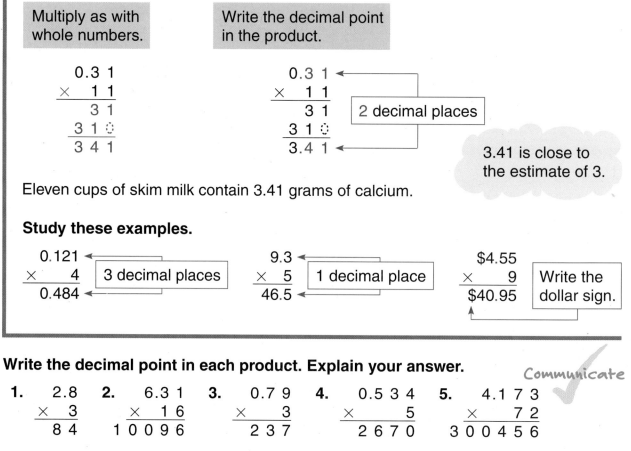

| Multiply as with whole numbers. | Write the decimal point in the product. |
|---|---|
| 0.3 1<br>× 1 1<br>3 1<br>3 1 0<br>3 4 1 | 0.3 1<br>× 1 1<br>3 1    2 decimal places<br>3 1 0<br>3.4 1 |

3.41 is close to the estimate of 3.

Eleven cups of skim milk contain 3.41 grams of calcium.

**Study these examples.**

| 0.121<br>× 4<br>0.484 | 3 decimal places | 9.3<br>× 5<br>46.5 | 1 decimal place | $4.55<br>× 9<br>$40.95 | Write the dollar sign. |
|---|---|---|---|---|---|

**Write the decimal point in each product. Explain your answer.**

*Communicate*

| 1. | 2.8<br>× 3<br>8 4 | 2. | 6.3 1<br>× 1 6<br>1 0 0 9 6 | 3. | 0.7 9<br>× 3<br>2 3 7 | 4. | 0.5 3 4<br>× 5<br>2 6 7 0 | 5. | 4.1 7 3<br>× 7 2<br>3 0 0 4 5 6 |

356

**Estimate. Then find the product.**

6.  0.6
    × 13

7.  0.8
    × 32

8.  0.53
    × 17

9.  0.49
    × 29

10. 0.83
    × 67

11. 0.169
    × 26

12. 0.834
    × 14

13. 0.479
    × 35

14. 1.5
    × 15

15. 9.2
    × 39

16. 3.05
    × 26

17. 4.98
    × 38

18. 5.052
    × 19

19. 9.152
    × 34

20. 7.891
    × 56

21. $0.74
    × 12

22. $0.93
    × 25

23. $5.87
    × 46

24. $8.39
    × 62

25. $14.55
    × 89

**Multiply.**

26. 3 × 0.4

27. 5 × 0.49

28. 9 × 0.019

29. 8 × 0.153

30. 7 × 2.7

31. 3 × 3.29

32. 4 × 2.013

33. 2 × 8.519

34. 13 × 15.9

35. 27 × 18.34

36. 35 × 35.02

37. 43 × 26.514

38. 9 × $0.49

39. 15 × $0.67

40. 36 × $1.03

41. 49 × $15.19

42. five times seven tenths

43. eight times seven hundredths

44. six times nineteen thousandths

45. two times five and two tenths

46. thirty-eight times thirteen and three hundredths

 **Choose a Computation Method**

**Solve and explain the method you used. Write whether you estimated or found an exact answer.**

47. One large banana contains 2.4 g of protein. How many grams of protein will a dozen large bananas contain?

48. One side of a square table measures 2.65 meters. Is the perimeter of the square table less than 10 meters?

49. Ms. Blake bought 3 lb of onions at $1.69 a pound, 2 lb of yams at $0.59 a pound, and 3 bunches of broccoli at $1.19 a bunch. Did she spend more than $10.00?

## 11-4 Multiplying Decimals by Decimals

Carla took 0.8 of the money from her savings account and spent 0.6 of it on books. How much of her savings did she spend on books?

To find how much of her savings, multiply: $0.6 \times 0.8 =$ __?__

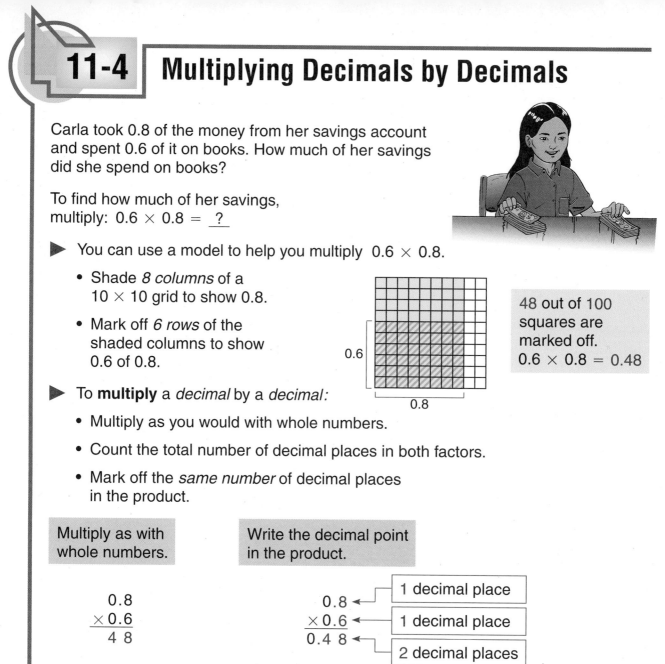

▶ You can use a model to help you multiply $0.6 \times 0.8$.

- Shade *8 columns* of a $10 \times 10$ grid to show 0.8.

- Mark off *6 rows* of the shaded columns to show 0.6 of 0.8.

48 out of 100 squares are marked off.
$0.6 \times 0.8 = 0.48$

▶ To **multiply** a *decimal* by a *decimal*:

- Multiply as you would with whole numbers.

- Count the total number of decimal places in both factors.

- Mark off the *same number* of decimal places in the product.

| Multiply as with whole numbers. | Write the decimal point in the product. |
|---|---|
| $\begin{array}{r} 0.8 \\ \times\,0.6 \\ \hline 4\,8 \end{array}$ | $\begin{array}{r} 0.8 \\ \times\,0.6 \\ \hline 0.4\,8 \end{array}$ |

In the right column:
0.8 ← 1 decimal place
× 0.6 ← 1 decimal place
0.4 8 ← 2 decimal places

Carla spent 0.48 of her savings on books.

**Study these examples.**

$\begin{array}{r} 6.5 \\ \times\,0.7\,3 \\ \hline 1\,9\,5 \\ 4\,5\,5\phantom{0} \\ \hline 4.7\,4\,5 \end{array}$

6.5 ← 1 decimal place
× 0.7 3 ← 2 decimal places
4.7 4 5 ← 3 decimal places

$\begin{array}{r} 4.2 \\ \times\,\phantom{0}1.7 \\ \hline 2\,9\,4 \\ 4\,2\phantom{0} \\ \hline 7.1\,4 \end{array}$

4.2 ← 1 decimal place
× 1.7 ← 1 decimal place
7.1 4 ← 2 decimal places

358

**Use the diagram to complete each statement.**

1.

$0.4 \times 0.7 = \underline{\ ?\ }$

2.

$\underline{\ ?\ } \times \underline{\ ?\ } = 0.54$

3.

$0.2 \times \underline{\ ?\ } = \underline{\ ?\ }$

**Use a 10 × 10 grid to find each product.**

**4.** $0.3 \times 0.9$  **5.** $0.8 \times 0.7$  **6.** $0.5 \times 0.6$  **7.** $0.9 \times 0.4$

**Write the decimal point in each product.**

**8.**
$$\begin{array}{r} 2.6 \\ \times\,0.4 \\ \hline 1\,0\,4 \end{array}$$

**9.**
$$\begin{array}{r} 3.5 \\ \times\,1.7 \\ \hline 5\,9\,5 \end{array}$$

**10.**
$$\begin{array}{r} 0.4\,2 \\ \times\quad 3.1 \\ \hline 1\,3\,0\,2 \end{array}$$

**11.**
$$\begin{array}{r} 9.1\,6 \\ \times\quad 0.3 \\ \hline 2\,7\,4\,8 \end{array}$$

**12.**
$$\begin{array}{r} 1\,1.6 \\ \times\,0.2\,3 \\ \hline 2\,6\,6\,8 \end{array}$$

**Multiply.**

**13.**
$$\begin{array}{r} 0.9 \\ \times\,0.4 \\ \hline \end{array}$$

**14.**
$$\begin{array}{r} 3.4 \\ \times\,0.8 \\ \hline \end{array}$$

**15.**
$$\begin{array}{r} 5.2 \\ \times\,0.6 \\ \hline \end{array}$$

**16.**
$$\begin{array}{r} 5.9 \\ \times\,0.03 \\ \hline \end{array}$$

**17.**
$$\begin{array}{r} 2.2 \\ \times\,0.16 \\ \hline \end{array}$$

**18.**
$$\begin{array}{r} 4.7 \\ \times\,2.6 \\ \hline \end{array}$$

**19.**
$$\begin{array}{r} 6.24 \\ \times\quad 0.9 \\ \hline \end{array}$$

**20.**
$$\begin{array}{r} 1.45 \\ \times\quad 0.5 \\ \hline \end{array}$$

**21.**
$$\begin{array}{r} 21.3 \\ \times\quad 1.5 \\ \hline \end{array}$$

**22.**
$$\begin{array}{r} 24.6 \\ \times\quad 2.3 \\ \hline \end{array}$$

**23.** $6.6 \times 4.83$  **24.** $4.8 \times 5.94$  **25.** $0.97 \times 65.8$  **26.** $3.17 \times 19.5$

**PROBLEM SOLVING**

**27.** Krissie is 1.43 m tall. Her mother is 1.2 times Krissie's height. How tall is Krissie's mother?

**28.** If Jack can run 8.53 km in one hour, how far can he run in 3.5 hours?

**Critical Thinking**

*Algebra*

**Tell whether each number in the pattern is multiplied by a *whole number* or by a *decimal*. Then complete the pattern.**

**29.** 50, 5, 0.5, $\underline{\ ?\ }$ , $\underline{\ ?\ }$

**30.** 2.5, 7.5, 22.5, $\underline{\ ?\ }$ , $\underline{\ ?\ }$

**31.** 1.2, 2.4, 4.8, $\underline{\ ?\ }$ , $\underline{\ ?\ }$

**32.** 20, 6, 1.8, $\underline{\ ?\ }$ , $\underline{\ ?\ }$

359

# Zeros in the Product

Sometimes you need to write zeros to the left of nonzero digits in the product in order to place the decimal point correctly.

Multiply: $0.3 \times 0.03 = $  ?

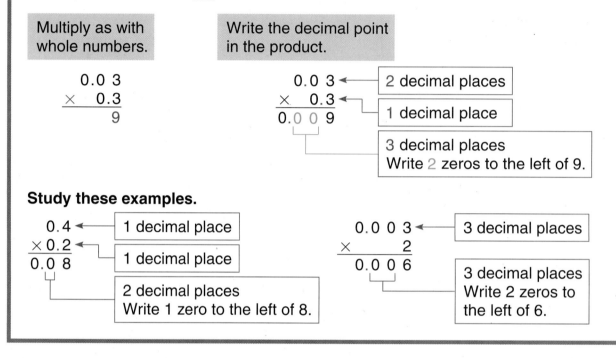

| Multiply as with whole numbers. | | Write the decimal point in the product. |
|---|---|---|

```
   0.0 3
 ×   0.3
       9
```

```
   0.0 3  ←  2 decimal places
 ×   0.3  ←  1 decimal place
 0.0 0 9
```
3 decimal places
Write 2 zeros to the left of 9.

**Study these examples.**

```
   0.4  ←  1 decimal place
 × 0.2  ←  1 decimal place
 0.0 8
```
2 decimal places
Write 1 zero to the left of 8.

```
 0.0 0 3  ←  3 decimal places
 ×     2
 0.0 0 6
```
3 decimal places
Write 2 zeros to the left of 6.

---

**Write the decimal point in the product.**
**Write in zeros where necessary.**

**1.**
```
   0.3
 ×0.2
     6
```

**2.**
```
 0.0 4
 ×   0.3
     1 2
```

**3.**
```
 0.3 4
 ×   0.2
     6 8
```

**4.**
```
     7.4
 ×0.0 1
     7 4
```

**5.**
```
 0.0 0 8
 ×       7
       5 6
```

**Multiply.**

**6.**
```
   0.2
 ×0.1
```

**7.**
```
 0.04
 × 0.2
```

**8.**
```
 0.03
 ×   9
```

**9.**
```
 0.003
 ×    3
```

**10.**
```
 0.002
 ×    4
```

**11.**
```
 0.16
 × 0.3
```

**12.**
```
 0.46
 × 0.2
```

**13.**
```
 0.19
 × 0.4
```

**14.**
```
 0.012
 ×    3
```

**15.**
```
 0.021
 ×    4
```

**16.**
```
   1.3
 ×0.03
```

**17.**
```
   1.1
 ×0.05
```

**18.**
```
   2.3
 ×0.04
```

**19.**
```
   6.7
 ×0.01
```

**20.**
```
   1.7
 ×0.04
```

**Find the product.**

**21.** $4 \times 0.003$      **22.** $0.06 \times 0.3$      **23.** $0.4 \times 0.02$      **24.** $0.03 \times 0.4$

**25.** $3.2 \times 0.02$      **26.** $0.7 \times 0.02$      **27.** $5.2 \times 0.01$      **28.** $0.13 \times 0.3$

**29.** $2 \times 0.021$      **30.** $0.3 \times 0.11$      **31.** $0.5 \times 0.05$      **32.** $1.2 \times 0.04$

**Write the letter of the correct answer.**

**33.** $8 \times 0.006 = \underline{\ ?\ }$      **a.** $4.8$      **b.** $0.48$      **c.** $0.048$

**34.** $0.05 \times 0.4 = \underline{\ ?\ }$      **a.** $0.002$      **b.** $0.02$      **c.** $0.2$

**35.** $0.3 \times 0.07 = \underline{\ ?\ }$      **a.** $0.021$      **b.** $0.21$      **c.** $2.1$

**PROBLEM SOLVING**

**36.** What is one and two tenths multiplied by seven hundredths?

**37.** Find the product of twenty-three hundredths times three tenths.

**38.** A clock uses 0.02 kilowatt hours of electricity a day. How much electricity does it use in 4 days?

**39.** A postcard weighs 0.004 kg. How many kilograms would six of these weigh?

**Calculator Activity**

Algebra ✓

**Find the products to discover a pattern.**

**40.** $0.25 \times 320$

$\boxed{0}\ \boxed{\cdot}\ \boxed{2}\ \boxed{5}\ \boxed{\times}\ \boxed{3}\ \boxed{2}\ \boxed{0}\ \boxed{=}$

**41.** $\frac{1}{4} \times 320$

$\boxed{1}\ \boxed{/}\ \boxed{4}\ \boxed{\times}\ \boxed{3}\ \boxed{2}\ \boxed{0}\ \boxed{=}$

**42.** $0.25 \times 32$      **43.** $\frac{1}{4} \times 32$      **44.** $0.25 \times 3.2$      **45.** $\frac{1}{4} \times 3.2$

**46.** Multiplying a number by 0.25 is the same as multiplying the number by the fraction $\underline{\ ?\ }$.

**47.** $0.2 \times 1500$      **48.** $0.2 \times 150$      **49.** $0.2 \times 15$      **50.** $0.2 \times 1.5$

**51.** $0.2 \times 2000$      **52.** $0.2 \times 200$      **53.** $0.2 \times 20$      **54.** $0.2 \times 2$

**55.** Multiplying a number by 0.2 is the same as multiplying the number by the fraction $\underline{\ ?\ }$.

# Dividing by 10, 100, and 1000

*Algebra* ✓

 ## Discover Together

**Materials Needed:** calculator, paper, pencil

Copy the given table. Use a calculator to complete it.
Look for patterns.

|    | n   | n ÷ 10 | n ÷ 100 | n ÷ 1000 |
|----|-----|--------|---------|----------|
| **1.** | 198 | ? | ? | ? |
| **2.** | 64  | ? | ? | ? |
| **3.** | 7   | ? | ? | ? |

Compare the position of the decimal point in *n* with the position
of the decimal point in *n* ÷ 10; in *n* ÷ 100; in *n* ÷ 1000.

**4.** What patterns do you notice in your complete table?
What happens to the decimal point when you divide
a decimal by 10? by 100? by 1000?

**5.** Examine the quotients in exercise 3. What happens
when there are not enough places to move the
decimal point as far to the left as needed?

Predict the quotients. Use the patterns to help you.

**6.** 4321 ÷ 10
4321 ÷ 100
4321 ÷ 1000

**7.** 765 ÷ 10
765 ÷ 100
765 ÷ 1000

**8.** 81 ÷ 10
81 ÷ 100
81 ÷ 1000

**9.** 6 ÷ 10
6 ÷ 100
6 ÷ 1000

Use a calculator to check your predictions.

**10.** Write a rule that you can use to divide a decimal
by 10, 100, and 1000.

Use your rule to find the quotients. Check by using a calculator.

**11.** 0.06 ÷ 10

**12.** 9 ÷ 10

**13.** 0.7 ÷ 100

**14.** 32 ÷ 100

**15.** 4 ÷ 1000

**16.** 5384 ÷ 1000

**17.** 2739.5 ÷ 100

**18.** 16,483 ÷ 1000

Now use your rule to predict if the divisor is 10, 100, or 1000.
Check using a calculator.

**19.** $2.08 \div \underline{\ ?\ } = 0.208$     **20.** $1.8 \div \underline{\ ?\ } = 0.018$     **21.** $59 \div \underline{\ ?\ } = 0.059$

**22.** $27.9 \div \underline{\ ?\ } = 0.279$     **23.** $865 \div \underline{\ ?\ } = 0.865$     **24.** $41.02 \div \underline{\ ?\ } = 4.102$

## Communicate

Discuss

**25.** Describe in your Math Journal the pattern formed by the number of zeros in 10, 100, and 1000 and the number of places the decimal point "moves" when you divide by these numbers.

**26.** When you divide a decimal by 10, 100, and 1000, why does the decimal point move to the left rather than to the right?

**27.** How is dividing a decimal by 10, 100, and 1000 the same as multiplying a decimal by 10, 100, and 1000? How is it different?

**28.** When and why do you need to write zeros in the quotient when dividing a decimal by 10, 100, and 1000?

**29.** Find the missing numbers. Explain your answers.

    **a.** $\underline{\ ?\ } \div 100 = 0.021$     **b.** $\underline{\ ?\ } \div 10 = 0.35$     **c.** $\underline{\ ?\ } \div 1000 = 0.024$

    **d.** $\underline{\ ?\ } \div 10 = 0.09$     **e.** $\underline{\ ?\ } \div 1000 = 2.006$     **f.** $\underline{\ ?\ } \div 100 = 0.012$

## Mental Math

**Write the output number.** Follow the steps for each machine.

| Input | ÷10 | ÷100 | Output |

**30.** 8591     **31.** 578     **32.** 69

| Input | ÷100 | ×1000 | Output |

**33.** 53.6     **34.** 6.2     **35.** 0.032

| Input | ×100 | ÷1000 | Output |

**36.** 354.9     **37.** 63.7     **38.** 9.5

| Input | ×100 | ×10 | Output |

**39.** 23.595     **40.** 4.13     **41.** 1.8

**Dividing Decimals by Whole Numbers**

Liam has 1.62 m of copper tubing that he cuts into 3 equal pieces. How long is each piece?

To find how long, divide: 1.62 ÷ 3 = __?__

> Think:
> What is 1.62 divided into 3 equal groups?

▶ You can use a model to help you divide 1.62 ÷ 3.

- Shade 1.62 on 10 × 10 grids.

- Cut the shaded grids apart as necessary to show 3 equal groups.

  1.62 ÷ 3 = 0.54

1.62   →   0.54

0.54

0.54

▶ To **divide** a *decimal* by a *whole number:*

| Write the decimal point of the quotient above the decimal point of the dividend. | Divide as you would with whole numbers. | Check. |
|---|---|---|

$$3\overline{)1.6\ 2}$$

$$\begin{array}{r} 0.5\ 4 \\ 3\overline{)1.6\ 2} \\ -1\ 5\phantom{0}\!\downarrow \\ \hline 1\ 2 \\ -1\ 2 \\ \hline 0 \end{array}$$

3 > 1  **Not enough** ones
3 < 16  **Enough** tenths
The quotient begins in the tenths place.

2 decimal places

$$\begin{array}{r} 0.5\ 4 \\ \times\phantom{0000}3 \\ \hline 1.6\ 2 \end{array}$$

2 decimal places

Each piece of copper tubing is 0.54 m long.

**Study these examples.**

$$\begin{array}{r} 1.1 \\ 8\overline{)8.8} \\ -8\phantom{0}\!\downarrow \\ \hline 0\ 8 \\ -8 \\ \hline 0 \end{array}$$

$$\begin{array}{r} 0.5\ 5\ 5 \\ 5\overline{)2.7\ 7\ 5} \\ -2\ 5\phantom{00}\!\downarrow \\ \hline 2\ 7 \\ -2\ 5\phantom{0}\!\downarrow \\ \hline 2\ 5 \\ -2\ 5 \\ \hline 0 \end{array}$$

$$\begin{array}{r} \$0.2\ 4 \\ 4\overline{)\$0.9\ 6} \\ -8\phantom{0}\!\downarrow \\ \hline 1\ 6 \\ -1\ 6 \\ \hline 0 \end{array}$$

Write the dollar sign and decimal point in the quotient.

**Use the diagram to complete each statement.**

**1.**

$\underline{\phantom{?}} \div 2 = \underline{\phantom{?}}$

**2.** $0.93 \div \underline{\phantom{?}} = \underline{\phantom{?}}$

**Divide and check.**

**3.** $3\overline{)1.2}$  **4.** $5\overline{)2.5}$  **5.** $7\overline{)4.2}$  **6.** $8\overline{)5.6}$  **7.** $9\overline{)5.4}$

**8.** $6\overline{)0.96}$  **9.** $5\overline{)0.75}$  **10.** $4\overline{)0.76}$  **11.** $2\overline{)0.84}$  **12.** $3\overline{)0.63}$

**13.** $7\overline{)0.861}$  **14.** $4\overline{)0.924}$  **15.** $9\overline{)2.214}$  **16.** $4\overline{)25.72}$  **17.** $5\overline{)23.55}$

**18.** $3\overline{)\$0.84}$  **19.** $8\overline{)\$0.96}$  **20.** $7\overline{)\$17.71}$  **21.** $6\overline{)\$55.56}$  **22.** $4\overline{)\$31.92}$

**23.** $76.8 \div 8$  **24.** $9.513 \div 7$  **25.** $5.337 \div 3$  **26.** $2.94 \div 6$

**27.** $\$44.45 \div 7$  **28.** $\$41.36 \div 8$  **29.** $\$5.13 \div 9$  **30.** $\$6.85 \div 5$

**31.** $\$97.44 \div 6$  **32.** $\$25.92 \div 3$  **33.** $\$364.50 \div 5$  **34.** $\$346.32 \div 9$

## PROBLEM SOLVING

**35.** A large carton of books weighs 24.9 lb. This is three times the weight of a smaller carton. How much does the smaller carton weigh?

**36.** Mr. Sotto traveled 232.5 km in 5 days. If he traveled the same distance each day, what was the distance traveled in one day?

**37.** Claire wants to divide a cost of $74.88 equally among 9 people. How much should each person pay?

## Share Your Thinking

Communicate ✓

**38.** How can a model help you divide a decimal by a whole number?

**39.** Why is it important to write the decimal point in the quotient before beginning to divide?

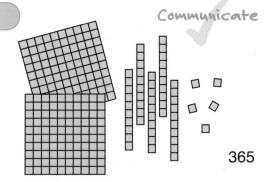

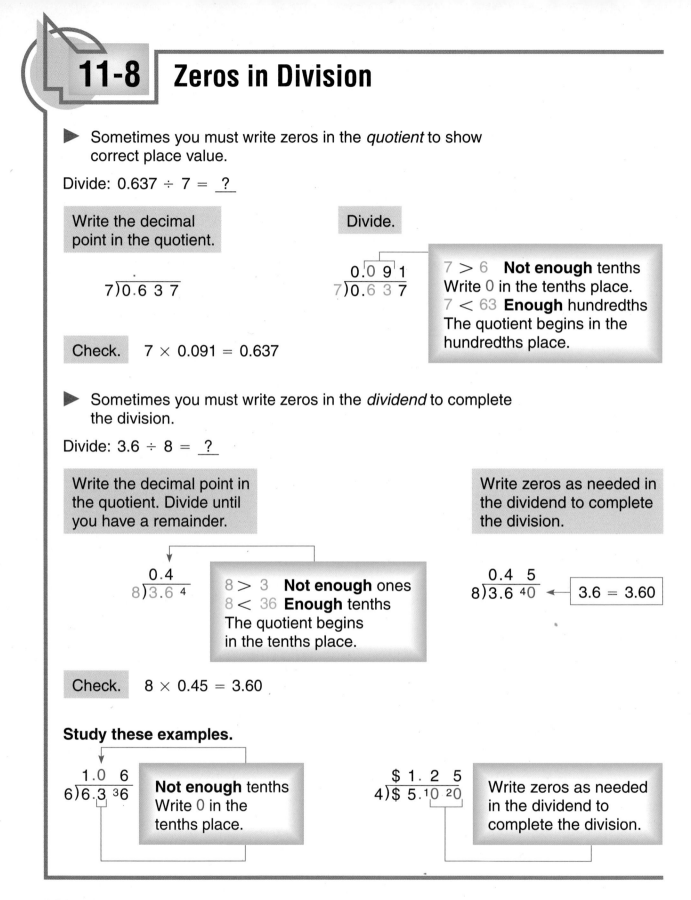

## 11-8 Zeros in Division

▶ Sometimes you must write zeros in the *quotient* to show correct place value.

Divide: 0.637 ÷ 7 = _?_

Write the decimal point in the quotient.

$$7\overline{)0.6\ 3\ 7}$$

Check.  7 × 0.091 = 0.637

Divide.

$$7\overline{)0.6\ 3\ 7}\quad 0.0\ 9\ 1$$

7 > 6  **Not enough** tenths
Write 0 in the tenths place.
7 < 63 **Enough** hundredths
The quotient begins in the hundredths place.

▶ Sometimes you must write zeros in the *dividend* to complete the division.

Divide: 3.6 ÷ 8 = _?_

Write the decimal point in the quotient. Divide until you have a remainder.

$$8\overline{)3.6\ ^4}\quad 0.4$$

8 > 3  **Not enough** ones
8 < 36 **Enough** tenths
The quotient begins in the tenths place.

Write zeros as needed in the dividend to complete the division.

$$8\overline{)3.6\ ^40}\quad 0.4\ 5$$

3.6 = 3.60

Check.  8 × 0.45 = 3.60

**Study these examples.**

$$6\overline{)6.3\ ^36}\quad 1.0\ 6$$

**Not enough** tenths
Write 0 in the tenths place.

$$4\overline{)\$\ 5.^10\ ^20}\quad \$1.2\ 5$$

Write zeros as needed in the dividend to complete the division.

366

**Divide and check.**

1. $2\overline{)0.014}$
2. $6\overline{)0.018}$
3. $8\overline{)8.24}$
4. $5\overline{)5.45}$
5. $9\overline{)27.81}$

6. $5\overline{)0.46}$
7. $8\overline{)0.44}$
8. $6\overline{)0.63}$
9. $6\overline{)6.15}$
10. $5\overline{)7.51}$

11. $4\overline{)0.424}$
12. $5\overline{)1.025}$
13. $9\overline{)2.745}$
14. $6\overline{)18.156}$
15. $7\overline{)21.364}$

16. $3\overline{)\$15.09}$
17. $8\overline{)\$16.72}$
18. $9\overline{)\$81.54}$
19. $4\overline{)\$13}$
20. $5\overline{)\$28}$

21. $16.2 \div 4$
22. $18.87 \div 6$
23. $33.32 \div 8$
24. $25.848 \div 6$

---

### More Zeros in the Dividend

For some divisions, writing zeros in the dividend does *not* complete the division. The quotient is rounded to a given place.

Divide: $4.4 \div 6 = \underline{\ ?\ }$

$$\begin{array}{r} 0.7\ 3\ 3\ 3\ \ldots = 0.733 \\ 6\overline{)4.4\ {}^20\ {}^20\ {}^20\ {}^2} \end{array}$$

rounded to the nearest thousandth

Divide: $3.56 \div 7 = \underline{\ ?\ }$

$$\begin{array}{r} 0.5\ 0\ 8\ 5\ \ldots = 0.509 \\ 7\overline{)3.5\ 6\ {}^60\ {}^40\ {}^5} \end{array}$$

rounded to the nearest thousandth

---

**Divide. Round the quotient to the nearest thousandth.**

25. $3\overline{)2.9}$
26. $7\overline{)1.5}$
27. $6\overline{)3.8}$
28. $9\overline{)1.83}$
29. $3\overline{)9.34}$

30. $6\overline{)0.64}$
31. $9\overline{)0.83}$
32. $3\overline{)0.95}$
33. $7\overline{)0.85}$
34. $6\overline{)0.59}$

### PROBLEM SOLVING

35. Jake rode 4 laps on his bike in 9.46 minutes. What was his average time for each lap?

36. Sofia bought 9 identical key chains for $27.72. How much did each key chain cost?

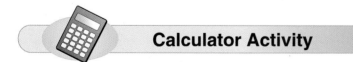

### Calculator Activity

**Divide. Round the quotient to the nearest thousandth.**

37. $36\overline{)1.842}$
38. $28\overline{)101.52}$
39. $16\overline{)48.756}$
40. $19\overline{)39.402}$

41. $27\overline{)25.659}$
42. $45\overline{)131.202}$
43. $32\overline{)467.563}$
44. $71\overline{)856.296}$

## 11-9 Estimating Decimal Quotients

On a bicycle trip, Marc plans to travel 226.85 km in 7 days. About how many kilometers a day will he travel if he travels the same distance each day?

To find about how many kilometers, estimate: 226.85 ÷ 7

▶ To **estimate** a *decimal quotient*:

- Write the decimal point in the quotient.

- Decide in which place the first nonzero digit of the quotient begins.

- Find the *first* nonzero digit of the quotient.

- Write zeros for the remaining digits.

$$\begin{array}{r} 3\ 0.0\ 0 \\ 7\overline{)2\ 2\ 6.8\ 5} \end{array}$$

7 > 2 **Not enough** hundreds
7 < 22 **Enough** tens
The quotient begins in the tens place.
About how many 7s in 22? 3

Marc will travel about 30 km a day.

### Study these examples.

$$\begin{array}{r} 0.4\ 0\ 0 \\ 8\overline{)3.2\ 4\ 8} \end{array}$$

8 > 3 **Not enough** ones
8 < 32 **Enough** tenths
About how many 8s in 32? 4

The quotient is close to 0.4.

$$\begin{array}{r} 0.0\ 5\ 0 \\ 6\overline{)0.3\ 1\ 4} \end{array}$$

6 > 3 **Not enough** tenths
6 < 31 **Enough** hundredths
About how many 6s in 31? 5

The quotient is greater than 0.05.

### Estimate the quotient.

1. 6)0.234
2. 7)0.244
3. 8)0.746
4. 3)0.997
5. 4)0.872

6. 7)6.566
7. 6)2.472
8. 3)2.976
9. 8)3.295
10. 5)3.315

11. 3)29.506
12. 9)36.279
13. 4)12.688
14. 5)39.719
15. 9)47.821

16. 7)36.494
17. 6)23.523
18. 8)38.344
19. 4)312.123
20. 9)286.391

**Estimate the quotient.**

**21.** $0.874 \div 5$     **22.** $0.855 \div 3$     **23.** $5.364 \div 4$     **24.** $9.088 \div 9$

**25.** $47.372 \div 9$     **26.** $23.018 \div 4$     **27.** $58.761 \div 8$     **28.** $38.554 \div 6$

---

### Using Compatible Numbers

To **estimate decimal quotients** using *compatible numbers:*

- Think of nearby numbers that are compatible.

- Divide.

Estimate: $8.316 \div 9$

> 9 and 81are compatible numbers.

$$9\overline{)8.3\ 1\ 6} \rightarrow \overset{0.9}{9\overline{)8.1\ 0}}$$

The quotient is about 0.9.

Estimate: $1.684 \div 42$

> 40 and 160 are compatible numbers.

$$42\overline{)1.6\ 8\ 4} \rightarrow \overset{4 \text{ hundredths}}{40\overline{)1\ 6\ 0}} \text{ hundredths}$$

The quotient is about 0.04.

---

**Estimate the quotient.** Use compatible numbers.

**29.** $4\overline{)2.302}$    **30.** $9\overline{)1.935}$    **31.** $7\overline{)4.351}$    **32.** $8\overline{)4.253}$    **33.** $6\overline{)3.756}$

**34.** $5\overline{)34.057}$    **35.** $7\overline{)29.361}$    **36.** $3\overline{)28.536}$    **37.** $8\overline{)63.016}$    **38.** $9\overline{)71.789}$

**39.** $1.339 \div 31$    **40.** $2.654 \div 53$    **41.** $3.128 \div 62$    **42.** $2.095 \div 38$

**43.** $62.158 \div 28$    **44.** $36.751 \div 61$    **45.** $461.651 \div 53$    **46.** $105.995 \div 19$

### PROBLEM SOLVING

**47.** Alan rode his bicycle 34.325 km in 5 hours. If he rode an equal distance each hour, about how many kilometers did he ride in one hour?

**48.** Beth can run 5.985 km in 21 minutes. About how many kilometers can she run in one minute?

 **Share Your Thinking**     Math Journal ✓

**49.** Write in your Math Journal at least three situations in which making an estimate is more useful or efficient than finding an exact answer.

**Estimating Money**

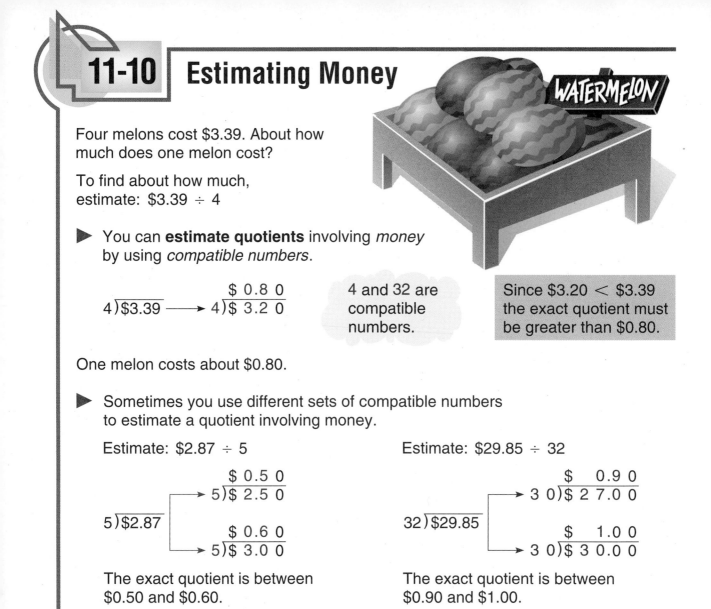

WATERMELON

Four melons cost $3.39. About how much does one melon cost?

To find about how much, estimate: $3.39 ÷ 4

▶ You can **estimate quotients** involving *money* by using *compatible numbers*.

$$4\overline{)\$3.39} \longrightarrow 4\overline{)\$\,3.2\,0}^{\;\$\,0.8\,0}$$

4 and 32 are compatible numbers.

Since $3.20 < $3.39 the exact quotient must be greater than $0.80.

One melon costs about $0.80.

▶ Sometimes you use different sets of compatible numbers to estimate a quotient involving money.

Estimate: $2.87 ÷ 5

$$5\overline{)\$2.87}$$

$$5\overline{)\$\,2.5\,0}^{\;\$\,0.5\,0}$$

$$5\overline{)\$\,3.0\,0}^{\;\$\,0.6\,0}$$

The exact quotient is between $0.50 and $0.60.

Estimate: $29.85 ÷ 32

$$32\overline{)\$29.85}$$

$$3\,0\overline{)\$\,2\,7.0\,0}^{\;\$\;\;0.9\,0}$$

$$3\,0\overline{)\$\,3\,0.0\,0}^{\;\$\;\;1.0\,0}$$

The exact quotient is between $0.90 and $1.00.

**Estimate the cost of one item.**

| | Item | Total Cost | Estimated Cost of One Item |
|---|---|---|---|
| **1.** | 6 bottles of apple juice | $ 2.49 | ? |
| **2.** | 9 tomatoes | $ 3.45 | ? |
| **3.** | 4 quarts of milk | $ 3.38 | ? |
| **4.** | 12 mugs | $59.76 | ? |
| **5.** | 23 oranges | $12.96 | ? |

**Write what compatible numbers you would use.**
**Then estimate the quotient.**

**6.** $3\overline{)\$1.06}$     **7.** $5\overline{)\$9.32}$     **8.** $9\overline{)\$8.25}$     **9.** $7\overline{)\$34.95}$     **10.** $6\overline{)\$13.79}$

**11.** $62\overline{)\$29.14}$   **12.** $54\overline{)\$37.84}$   **13.** $92\overline{)\$82.88}$   **14.** $31\overline{)\$59.96}$   **15.** $28\overline{)\$56.65}$

**16.** $\$149.50 \div 15$    **17.** $\$231.25 \div 42$    **18.** $\$412.18 \div 83$    **19.** $\$186.62 \div 31$

---

### Rounding to the Nearest Cent

Four mugs cost $9.89. How much does one mug cost?
Round the amount to the nearest cent.

To find how much,       $\$9.89 \div 4 = \underline{\ ?\ }$

| Write the decimal point in the quotient. Divide. | | Round the amount to the nearest cent. |
|---|---|---|

$$\begin{array}{r} \$\,2.\,4\ 7\ 2 \\ \hline 4\,)\$\ 9.^{1}8\ ^{2}9\ ^{1}0 \end{array}$$

> Add a zero in the dividend.
> $\$9.89 = \$9.890$

$$\begin{array}{r} \$\,2.\,4\ \overset{\frown}{7}\ 2 \\ \hline 4\,)\$\ 9.^{1}8\ ^{2}9\ ^{1}0 \end{array} = \$2.47$$

> $2 < 5$
> Round **down** to $2.47.

One mug costs about $2.47.

---

**Divide. Round the quotient to the nearest cent.**

**20.** $8\overline{)\$1.24}$    **21.** $5\overline{)\$3.78}$    **22.** $2\overline{)\$1.11}$    **23.** $3\overline{)\$5.29}$    **24.** $6\overline{)\$8.20}$

**25.** $6\overline{)\$33.32}$   **26.** $4\overline{)\$22.61}$   **27.** $5\overline{)\$26.12}$   **28.** $9\overline{)\$51.09}$   **29.** $7\overline{)\$28.46}$

### PROBLEM SOLVING

**30.** Ruby earns $52.50 in 6 hours. About how much does she earn in one hour?

**31.** A set of 35 identical books costs $236.25. About how much does one book cost?

### Challenge

**32.** A monthly pass for a commuter train costs $84. A single ticket costs $2.75. Rose rides the train an average of 44 times a month. About how much does she save per ride if she buys a monthly pass instead of single tickets?

## 11-11 Problem Solving: Write a Number Sentence

*Algebra* ✓

| Number Sentences | |
|---|---|
| **Addition Sentences** | **Multiplication Sentences** |
| Let $a$ and $b$ represent addends, and $c$ represent the sum.<br>$a + b = c$    $b + a = c$ | Let $m$ and $n$ represent factors, and $r$ represent the product.<br>$m \times n = r$    $n \times m = r$ |
| **Related Subtraction Sentences**<br>$c - b = a$    $c - a = b$ | **Related Division Sentences**<br>$r \div n = m$    $r \div m = n$ |

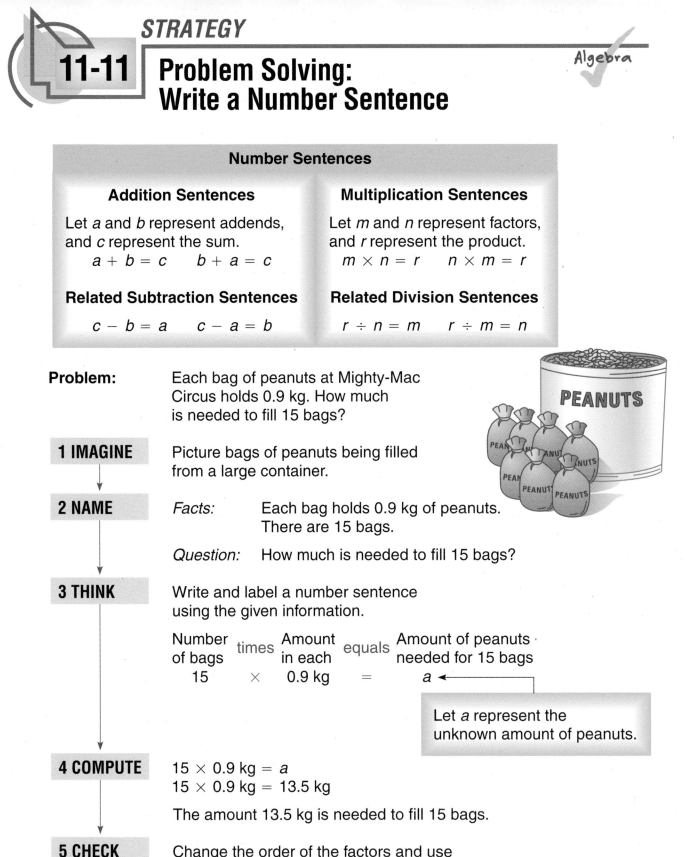

**Problem:** Each bag of peanuts at Mighty-Mac Circus holds 0.9 kg. How much is needed to fill 15 bags?

**1 IMAGINE** Picture bags of peanuts being filled from a large container.

**2 NAME** *Facts:* Each bag holds 0.9 kg of peanuts. There are 15 bags.

*Question:* How much is needed to fill 15 bags?

**3 THINK** Write and label a number sentence using the given information.

| Number of bags | times | Amount in each | equals | Amount of peanuts needed for 15 bags |
|---|---|---|---|---|
| 15 | $\times$ | 0.9 kg | $=$ | $a$ |

Let $a$ represent the unknown amount of peanuts.

**4 COMPUTE** $15 \times 0.9 \text{ kg} = a$
$15 \times 0.9 \text{ kg} = 13.5 \text{ kg}$

The amount 13.5 kg is needed to fill 15 bags.

**5 CHECK** Change the order of the factors and use a calculator to check your computations.

**Write a number sentence to solve each problem.**

Algebra ✓

1. A bicyclist travels 36.3 miles in 2 hours.
   What is her rate of speed in miles per hour?

**IMAGINE**    Put yourself in the problem.

**NAME**    *Facts:*    distance—36.3 miles
                        time—2 hours

           *Question:*   How many miles per hour
                         did the bicyclist travel?

**THINK**    Write a word sentence for the problem, then
             write a number sentence. Let *r* represent rate.
             Divide to find the missing factor.

             Distance equals rate of speed times time.
             36.3 mi    =    *r* mph    ×    2 h

Write the related division sentence.

**COMPUTE** ⟶ **CHECK**

2. Devon lives 8.25 km from the river. In the morning he
   walks 5.7 km toward the river. How much farther does
   he need to walk to reach the river?

3. The length of a river is 27.6 mi. Joan
   kayaked half the length of the river.
   How many miles did Joan kayak?

4. A swimmer took 2.75 h to swim upstream
   and 1.8 h to swim downstream. How
   long did it take the swimmer to cover
   the entire distance?

5. A set of 32 new fifth grade math books
   costs $468.96. About how much does
   each math book cost?

6. Ninety books weigh 720.9 lb. What is the
   weight of one book if they all weigh the same amount?

7. Mr. Brophy traveled 232.5 km in 5 days. If he traveled the same
   distance each day, what was the distance he traveled in one day?

## 11-12 | Problem-Solving Applications

### Connections: Science

**Solve and explain the method you used.**

1. In a science experiment one lens is positioned 0.25 m from a light source, and a second lens is positioned ten times farther. How far is the second lens from the light source?

2. Adam discovers that the first lens has a focal length of 1.4 m. The second lens has a focal length 3 times greater. What is the focal length of the second lens?

3. The radius of Sara's lens is 0.235 dm. What is the diameter of her lens?

4. Carlotta divides 0.09 L of bleach equally among 3 beakers. How much bleach is in each beaker?

5. A set of 8 tuning forks costs $19.84. How much does each tuning fork cost?

6. The largest tuning fork is 4 times the size of the smallest. The largest tuning fork is 31.48 cm long. How long is the smallest tuning fork?

7. In each of four experiments Marta uses 23.2 mL, 20.8 mL, 17.3 mL, and 19.7 mL of distilled water. About how many milliliters does she use in all?

8. It took Adam 100.3 s to light a candle using a small lens. It took one half as long using a large lens. How long did it take to light a candle using a large lens?

9. The tone of the largest tuning fork lasts for 125.75 s. The tone of the smallest tuning fork lasts two tenths of this time. How long does the tone of the smallest fork last?

10. A solution's temperature increased 11.3°C in 5 minutes. What was the average temperature increase per minute?

Choose a strategy from the list or use
another strategy you know to solve
each problem.

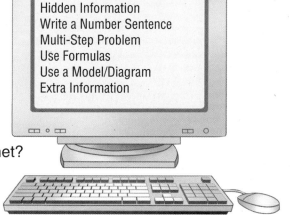

USE THESE STRATEGIES
Hidden Information
Write a Number Sentence
Multi-Step Problem
Use Formulas
Use a Model/Diagram
Extra Information

**11.** Tyrell uses 14 magnets and 6 batteries
in a physics experiment. Each magnet
costs $1.29. How much does Tyrell
spend on magnets?

**12.** The rectangular magnets are 5.75 cm by
3.4 cm. What is the perimeter of each magnet?

**13.** Each magnet can lift 0.542 kg. Can
fourteen magnets together lift a 6.5-kg
metal box?

**14.** Each magnet has a mass of 95.5 g. Kim uses 9 magnets to lift a
4.5-kg box. What is the total mass of the magnets and the box?

**15.** A tank holds 0.38 cubic meters. Vicki fills 0.1 of the tank
with gravel. How many cubic meters of water does she
need to fill the tank?

**16.** Joni sets a 24-hour timer so that a heat lamp shines on her plants for
3.75 h every day. An incandescent bulb shines on the plants for twice
as long every day. How long will the incandescent bulb be on today?

**17.** Each of these test tubes can hold 0.015 L.
Mr. Henry pours out half of the water in test
tube B. How much water is left in test tube B?

**18.** Ms. Cooper fills the rest of test tube A
with an acid. How much acid does she use?

**19.** Mr. Henry uses 0.3 of the water from
test tube C. Now can he add 0.006 L
of bleach to test tube C?

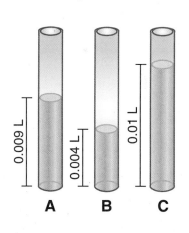

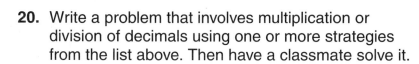

**Make Up Your Own**

Communicate

**20.** Write a problem that involves multiplication or
division of decimals using one or more strategies
from the list above. Then have a classmate solve it.

# Chapter Review and Practice

**Find the missing number.** (See pp. 352–353, 362–363.)

1. __?__ × 6.1 = 61

2. __?__ × 42.3 = 4230

3. __?__ × 6.23 = 6230

4. 43.7 ÷ __?__ = 4.37

5. 2.7 ÷ __?__ = 0.027

6. 25 ÷ __?__ = 0.025

7. 10 × __?__ = 14.3

8. 1000 × __?__ = 593

9. 100 × __?__ = 74.6

10. __?__ ÷ 10 = 7.914

11. __?__ ÷ 100 = 4.567

12. __?__ ÷ 1000 = 0.009

**Estimate each product. Then tell whether the actual product is *greater than* or *less than* the estimated product.** (See pp. 354–355.)

13.
$$\begin{array}{r} 3.396 \\ \times\ \ \ 7.4 \\ \hline \end{array}$$

14.
$$\begin{array}{r} 14.87 \\ \times\ \ 0.73 \\ \hline \end{array}$$

15.
$$\begin{array}{r} 8.147 \\ \times\ \ \ 6.3 \\ \hline \end{array}$$

16.
$$\begin{array}{r} 25.423 \\ \times\ \ \ 0.58 \\ \hline \end{array}$$

**Use a 10 × 10 grid to find the product or quotient.** (See pp. 358–359, 364–365.)

17. 0.4 × 0.6

18. 0.9 × 0.5

19. 1.98 ÷ 2

20. 0.87 ÷ 3

**Multiply.** (See pp. 356–361.)

21. 6 × 0.43

22. 3.8 × 0.6

23. 0.64 × 0.4

24. 7 × 1.4

25. 0.18 × 0.4

26. 4.3 × 0.2

**Divide and check.** (See pp. 364–367.)

27. 0.546 ÷ 2

28. 0.4 ÷ 8

29. 6.4 ÷ 4

30. 7.56 ÷ 3

31. 9.5 ÷ 5

32. 0.49 ÷ 7

**Estimate the quotient. Use compatible numbers.** (See pp. 368–371.)

33. $8\overline{)1.754}$

34. $6\overline{)4.159}$

35. $7\overline{)29.543}$

36. $21\overline{)43.359}$

37. $9\overline{)\$28.53}$

38. $4\overline{)\$37.34}$

39. $3\overline{)\$19.97}$

40. $43\overline{)\$89.15}$

## PROBLEM SOLVING

(See pp. 358–359, 370–371, 372–374.)

41. Mr. Burton ordered 12.4 cases of fruit for his store. Three tenths of the cases were damaged. How many cases of fruit were damaged?

42. Nine part-time workers earned $463.15. About how much did each worker get if the money was divided equally?

(See *Still More Practice*, p. 486.)

## FRACTIONS TO DECIMALS

Every fraction is *equivalent to* or can be *renamed as* a decimal.

To rename a fraction as a decimal:

▶ Find an equivalent fraction whose denominator is a power of 10.

> Powers of 10 are:
> 1, 10, 100, 1000,...

$$\frac{1}{2} = \frac{1 \times 5}{2 \times 5} = \frac{5}{10} = 0.5 \qquad \frac{3}{4} = \frac{3 \times 25}{4 \times 25} = \frac{75}{100} = 0.75$$

power of 10

or

▶ Divide the numerator by the denominator.

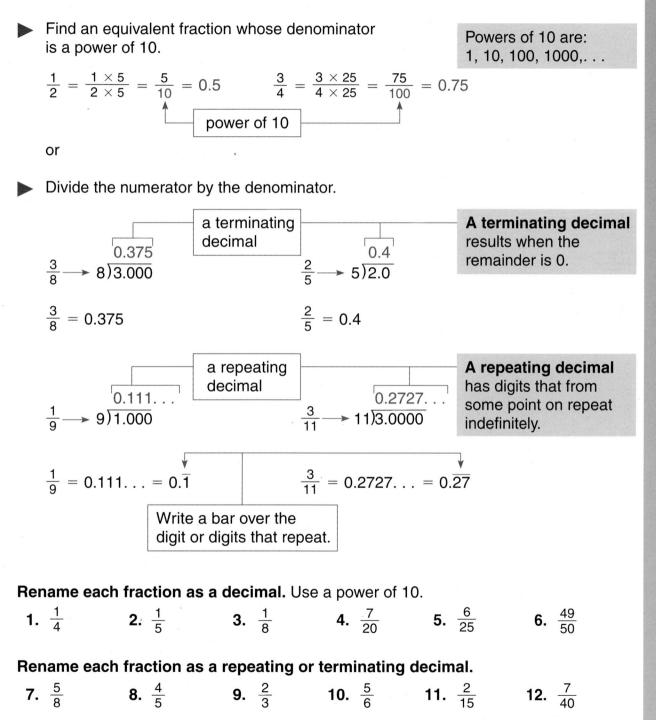

a terminating decimal

$$\frac{3}{8} \longrightarrow 8\overline{)3.000} \quad (0.375) \qquad \frac{2}{5} \longrightarrow 5\overline{)2.0} \quad (0.4)$$

**A terminating decimal** results when the remainder is 0.

$$\frac{3}{8} = 0.375 \qquad \frac{2}{5} = 0.4$$

a repeating decimal

$$\frac{1}{9} \longrightarrow 9\overline{)1.000} \quad (0.111...) \qquad \frac{3}{11} \longrightarrow 11\overline{)3.0000} \quad (0.2727...)$$

**A repeating decimal** has digits that from some point on repeat indefinitely.

$$\frac{1}{9} = 0.111... = 0.\overline{1} \qquad \frac{3}{11} = 0.2727... = 0.\overline{27}$$

Write a bar over the digit or digits that repeat.

**Rename each fraction as a decimal.** Use a power of 10.

1. $\frac{1}{4}$    2. $\frac{1}{5}$    3. $\frac{1}{8}$    4. $\frac{7}{20}$    5. $\frac{6}{25}$    6. $\frac{49}{50}$

**Rename each fraction as a repeating or terminating decimal.**

7. $\frac{5}{8}$    8. $\frac{4}{5}$    9. $\frac{2}{3}$    10. $\frac{5}{6}$    11. $\frac{2}{15}$    12. $\frac{7}{40}$

# Check Your Mastery

## Performance Assessment

**Write the missing output and write a rule for each table.**

**1.**

| Input | 7.8 | 0.13 | 0.6 |
|---|---|---|---|
| Output | 0.78 | 0.013 | ? |

**2.**

| Input | 0.065 | 1.03 | 0.008 |
|---|---|---|---|
| Output | 6.5 | 103 | ? |

**Make up an input-output table for each rule.**

**3.** rule: input times 1000

**4.** rule: input divided by 100

**Estimate each product. Then tell whether the actual product is _greater than_ or _less than_ the estimated product.**

**5.**  $\begin{array}{r} 5.65 \\ \times\ \ 3.4 \\ \hline \end{array}$

**6.**  $\begin{array}{r} 8.436 \\ \times\ \ \ 7.6 \\ \hline \end{array}$

**7.**  $\begin{array}{r} 18.76 \\ \times\ \ 0.44 \\ \hline \end{array}$

**8.**  $\begin{array}{r} 26.877 \\ \times\ \ \ \ 0.47 \\ \hline \end{array}$

**Use a 10 × 10 grid to find the product or quotient.**

**9.** 0.8 × 0.7

**10.** 0.2 × 0.9

**11.** 2.42 ÷ 2

**12.** 3.16 ÷ 4

**Multiply.**

**13.** 6 × 4.2

**14.** 0.72 × 0.3

**15.** 0.07 × 0.6

**16.** 9.6 × 0.85

**17.** 0.7 × 9.19

**18.** 8 × 0.43

**Divide and check.**

**19.** 0.38 ÷ 2

**20.** 0.24 ÷ 6

**21.** 1.4 ÷ 4

**22.** 0.63 ÷ 7

**23.** 4.12 ÷ 5

**24.** 1.908 ÷ 3

**Estimate the quotient.** Use compatible numbers.

**25.** $6\overline{)29.457}$

**26.** $8\overline{)41.053}$

**27.** $72\overline{)139.125}$

**28.** $7\overline{)\$22.98}$

**29.** $4\overline{)\$35.97}$

**30.** $54\overline{)\$295.72}$

**PROBLEM SOLVING**  *Use a strategy you have learned.*

**31.** Tom paid a total of $67.95 for 3 identical swimsuits. How much did he pay for each?

**32.** Ron bought 15 keychains for $46.35. About how much did each keychain cost?

**33.** Martha worked 7.4 h one day. Eight tenths of this time was spent at the cash register. How much time did Martha work at the cash register?

# Cumulative Review IV

**Choose the best answer.**

---

**1.** In the number 12,345,678,000, which digit is in the billions place?

   **a.** 1     **b.** 2
   **c.** 3     **d.** 5

**2.** $406 \times \$17.98$

   **a.** \$827.08     **b.** \$7299.88
   **c.** \$7479.68     **d.** not given

---

**3.** Which number is divisible by 3 and by 6 but not divisible by 9?

   **a.** 20,007     **b.** 72,000
   **c.** 72,111     **d.** 73,110

**4.** Which is the least common denominator of

$$\frac{3}{5}, \quad \frac{1}{2}, \quad \frac{5}{6} \ ?$$

   **a.** 18     **b.** 30
   **c.** 60     **d.** 16

---

**5.** How much greater than

$8\frac{1}{3} - 6\frac{1}{4}$ is

$8\frac{1}{3} + 6\frac{1}{4}$ ?

   **a.** 12     **b.** $12\frac{1}{12}$
   **c.** $12\frac{5}{24}$     **d.** $12\frac{1}{2}$

**6.** $2\frac{2}{5} \div 1\frac{1}{7}$

   **a.** $1\frac{2}{35}$     **b.** $2\frac{1}{10}$
   **c.** $2\frac{26}{35}$     **d.** $1\frac{1}{10}$

---

**7.** Use the data. Which is greatest: range, mean, median, mode?

**School Enrollment**

| Grade | 3 | 4 | 5 | 6 |
|---|---|---|---|---|
| Number of Students | 83 | 79 | 87 | 79 |

   **a.** range     **b.** mean
   **c.** median     **d.** mode

**8.** Which is true about the measure of the angle?

   **a.** 90°
   **b.** greater than 90°
   **c.** less than 90°
   **d.** cannot tell

---

**9.** Estimate.

$31.09 + 7.86$

   **a.** 11.5
   **b.** 23.9
   **c.** 39
   **d.** 45

**10.** $45.8 - 4.294$

   **a.** 0.286
   **b.** 41.606
   **c.** 41.694
   **d.** 41.506

---

**11.** $8.1 \times 7.56$

   **a.** 61.236
   **b.** 68.04
   **c.** 612.36
   **d.** not given

**12.** $3.612 \div 4$

   **a.** 0.903
   **b.** 0.93
   **c.** 9.03
   **d.** 9.3

# Ongoing Assessment IV

## For Your Portfolio

**Solve each problem. Explain the steps and the strategy or strategies you used for each. Then choose one from problems 1–4 for your Portfolio.**

1. Rosalie made a rectangular afghan 6.25 ft wide and 7.5 ft long. What is its perimeter?

2. Luz multiplied a number by 3 and then divided the result by 5. The answer was 0.36. Find Luz's original number.

3. A circular magnifying glass has a radius that is $\frac{2}{3}$ in. Estimate the circumference of the magnifying glass.

4. A scale model train is 14.2 cm long. Each centimeter represents 87 m on the actual train. How long is the actual train?

**Tell about it.**

5. What strategy did you use to solve problem 3? problem 4?

*Communicate* ✓

6. Explain how the Use Simpler Numbers strategy helps you solve problems 3 and 4.

---

## For Rubric Scoring

**Listen for information on how your work will be scored.**

7. An **arithmetic sequence** is a pattern in which the same number is **added** to the previous number to obtain the next number.

   What number is added?     0.2   0.5   0.8   1.1   1.4

8. A **geometric sequence** is a pattern in which the same number is **multiplied** to the previous number to obtain the next number.

   What number is multiplied?     0.2   0.4   0.8   1.6   3.2

9. **a.** Write other examples of arithmetic and geometric sequences.

   **b.** See if you can find a sequence that is both arithmetic and geometric. (*Hint:* Try adding 0 and multiplying by 1.)

# Metric Measurement, Area, and Volume

## 12

The Mississippi River is about 4,000 kilometers long.

An M&M is about 1 centimeter long.

There are 100 centimeters in a meter, and 1,000 meters in a kilometer.

Estimate how many M&Ms it would take to measure the length of the Mississippi River.

From *Math Curse* by *Jon Scieszka*

**In this chapter you will:**

Investigate metric units of length, capacity, and mass
Use area formulas
Classify space figures
Learn about cubic measure and volume
Explore electronic spreadsheets
Solve problems by drawing a picture

**Critical Thinking/ Finding Together**

You have one piece of pipe 1.3 m long and another piece 30 cm long. How can you use these two pieces of pipe to measure 2 m on a third piece of pipe?

## 12-1 Metric Measurement

The **metric system** is a *decimal* system of measurement. Decimal means that the system is based on tens.

▶ The standard metric units are:
the **meter (m)**, which is used to measure length, the **liter (L)**, which is used to measure capacity, and the **gram (g)**, which is used to measure mass.

▶ You can use :
a **meterstick** to measure length, a **graduated cylinder** to measure capacity, and a **metric balance** to measure mass.

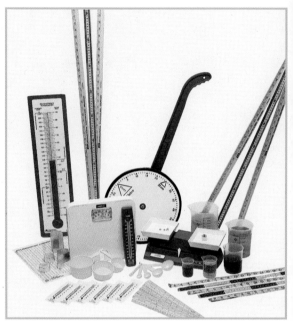

The baseball bat is 1 meter long.

1 liter

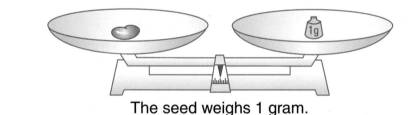

The seed weighs 1 gram.

---

**Which metric unit is used to measure each? Write *m*, *L*, or *g*.**

1. milk in a carton
2. mass of a football
3. height of a building
4. capacity of a water cooler
5. distance of a softball throw
6. mass of potatoes
7. string of a kite
8. juice in a pitcher

9. Explain in your Math Journal how the metric system of measurement differs from the customary system of measurement.

Math Journal

## Some Metric Relationships

The chart below shows how the metric units of length, capacity, or mass are related to the standard metric units and to each other.

| Metric Units of Length | $(1 \times 1000)$ m $= 1000$ m | | | $(1 \div 10)$ m $= 0.1$ m | $(1 \div 100)$ m $= 0.01$ m | $(1 \div 1000)$ m $= 0.001$ m |
|---|---|---|---|---|---|---|
| | 1 kilometer | | 1 meter | 1 decimeter | 1 centimeter | 1 millimeter |
| | (km) | | (m) | (dm) | (cm) | (mm) |
| Metric Units of Capacity | $(1 \times 1000)$ L $= 1000$ L | | | $(1 \div 10)$ L $= 0.1$ L | $(1 \div 100)$ L $= 0.01$ L | $(1 \div 1000)$ L $= 0.001$ L |
| | 1 kiloliter | | 1 liter | 1 deciliter | 1 centiliter | 1 milliliter |
| | (kL) | | (L) | (dL) | (cL) | (mL) |
| Metric Units of Mass | $(1 \times 1000)$ g $= 1000$ g | | | $(1 \div 10)$ g $= 0.1$ g | $(1 \div 100)$ g $= 0.01$ g | $(1 \div 1000)$ g $= 0.001$ g |
| | 1 kilogram | | 1 gram | 1 decigram | 1 centigram | 1 milligram |
| | (kg) | | (g) | (dg) | (cg) | (mg) |

**Write the letter of the smaller unit of measure.**

10. **a.** milliter
    **b.** liter

11. **a.** meter
    **b.** decimeter

12. **a.** gram
    **b.** kilogram

13. **a.** centimeter
    **a.** millimeter

**Write the letter of the best metric unit to be used for each.**

14. mass of an apple     **a.** gram     **b.** kilogram     **c.** milligram

15. thickness of a nickel     **a.** kilometer     **b.** meter     **c.** millimeter

16. a drop of water     **a.** liter     **b.** milliliter     **c.** kiloliter

17. width of a cup     **a.** meter     **b.** centimeter     **c.** kilometer

## Skills to Remember

**Find the products.**

18.   $10 \times 35$
    $100 \times 35$
    $1000 \times 35$

19.   $10 \times 430$
    $100 \times 430$
    $1000 \times 430$

20.   $0.1 \times 17$
    $0.01 \times 17$
    $0.001 \times 17$

21.   $0.1 \times 890$
    $0.01 \times 890$
    $0.001 \times 890$

**Find the quotients.**

22. $9 \div 10$
    $9 \div 100$
    $9 \div 1000$

23. $34 \div 10$
    $34 \div 100$
    $34 \div 1000$

24. $450 \div 10$
    $450 \div 100$
    $450 \div 1000$

25. $6700 \div 10$
    $6700 \div 100$
    $6700 \div 1000$

# Renaming Metric Units

To **rename metric units**, use the relations between the units.

| | | |
|---|---|---|
| 1 km = 1000 m | 1 kL = 1000 L | 1 kg = 1000 g |
| 1 m = 10 dm | 1 L = 10 dL | 1 g = 10 dg |
| 1 m = 100 cm | 1 L = 100 cL | 1 g = 100 cg |
| 1 m = 1000 mm | 1 L = 1000 mL | 1 g = 1000 mg |
| 1 dm = 10 cm | 1 dL = 10 cL | 1 dg = 10 cg |
| 1 dm = 100 mm | 1 dL = 100 mL | 1 dg = 100 mg |
| 1 cm = 10 mm | 1 cL = 10 mL | 1 cg = 10 mg |

▶ *Multiply* to rename larger units as smaller units.

25 m = __?__ dm
25 m = (25 × 10) dm
25 m = 250 dm

Think:
1 m = 10 dm

4 kg = __?__ g
4 kg = (4 × 1000) g
4 kg = 4000 g

Think:
1 kg = 1000 g

▶ *Divide* to rename smaller units as larger units.

Think:
1000 mm = 1 m

600 cL = __?__ L
600 cL = (600 ÷ 100) L
600 cL = 6 L

Think:
100 cL = 1 L

800 000 mm = __?__ m
800 000 mm = (800 000 ÷ 1000)
800 000 mm = 800 m

**Study these examples.**

7.48 dg = __?__ g
7.48 dg = (7.48 × 10) g
7.48 dg = 74.8 g

638 L = __?__ kL
638 L = (638 ÷ 1000) kL
638 L = 0.638 kL

## Write the letter of the correct answer.

1. 9 cm = __?__ mm      **a.** 0.9      **b.** 90      **c.** 900      **d.** 9000

2. 4 kg = __?__ g      **a.** 0.4      **b.** 40      **c.** 400      **d.** 4000

3. 375 mL = __?__ L      **a.** 0.375      **b.** 3.75      **c.** 37.5      **d.** 3750

4. 84.6 dm = __?__ m      **a.** 0.846      **b.** 8.46      **c.** 846      **d.** 8460

5. 9.8 g = __?__ mg      **a.** 0.098      **b.** 0.98      **c.** 0.0098      **d.** 9800

**Copy and complete.**

**6.** 5 km = _?_ m  **7.** 9 m = _?_ cm  **8.** 4 dm = _?_ cm

**9.** 7 m = _?_ mm  **10.** 800 mm = _?_ cm  **11.** 8 L = _?_ dL

**12.** 84 g = _?_ cg  **13.** 4000 cL = _?_ L  **14.** 16 000 g = _?_ kg

**15.** 11.5 dm = _?_ m  **16.** 25 300 m = _?_ km  **17.** 50 dL = _?_ L

**18.** 3.78 cm = _?_ mm  **19.** 40.3 kL = _?_ L  **20.** 734 g = _?_ kg

**21.** 585 m = _?_ km  **22.** 836 mm = _?_ m  **23.** 479 cg = _?_ g

## PROBLEM SOLVING

**24.** Sergey Bubka's Olympic gold-medal-winning pole vault in 1988 was 5.90 m. Would a vault of 595 cm be higher or lower than Bubka's jump?

**25.** Isabel needs 350 mL of milk to make a loaf of bread. How many liters of milk does she need to make 8 loaves of bread?

**26.** An orange contains about 0.07 g of vitamin C. About how many milligrams of vitamin C does it contain?

**27.** Marco was running in the 600-m race. He had run 45 000 cm. How many meters farther did he have to run to complete the race?

## Calculator Activity

*Algebra*

**Find the missing number to discover a pattern in each row.**

**28.** 18.5 m = _?_ dm  **29.** 185 dm = _?_ cm  **30.** 1850 cm = _?_ mm

**31.** 173 L = _?_ dL  **32.** 1730 dL = _?_ cL  **33.** 17 300 cL = _?_ mL

**34.** 2500 mm = _?_ cm  **35.** 250 cm = _?_ dm  **36.** 25 dm = _?_ m

**37.** 68 000 mg = _?_ cg  **38.** 6800 cg = _?_ dg  **39.** 680 dg = _?_ g

**40.** To which direction, right or left, is the decimal point moved when renaming a larger metric unit as a smaller metric unit? a smaller metric unit as a larger metric unit?

*Communicate*

**41.** In the metric system, how many of a unit does it take to make one of the next larger unit? What part of a unit is the next smaller unit?

# 12-3 Relating Metric Units of Length

## Discover Together

**Materials Needed:** metric ruler or meterstick, paper, pencil, calculator

The **millimeter (mm)**, **centimeter (cm)**, **decimeter (dm)**, **meter (m)**, and **kilometer (km)** are metric units of length.

| | | |
|---|---|---|
| 1 m | = | 1000 mm |
| 1 m | = | 100 cm |
| 1 m | = | 10 dm |
| 1 km | = | 1000 m |

1. Which units are smaller than a meter? larger than a meter?

2. Which unit would you use to measure the height of your desk? Explain why you think your choice is reasonable.

3. What objects in your classroom would you measure in meters? Explain why your choices are reasonable.

4. What unit would you use to measure the distance between two cities? Explain why you think your choice is reasonable.

5. What unit would you use to measure the length of an ant? Explain why you think your choice is reasonable.

You can use a metric ruler or a meterstick to measure the length of an object. A meterstick usually shows decimeters, millimeters, and centimeters.

6. Find the marks that represent each unit on your metric ruler.

1 mm    1 cm                                    1 dm

7. How many millimeters long is your metric ruler? How many centimeters? How many decimeters?

8. How many millimeters long is a meterstick? How many centimeters? How many decimeters?

9. Use your metric ruler to measure the length of your math book. What is the length of your math book in millimeters? in centimeters? in decimeters?

**10.** Name 3 objects you would measure in millimeters; in centimeters; in decimeters; in meters; in kilometers.

Sometimes it is necessary to take precise measurements. The smaller the unit of measure you use, the more precise your measurement will be. When you measure an object, you measure to the nearest unit of that measure.

**11.** Use your metric ruler as shown to measure the length of the given ribbon.

What is the length of the ribbon to the nearest mm? the nearest cm? the nearest dm?

**12.** Estimate. Then measure each to the nearest mm, nearest cm, and nearest dm.

    **a.** length of your pen    **b.** diameter of a coin    **c.** height of the board

## Communicate

Discuss ✓

**13.** What is the smallest metric unit of length? the largest metric unit of length?

**14.** Which is the most precise unit of measure to use: meter, decimeter, centimeter, or millimeter? Why?

**15.** At the hardware store Alex asked for an extension cord that was 4 km long. Was this an appropriate length to ask for? If not, what length do you think he should have asked for?

### Choose a Computation Method

**Find the missing unit. Use mental math, paper and pencil, or a calculator. Explain the method you used.**

**16.** 9.5 dm = 950 ___?___

**17.** 4 cm = 0.04 ___?___

**18.** 2.5 mm = 0.25 ___?___

**19.** 1200 m = 1.2 ___?___

**20.** 2.5 m = 2500 ___?___

**21.** 0.34 km = 34 000 ___?___

# 12-4 Relating Units of Capacity

The **milliliter (mL)**, **centiliter (cL)**, **deciliter (dL)**,
**liter (L)**, and **kiloliter (kL)**, are metric units of capacity.

| | | |
|---|---|---|
| 1 L | = | 1000 mL |
| 1 L | = | 100 cL |
| 1 L | = | 10 dL |
| 1 kL | = | 1000 L |

▶ The liter, milliliter, and kiloliter are the most commonly
used metric units of capacity.

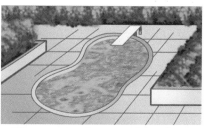

A tall thermos
holds about 1 L.

A medicine dropper
holds about 0.5 mL.

The water in a swimming
pool is measured in kL.

▶ You can use graduated cylinders of various sizes
to measure liquid capacity.

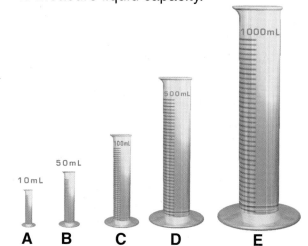

**A    B    C    D    E**

- Cylinder *A* holds 10 mL or 1 cL.
- Cylinder *B* holds 50 mL.
- Cylinder *C* holds 100 mL or 1 dL.
- Cylinder *D* holds 500 mL.
- Cylinder *E* holds 1000 mL or 1 L.

**Study these examples.**

15 L  ?  1500 mL

15 L = (15 × 1000) mL
15 L = 15 000 mL
15 000 mL > 1500 mL

So 15 L > 1500 mL.

Think:
1 L = 1000 mL

360 L  ?  3.6 kL

360 L = (360 ÷ 1000) kL
360 L = 0.36 kL
0.36 kL < 3.6 kL

So 360 L < 3.6 kL.

Think:
1000 L = 1 kL

**Which metric unit is best to measure the capacity of each?**
**Write *mL*, *L*, or *kL*.**

1. a fish tank
2. an oil tanker
3. an ice tray
4. a milk truck
5. a baby bottle
6. a washing machine

**Compare. Write <, =, or >.**

7. 2 L ? 250 cL
8. 13 L ? 130 mL
9. 36 kL ? 36 000 L
10. 52 L ? 515 dL
11. 2600 L ? 26 kL
12. 35 dL ? 4 L
13. 760 cL ? 75 L
14. 12 L ? 12 000 mL
15. 173 L ? 1730 cL
16. 860 mL ? 8.6 L
17. 17.3 kL ? 1730 L
18. 2.5 L ? 25 dL

**PROBLEM SOLVING**

19. Rhoda wants to add a small amount of food coloring to the pie she is making. What metric unit of capacity should she use to measure the food coloring?

20. Mr. Navarro has 28 students in his science class. Each student in his class needs 250 mL of salt solution to do one experiment. How many liters of salt solution does the class need for the experiment?

21. Ms. Haraguchi made fruit punch for her party. To make the punch, she used 1.5 L of orange juice, 300 cL of ginger ale, 5 dL of lemon juice, and 1 L of club soda. How many deciliters of punch did Ms. Haraguchi make?

**Project**

*Communicate* ✓

**Choose 4 empty containers of different sizes and shapes.**

22. Use a small paper cup as your unit of measure.
    • Estimate how many times you would have to fill the paper cup with water to fill each of the 4 empty containers.
    • Use the paper cup and water to measure the actual capacity of each container.

23. Use a graduated cylinder to measure the capacity of each container in milliliters. Then tell whether each container holds less than, equal to, or greater than one liter.

24. Report to your class on your experiment.

## 12-5 | Relating Metric Units of Mass

The **milligram (mg)**, **centigram (cg)**, **decigram (dg)**, **gram (g)**, **kilogram (kg)**, and **metric ton (t)** are metric units of mass.

The most commonly used metric units of mass are the milligram, gram, kilogram, and metric ton.

| | | |
|---|---|---|
| 1 g | = | 1000 mg |
| 1 g | = | 100 cg |
| 1 g | = | 10 dg |
| 1 kg | = | 1000 g |
| 1 t | = | 1000 kg |

 **Discover Together**

**Materials Needed:** metric balance, gram masses, nickel, paper, pencil

1. Which units are smaller than a gram? larger than a gram?

2. A grain of salt has a mass of about one milligram. Name other objects that have a mass of about 1 mg.

3. What objects would you use to measure mass in milligrams?

4. A standard paper clip has a mass of about one gram. Name other objects that have a mass of about 1 g.

5. Estimate the mass of a nickel by comparing it with the mass of a standard paper clip. How many standard paper clips do you think are equal to the mass of a nickel?

6. About how many grams do you think a nickel would weigh?

7. Use a metric balance to find the actual mass of a nickel. Then compare the mass with your estimate. How does your estimate compare with the mass?

Now choose 5 classroom objects of different sizes and mass.

8. Estimate the mass of each object. Then use a metric balance to find the mass in grams. Record your answers in a table like the one shown.

9. How does each estimate in your table compare with the actual measurement?

| Object | Estimate | Mass in Grams |
|---|---|---|
| | | |
| | | |
| | | |

10. Estimate the mass of a hardcover dictionary by comparing it with the mass of a bag of 1000 standard paper clips. About how many grams do you think a hardcover dictionary would weigh?

11. If 1000 g = 1 kg, about how many kilograms do you think a hardcover dictionary would weigh?

12. Name some objects you know that have their mass measured in kilograms?

The mass of extremely heavy objects is expressed in metric tons. A camper has a mass of about 3 t.

13. Name some objects you know that have their mass measured in metric tons.

14. How many grams are in one metric ton?

15. Why are you less likely to use the metric ton than the gram, the milligram, or the kilogram as a unit of mass in your everyday life?

> 1 metric ton (t) = 1000 kilograms (kg)

16. Which is a greater mass: 3 g or 300 mg? 400 g or 4.5 kg? 2.75 t or 2000 kg? Explain your answers.

## Communicate

Discuss

17. What is the smallest metric unit of mass? the largest metric unit of mass?

18. What unit would you use to measure the mass of a small leaf? a loaf of bread? an automobile? a table? Explain your answers.

19. You are cooking chicken for dinner. The recipe calls for a large chicken. Will you buy a chicken that is about 4 g or 4 kg? Why?

### Mental Math

20. Express in cm:   5 dm, 10 dm, 15 dm, 100 mm, 150 mm, 200 mm

21. Express in m:   8 km, 6 km, 7 km, 50 dm, 70 dm, 400 dm

22. Express in g:   2 kg, 4 kg, 9 kg, 70 dg, 80 dg, 600 dg

23. Express in L:   3 kL, 5 kL, 8 kL, 40 dL, 90 dL, 700 dL

# 12-6 | Square Measure

The **area** of a figure is the number of square units that cover its surface.

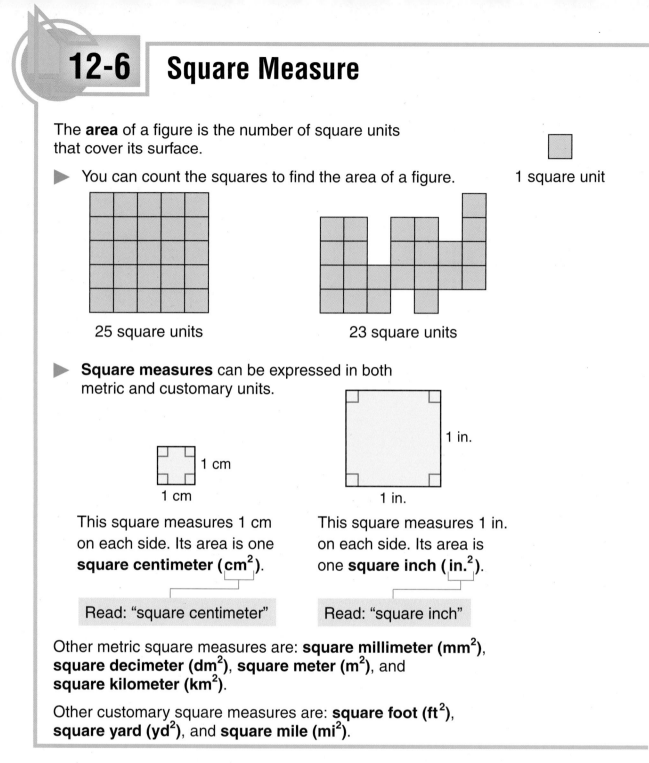

▶ You can count the squares to find the area of a figure.

1 square unit

25 square units

23 square units

▶ **Square measures** can be expressed in both metric and customary units.

1 cm
1 cm

1 in.
1 in.

This square measures 1 cm on each side. Its area is one **square centimeter ($cm^2$)**.

This square measures 1 in. on each side. Its area is one **square inch ($in.^2$)**.

Read: "square centimeter"

Read: "square inch"

Other metric square measures are: **square millimeter ($mm^2$)**, **square decimeter ($dm^2$)**, **square meter ($m^2$)**, and **square kilometer ($km^2$)**.

Other customary square measures are: **square foot ($ft^2$)**, **square yard ($yd^2$)**, and **square mile ($mi^2$)**.

**Find the area of each figure.**

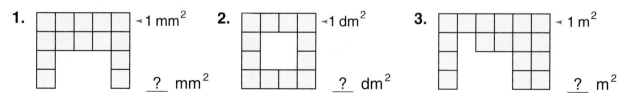

1. ◄ 1 $mm^2$
   ___ $mm^2$

2. ◄ 1 $dm^2$
   ___ $dm^2$

3. ◄ 1 $m^2$
   ___ $m^2$

**Find the area of each figure.**

4. 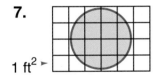 ◄ 1 ft²

    ___?___ ft²

5. ◄ 1 yd²

    ___?___ yd²

6. ◄ 1 mi²

    ___?___ mi²

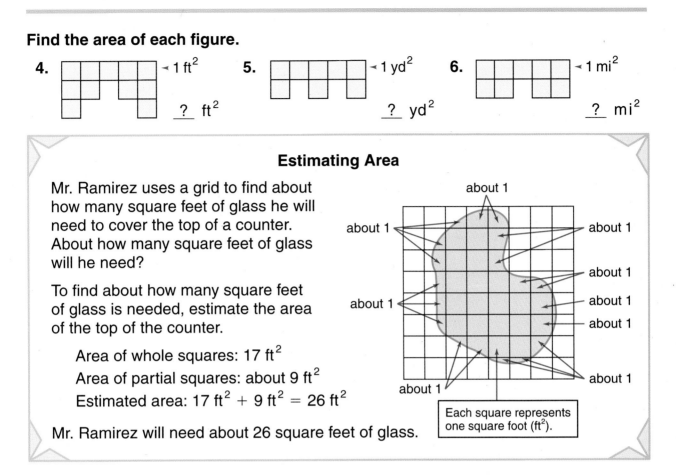

## Estimating Area

Mr. Ramirez uses a grid to find about how many square feet of glass he will need to cover the top of a counter. About how many square feet of glass will he need?

To find about how many square feet of glass is needed, estimate the area of the top of the counter.

Area of whole squares: 17 ft²
Area of partial squares: about 9 ft²
Estimated area: 17 ft² + 9 ft² = 26 ft²

Mr. Ramirez will need about 26 square feet of glass.

about 1
about 1
about 1
about 1
about 1
about 1
about 1
about 1
about 1
about 1

Each square represents one square foot (ft²).

**Estimate the area of each figure.**

7.

1 ft² ►

8.

1 m² ►

9.

1 yd² ►

## PROBLEM SOLVING

10. Karina is making a design by using a grid as shown. About how many square feet is her design if each square in the grid represents one square foot?

1 ft² ►

### Share Your Thinking

11. Use grid paper to make a design like Karina's in exercise 10. Tell a classmate how you planned your design. Work together to estimate the number of square feet used in your design.

Communicate ✓

# 12-7    Areas of Rectangles and Squares

▶ The rectangle on the right contains 45 squares, or 9 rows of 5 squares each.

The area of the rectangle is found by *multiplying* the *length by* the *width*. So, the *formula* for finding the **area of a rectangle** is:

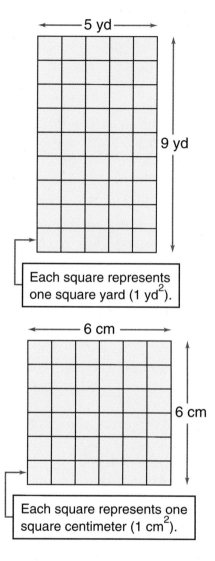

5 yd

9 yd

Each square represents one square yard ($1 \text{ yd}^2$).

$$\begin{array}{ccccc} \textbf{Area} & & \textbf{length} & & \textbf{width} \\ \downarrow & & \downarrow & & \downarrow \\ A & = & \ell & \times & w \\ A & = & 9 \text{ yd} & \times & 5 \text{ yd} \\ A & = & 45 \text{ yd}^2 \end{array}$$

The area of the rectangle is 45 square yards.

▶ The square on the right contains 36 squares, or 6 rows of 6 squares each.

The area of the square is found by *multiplying* the *side by* the *side*. So, the *formula* for finding the **area of a square** is:

6 cm

6 cm

Each square represents one square centimeter ($1 \text{ cm}^2$).

$$\begin{array}{ccccccc} \textbf{Area} & & \textbf{side} & & \textbf{side} & & \\ \downarrow & & \downarrow & & \downarrow & & \\ A & = & s & \times & s & = & s^2 \\ A & = & 6 \text{ cm} & \times & 6 \text{ cm} & & \\ A & = & 36 \text{ cm}^2 & & & & \end{array}$$

Read: "s squared"

The area of the square is 36 square centimeters.

## Study these examples.

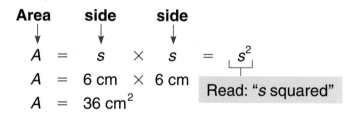

$4\frac{1}{2}$ in.

$5\frac{1}{3}$ in.

$$A = \ell \times w$$
$$A = 5\frac{1}{3} \text{ in.} \times 4\frac{1}{2} \text{ in.}$$
$$A = \frac{\overset{8}{\cancel{16}}}{\underset{1}{\cancel{3}}} \text{ in.} \times \frac{\overset{3}{\cancel{9}}}{\underset{1}{\cancel{2}}} \text{ in.}$$
$$A = 24 \text{ in.}^2$$

7.2 m

7.2 m

$$A = s \times s$$
$$A = 7.2 \text{ m} \times 7.2 \text{ m}$$
$$A = 51.84 \text{ m}^2$$

**Find the area of each rectangle.**

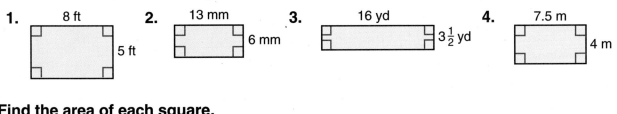

**1.** 8 ft    5 ft

**2.** 13 mm    6 mm

**3.** 16 yd    $3\frac{1}{2}$ yd

**4.** 7.5 m    4 m

**Find the area of each square.**

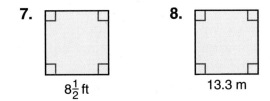

**5.** 3 yd

**6.** 9 dm

**7.** $8\frac{1}{2}$ ft

**8.** 13.3 m

**Copy and complete each table.**

| Rectangle | | |
|---|---|---|
| $\ell$ | $w$ | $A = \ell \times w$ |
| **9.** 7.3 cm | 3.1 cm | ? |
| **10.** $13\frac{1}{3}$ ft | $3\frac{3}{4}$ ft | ? |

| Square | |
|---|---|
| $s$ | $A = s \times s$ |
| **11.** 4.5 cm | ? |
| **12.** $4\frac{1}{3}$ in. | ? |

**Use your centimeter ruler to measure the sides to the nearest millimeter. Then find the area.**

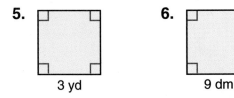

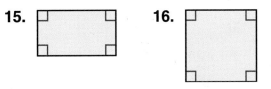

**13.**    **14.**    **15.**    **16.**

**PROBLEM SOLVING**

**17.** Which has a greater area, a rectangle that has a length of 80 cm and a width of 20 cm, or a square that measures 40 cm on each side?

**18.** How many cans of paint are needed to paint 2 walls that are each 8 ft high and 18 ft long if one can of paint covers an area of 100 square feet?

**Critical Thinking**

**How many different rectangles with whole number dimensions can you make for each given area?** Use grid paper to construct each figure.

**19.** 7 square units    **20.** 10 square units    **21.** 8 square units

# 12-8 | Areas of Parallelograms and Triangles

 **Discover Together**

**Materials Needed:** grid paper, pencil, ruler, scissors

You can use what you know about finding the area of a rectangle to help you find the area of other polygons.

Look at the parallelograms below.

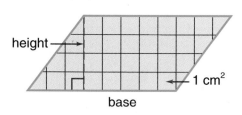

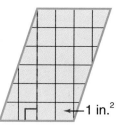

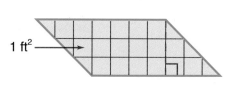

Any side of a parallelogram can serve as the *base*. The *height* is the length of the perpendicular segment from the base to the opposite vertex.

1. Find and record the base (*b*) and height (*h*) of each parallelogram.

2. How would you find the height of each parallelogram if it was not marked with a dotted line?

3. On grid paper copy and then cut out each parallelogram along each dotted line. Place the two pieces of each parallelogram together to form a rectangle.

4. Were you able to form a rectangle from each parallelogram?

5. What is the area of each rectangle formed?

6. How do the base and height of each parallelogram relate to the length and width of its related rectangle?

7. What is the area of each parallelogram? How does the area of each parallelogram compare with the area of its related rectangle?

8. What formula would you use to find the area of a parallelogram with base *b* and height *h*?

Now look at the parallelograms below.

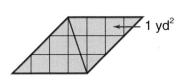

9. Find and record the base (b) and height (h) of each parallelogram. Then find its area.

10. On grid paper copy and cut out each parallelogram. Then cut along each diagonal to make two triangles.

    Are the two triangles of each parallelogram congruent?

11. How do the base and height of each triangle relate to the base and height of its related parallelogram?

12. How does the area of each triangle compare with the area of its related parallelogram? What is the area of each of the triangles?

13. What formula would you use to find the area of a triangle with base b and height h?

## Communicate

Discuss

14. What two measurements are needed for finding the area of parallelograms and of triangles?

15. Write in your Math Journal the formulas for finding the area of parallelograms and of triangles. Give an example using each formula.

Math Journal

### Challenge

Communicate

16. In the given figure, ABCD is a parallelogram. If $\overline{DM}$ and $\overline{CM}$ are the same length, how does the area of triangle ABM relate to the area of parallelogram ABCD? Use grid paper to model and explain your answer.

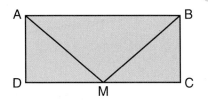

# 12-9 Space Figures

**Space figures** or **solids** are three-dimensional figures. Some of their parts are in more than one plane.

▶ **Prisms** are space figures with two faces, called **bases**, bounded by polygons that are parallel and congruent.

A **cube** is a special kind of prism with 6 square faces.

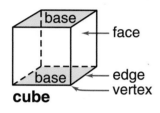
**cube**

A **face** is a flat surface of a space figure.
An **edge** is a segment where 2 faces meet.
A **vertex** is a point where 2 or more edges meet.

Other prisms are named by their bases.

 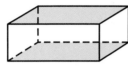  

**triangular prism**     **rectangular prism**     **pentagonal prism**     **hexagonal prism**

▶ **Pyramids** are space figures with triangular faces that meet at a common vertex. Pyramids are also named by their bases.

**square pyramid**     **rectangular pyramid**     **triangular pyramid**

**pentagonal pyramid**     **hexagonal pyramid**

▶ Some space figures have curved surfaces.

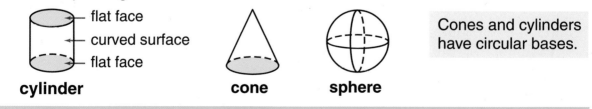

flat face
curved surface
flat face
**cylinder**          **cone**          **sphere**

Cones and cylinders have circular bases.

**Write the name of the space figure each is most like.**

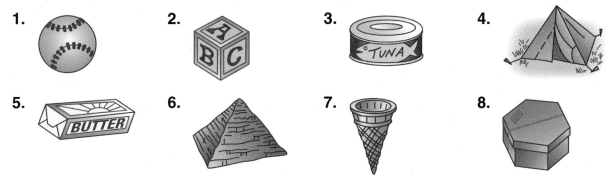

1.
2.
3.
4.

5.
6.
7.
8.

**Copy and complete the table.**

| | Space Figure | Number of | | |
|---|---|---|---|---|
| | | Faces | Vertices | Edges |
| 9. | rectangular prism | ? | ? | ? |
| 10. | triangular prism | ? | ? | ? |
| 11. | pentagonal prism | ? | ? | ? |
| 12. | hexagonal prism | ? | ? | ? |
| 13. | square pyramid | ? | ? | ? |
| 14. | rectangular pyramid | ? | ? | ? |
| 15. | triangular pyramid | ? | ? | ? |
| 16. | pentagonal pyramid | ? | ? | ? |
| 17. | hexagonal pyramid | ? | ? | ? |

**Write *True* or *False* for each statement. Explain your answer.**

Communicate ✓

18. Cylinders, cones, and spheres have no edges or vertices.

19. Cylinders and cones have flat surfaces.

20. A sphere has no flat surfaces.

21. A cone has more than one base.

**Critical Thinking**

**Write the space figure that can be made from each net.**
The dashed lines indicate folds.

22.
23.
24.
25.

399

# Cubic Measure

The volume of space figures is measured in cubic units.

▶ You can count the cubes to find the **cubic measure** or volume of a space figure.

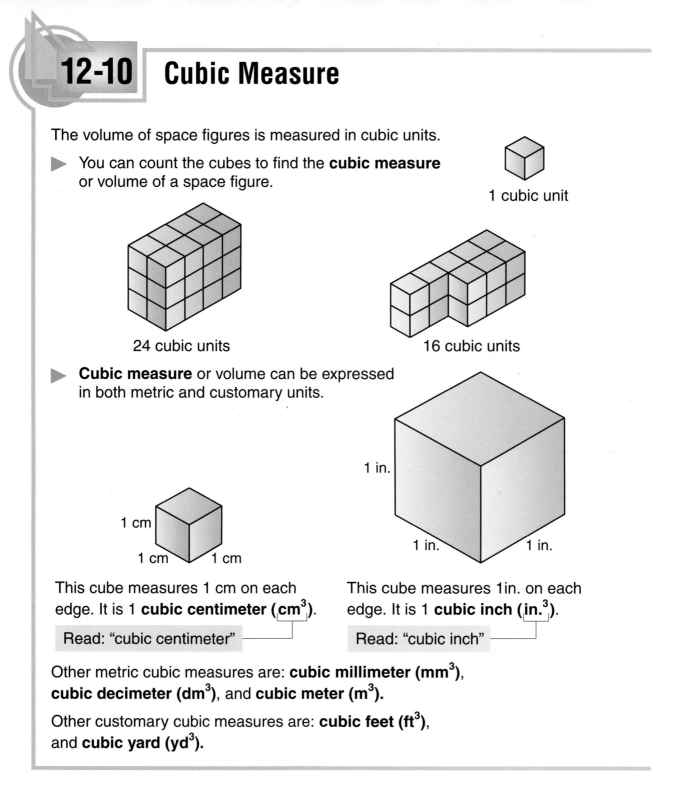

1 cubic unit

24 cubic units

16 cubic units

▶ **Cubic measure** or volume can be expressed in both metric and customary units.

1 cm
1 cm   1 cm

1 in.
1 in.   1 in.

This cube measures 1 cm on each edge. It is 1 **cubic centimeter ($cm^3$)**.

Read: "cubic centimeter"

This cube measures 1 in. on each edge. It is 1 **cubic inch ($in.^3$)**.

Read: "cubic inch"

Other metric cubic measures are: **cubic millimeter ($mm^3$)**, **cubic decimeter ($dm^3$)**, and **cubic meter ($m^3$)**.

Other customary cubic measures are: **cubic feet ($ft^3$)**, and **cubic yard ($yd^3$)**.

**Find the cubic measure of each by counting the cubes.**

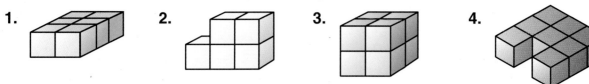

1.  2.  3.  4.

**Find the cubic measure of each.**

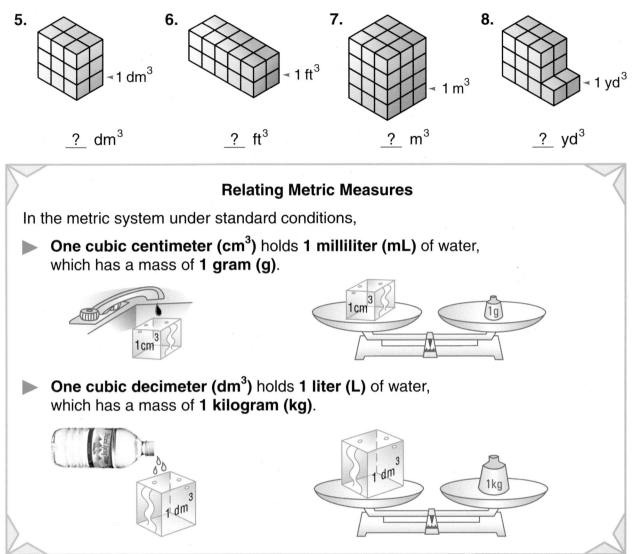

5. ◄ 1 dm³

___?___ dm³

6. ◄ 1 ft³

___?___ ft³

7. ◄ 1 m³

___?___ m³

8. ◄ 1 yd³

___?___ yd³

### Relating Metric Measures

In the metric system under standard conditions,

▶ **One cubic centimeter (cm³)** holds **1 milliliter (mL)** of water, which has a mass of **1 gram (g)**.

1cm³      1cm³      1g

▶ **One cubic decimeter (dm³)** holds **1 liter (L)** of water, which has a mass of **1 kilogram (kg)**.

1 dm³      1 dm³      1kg

**Copy and complete the table.**

|     | Cubic Measure | Capacity of Water | Mass of Water |
|-----|---------------|-------------------|---------------|
| 9.  | 3 cm³         | 3 mL              | ?             |
| 10. | 5 dm³         | ?                 | 5 kg          |
| 11. | ?             | 2 mL              | 2 g           |

**PROBLEM SOLVING**

12. What cubic measure can hold 25 mL of water?

13. What cubic measure can hold 8 kg of water?

401

# Volume

Find the volume of a rectangular prism that measures 4 cm long, 2 cm wide, and 3 cm high.

▶ The **volume** of a space figure is its cubic measure, or the number of cubic units it contains.

You can find the volume of the prism by *counting the cubes* it contains:

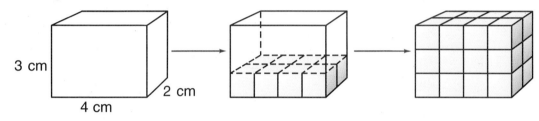

There are $4 \times 2$, or 8, cubes in each layer and there are 3 layers of cubes. Thus, $8 \times 3$, or 24 cubes fill the prism.

or

You can use the *formula* to find the
**volume of a rectangular prism**:

| **Volume** | **length** | **width** | **height** |
|:---:|:---:|:---:|:---:|
| ↓ | ↓ | ↓ | ↓ |

$$V = \ell \times w \times h$$
$$V = 4 \text{ cm} \times 2 \text{ cm} \times 3 \text{ cm}$$
$$V = 24 \text{ cm}^3$$

Read: "24 cubic centimeters"

The volume of the rectangular prism is $24 \text{ cm}^3$.

**Find the length, width, and height of each rectangular prism. Then use the formula to find the volume.**

**1.**

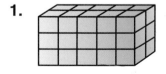

$\ell =$ _?_ units; $w =$ _?_ units

$h =$ _?_ units; $V =$ _?_ cubic units

**2.**

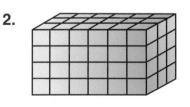

$\ell =$ _?_ units; $w =$ _?_ units

$h =$ _?_ units; $V =$ _?_ cubic units

**Find the volume of each rectangular prism.**

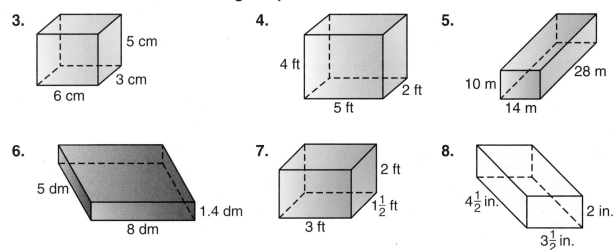

**3.** 5 cm, 3 cm, 6 cm

**4.** 4 ft, 5 ft, 2 ft

**5.** 10 m, 28 m, 14 m

**6.** 5 dm, 8 dm, 1.4 dm

**7.** 2 ft, $1\frac{1}{2}$ ft, 3 ft

**8.** $4\frac{1}{2}$ in., 2 in., $3\frac{1}{2}$ in.

**Copy and complete the table.**

| Rectangular Prism | | | |
|---|---|---|---|
| $\ell$ | $w$ | $h$ | $V = \ell \times w \times h$ |
| **9.** $2\frac{1}{4}$ yd | $\frac{2}{3}$ yd | 5 yd | ? |
| **10.** $8\frac{1}{3}$ ft | $2\frac{2}{5}$ ft | $1\frac{3}{4}$ ft | ? |
| **11.** 0.8 m | 0.3 m | 2 m | ? |
| **12.** 7.5 dm | 3 dm | 5.1 dm | ? |

**Use your centimeter ruler to measure the length, width, and height of each rectangular prism to the nearest millimeter. Then find the volume.**

**13.**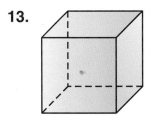

**14.**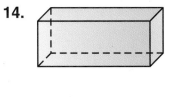

**15.**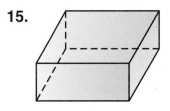

## PROBLEM SOLVING

**16.** A sandbox measures 6 feet long, 5 feet wide, and 3 feet deep. How many cubic feet of sand are needed to fill it?

**17.** Find the volume of a gift box that measures 8 inches long, $5\frac{1}{2}$ inches wide, and 2 inches high.

## 12-12 Estimating Volume

Marco wants to build a cube-shaped box large enough to hold a baseball he caught at the stadium. He is deciding whether to build a box with a volume of 1 cubic centimeter or a box with a volume of 1 cubic decimeter. Which size is more reasonable for the baseball?

 **Hands-On Understanding**

**Materials Needed:** centimeter grid paper, tape, scissors, pencil, ruler, base ten blocks

**Step 1**    Draw the net at the right on centimeter grid paper.

**Step 2**    Draw a second net with each face as a square of 1 decimeter on each side on centimeter grid paper.

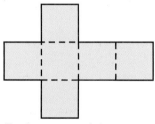

Each square of the net is 1 cm on each side.

**Step 3**    Cut out the outline of each net. Then fold and tape each net to form a box.

What is the volume of each box? Which of these boxes is a more reasonable size to hold a baseball?

1. What objects do you know that would fit into a cube-shaped box with a volume of 1 cm$^3$? with a volume of 1 dm$^3$?

2. How many centimeter cubes would you need to fill a decimeter cube? What is the volume of a cubic decimeter box in cubic centimeters?

3. How many decimeter cubes would you need to fill a meter cube What is the volume of a cubic meter box in cubic decimeters? in cubic centimeters?

**Which size, $cm^3$ or $dm^3$, is a reasonable size to hold each object?**

**4.** a sunflower seed      **5.** a tennis ball      **6.** a miniature car

**7.** a ring      **8.** a cat's-eye marble      **9.** a Ping-Pong ball

**10.** Find or make a cube-shaped box that has a volume of about 1 in.$^3$ Then use this as a model to find larger objects, such as boxes, that are about 12 times the length, width, and height of a cubic inch.

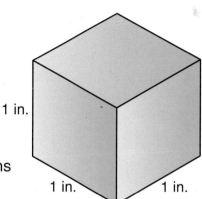

1 in.

1 in.      1 in.

**11.** What would be the length, width, and height of each of these objects?

**12.** What other unit of length can you use for the dimensions of these objects besides inches? Why?

**13.** What would be the approximate volume of each object?

**14.** What is the volume of a cubic foot box in cubic inches?

**Write the letter of the best estimate of volume for each.**

**15.** crayon box      **a.** 500 m$^3$      **b.** 500 dm$^3$      **c.** 500 cm$^3$

**16.** tissue box      **a.** 90 in.$^3$      **b.** 90 ft$^3$      **c.** 90 yd$^3$

**17.** cassette tape      **a.** 140 mm$^3$      **b.** 140 cm$^3$      **c.** 140 m$^3$

## Communicate

Discuss

**18.** Which is larger: 10 cm$^3$ or 1 dm$^3$? 100 dm$^3$ or 1 m$^3$? 12 in.$^3$ or 1 ft$^3$? Explain your answers.

**19.** Can rectangular prisms look different but have the same volume? Explain your answer.

### Challenge

Communicate

**20.** Choose 3 classroom objects that are shaped like rectangular prisms. Figure out a way to estimate the volume of each object. Explain the method you used.

# TECHNOLOGY

## Electronic Spreadsheets

An **electronic spreadsheet** is a computer-based software program that arranges data and formulas in a column-and-row format.

Each entry in a spreadsheet goes into its own **cell**. A cell can contain data, labels, or formulas. A cell is identified by the intersection of a column (A, B, C, . . .) and a row (1, 2, 3, . . .).

column C

|  | A | B | C | D | E | F |
|---|---|---|---|---|---|---|
| 1 | GAME | ORIGINAL | SALE | DISCOUNT | NUMBER | TOTAL |
| 2 |  | PRICE | PRICE |  | SOLD | AMOUNT |
| 3 | ROCKET | $59.95 | $39.95 | $20.00 | 56 | $2237.20 |
| 4 | WORDZ | $49.95 | $44.95 | $5.00 | 144 |  |
| 5 | FLAG | $38.95 | $29.95 | $9.00 | 760 |  |
| 6 | OPEN DOOR | $119.95 | $88.95 | $31.00 | 321 |  |

row 4 → 4

▶ What information is in cell C4?
Cell C4 is in column C, row 4.
The sale price ($44.95) for the video game WordZ is in cell C4.

A **formula** for a cell assigns a result that will appear in that cell.

▶ What formula can be entered in cell D3 to obtain the amount of discount for the video game Rocket?

To find the discount,
subtract: original price − sale price
          B3          C3

Locate the cell for each.

A formula must begin with a number or a sign (+/−). If a formula begins with a cell location, it must be preceded by a + sign or be enclosed with parentheses.

So the formula can be +B3−C3 or (B3−C3).

**Use the spreadsheet on page 406 for problems 1–9.**

1. What information can a cell of a spreadsheet contain?

2. What information is in cell E5?

3. In which cell can you find the original price for the video game Open Door?

4. If you wanted to enter the formula C3 + D3, what would you type?

5. What formula would you enter in cell D4? D5? D6?

6. What formula would you use to find the total number of games sold?

7. Column F represents the total amount sold at each video game's sale price. What formula was entered in cell F3?

8. What formula would you enter in cell F4? F5? F6?

9. What is the total amount sold for each video game?

|   | A | B | C | D | E | F | G |
|---|---|---|---|---|---|---|---|
| 1 | FIGURE | LENGTH | WIDTH | HEIGHT | PERIMETER | AREA | VOLUME |
| 2 | SQUARE |  | 8 CM |  |  |  |  |
| 3 | RECTANGLE ABCD | 9 CM | 4 CM |  |  |  |  |
| 4 | RECTANGLE WXYZ | 8 CM | 6 CM |  |  |  |  |
| 5 | RECTANGLE MNOP | 4 CM | 11 CM |  |  |  |  |
| 6 | EQUILATERAL TRIANGLE |  |  |  | 12 CM |  |  |
| 7 | RECTANGULAR PRISM | 8 CM | 5 CM | 3 CM |  |  |  |

**Use the spreadsheet above for problems 10–15.**

10. What number should be entered in cell B2?

11. What formula would you enter in each cell in column E?

12. What formula would you enter in each cell in column F?

13. What must be the length of each side of the equilateral triangle?

14. What are the perimeter and area of the square and each rectangle?

15. What formula would you enter in cell G7?

# 12-14 | Problem Solving: Draw a Picture

**Problem:** Marlene cut a mat for a picture from a sheet 24 inches by 15 inches. If the mat was 2 inches wide, what was the area of the mat she used?

15 in.

24 in.

**1 IMAGINE** Create a mental picture of the matting.

**2 NAME**

*Facts:* paper—24 in. by 15 in.
width of mat—2 in.

*Question:* What is the area of the mat?

**3 THINK** Draw the picture you imagined.

First find the length and width of the inside rectangle. Subtract $2 \times 2$ in., or 4 inches, from each side.

Next use the area formula to find the area of each rectangle.

Then subtract the smaller area from the larger to find the area of the mat.

15 in.
2 in.
24 in.
2 in.
2 in.
2 in.

**4 COMPUTE**

**Smaller Rectangle**

$\ell$ = 24 in. − 4 in. = 20 in.
$w$ = 15 in. − 4 in. = 11 in.

$A = \ell \times w$
$= 20$ in. $\times 11$ in.
$= 220$ sq in.

**Larger Rectangle**

$A = \ell \times w$
$= 24$ in. $\times 15$ in.
$= 360$ sq in.

Difference ⟶ 360 sq in. − 220 sq in. = 140 sq in.

The area of the mat is 140 sq in.

**5 CHECK** You can draw the picture on grid paper and count the number of square units of mat.

Use a calculator to check your computations.

**Draw a picture to solve each problem.**

1. Daryl drew a right triangle on grid paper. The length of its base
   was double the length of its height. Its area was 16 square units.
   If both dimensions were whole numbers, find its height and base.

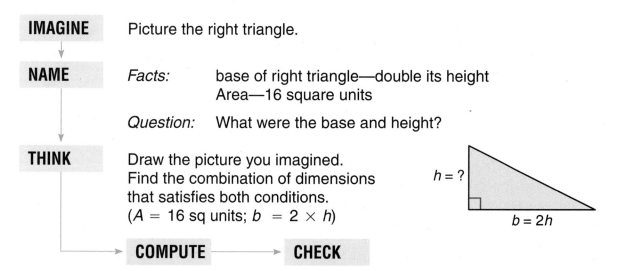

| IMAGINE | Picture the right triangle. |
|---|---|

**NAME**

*Facts:*     base of right triangle—double its height
                     Area—16 square units

*Question:*    What were the base and height?

**THINK**

Draw the picture you imagined.
Find the combination of dimensions
that satisfies both conditions.
($A = 16$ sq units; $b = 2 \times h$)

$h = ?$

$b = 2h$

**COMPUTE**   →   **CHECK**

2. Kate made a cube that has a volume of 27 cubic units.
   She painted each of the 3 sets of parallel faces the
   same color: red, blue, or yellow. What part of the
   cubic units has all 3 colors?

3. What is the least perimeter Jason can make by joining
   5 regular hexagons side to side if each side is 2.5 cm?
   What is the greatest perimeter?

4. A right triangle has an area of 9 cm². The base and height
   are whole numbers. What are two possible lengths?

5. Kelly made a design by pasting an isosceles right triangle
   in the center of a square 10 cm on each side. If the length
   of each perpendicular side of the triangle is 5.2 cm, what
   is the area of the square that is still showing?

**Make Up Your Own**

*Discuss* ✓

6. Draw 3 different polygons that have an area of 9 sq cm.
   Which polygon has the greatest perimeter? the least?
   Share your work with a classmate.

**Solve each problem and explain the method you used.**

1. A giant fold-out greeting card is 48.5 cm long. How much shorter than a meter is the card?

2. A musical card is 1.65 dm long and 1.1 dm wide. Its envelope is 0.2 cm longer on each side. What are the length and width of the envelope?

3 A special pop-up birthday card has a mass of 12.5 g. The card store sells these cards in a pack that weighs about 1 kg. About how many pop-up cards are in each pack?

4. Each holder on the postcard rack can take up to 10 centimeters of cards. Postcards are printed on 2-mm thick paper. How many postcards can fit in one holder?

5. Each perfumed card uses 0.5 mL of perfume. How many cards can be made with a liter of perfume?

6. Each colored square of this greeting card represents 1 cm². What is the area of the front of the card? of the word?

7. Whimsical Greeting Cards come in odd shapes. One greeting card is a 12-cm square. What is the area of this card in square centimeters?

8. A box contains cards with a hologram on the front. Each hologram is 53.2 mm wide and 81.5 mm tall. What is the area of each hologram?

9. A right-triangular birthday pennant has a base of 7.2 dm and a height of 2.6 dm. What is its area?

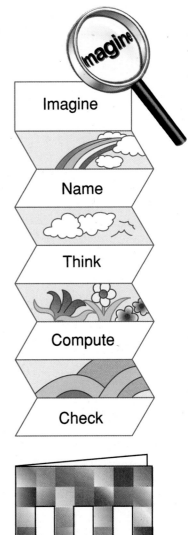

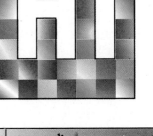

**Choose a strategy from the list or use another strategy you know to solve each problem.**

10. A card shaped like a regular pentagon has a perimeter of 35 decimeters. How many centimeters long is each side?

11. A rectangular greeting card has an area of 176 cm². One side is 16 cm. How long is the other side?

12. One birthday card comes with 2 g of confetti inside. Can 195 cards be made with 385 g of confetti?

13. Ron, Yvonne, and Fran tried to guess the age of their grandmother. Their guesses were 68, 75, and 70. One guess was incorrect by 4 years, one by 3 years, and one by 2 years. How old is their grandmother?

14. A giant right-triangular card has an area of 210 cm². The height of the triangle is 28 cm. How long is the base of this card?

15. A clerk is arranging 192 cubic units that are 1 decimeter on each edge in a display. If the display's height cannot exceed 8 dm, what might the clerk use as the length and width of the display?

16. What is the circumference of the largest circle you can cut from a piece of paper 2.15 dm by 2.8 dm?

**Use the diagram for problems 17–20.**
**Tell whether each statement is *True* or *False*.**

17. No birthday cards are pop-up cards.

18. All postcards are rectangular.

19. All triangular cards are birthday cards.

20. Some pop-up cards are rectangular birthday cards.

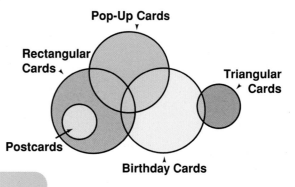

Pop-Up Cards

Rectangular Cards

Triangular Cards

Postcards

Birthday Cards

### Make Up Your Own

21. Write a problem that uses the information in the diagram. Have someone solve it.

Communicate

# Chapter Review and Practice

**Which metric unit is used to measure each? Write *m*, *L*, or *g*.**    *(See pp. 382–383.)*

1. length of a school cafeteria
2. capacity of an aquarium
3. mass of a typewriter
4. width of a yard

**Complete.**    *(See pp. 384–391.)*

5. $5\,L = \underline{\ ?\ }\ mL$
6. $70\,mm = \underline{\ ?\ }\ cm$
7. $3000\,mg = \underline{\ ?\ }\ g$

8. $2.8\,cm = \underline{\ ?\ }\ mm$
9. $20.5\,mg = \underline{\ ?\ }\ cg$
10. $2.96\,km = \underline{\ ?\ }\ m$

11. $1.2\,m = \underline{\ ?\ }\ dm$
12. $2.65\,kg = \underline{\ ?\ }\ g$
13. $3.9\,L = \underline{\ ?\ }\ mL$

**Estimate the area of each figure.**    *(See pp. 392–393.)*

14.

15.

$1\,yd^2$ ►

$1\,m^2$ ►

**Find the area of each figure.**    *(See pp. 392–397.)*

16.

17.    6.2 cm

8.6 cm

18.    $1\frac{1}{2}$ ft

$1\frac{1}{2}$ ft

19.    3 in.

8 in.

20.    5 in.

$10\frac{1}{2}$ in.

21.    35 cm ►

30 cm

**Write the name of the space figure each object is most like.**    *(See pp. 398–399.)*

22.

23.

24.

## PROBLEM SOLVING    *(See pp. 400–405, 408–411.)*

25. How many cubic centimeters will 65 mL of water fill?

26. A doghouse is 3 ft by 4 ft by 4 ft. Is the volume of the doghouse more or less than a doghouse with a volume of $1\,yd^3$?

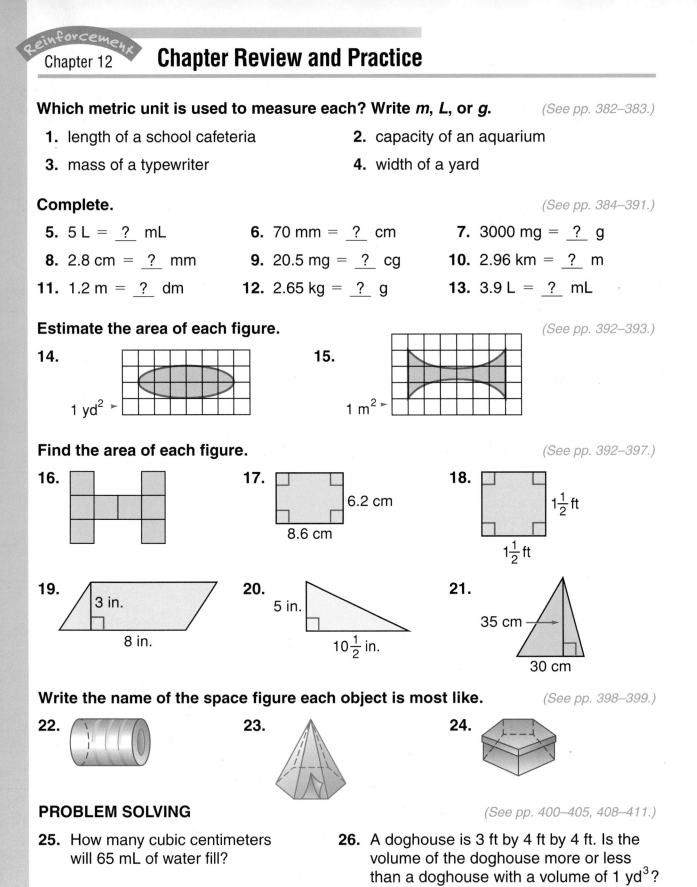

*(See Still More Practice, p. 487.)*

## SURFACE AREA

The **surface area** (*S*) of a space figure is the sum of the areas of all its faces.

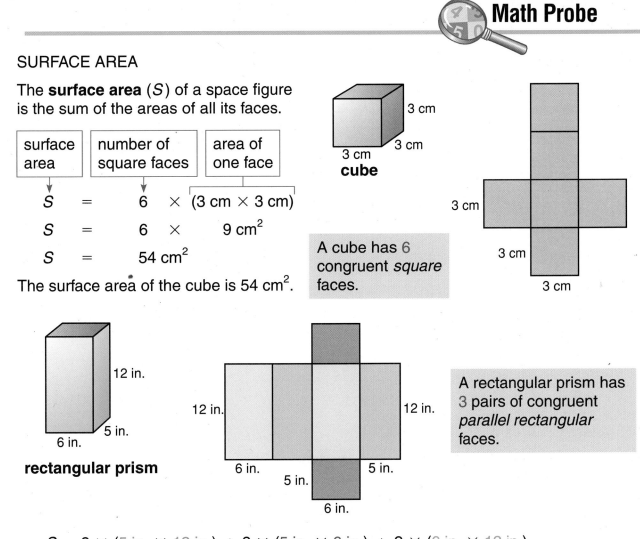

| surface area | number of square faces | area of one face |
|---|---|---|

$S = 6 \times (3 \text{ cm} \times 3 \text{ cm})$

$S = 6 \times 9 \text{ cm}^2$

$S = 54 \text{ cm}^2$

The surface area of the cube is 54 cm².

**cube**

A cube has 6 congruent *square* faces.

A rectangular prism has 3 pairs of congruent *parallel rectangular* faces.

**rectangular prism**

$S = 2 \times (5 \text{ in.} \times 12 \text{ in.}) + 2 \times (5 \text{ in.} \times 6 \text{ in.}) + 2 \times (6 \text{ in.} \times 12 \text{ in.})$

$S = 2 \times 60 \text{ in.}^2 + 2 \times 30 \text{ in.}^2 + 2 \times 72 \text{ in.}^2$

$S = 120 \text{ in.}^2 + 60 \text{ in.}^2 + 144 \text{ in.}^2$

$S = 324 \text{ in.}^2$

The surface area of the rectangular prism is 324 in.².

**Find the surface area of each figure.**

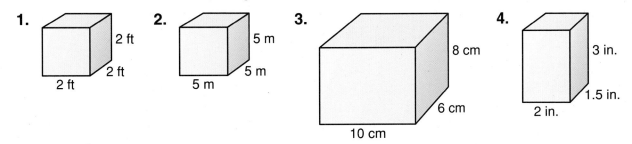

1. 2 ft, 2 ft, 2 ft
2. 5 m, 5 m, 5 m
3. 8 cm, 6 cm, 10 cm
4. 3 in., 1.5 in., 2 in.

413

# Check Your Mastery

## Performance Assessment

**Use the stamps.**

1. Measure the length and width of the rectangular stamp, then find its area.

2. A square stamp is $\frac{7}{8}$ in. on each side. A card contains 8 stamps. What is its area?

**Write the letter of the best estimate.**

3. length of a bicycle        **a.** 2 mm    **b.** 2 cm    **c.** 2 m

4. mass of an envelope        **a.** 2 mg    **b.** 2 g    **c.** 1 kg

5. capacity of a thimble       **a.** 3 mL    **b.** 30 mL    **c.** 3 L

**Compare. Write $<$, $=$, or $>$.**

6. 7.3 km __?__ 7000 m     7. 940 mL __?__ 9.4 L     8. 8.4 kg __?__ 8400 g

**Estimate the area of each figure.**

9.
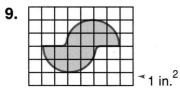
◄ 1 in.$^2$

10.
◄ 1 m$^2$

**Find the area of each figure.**

11.
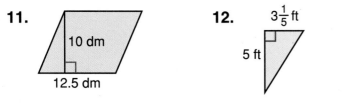
10 dm
12.5 dm

12. $3\frac{1}{5}$ ft

5 ft

13.
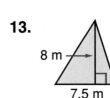
8 m
7.5 m

**Write the name of the space figure each object is most like.**

14.

15.

16.

**PROBLEM SOLVING**    *Use a strategy you have learned.*

17. How many cubic centimeters will 35 mL of water fill?

18. A birdfeeder is 16 cm by 20 cm by 12 cm. A sack of birdseed has a volume of 4 dm$^3$. Is this enough birdseed to fill the feeder?

# Ratio, Proportion, and Percent

# 13

**In this chapter you will:**

Relate ratios to fractions
Use proportion in scale drawings and maps
Relate fractions and decimals to percents
Find the percent of a number
Solve problems by combining strategies

**Critical Thinking/Finding Together**

The cashier gave you 9 coins in change, totaling one dollar. The largest coin was a quarter and the smallest coin was a nickel. How many of each kind of coin did you receive?

## Smart

My dad gave me one dollar bill
'Cause I'm his smartest son,
And I swapped it for two shiny quarters
'Cause two is more than one!

And then I took the quarters
And traded them to Lou
For three dimes—I guess he don't know
That three is more than two!

Just then, along came old blind Bates
And just 'cause he can't see
He gave me four nickels for my three dimes,
And four is more than three!

And I took the nickels to Hiram Coombs
Down at the seed-feed store,
And the fool gave me five pennies for them,
And five is more than four!

And then I went and showed my dad,
And he got red in the cheeks
And closed his eyes and shook his head—
Too proud of me to speak!

*Shel Silverstein*

415

# 13-1 Ratios as Fractions

A number of balls are on display in the sports store window. What is the ratio of the number of baseballs to the number of soccer balls?

A **ratio** is a way of comparing two numbers or quantities by division.

The ratio of the number of baseballs to the number of soccer balls is 5 to 3.

There are three ways to write a ratio:

$$5 \text{ to } 3 \quad \text{or} \quad 5 : 3 \quad \text{or} \quad \frac{5}{3}$$

▶ Some ratios can be written in simplest form.

The ratio of the number of soccer balls to the number of tennis balls is:

$$3 \text{ to } 6 = 1 \text{ to } 2 \quad \text{or} \quad 3 : 6 = 1 : 2 \quad \text{or} \quad \frac{3}{6} = \frac{1}{2}.$$

ratios in simplest form

▶ 3 to 2 and 2 to 3 are two different ratios.

The ratio of the number of soccer balls to basketballs is:

$$3 \text{ to } 2 \quad \text{or} \quad 3 : 2 \quad \text{or} \quad \frac{3}{2}.$$

The ratio of the number of basketballs to soccer balls is:

$$2 \text{ to } 3 \quad \text{or} \quad 2 : 3 \quad \text{or} \quad \frac{2}{3}.$$

$$3 \text{ to } 2 \neq 2 \text{ to } 3 \quad \text{or} \quad 3 : 2 \neq 2 : 3 \quad \text{or} \quad \frac{3}{2} \neq \frac{2}{3}$$

$\neq$ means "is not equal to."

**Write each ratio in 3 ways.**

1. gloves to bats
2. gloves to caps
3. bats to caps
4. balls to bats

**Write each ratio in simplest form.**

5. 4 to 6

6. 9 : 27

7. $\dfrac{14}{21}$

8. 12 to 24

9. 13 : 25

10. 16 to 4

11. $\dfrac{26}{39}$

12. 24 : 36

13. 100 : 125

14. $\dfrac{5}{33}$

---

### Equal Ratios

*Equal ratios* have the same value. Equal ratios can be written as *equivalent fractions*.

To write an equal ratio:

- Write the given ratio as a fraction.

$\dfrac{6}{10}$

- Multiply or divide both the numerator and the denominator by the same number.

$\dfrac{6 \times 3}{10 \times 3} = \dfrac{18}{30}$ or $\dfrac{6 \div 2}{10 \div 2} = \dfrac{3}{5}$

- Express the result as a fraction.

$\dfrac{6}{10} = \dfrac{18}{30} = \dfrac{3}{5}$ ← equal ratios

---

**Complete.**

Algebra

15. $\dfrac{1}{5} = \dfrac{?}{10}$

16. $\dfrac{3}{4} = \dfrac{?}{12}$

17. $\dfrac{2}{3} = \dfrac{?}{15}$

18. $\dfrac{2}{5} = \dfrac{?}{10}$

19. $\dfrac{6}{16} = \dfrac{?}{8}$

20. $\dfrac{9}{30} = \dfrac{?}{10}$

21. $\dfrac{8}{12} = \dfrac{?}{3}$

22. $\dfrac{25}{35} = \dfrac{?}{7}$

### PROBLEM SOLVING

23. During one baseball season, Glenn was at bat 25 times and had 13 hits. What is the ratio of hits to times at bat?

24. Sally took a 30-question grammar test. She had 23 answers correct. What is the ratio of the number of correct answers to the number of incorrect answers?

### Share Your Thinking

Communicate

25. Explain why the order of the numbers is important when you read and write a ratio.

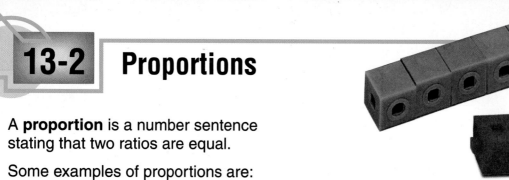

A **proportion** is a number sentence stating that two ratios are equal.

Some examples of proportions are:

2 is to 5 as 6 is to 15          $2 : 5 = 6 : 15$          $\frac{2}{5} = \frac{6}{15}$

$$\frac{1 \text{ liter}}{4 \text{ glasses}} = \frac{2 \text{ liters}}{8 \text{ glasses}} \qquad \frac{30 \text{ miles}}{20 \text{ minutes}} = \frac{15 \text{ miles}}{10 \text{ minutes}}$$

There are two ways to find out if two ratios form a proportion.

- Write the ratios as fractions in simplest form. Two ratios form a proportion if they can be simplified to give the same fraction.

$$\frac{8}{12} \overset{?}{=} \frac{6}{9} \longrightarrow \frac{8 \div 4}{12 \div 4} = \frac{2}{3} \quad \text{and} \quad \frac{6 \div 3}{9 \div 3} = \frac{2}{3}$$

$\frac{8}{12} = \frac{6}{9}$ is a proportion.

Fractions are the same.

- Use the *cross-products* rule. Two ratios form a proportion if their cross products are equal.

first $\frac{1}{3}$ $\times$ $\frac{3}{9}$ third          $1 \times 9 = 3 \times 3$
second $\qquad$ fourth

$\qquad\qquad\qquad\qquad 9 \qquad 9$

$\frac{1}{3} = \frac{3}{9}$ is a proportion.

The product of the first and fourth numbers and the product of the second and third numbers are equal.

**Explain the way you used to find out if each pair of fractions forms a proportion.**

Communicate

1. $\frac{1}{6}, \frac{3}{18}$     2. $\frac{2}{3}, \frac{4}{9}$     3. $\frac{4}{5}, \frac{8}{15}$     4. $\frac{12}{10}, \frac{5}{6}$     5. $\frac{2}{7}, \frac{6}{21}$

**Use the cross-products rule to find out which of these are proportions. Write *Yes* or *No*.**

6. $\frac{5}{7} \overset{?}{=} \frac{10}{14}$     7. $\frac{8}{5} \overset{?}{=} \frac{40}{25}$     8. $\frac{2}{11} \overset{?}{=} \frac{14}{22}$     9. $\frac{5}{3} \overset{?}{=} \frac{39}{16}$

## Missing Number in a Proportion

To find the missing number in a proportion:

- Use equal ratios.

  Two cups of rice serve 6 people.
  How many people do 3 cups of rice serve?

  $$\frac{2 \text{ cups rice}}{3 \text{ cups rice}} = \frac{6 \text{ people}}{n \text{ people}} \longrightarrow \frac{2}{3} = \frac{6}{n} \longrightarrow \frac{2 \times 3}{3 \times 3} = \frac{6}{9}, \, n = 9$$

  Three cups of rice serve 9 people.

- Use the cross-products rule.

  $$\frac{1}{4} \diagdown \frac{3\frac{3}{4}}{n} \longrightarrow 1 \times n = 4 \times 3\frac{3}{4} \longrightarrow n = 4 \times 3\frac{3}{4} = \frac{\overset{1}{\cancel{4}}}{1} \times \frac{15}{\underset{1}{\cancel{4}}} = 15$$

**Find the missing number in the proportion.**

*Algebra* ✓

**10.** $\frac{3}{4} = \frac{12}{n}$

**11.** $\frac{12}{14} = \frac{n}{28}$

**12.** $\frac{16}{n} = \frac{4}{5}$

**13.** $\frac{n}{15} = \frac{6}{10}$

**14.** $\frac{1}{2} = \frac{2\frac{1}{2}}{n}$

**15.** $\frac{1}{8} = \frac{1\frac{1}{8}}{n}$

**16.** $\frac{2\frac{1}{4}}{n} = \frac{1}{4}$

**17.** $\frac{n}{2} = \frac{10}{1}$

**18.** $\frac{2 \text{ oz cheese}}{6 \text{ oz cheese}} = \frac{4 \text{ sandwiches}}{n \text{ sandwiches}}$

**19.** $\frac{1 \text{ box}}{3 \text{ boxes}} = \frac{16 \text{ crayons}}{n \text{ crayons}}$

## PROBLEM SOLVING

**20.** If 2 apples cost 40¢, how much will 4 apples cost?

**21.** If 3 oranges cost 75¢, how many oranges could you buy for 25¢?

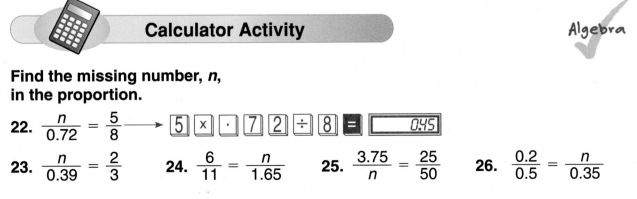

## Calculator Activity

*Algebra* ✓

**Find the missing number, $n$, in the proportion.**

**22.** $\frac{n}{0.72} = \frac{5}{8} \longrightarrow$ [5] [×] [.] [7] [2] [÷] [8] [=] [ 0.45 ]

**23.** $\frac{n}{0.39} = \frac{2}{3}$

**24.** $\frac{6}{11} = \frac{n}{1.65}$

**25.** $\frac{3.75}{n} = \frac{25}{50}$

**26.** $\frac{0.2}{0.5} = \frac{n}{0.35}$

# 13-3 Scale and Maps

A **scale drawing** of something is accurate, but *different* in size.

A **scale** is the ratio of the pictured measure to the actual measure.

The scale distance between San Antonio and Houston is $1\frac{5}{8}$ in.

To find the actual distance between San Antonio and Houston:

- Set up a proportion.

$$\frac{\text{Scale measure}}{\text{Actual measure}} = \frac{\text{Scale distance}}{\text{Actual distance}}$$

- Substitute.

$$\frac{1 \text{ in.}}{120 \text{ miles}} = \frac{1\frac{5}{8} \text{ in.}}{n \text{ miles}}$$

- Use the cross-products rule to solve.

$$\frac{1}{120} \bowtie \frac{1\frac{5}{8}}{n} \rightarrow 1 \times n = 120 \times 1\frac{5}{8} \rightarrow n = \frac{\overset{15}{\cancel{120}}}{1} \times \frac{13}{\underset{1}{\cancel{8}}} = \frac{15 \times 13}{1 \times 1} = 195$$

The actual distance between San Antonio and Houston is about 195 miles.

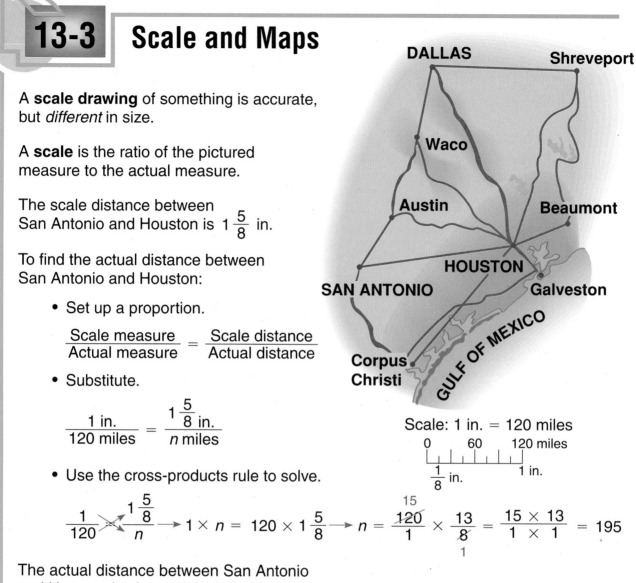

Scale: 1 in. = 120 miles

0    60    120 miles

$\frac{1}{8}$ in.    1 in.

Copy and complete the table. Measure the scale distance on the map above to the nearest $\frac{1}{8}$ in.

|     | Between Cities | Scale Distance (in.) | Actual Distance (mi) |
|-----|----------------|----------------------|----------------------|
| 1.  | Houston—Beaumont | $\frac{5}{8}$ in. | ? |
| 2.  | Dallas—Shreveport | ? | ? |
| 3.  | Austin—San Antonio | ? | ? |
| 4.  | Waco—Dallas | ? | ? |
| 5.  | Corpus Christi—Galveston | ? | ? |

**Use the scale 1 in. = 8 mi to complete each table.**

| To go from: | Scale Distance | Actual Distance |
|---|---|---|
| **6.** Dunes Club to Far Park | 2 in. | ? |
| **7.** Hotel to Sandy Beach | $2\frac{3}{4}$ in. | ? |
| **8.** Lighthouse to Park | $3\frac{1}{2}$ in. | ? |

| To go from: | Scale Distance | Actual Distance |
|---|---|---|
| **9.** Atlantic City to Hotel | $1\frac{1}{2}$ in. | ? |
| **10.** Sandy Beach to Lighthouse | $2\frac{1}{2}$ in. | ? |
| **11.** Camp Grounds to Atlantic City | $4\frac{3}{4}$ in. | ? |

**TREASURE ISLAND**

**Measure the scale distance to the nearest centimeter. Then estimate the distance from the treasure to each place.**

**12.** Rockaway Cove

**13.** Town

**14.** West Mount

**15.** Old Oak Tree

**16.** The scale distance between Watch Tower and East Mount is about 7 centimeters. Estimate the distance.

**17.** The distance between Sandy Beach and Sleepy Lagoon is about 50 kilometers. About how many centimeters is the scale distance?

**18.** Estimate the distance between Watch Tower and Sandy Beach.

**19.** Create a small map of your school yard. Explain in your Math Journal why a scale is needed when making a map.

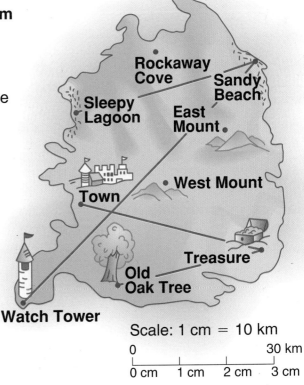

Scale: 1 cm = 10 km

0     30 km

0 cm   1 cm   2 cm   3 cm

Math Journal

## Skills to Remember

**Write in simplest form.**

**20.** $\frac{2}{10}$    **21.** $\frac{6}{10}$    **22.** $\frac{5}{10}$    **23.** $\frac{25}{100}$    **24.** $\frac{80}{100}$    **25.** $\frac{16}{100}$

# Relating Fractions to Percents

*Discover Together*

**Materials Needed:** 10 × 10 grids, colored pencils or crayons, large sheet of paper

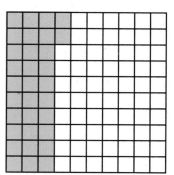

A fraction with a denominator of 100 can be written as a **percent**. Percent means the part of each hundred. The symbol for percent is **%**.

**1.** Shade 32 squares in a 10 × 10 grid as shown in the above grid.

**2.** What fractional part of the grid is shaded? Write it as a fraction with a denominator of 100.

Another way to describe the amount shaded is with a percent. Thirty-two percent of the grid is shaded.

**3.** Write the amount shaded as a percent.

Look at the grids below. Write the fraction with a denominator of 100 and the percent that tells what part is shaded.

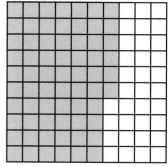

**4.**

**5.**

**6.**

Now shade a 10 × 10 grid to model each fraction.
Then write the fraction as a percent.

**7.** $\frac{16}{100}$    **8.** $\frac{9}{100}$    **9.** $\frac{95}{100}$    **10.** $\frac{44}{100}$    **11.** $\frac{30}{100}$    **12.** $\frac{89}{100}$

**13.** $\frac{15}{100}$    **14.** $\frac{57}{100}$    **15.** $\frac{88}{100}$    **16.** $\frac{75}{100}$    **17.** $\frac{1}{100}$    **18.** $\frac{100}{100}$

Look at the percents below. Write each as a fraction with a denominator of 100. Then rename the fraction in simplest form.

**19.** 65%    **20.** 8%    **21.** 50%    **22.** 10%    **23.** 55%    **24.** 4%

**25.** 98%    **26.** 25%    **27.** 90%    **28.** 75%    **29.** 46%    **30.** 100%

**31.** What percent in exercises 19–30 means one fourth? one half? three fourths? four fourths? Explain the percent pattern they form.

**32.** Give equivalent fractions for $\frac{1}{4}$, $\frac{1}{2}$, $\frac{3}{4}$, $\frac{4}{4}$. Do these equivalent fractions form the percent pattern in exercise 31? Explain your answer.

Now shade a 10 × 10 grid to model each percent. Then write as a fraction in simplest form.

**33.** 20%    **34.** 5%    **35.** 80%    **36.** 30%    **37.** 4%    **38.** 90%

**39.** 40%    **40.** 2%    **41.** 70%    **42.** 85%    **43.** 13%    **44.** 35%

**45.** What percent is modeled on a 10 × 10 grid if all squares of the grid are shaded? if none of the squares are shaded?

## Communicate

Discuss

**46.** Explain why $\frac{7}{10}$ is not 7%. What is the correct percent for $\frac{7}{10}$?

**47.** What percent means the whole? Write three fractions that are equivalent to a whole.

**48.** Is it possible to shade a 10 × 10 grid so that it is 15% blue, 75% red, and 20% green? Explain your answer.

**49.** Write a rule in your Math Journal on how to rename a percent as a fraction.

Math Journal

## Project

**50.** Find at least 10 examples of how percents are used in advertisements. Use a 10 × 10 grid to color the percent in each example. Paste the colored grids and the advertisements on a large sheet of paper. Make a classroom display showing these and other uses of percent.

# 13-5 Relating Percents to Decimals

Percents can also be written as decimals.

▶ To write a percent as a decimal:

- Drop the percent symbol (%).

- Move the decimal point *two* places to the left. [To do this, you may have to write zero(s).]

| Percent | Decimal |
|---------|---------|
| 45% ⟶ .45. ⟶ 0.45 | |

| Percent | Decimal |
|---------|---------|
| 5% ⟶ .05. ⟶ 0.05 | |

Write zero.

▶ To write a decimal as a percent:

- Move the decimal point *two* places to the right. [To do this, you may have to write zero(s).]

- Write the percent symbol (%).

| Decimal | Percent |
|---------|---------|
| 0.59 ⟶ 0.59. ⟶ 59% | |

| Decimal | Percent |
|---------|---------|
| 0.4 ⟶ 0.40. ⟶ 40% | |

Write zero.

**Write as a decimal.**

| 1. 65% | 2. 83% | 3. 12% | 4. 28% | 5. 10% | 6. 3% |
|--------|--------|--------|--------|--------|-------|
| 7. 78% | 8. 47% | 9. 7% | 10. 50% | 11. 23.6% | 12. 12.7% |

**Write as a percent.**

| 13. 0.15 | 14. 0.73 | 15. 0.08 | 16. 0.99 | 17. 0.57 | 18. 0.4 |
|----------|----------|----------|----------|----------|---------|
| 19. 0.62 | 20. 0.6 | 21. 0.93 | 22. 0.1 | 23. 0.123 | 24. 1.85 |

## Money as Percent of a Dollar

Coins can be expressed as a **percent** of a dollar.

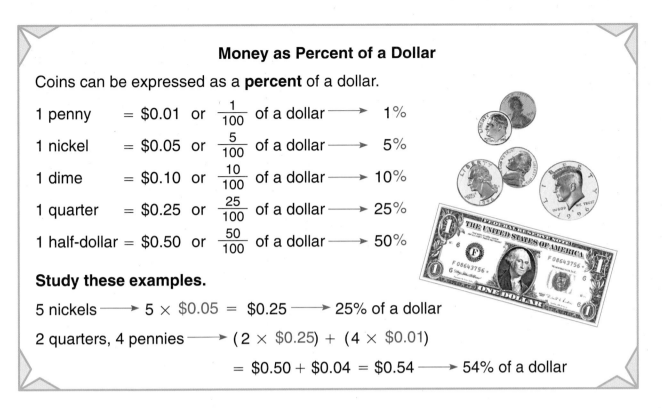

1 penny   = $0.01  or  $\frac{1}{100}$ of a dollar ⟶ 1%

1 nickel   = $0.05  or  $\frac{5}{100}$ of a dollar ⟶ 5%

1 dime   = $0.10  or  $\frac{10}{100}$ of a dollar ⟶ 10%

1 quarter   = $0.25  or  $\frac{25}{100}$ of a dollar ⟶ 25%

1 half-dollar = $0.50  or  $\frac{50}{100}$ of a dollar ⟶ 50%

**Study these examples.**

5 nickels ⟶ 5 × $0.05 = $0.25 ⟶ 25% of a dollar

2 quarters, 4 pennies ⟶ ( 2 × $0.25) + ( 4 × $0.01)

= $0.50 + $0.04 = $0.54 ⟶ 54% of a dollar

### Write as a percent of a dollar.

**25.** 9 nickels    **26.** 7 pennies    **27.** 3 dimes    **28.** 2 quarters

**29.** 58 pennies   **30.** 4 nickels    **31.** 5 dimes    **32.** 3 quarters

**33.** 1 quarter, 5 pennies       **34.** 2 nickels, 3 pennies

**35.** 2 quarters, 1 dime         **36.** 4 dimes, 3 nickels

**37.** 1 half-dollar, 2 pennies   **38.** 1 half-dollar, 2 dimes

### PROBLEM SOLVING

**39.** Mario had $1.00. He spent 65¢. What percent of his money did he spend?

**40.** Nelda needs 0.02 liter of acid for her experiment. What percent of a liter does she need?

### Critical Thinking

*Communicate*

### Compare. Write <, =, or >. Explain your answer.

**41.** 0.75 _?_ 75%       **42.** 0.13 _?_ 1.3%       **43.** 2.5 _?_ 25%

**44.** 0.06 _?_ 60%       **45.** 0.27 _?_ 27%        **46.** 0.032 _?_ 3.2%

# 13-6 Finding the Percent of a Number

There are 60 questions on a social studies exam. Twenty-five percent of the questions are about map skills. How many of the questions are about map skills?

To find how many of the questions are about map skills, find the percent of a number:

25% of 60 = ___?___

**To find the percent of a number:**

- Write the percent as a decimal.
- Multiply.

or

- Write the percent as a fraction.
- Multiply.

There are 15 questions about map skills.

25% of 60 = ___?___

$$25\% = 0.25$$

$$0.25 \times 60 = 15.00$$

$$25\% = \frac{25}{100} = \frac{1}{4}$$

$$\frac{1}{\overset{}{\underset{1}{4}}} \times \overset{15}{\cancel{60}} = 15$$

**Find the percent of the number by writing the percent as a decimal.**

**1.** 50% of 32

**2.** 25% of 16

**3.** 10% of 50

**4.** 20% of 25

**5.** 35% of 20

**6.** 40% of 20

**7.** 15% of 40

**8.** 6% of 200

**9.** 4% of 250

**Find the percent of the number by writing the percent as a fraction.**

**10.** 10% of 120

**11.** 50% of 46

**12.** 25% of 224

**13.** 75% of 48

**14.** 20% of 325

**15.** 30% of 80

**16.** 80% of 240

**17.** 15% of 180

**18.** 60% of 315

**19.** 40% of 300

**20.** 90% of 200

**21.** 35% of 120

**Find the percent of the number.**

**22.** 55% of 800

**23.** 50% of 726

**24.** 45% of 120

**25.** 75% of 340

**26.** 90% of 630

**27.** 15% of 460

**28.** 20% of 760

**29.** 18% of 500

**30.** 24% of 700

**Compare. Use <, =, or >.**

**31.** 10% of 20 _?_ 20% of 40

**32.** 30% of 60 _?_ 40% of 20

**33.** 15% of 60 _?_ 25% of 60

**34.** 20% of 150 _?_ 20% of 180

**35.** 30% of 40 _?_ 60% of 20

**36.** 45% of 300 _?_ 65% of 200

**PROBLEM SOLVING  Use the percent chart.**

**37.** Five percent of 80 fifth graders have red hair. How many fifth graders have red hair?

**38.** Ten percent of the 150 new cars that are on display at the Auto-Rama are minivans. How many minivans are on display?

**39.** At Irwin School, 75% of the 348 students ride the bus to school. How many students ride the bus to school?

| Percent Chart | | |
|---|---|---|
| **Percent** | **Decimal** | **Fraction** |
| 1% | 0.01 | $\frac{1}{100}$ |
| 5% | 0.05 | $\frac{5}{100} = \frac{1}{20}$ |
| 10% | 0.10 | $\frac{10}{100} = \frac{1}{10}$ |
| 25% | 0.25 | $\frac{25}{100} = \frac{1}{4}$ |
| 50% | 0.50 | $\frac{50}{100} = \frac{1}{2}$ |
| 75% | 0.75 | $\frac{75}{100} = \frac{3}{4}$ |

**40.** Molly's stamp collection totals 236 stamps. Twenty-five percent of the stamps are from France. How many stamps are from France?

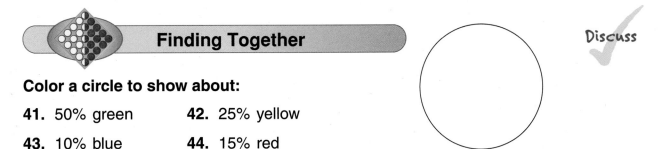

**Finding Together**

Discuss

**Color a circle to show about:**

**41.** 50% green

**42.** 25% yellow

**43.** 10% blue

**44.** 15% red

# Using Percent

At Kennedy School, 180 students take Allied Arts courses. How many students take Fine Arts?

The **circle graph** at the right shows the percent of students taking each Allied Arts course.

**Allied Arts Courses**

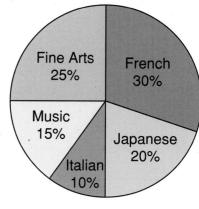

To find how many students take Fine Arts, find the percent of a number: 25% of 180 = _?_

|  | Percent | × | Total | | Number of Students |
|--|---------|---|-------|--|------------------|

$$25\% \quad \text{of} \quad 180 \quad = \quad ?$$

$$\frac{1}{\overset{1}{\cancel{4}}} \times \overset{45}{\cancel{180}} = 45 \quad \text{or}$$

$$\begin{array}{r} 1\,8\,0 \\ \times\ 0.2\,5 \\ \hline 9\,0\,0 \\ 3\,6\,0 \\ \hline 4\,5.0\,0 \end{array}$$

There are 45 students taking Fine Arts.

**Copy and complete the table.
Use the circle graph above.**

|  | Subject | Percent | Number of Students |
|--|---------|---------|--------------------|
| 1. | Music | ? | ? |
| 2. | Italian | ? | ? |
| 3. | Japanese | ? | ? |
| 4. | French | ? | ? |

**Use the circle graph.**

Mr. Smith's monthly income is $3500.
How much is his budget for:

5. education?

6. food?

7. shelter?

8. clothing?

9. recreation?

10. savings?

**Mr. Smith's Monthly Budget**

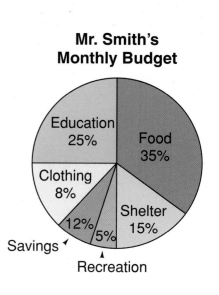

## Finding Discount

During a sale, LP Electronics offers a discount of 20% on a video recorder with a regular price of $500. How much is the discount?

A **discount** is a savings on the regular price of an item. The **rate of discount** is given as a percent.

To find the discount, find:  20% of $500 = __?__

| Rate of Discount | × | Regular Price | = | Discount |
|:---:|:---:|:---:|:---:|:---:|
| ↓ | ↓ | ↓ | | ↓ |
| 20% | of | $500 | = | ? |
| 0.20 | × | $500 | = | $100 |

The discount is $100.

## Copy and complete the table.

| | Item | Regular Price | Rate of Discount | Discount |
|---|---|:---:|:---:|:---:|
| **11.** | towel | $14 | 25% | ? |
| **12.** | tablecloth | $30 | 15% | ? |
| **13.** | bed sheets | $200 | 30% | ? |
| **14.** | shower curtain | $25 | 5% | ? |

## PROBLEM SOLVING

**15.** Bikes with a regular price of $120 are offered at a 35% discount. What is the discount?

**16.** Beach chairs with a regular price of $30 are on sale at a 15% discount. What is the discount?

**17.** Explain in your Math Journal why stores advertise percent off rather than dollars off.

*Math Journal*

## Make Up Your Own

**18.** Use some of the given information and supply any missing information to write and solve your own problem.

FALL CLEARANCE
40% OFF — Leather Handbags, Men's & Ladies' Shoes

# 13-8 Problem Solving: Combining Strategies

**Problem:** Tasha decides to save some money. The first day she puts a nickel in a bank. Each day she plans to double the amount she put in the day before. How much money will she have saved in a week?

**1 IMAGINE** Create a mental picture.

**2 NAME** *Facts:* First day—Tasha saves a nickel.

Each day following, she doubles the amount she puts in the bank.

*Question:* How much money will Tasha have saved in a week?

**3 THINK** Some problems are easier to solve by combining strategies.

Is there hidden information? Yes.

1 nickel = $0.05 and 1 week = 7 days

Make a table to record the amount saved each day.

Find a pattern.

**4 COMPUTE**

|  | 1st | 2nd | 3rd | 4th | 5th | 6th | 7th |
|---|---|---|---|---|---|---|---|
| **Saved** | $0.05 | $0.10 | $0.20 | $0.40 | $0.80 | $1.60 | $3.20 |
| **Total** | $0.05 | $0.15 | $0.35 | $0.75 | $1.55 | $3.15 | $6.35 |

**5 CHECK** In 1 week Tasha saved $6.35.

Use a calculator to check your computation.

**Combine strategies to solve each problem.**

1. Caren bought some greeting cards. She gave 5 cards to her sister. After sending 3 of the remaining cards, Caren had 2 left. What percent of the cards Caren bought does she have left?

| | |
|---|---|
| **IMAGINE** | Put yourself in the problem. |

| | | |
|---|---|---|
| **NAME** | *Facts:* | bought some cards<br>gave 5 cards away<br>sent 3 cards<br>had 2 cards left |
| | *Question:* | What percent of the bought cards are left? |

Cards

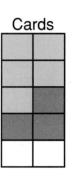

| | |
|---|---|
| **THINK** | First find the number of cards Caren bought by *working backwards.* $2 + 3 + 5 = \underline{\ ?\ }$ |
| | Then find the percent by *writing a number sentence* or *using drawings.* |

**COMPUTE** ⟶ **CHECK**

2. In a box of 40 assorted cards, 12 were birthday cards, 10 were anniversary cards, 6 were get-well cards, and the rest were all-occasion cards. What percent of the box of cards were all-occasion cards?

3. Two out of every seven pieces of mail the Zimmer family receives are bills. If they received a half-dozen bills last week, how many pieces of mail did they receive?

4. Three out of every 5 thank-you cards Diane wrote were to her family. The rest were to her friends. If Diane wrote 8 cards to her friends, how many thank-you cards did she have?

5. Mary has 162 cards to put into 15 boxes. Some boxes hold 10 cards; others hold a dozen. Fifty cards are yellow. How many of each size box will Mary use?

431

# 13-9 | Problem-Solving Applications

**Solve each problem and explain the method you used.**

1. The stationery store is having a spring sale. For every 5 pencils you buy, you get 2 free. If Arnie pays for 15 pencils, how many does he get free?

2. The store clerk notices that he sold pens and pencils in a ratio of 4 : 9. He sold 24 pens. How many pencils did he sell?

3. Two out of every 5 customers bought markers. What percent did *not* buy markers?

4. The store earns $.12 on every $.49 eraser it sells. How much money will the store earn on the sale of 2 dozen erasers?

5. This week eight tenths of the stationery items are on sale. What percent of the stationery items are *not* on sale?

6. A book bag usually costs $15, but during the sale its price is reduced by 30%. How much will be saved?

7. The list price of a dictionary is $24.00. Helen saved $6.00 when she bought it at the sale. What percent of the list price did she save?

8. Which is less expensive during the sale: a $12 sweatshirt reduced by 25% or a $15 sweatshirt reduced by 45%?

9. In a brochure the scale for a picture of a computer is 1 cm = 4 cm. The computer screen has a scale length of 7 cm and a scale width of 5 cm. What are its actual dimensions?

10. The scale length of the keyboard is 12 cm. Its actual width is 37.5% of its length. What are its length and width?

11. For every $20 spent, a customer pays a $1.20 sales tax. Lori bought 3 pen-and-pencil sets and paid $2.16 in sales tax. How much did she spend?

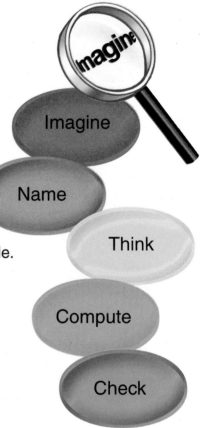

Imagine

Name

Think

Compute

Check

**Choose a strategy from the list or use another strategy you know to solve each problem.** You may combine strategies.

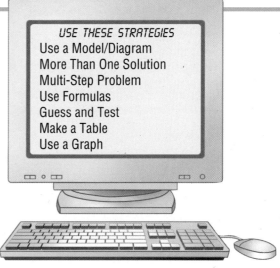

USE THESE STRATEGIES
Use a Model/Diagram
More Than One Solution
Multi-Step Problem
Use Formulas
Guess and Test
Make a Table
Use a Graph

12. Angela bought a ream of paper listed at $20 for 10% less. How much money did she save? She was charged an additional $1.08 in sales tax. How much did she pay for the paper?

13. A $7 T–shirt at the bookstore is reduced by 50%. What is the final cost, including $0.21 sales tax?

14. Each day the price of a school umbrella will be reduced by another 10% until all the umbrellas have been sold. The original price of each umbrella is $10. What is the price on the 5th day of the sale?

15. Greg buys a sheet of paper 24 in. by 18 in. First he folds it in half vertically, then horizontally. What is the perimeter of the final rectangle?

| Model | Original |
|-----------|----------|
| Mini-Max | $7.30 |
| Midi-Max | $9.10 |
| Super Sum | $8.10 |
| Turbo Plus | $9.50 |

16. This table shows the original prices of calculators on sale for 30% off. Kirk spent less than $11. Which 2 calculators did he buy?

17. Mercedes bought 2 calculators from the table at the 30% discount. She spent $11.76. Which 2 calculators did she buy?

**Use the circle graph for problems 18–21.**

18. Which 3 items represent about 50% of the profits?

19. The bookstore's profits were $1470 last week. What was the profit from sales of writing tools?

20. How much more profit was there on paper supplies than on clothing?

**Last Week's Profits**

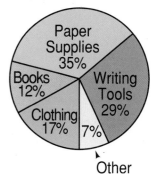

21. Write a multi-step problem that uses the data from the circle graph. Have a classmate solve it.

**Make Up Your Own**

Communicate ✓

**Write in 3 ways the ratio of the number of:** *(See pp. 416–417.)*

1. kites to balls

2. cars to kites

3. balls to cars

**Find the missing number in the proportion.** *(See pp. 418–419.)*

4. $\dfrac{3}{4} = \dfrac{n}{12}$

5. $\dfrac{6}{7} = \dfrac{18}{n}$

6. $\dfrac{n}{16} = \dfrac{7}{8}$

7. $\dfrac{5}{n} = \dfrac{7}{35}$

**Find the actual measurements.** *(See pp. 420–421.)*

8. What is the length of the soccer field?

9. What is the width of the soccer field?

1 in.

$1\dfrac{3}{8}$ in.     Scale: 1 in. = 80 yd

**Write as a percent.** *(See pp. 422–425.)*

10. $\dfrac{27}{100}$

11. $\dfrac{65}{100}$

12. 0.83

13. 0.52

14. 0.3

**Shade a 10 × 10 grid to model each percent.
Then write as a fraction in simplest form.** *(See pp. 422–423.)*

15. 20%

16. 45%

17. 16%

18. 70%

19. 81%

**Write as a decimal.** *(See pp. 424–425.)*

20. 46%

21. 68%

22. 5%

23. 9%

24. 76%

**Find the percent of the number.** *(See pp. 426–427.)*

25. 20% of 200

26. 50% of 136

27. 25% of 120

## PROBLEM SOLVING

*(See pp. 418–419, 426–427, 430–433.)*

28. William paid $1.20 for 2 hot dogs. He also paid $.75 for a soda and $1.09 for french fries. How much would he pay for 6 hot dogs?

29. Ten percent of a $25 gas bill is tax. How much is the tax?

(See *Still More Practice*, p. 488.)

## PERCENT PATTERNS

Study the pattern for these percents and their equivalent fractions.

$$2\% = \frac{2}{100} = \frac{1}{50}$$ $$4\% = \frac{4}{100} = \frac{1}{25}$$

$$4\% = 2 \times 2\% = 2 \times \frac{1}{50} = \frac{2}{50}$$ $$12\% = 3 \times 4\% = 3 \times \frac{1}{25} = \frac{3}{25}$$

$$6\% = 3 \times 2\% = 3 \times \frac{1}{50} = \frac{3}{50}$$ $$20\% = 5 \times 4\% = 5 \times \frac{1}{25} = \frac{5}{25}$$

$$8\% = 4 \times 2\% = 4 \times \frac{1}{50} = \frac{4}{50}$$ $$28\% = 7 \times 4\% = 7 \times \frac{1}{25} = \frac{7}{25}$$

$$5\% = \frac{5}{100} = \frac{1}{20}$$ $$10\% = \frac{10}{100} = \frac{1}{10}$$

$$25\% = 5 \times 5\% = 5 \times \frac{1}{20} = \frac{5}{20}$$ $$30\% = 3 \times 10\% = 3 \times \frac{1}{10} = \frac{3}{10}$$

$$45\% = 9 \times 5\% = 9 \times \frac{1}{20} = \frac{9}{20}$$ $$50\% = 5 \times 10\% = 5 \times \frac{1}{10} = \frac{5}{10}$$

$$65\% = 13 \times 5\% = 13 \times \frac{1}{20} = \frac{13}{20}$$ $$70\% = 7 \times 10\% = 7 \times \frac{1}{10} = \frac{7}{10}$$

**Find the equivalent fractions.** Look for a pattern.

**1.** 10%, 20%, 30%, 40%

**2.** 20%, 40%, 60%, 80%

**3.** 25%, 50%, 75%, 100%

**4.** 15%, 30%, 45%, 60%

**5.** 12%, 24%, 36%, 48%

**6.** 8%, 16%, 32%, 64%

**7.** 5%, 20%, 35%, 50%

**8.** 4%, 32%, 60%, 88%

## PROBLEM SOLVING

**9.** If $\frac{1}{8} = 12.5\%$, then what percent is equivalent to $\frac{3}{8}$?

**10.** If $\frac{1}{3} = 33\frac{1}{3}\%$, then what percent is equivalent to $\frac{2}{3}$?

**11.** If $\frac{1}{9} = 11\frac{1}{9}\%$, then what percent is equivalent to $\frac{7}{9}$?

**12.** If $\frac{1}{7} = 14\frac{2}{7}\%$, then what percent is equivalent to $\frac{3}{7}$?

**13.** If $\frac{1}{6} = 16\frac{2}{3}\%$, then what percent is equivalent to $\frac{5}{6}$?

# Check Your Mastery

## Performance Assessment

**Show all the ratios in exercises 1–3 on one fraction strip.**

1. Color the fraction strip so that the ratio of:
   **a.** red to blue is 2 to 5        **b.** yellow to blue is 3 to 5
   **c.** not yellow to red is 9 to 2

2. Write each ratio in exercise 1 in 2 other ways.

3. Describe what the ratio of 2 to 2 represents.

**Which are proportions? Write = or ≠.**

4. $\dfrac{5}{21}$ ? $\dfrac{10}{40}$

5. $\dfrac{4}{9}$ ? $\dfrac{8}{16}$

6. $\dfrac{16}{2}$ ? $\dfrac{32}{4}$

**Solve for *n*.**

7. $\dfrac{5}{n} = \dfrac{25}{3}$

8. $\dfrac{7}{9} = \dfrac{n}{81}$

9. $\dfrac{n}{12} = \dfrac{7}{4}$

**Write as a percent.**

10. $\dfrac{42}{100}$

11. $\dfrac{57}{100}$

12. $\dfrac{19}{100}$

13. $\dfrac{23}{100}$

14. 0.26

15. 0.31

16. 0.6

17. 0.03

**Shade a 10 × 10 grid to model each percent.
Then write as a fraction in simplest form.**

18. 40%

19. 51%

20. 75%

21. 14%

**Write as a decimal.**

22. 19%

23. 90%

24. 7%

25. 4%

**Find the percent of the number.**

26. 4% of 120

27. 30% of 250

28. 90% of 300

29. 75% of 150

**PROBLEM SOLVING**    *Use a strategy you have learned.*

30. On a map 1 cm represents 6 m. What does 7 cm represent?

31. Beach towels with a regular price of $15 are offered at a 20% discount. What is the discount?

# Cumulative Test II

**Choose the best answer.**

1. Which shows 5 billion in expanded form?
   a. $5 \times 1{,}000{,}000$
   b. $5 \times 1{,}000{,}000{,}000$
   c. $5 \times 10{,}000{,}000{,}000$
   d. $5 \times 100{,}000{,}000$

2. Estimate.

   $221 \times 4632$

   a. 800,000
   b. 1,400,000
   c. 8,000,000
   d. 1,000,000

3. Which number is divisible by 2, 3, 5, 6, 9, and 10?
   a. 135
   b. 600
   c. 1620
   d. 2025

4. Which shows the prime factorization of 84?
   a. $2 \times 42$
   b. $2 \times 3 \times 7$
   c. $2 \times 2 \times 3 \times 7$
   d. $3 \times 4 \times 7$

5. $18 - 1\frac{1}{8}$
   a. $17\frac{1}{8}$
   b. $17\frac{7}{8}$
   c. $19\frac{1}{8}$
   d. not given

6. $18 \div 1\frac{1}{8}$
   a. 4
   b. 16
   c. $20\frac{1}{4}$
   d. not given

7. Which graph shows how a whole is divided into fractional parts?
   a. bar graph
   b. circle graph
   c. pictograph
   d. line graph

8. Which of the following is *not* a quadrilateral?
   a. trapezoid
   b. rhombus
   c. parallelogram
   d. hexagon

9. Rename.

   $10 \text{ qt} = \underline{\ ?\ }$

   a. 5 c
   b. 5 pt
   c. 16 pt
   d. $2\frac{1}{2}$ gal

10. Round 7.248 to the nearest hundredth.
    a. 0.725
    b. 7.24
    c. 7.25
    d. 7.3

11. Find the area.

    2.5 m
    2.5 m

    a. $2.5 \text{ m}^2$
    b. $5 \text{ m}^2$
    c. $6.25 \text{ m}^2$
    d. $10 \text{ m}^2$

12. Which space figure is shown?

    a. triangular pyramid
    b. rectangular prism
    c. rectangular pyramid
    d. square prism

13. Rename 48% as a fraction in lowest terms.
    a. $\frac{12}{25}$
    b. $\frac{24}{50}$
    c. $\frac{48}{100}$
    d. $\frac{4}{15}$

14. Find the missing number.

    $\frac{4}{9} = \frac{12}{n}$

    a. 8
    b. 18
    c. 27
    d. 36

**15.** Make a line graph for the data.

**Daily Temperature (in °F)**

| Day | S | M | T | W | Th | F | S |
|---|---|---|---|---|---|---|---|
| Temperature | 26 | 23 | 32 | 35 | 27 | 31 | 27 |

**16.** Find the perimeter and the area.

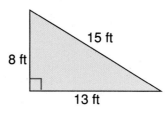

15 ft

8 ft

13 ft

**Compute or find the answer.**

**17.** $6.56 + 3.36 + 2.131$

**18.** $0.83 - 0.612$

**19.** $0.35 \times 16.9$

**20.** $3.268 \div 4$

**21.** $5\overline{)5.32}$

**22.** 24% of 500

## PROBLEM SOLVING

**23.** Lois has 3 packages. Each weighs 2 lb 10 oz. Find the total weight of the three packages.

**24.** Ari divided 12.5 lb of trail mix equally into 10 boxes. How much trail mix is in each box?

**25.** A bag contains 1 red, 1 green, 1 blue, and 1 yellow marble. Pick a marble from the bag without looking and put it back. Then pick another marble. How many different outcomes are possible?

**26.** Terry drew a scale drawing of his school building and made the scale 1 cm = 12 m. If the drawing is 9.5 cm high, what is the height of his school?

## For Rubric Scoring

**Listen for information on how your work will be scored.**

**27.** Examine the statement in the box.

- If you think it is *always* true, explain why.
- If you think it is *never* true, explain why.
- If you think it is *sometimes* true, draw examples to show this.

> Any rectangular sheet of paper that has twice the length and twice the width of another rectangular sheet of paper will also have twice the area.

**28.** Locate on a geoboard or draw on dot paper some pairs of rectangles where the area of one is exactly twice the area of the other. What do you notice about the width and the length of each pair?

# Moving On: Algebra

# 14

## Exit x

Let $x$ be this
and $y$ be that,
my teacher says. And I
expecting $x$ to be complex
enough, put wily $y$
to work. If $vex$
is $x^2$, $rex$
will equal one-no-three.
But that's not why
$x$ over my
right shoulder laughs at me.

*David McCord*

**In this chapter you will:**

Simplify expressions
Learn about function tables and
  coordinate geometry
Solve addition and multiplication
  equations
Solve problems by writing an equation

**Critical Thinking/Finding Together**

If $x$ and $y$ in the equations below stand for
different digits, but are the same in every
equation, what are their values?

$$x + y = 12 \qquad y \times y = x$$
$$x - y = 6 \qquad 27 \div x = y$$

# Moving On: Algebra

## 14-1 Expressions: Addition and Subtraction

A **mathematical expression** is a name for a number.

A mathematical expression may involve addition or subtraction.

34 →
20 + 14    18 + 16    (addition expressions)
40 − 6     38 − 4     (subtraction expressions)

| | Definition | Examples |
|---|---|---|
| **Numerical Expression** | A mathematical expression that contains only numbers | 11 + 1<br>15 − 3     expressions<br>4 + 4 + 4   that name 12 |
| **Variable** | A letter used to represent a number that is unknown | Let $t$ represent the temperature.<br>Let $n$ represent the number of students.<br>Let $x$ represent your age. |
| **Algebraic Expression** | A mathematical expression that contains one or more variables | $t - 20$<br>$x$<br>$c + d$ |

---

Write an addition and a subtraction expression for each number.

**1.** 10       **2.** 25       **3.** 1       **4.** 0       **5.** 18

**6.** 16       **7.** 100      **8.** 200      **9.** $\frac{1}{2}$      **10.** $\frac{3}{4}$

**11.** 150      **12.** 650      **13.** 1000     **14.** $\frac{2}{3}$      **15.** $\frac{4}{5}$

Label each expression *numerical* or *algebraic.*

**16.** 7 + 2          **17.** $n - 5$          **18.** $y$          **19.** $x + y$

**20.** 9 − 9          **21.** 3 + 3 + 3          **22.** 20 − 5          **23.** $80 - r$

**24.** 15 − 2 − 3     **25.** $b + 10$          **26.** $30 + c$          **27.** $m + n + o$

## Evaluating Algebraic Expressions

To find the value of an algebraic expression:

- Substitute number(s) for the variable(s).
- Compute.

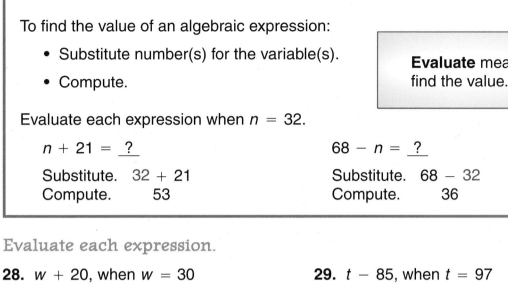

> **Evaluate** means find the value.

Evaluate each expression when $n = 32$.

$n + 21 = \underline{\ ?\ }$

Substitute. $32 + 21$
Compute. $\quad\quad 53$

$68 - n = \underline{\ ?\ }$

Substitute. $68 - 32$
Compute. $\quad\quad 36$

*Evaluate each expression.*

**28.** $w + 20$, when $w = 30$

**29.** $t - 85$, when $t = 97$

**30.** $36 + k$, when $k = 24$

**31.** $212 - d$, when $d = 107$

**32.** $s - 3\frac{1}{3}$, when $s = 9\frac{2}{3}$

**33.** $15\frac{1}{2} + z$, when $z = 17\frac{1}{2}$

**34.** $x + y$, when $x = 13$ and $y = 9$

**35.** $a - b$, when $a = 900$ and $b = 263$

*Copy and complete the table.*

| | $n$ is equal to | Expression | Value |
|---|---|---|---|
| **36.** | 12 | $n + 3$ | ? |
| **37.** | 20 | $n - 8$ | ? |
| **38.** | ? | $n + n$ | 16 |
| **39.** | ? | $36 - n$ | 30 |

 **Connections: Geometry**

**40.** The length and width, in meters, of a rectangle are represented by $x + 9$ and $x + 5$. Find the area of the rectangle when $x = 8$.

**41.** The length, in inches, of the sides of a triangle are represented by $y$, $y + 3$, and $y - 2$. Find the perimeter of the triangle when $y = 10$.

# 14-2 Expressions: Multiplication and Division

A mathematical expression may also involve multiplication or division.

$$24 \begin{cases} 4 \times 6 \quad 4 \cdot 6 \quad 4(6) \quad \text{(multiplication expressions)} \\ \\ 72 \div 3 \quad \dfrac{48}{2} \quad \text{(division expressions)} \end{cases}$$

| | Numerical Expressions | Algebraic Expressions | |
|---|---|---|---|
| **Multiplication** | $15 \times 3$ | $6 \times a$ | "times" sign |
| | $15 \cdot 3$ | $6 \cdot a$ | raised dot |
| | $15(3)$ or $(15)(3)$ | $6(a)$ or $(6)(a)$ | parentheses |
| | | $6a$ | no sign |
| **Division** | $12 \div 3$ | $a \div 9$ | "divided by" sign |
| | $\dfrac{12}{3}$ | $\dfrac{a}{9}$ | fraction form |

Study these examples.

Evaluate each expression when $x = 7$.

$8 \cdot x = \underline{\ ?\ }$

Substitute. $8 \cdot 7$

Compute. $\quad 56$

$\dfrac{49}{x} = \underline{\ ?\ }$

Substitute. $\dfrac{49}{7}$

Compute. $\quad 7$

---

Write a multiplication and a division expression for each number.

**1.** 14     **2.** 18     **3.** 11     **4.** 3     **5.** 112

**6.** 138     **7.** 324     **8.** 568     **9.** $\dfrac{1}{3}$     **10.** $\dfrac{1}{2}$

Evaluate each expression.

**11.** $212 \times d$, when $d = 10$     **12.** $13 \cdot n$, when $n = 5$

**13.** $36 \div k$, when $k = 6$     **14.** $x \div 11$, when $x = 297$

Name the operation shown in each expression.

**15.** $7 + m$

**16.** $14 \div n$

**17.** $88 - w$

**18.** $2148 \times h$

**19.** $3g$

**20.** $\dfrac{x}{90}$

**21.** $2\ell + 2w$

**22.** $3x - 8$

Copy and complete the table.

| $n$ is equal to | Expression | Value |
|---|---|---|
| **23.** 16 | $2n$ | ? |
| **24.** 12 | $\dfrac{36}{n}$ | ? |
| **25.** ? | $3n$ | 75 |
| **26.** ? | $\dfrac{n}{6}$ | 8 |

Choose the correct answer.
Which expression is equal to:

**27.** 682, when $x = 62$?

   **a.** $\dfrac{682}{x}$  **b.** $740 - x$  **c.** $11 \cdot x$

**28.** 48, when $y = 10$?

   **a.** $4y$  **b.** $68 - y$  **c.** $\dfrac{480}{y}$

**29.** 95, when $z = 32$?

   **a.** $6z$  **b.** $3z - 1$  **c.** $\dfrac{320}{z}$

Express hours as minutes. Use the expression $60 \times h$.
Let h represent the number of hours.

**30.** 10 hours

**31.** 25 hours

**32.** $3\frac{1}{3}$ hours

**33.** $4\frac{1}{5}$ hours

## PROBLEM SOLVING

**34.** The temperature, in °F, is represented by $\dfrac{9}{5}$ C + 32. Find the temperature when C = 15°.

**35.** The base and height, in inches, of a triangle are represented by $3a + 1$ and $2a$. Find the area of the triangle when a = 5.

### Make Up Your Own

*Communicate* ✓

**36.** Make up an algebraic expression for your name.
- Count the letters in your first name.
- Create an algebraic expression for your name.
- Assign a value to your variable.
- Evaluate your algebraic expression.

Jessica = 7 letters
Jessica ⟶ $x + 10$
$x = 7$
$x + 10 = 7 + 10 = 17$

# 14-3 Function Tables

The table shows the charges for an overdue library book.

| Days Late | 1 | 2 | 3 | 4 | 5 | d |
|---|---|---|---|---|---|---|
| Charges (in cents) | 5 | 10 | 15 | 20 | 25 | ? |

Think: The charges are 5 cents *times* the number of days the book is late.

Charges for 1 day:   5¢ · 1   or   5 cents
2 days:   5¢ · 2   or   10 cents
*d* days:   5¢ · *d*   or   5*d* cents

A **rule** for the table above is 5 *times* *d* or 5*d* cents, where *d* represents the number of days a book is late.

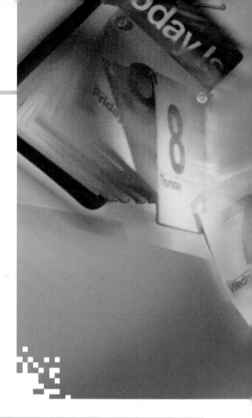

---

Use the table above to find the charge for each overdue book.

**1.** 3 days late      **2.** 5 days late      **3.** 4 days late

**4.** 6 days late      **5.** 10 days late      **6.** 0 days late

Use the table above to find the number of days each book is late.

**7.** $0.10     **8.** $0.20     **9.** $0.05     **10.** $0.15

**11.** $0.25     **12.** $0.50     **13.** $0.30     **14.** $0.45

Copy and complete the table for two late books.

| | d (days) | 10d | Charges (in cents) |
|---|---|---|---|
| **15.** | 1 | 10 · 1 | ? |
| **16.** | 2 | 10 · 2 | ? |
| **17.** | 5 | 10 · 5 | ? |
| **18.** | 8 | ? | ? |
| **19.** | 9 | ? | ? |
| **20.** | 10 | ? | ? |

OVERDUE
BOOKS

Copy and complete each table.

**21.** Let $n$ = number of videos rented. Let $3n$ = total cost.

| $n$ | 1 | 2 | 3 | 4 | 5 |
|---|---|---|---|---|---|
| $3n$ | $3 | $6 | ? | ? | ? |

**22.** Let $f$ = age now. Let $f + 6$ = age in 6 years.

| $f$ | 16 | 33 | 50 | 67 | 74 |
|---|---|---|---|---|---|
| $f + 6$ | ? | ? | ? | ? | ? |

**23.** Let $a$ = number of days. Let $\dfrac{a}{7}$ = number of weeks.

| $a$ | 7 | 14 | 35 | 77 | $3\frac{1}{2}$ |
|---|---|---|---|---|---|
| $\dfrac{a}{7}$ | ? | ? | ? | ? | ? |

Write the rule for each table.

**24.**

| $x$ | ? |
|---|---|
| 2 | 6 |
| 3 | 9 |
| 4 | 12 |
| 5 | 15 |

Think:

$3 \cdot 2 = 6$

$3 \cdot 3 = 9$

$\vdots$

So $3 \cdot x$ or $3x$ is the rule.

**25.**

| $m$ | ? |
|---|---|
| 20 | 4 |
| 21 | 5 |
| 22 | 6 |
| 23 | 7 |

**26.**

| $y$ | ? |
|---|---|
| $\frac{1}{2}$ | $\frac{1}{4}$ |
| $\frac{1}{3}$ | $\frac{1}{6}$ |
| $\frac{1}{4}$ | $\frac{1}{8}$ |
| $\frac{1}{5}$ | $\frac{1}{10}$ |

**27.**

| $n$ | ? |
|---|---|
| $\frac{1}{2}$ | $\frac{1}{3}$ |
| $\frac{1}{4}$ | $\frac{1}{6}$ |
| $\frac{1}{6}$ | $\frac{1}{9}$ |
| $\frac{1}{8}$ | $\frac{1}{12}$ |

## Critical Thinking

Match the rule with the correct table.

Remember: Use the order of operations.

**28.** $(2 \cdot x) + 1$   **29.** $(3 \cdot x) - 1$   **30.** $(2 \cdot x) + 2$

**a.**

| $x$ | 2 | 3 | 4 | 5 |
|---|---|---|---|---|
| ? | 5 | 7 | 9 | 11 |

**b.**

| $x$ | 2 | 3 | 4 | 5 |
|---|---|---|---|---|
| ? | 6 | 8 | 10 | 12 |

**c.**

| $x$ | 2 | 3 | 4 | 5 |
|---|---|---|---|---|
| ? | 5 | 8 | 11 | 14 |

445

# Moving On: Algebra

## 14-4 Addition Equations

An **equation** is a statement indicating that two mathematical expressions are equal.

To **solve** an *addition equation* with a variable, use Guess and Test:

- Guess—Substitute numbers for the variable.
- Test—Check to see if you get a true statement.

| Addition Equations | |
|---|---|
| $12 + 4 = 16$ | $x + 9 = 15$ |
| $3 + 5 = 2 + 6$ | $18 = c + 10$ |

Solve: $x + 9 = 15$

$$x + 9 = 15$$

Try 4. $4 + 9 = 15$   No, $13 \neq 15$

Try 7. $7 + 9 = 15$   No, $16 \neq 15$

Try 6. $6 + 9 = 15$   Yes, $15 = 15$

So $x = 6$.

Solve: $0.6 = 0.2 + c$

$$0.6 = 0.2 + c$$

Try 0.1. $0.6 \stackrel{?}{=} 0.2 + 0.1$   No, $0.6 \neq 0.3$

Try 0.3. $0.6 \stackrel{?}{=} 0.2 + 0.3$   No, $0.6 \neq 0.5$

Try 0.4. $0.6 \stackrel{?}{=} 0.2 + 0.4$   Yes, $0.6 = 0.6$

So $c = 0.4$.

---

Choose the correct solution.

**1.** $x + 2 = 9$

    **a.** $x = 5$    **b.** $x = 6$    **c.** $x = 7$

**2.** $5 + m = 14$

    **a.** $m = 6$    **b.** $m = 9$    **c.** $m = 11$

**3.** $17 = n + 5$

    **a.** $n = 22$    **b.** $n = 12$    **c.** $n = 11$

**4.** $21 = 8 + d$

    **a.** $d = 13$    **b.** $d = 17$    **c.** $d = 29$

**5.** $0.4 + p = 0.7$

    **a.** $p = 0.2$    **b.** $p = 0.3$    **c.** $p = 0.5$

**6.** $r + 0.5 = 0.7$

    **a.** $r = 0.2$    **b.** $r = 0.4$    **c.** $r = 0.7$

**7.** $3.1 = 1.1 + h$

    **a.** $h = 2$    **b.** $h = 2.1$    **c.** $h = 1.2$

**8.** $6.4 = t + 3.2$

    **a.** $t = 2.2$    **b.** $t = 4.2$    **c.** $t = 3.2$

Solve the equation.

**9.** $t + 3 = 7$

**10.** $m + 7 = 13$

**11.** $g + 2 = 21$

**12.** $a + 9 = 17$

**13.** $w + 15 = 20$

**14.** $x + 20 = 40$

**15.** $25 + b = 50$

**16.** $40 + c = 43$

**17.** $18 = d + 12$

**18.** $9 = x + 6$

**19.** $79 = 70 + p$

**20.** $86 = 59 + f$

**21.** $y + 0.1 = 1$

**22.** $0.2 = c + 0.1$

**23.** $0.4 = z + 0$

Solve for the variable.

**24.** Perimeter $= 24$ in.

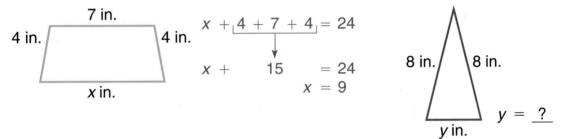

$x + \underbrace{4 + 7 + 4} = 24$

$x + \qquad 15 \qquad = 24$

$x = 9$

**25.** Perimeter $= 20$ in.

8 in.   8 in.

$y$ in.

$y = \underline{\ ?\ }$

**26.** Perimeter $= 18$ cm

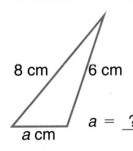

8 cm    6 cm

$a$ cm    $a = \underline{\ ?\ }$

**27.** Perimeter $= 47$ cm

9 cm   9 cm

10 cm    10 cm

$d$ cm    $d = \underline{\ ?\ }$

**28.** Perimeter $= 31$ ft

$c$ ft

7 ft    10 ft

8 ft    $c = \underline{\ ?\ }$

 **Mental Math**

Solve. Use missing addends.

Think: $8 + \underline{\ ?\ } = 17$
$\underline{\ ?\ } = 9$

$8 + m = 17$
  So $m = 9$.

**29.** $r + 9 = 16$

**30.** $8 + s = 8$

**31.** $14 = n + 5$

# 14-5 Multiplication Equations

Equations may also involve multiplication.

To **solve** a *multiplication equation* with a variable, use Guess and Test:

- Guess—Substitute numbers for the variable.
- Test—Check to see if you get a true statement.

| Multiplication Equations | |
| --- | --- |
| $12 \times 3 = 36$ | $6 \times n = 12$ |
| $4 \times 5 = 10 \times 2$ | $7 \cdot c = 35$ |
| $7(8) = 56$ | $9(a) = 54$ |
| $96 = (8)(12)$ | $32 = (4)(y)$ |
| $18 \cdot 2 = (9)(4)$ | $5x = 125$ |

Solve:     $8x = 32$

$8x = 32$

Try 5.   $8 \cdot 5 \stackrel{?}{=} 32$   No, $40 \neq 32$

Try 3.   $8 \cdot 3 \stackrel{?}{=} 32$   No, $24 \neq 32$

Try 4.   $8 \cdot 4 \stackrel{?}{=} 32$   Yes, $32 = 32$

So $x = 4$.

Solve:   $9 = \frac{1}{6} \cdot y$

$9 = \frac{1}{6} \cdot y$

Try 36.   $9 \stackrel{?}{=} \frac{1}{6} \cdot 36$   No, $9 \neq 6$

Try 48.   $9 \stackrel{?}{=} \frac{1}{6} \cdot 48$   No, $9 \neq 8$

Try 54.   $9 \stackrel{?}{=} \frac{1}{6} \cdot 54$   Yes, $9 = 9$

So $y = 54$.

---

Choose the correct solution.

**1.** $5x = 60$

   **a.** $x = 10$  **b.** $x = 11$  **c.** $x = 12$

**2.** $10w = 150$

   **a.** $w = 15$  **b.** $w = 10$  **c.** $w = 20$

**3.** $16 = 48(n)$

   **a.** $n = \frac{1}{2}$  **b.** $n = \frac{1}{3}$  **c.** $n = \frac{1}{6}$

**4.** $35 = (105)(y)$

   **a.** $y = \frac{1}{3}$  **b.** $y = \frac{1}{6}$  **c.** $y = \frac{1}{5}$

**5.** $\frac{1}{7} \cdot m = 6$

   **a.** $m = 42$  **b.** $m = 35$  **c.** $m = 48$

**6.** $\frac{2}{3} \cdot d = 8$

   **a.** $d = 6$  **b.** $d = 12$  **c.** $d = 16$

Solve the equation.

**7.** $3t = 21$      **8.** $4y = 8$      **9.** $2x = 18$      **10.** $5a = 20$

**11.** $4w = 44$      **12.** $6s = 18$      **13.** $6r = 42$      **14.** $25q = 75$

**15.** $35 = 7z$      **16.** $20 = 2r$      **17.** $48 = 8k$      **18.** $150 = 10d$

**19.** $108 = 9(t)$      **20.** $38 = 19(v)$      **21.** $18 = \frac{1}{2} \cdot b$      **22.** $39 = \frac{1}{3} \cdot c$

Solve for the variable.

**23.** Area $= 18$ in.$^2$

$6w = 18$

$w$ in.    $w = 3$

6 in.

**24.** Area $= 96$ in.$^2$

8 in.

$\ell$ in.

$\ell = \underline{\ ?\ }$

**25.** Volume $= 64$ cm$^3$

2 cm

$w$ cm    $w = \underline{\ ?\ }$

8 cm

**26.** Volume $= 12$ cm$^3$

$h$ cm

$h = \underline{\ ?\ }$

2 cm

3 cm

## Challenge

Solve the equation.

**27.** $x - 6 = 18$

Try 20.    $20 - 6 \stackrel{?}{=} 18$    No, $14 \neq 18$

Try 24.    $24 - 6 = 18$    Yes, $18 = 18$

So $x = 24$.

**28.** $\frac{a}{2} = 8$

Try 18.    $\frac{18}{2} \stackrel{?}{=} 8$    No, $9 \neq 8$

Try 16.    $\frac{16}{2} = 8$    Yes, $8 = 8$

So $a = 16$.

**29.** $c - 9 = 20$      **30.** $27 - b = 14$      **31.** $\frac{m}{5} = 4$      **32.** $\frac{n}{3} = 9$

**33.** $a - 11 = 15$      **34.** $38 - d = 13$      **35.** $6 = \frac{w}{12}$      **36.** $7 = \frac{v}{13}$

449

# 14-6 Fractions in Algebra

Some equations can be solved using the properties of addition or multiplication.

Solve for n.

| Equation | Property | Solution |
|---|---|---|
| $n + \frac{1}{2} = \frac{1}{2}$ | Identity Property of Addition | $n = 0$ |
| $\frac{5}{7} + \frac{1}{7} = \frac{1}{7} + n$ | Commutative Property of Addition | $n = \frac{5}{7}$ |
| $\frac{1}{3} \cdot n = \frac{1}{3}$ | Identity Property of Multiplication | $n = 1$ |
| $\left(3 \cdot \frac{1}{2}\right) \cdot n = 3 \cdot \left(\frac{1}{2} \cdot \frac{3}{5}\right)$ | Associative Property of Multiplication | $n = \frac{3}{5}$ |
| $n \cdot \frac{1}{4} = 0$ | Zero Property of Multiplication | $n = 0$ |

Solve for a. Use the properties to help you.

**1.** $\frac{2}{3} \cdot a = 4 \cdot \frac{2}{3}$     $a = 4$
Commutative Property of Multiplication

**2.** $\frac{3}{4} + a = \frac{3}{4}$

**3.** $a + 0 = \frac{3}{5}$

**4.** $\frac{7}{8} \cdot a = \frac{1}{3} \cdot \frac{7}{8}$

**5.** $a + \frac{1}{2} = \frac{1}{2} + \frac{2}{3}$

**6.** $\frac{2}{5} + a = \frac{1}{5} + \frac{2}{5}$

**7.** $\frac{6}{7} \cdot a = \frac{6}{7}$

**8.** $a \cdot \frac{1}{9} = 0$

**9.** $\frac{2}{5} \cdot a = 0$

**10.** $\frac{3}{8} + a = \frac{3}{8}$

**11.** $1 \cdot a = \frac{2}{3}$

**12.** $a \cdot \frac{1}{6} = \frac{1}{6}$

**13.** $\frac{6}{11} + \frac{3}{10} = a + \frac{6}{11}$

**14.** $\frac{3}{4} + \left(\frac{1}{2} + \frac{3}{5}\right) = \left(\frac{3}{4} + \frac{1}{2}\right) + a$

**15.** $\frac{5}{9} + \left(a + \frac{2}{3}\right) = \left(\frac{5}{9} + \frac{1}{6}\right) + \frac{2}{3}$

**16.** $\frac{1}{4} \times \left(a + \frac{1}{5}\right) = \left(\frac{1}{4} \times \frac{1}{3}\right) + \left(\frac{1}{4} \times \frac{1}{5}\right)$

## Solving Equations with Fractions

Solve the equations.

$$n + \frac{2}{7} = \frac{6}{7}$$

Try $\frac{2}{7}$.  $\frac{2}{7} + \frac{2}{7} \overset{?}{=} \frac{6}{7}$  No, $\frac{4}{7} \neq \frac{6}{7}$

Try $\frac{4}{7}$.  $\frac{4}{7} + \frac{2}{7} \overset{?}{=} \frac{6}{7}$  Yes, $\frac{6}{7} = \frac{6}{7}$

So $n = \frac{4}{7}$.

$$\frac{1}{3} \cdot x = \frac{1}{12}$$

Try $\frac{1}{2}$.  $\frac{1}{3} \cdot \frac{1}{2} \overset{?}{=} \frac{1}{12}$  No, $\frac{1}{6} \neq \frac{1}{12}$

Try $\frac{1}{4}$.  $\frac{1}{3} \cdot \frac{1}{4} \overset{?}{=} \frac{1}{12}$  Yes, $\frac{1}{12} = \frac{1}{12}$

So $x = \frac{1}{4}$.

Solve the equation. Use Guess and Test.

**17.** $y + \frac{1}{3} = \frac{2}{3}$

**18.** $a + \frac{1}{5} = \frac{4}{5}$

**19.** $\frac{1}{2} \cdot b = \frac{1}{4}$

**20.** $\frac{1}{3} \cdot m = \frac{1}{6}$

**21.** $\frac{5}{6} = c + \frac{1}{6}$

**22.** $\frac{1}{8} = \frac{1}{2} \cdot d$

**23.** $\frac{2}{7} + n = \frac{9}{14}$

**24.** $\frac{2}{5} \cdot n = \frac{1}{3}$

**25.** $7\frac{4}{9} = x + \frac{1}{3}$

Solve the equation. Use *properties* or Guess and Test.

**26.** $\frac{2}{3} \cdot y = \frac{2}{3}$

**27.** $\frac{4}{9} + z = \frac{4}{9}$

**28.** $\frac{3}{4} = t + \frac{1}{4}$

**29.** $\frac{1}{5} \cdot \frac{1}{2} = \frac{1}{2} \cdot n$

**30.** $\frac{1}{14} = \frac{1}{7} \cdot m$

**31.** $\frac{5}{9} + \frac{1}{3} = \frac{1}{3} + a$

**32.** $\left(\frac{1}{2} + \frac{1}{3}\right) + \frac{1}{4} = \frac{1}{2} + \left(\frac{1}{3} + r\right)$

**33.** $\frac{3}{5} \cdot \left(\frac{1}{2} \cdot s\right) = \left(\frac{3}{5} \cdot \frac{1}{2}\right) \cdot \frac{1}{10}$

## Skills to Remember

Write the number or letter for each point.

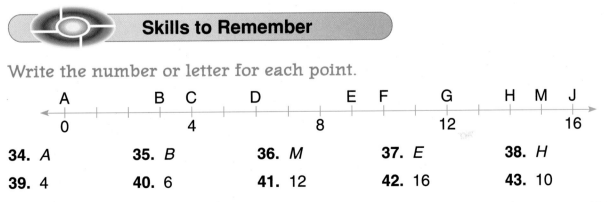

**34.** A

**35.** B

**36.** M

**37.** E

**38.** H

**39.** 4

**40.** 6

**41.** 12

**42.** 16

**43.** 10

# Moving On: Algebra

## 14-7 Coordinate Geometry

**Ordered pairs** are used to locate points on a grid. The numbers that are used to represent a point are called **coordinates**.

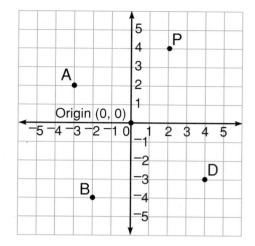

(2, 4) are the coordinates of point *P*.

1st coordinate
2nd coordinate

A grid can be divided into four sections. The point where the two scales meet is called the **origin**.

The coordinates of the origin are (0, 0).

To locate a point on a grid:

• Read the value of the first coordinate of the ordered pair.

• Start at 0.

• Move the number of units on the horizontal scale indicated by the first coordinate. The ⁻ sign tells you to move left.

• Read the value of the second coordinate.

• Move the number of units on the vertical scale indicated by that value. The ⁻ sign tells you to move down.

(⁻3, 2) locates the point *A*.

Move *up* 2 units.

Start at 0. Move *left* 3 units.

Study these examples.

(⁻2, ⁻4) locates the point *B*.

Start at 0. Move *left* 2 units.

Move *down* 4 units.

(4, ⁻3) locates the point *D*.

Start at 0. Move *right* 4 units.

Move *down* 3 units.

452

Use the grid at the right for exercises 1–6.
Name the point for each set of coordinates.

**1.** (⁻2, 2)  **2.** (0, 0)  **3.** (2, 2)

**4.** (0, ⁻3)  **5.** (2, ⁻4)  **6.** (⁻2, ⁻2)

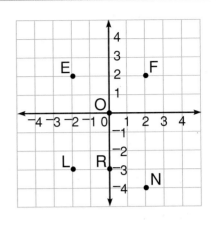

Use the grid at the right for exercises 7–17.
Copy and complete the chart.

| | Point | Coordinates |
|---|---|---|
| **7.** | ? | (⁻5, 1) |
| **8.** | ? | (⁻1, 1) |
| **9.** | ? | (⁻1, ⁻2) |
| **10.** | ? | (⁻5, ⁻2) |
| **11.** | ? | (3, 2) |
| **12.** | ? | (5, 0) |
| **13.** | ? | (5, ⁻2) |
| **14.** | ? | (3, ⁻4) |
| **15.** | ? | (1, ⁻2) |
| **16.** | ? | (1, 0) |

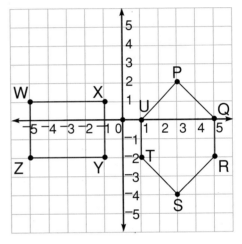

**17.** Classify figure *WXYZ* and figure *PQRSTU*.

Use a grid to locate the points. Then connect them.

**18.** *A* (1, 4); *M* (4, 4); *H* (1, 8)   **19.** *P* (⁻2, 8); *S* (⁻8, 8); *T* (⁻2, 16)

**20.** *B* (3, ⁻1); *C* (5, ⁻1); *D* (5, ⁻3); *E* (3, ⁻3)

**21.** *W* (⁻7, ⁻2); *X* (⁻11, ⁻2); *Y* (⁻11, ⁻6); *Z* (⁻7, ⁻6)

**22.** What figures have you made? What did you discover?

**Project**

Discuss ✓

**23.** On a grid, draw a simple picture made up of line segments.
Make a list of coordinates that must be connected to make
the picture. Display your work in your classroom.

# Moving On: Algebra

## 14-8 Introduction to Integers

Kyle earned $8 running errands for neighbors.
He spent $3.

You can write these numbers as **integers**.

*earned* $8    $^+8$ dollars    ← Read: "positive eight dollars"

*spent* $3    $^-3$ dollars    ← Read: "negative three dollars"

▶ Integers are the whole numbers and their opposites.
They are either positive, negative, or zero.
They can be shown on a number line.

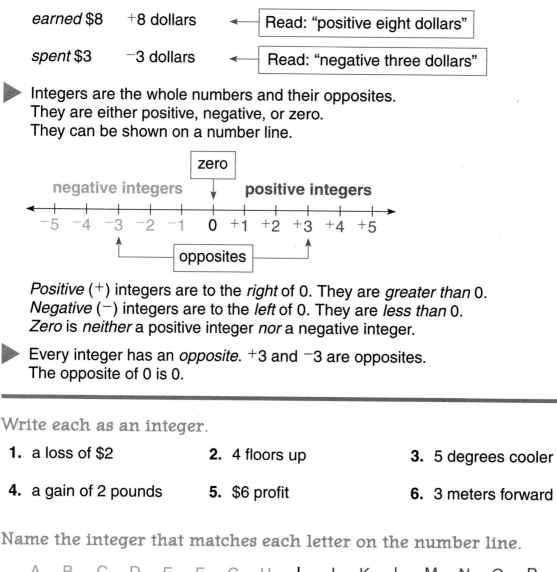

Positive ($^+$) integers are to the *right* of 0. They are *greater than* 0.
Negative ($^-$) integers are to the *left* of 0. They are *less than* 0.
Zero is *neither* a positive integer *nor* a negative integer.

▶ Every integer has an *opposite*. $^+3$ and $^-3$ are opposites.
The opposite of 0 is 0.

---

Write each as an integer.

**1.** a loss of $2      **2.** 4 floors up      **3.** 5 degrees cooler

**4.** a gain of 2 pounds      **5.** $6 profit      **6.** 3 meters forward

Name the integer that matches each letter on the number line.

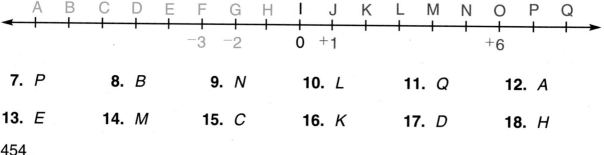

**7.** *P*     **8.** *B*     **9.** *N*     **10.** *L*     **11.** *Q*     **12.** *A*

**13.** *E*     **14.** *M*     **15.** *C*     **16.** *K*     **17.** *D*     **18.** *H*

For each integer, name the integer that is
just *before* and just *after* it on a number line.

**19.** $^+9$     **20.** $^-17$     **21.** $^-6$     **22.** 0     **23.** $^-10$     **24.** $^+1$

**25.** $^-1$     **26.** $^+13$     **27.** $^-26$     **28.** $^-8$     **29.** $^+4$     **30.** $^-2$

Write the opposite of each integer.

**31.** $^+5$     **32.** $^+8$     **33.** $^-6$     **34.** $^+9$     **35.** $^-17$     **36.** $^-3$

**37.** $^-11$     **38.** $^+88$     **39.** $^-1$     **40.** 0     **41.** $^-67$     **42.** $^+49$

**43.** $^+14$     **44.** $^-63$     **45.** $^+70$     **46.** $^+105$     **47.** $^-213$     **48.** $^+300$

## PROBLEM SOLVING

**49.** If you record a deposit of eighteen dollars as $^+$\$18, how would you record a withdrawal of eighteen dollars?

**50.** In a game the card for $^+7$ says "Go Ahead 7 Steps." What would the card for $^-7$ say?

**51.** Begin at 0. What happens if you go up 6 steps ($^+6$) and then down 6 steps ($^-6$)?

**52.** On a vertical number line, are the numbers above zero positive or negative?

**53.** If 0 is sea level, how would twenty-five feet below sea level be written?

**54.** If 0 is sea level, how would forty-seven feet above sea level be written?

**55.** List real-life situations in which positive and negative integers are used.

0 ft sea level

*Math Journal*

## Finding Together

*Discuss*

Name each integer on a horizontal number line.

**56.** six to the right of negative three

**57.** four to the left of one

# Moving On: Algebra

## 14-9 Compare and Order Integers

You can use a number line to compare and order integers.

$$\begin{array}{ccccccccccccccccc} & -8 & -7 & -6 & -5 & -4 & -3 & -2 & -1 & 0 & +1 & +2 & +3 & +4 & +5 & +6 & +7 & +8 \end{array}$$

▶ To **compare integers** using a horizontal number line, any integer is greater than an integer to its left.

| | | |
|---|---|---|
| $+4 > +2$ since $+2$ is left of $+4$. | $+2 < +4$ | A positive integer is greater than any negative integer. |
| $-3 < +1$ since $+1$ is right of $-3$. | $+1 > -3$ | |
| $-3 > -5$ since $-5$ is left of $-3$. | $-5 < -3$ | |

Order $+5, -4, 0.$

▶ To **order integers** using a horizontal number line:

• Least to greatest — Begin with the integer farthest to the *left*.

• Greatest to least — Begin with the integer farthest to the *right*.

> Think: $-4$ is farthest to the left and $+5$ is farthest to the right on the number line; 0 is between $+5$ and $-4$.

The order from least to greatest is: $-4, 0, +5$

The order from greatest to least is: $+5, 0, -4$

---

Choose the greater integer.

**1.** $+3, +5$      **2.** $0, -6$      **3.** $+4, +1$      **4.** $0, +3$

**5.** $+2, -2$      **6.** $+1, 0$      **7.** $-2, +4$      **8.** $+5, +6$

**9.** $+2, +4$      **10.** $-3, -7$      **11.** $-1, +1$      **12.** $-6, -2$

Compare. Write < or >.

**13.** $-1 \underline{\ ?\ } +1$    **14.** $+6 \underline{\ ?\ } -5$    **15.** $+4 \underline{\ ?\ } +1$    **16.** $-11 \underline{\ ?\ } -14$

**17.** $-2 \underline{\ ?\ } +6$    **18.** $+12 \underline{\ ?\ } -10$    **19.** $-6 \underline{\ ?\ } 0$    **20.** $+9 \underline{\ ?\ } 0$

456

Arrange in order from least to greatest.

**21.** ⁻5, 0, ⁻4

**22.** ⁺5, ⁺3, ⁻7

**23.** ⁻1, ⁻9, ⁺2

**24.** ⁺14, ⁻6, ⁻1

**25.** ⁻8, ⁺5, 0

**26.** ⁺9, ⁺8, ⁻1

**27.** ⁻6, ⁻9, ⁻3

**28.** ⁻2, ⁺7, ⁻1

**29.** ⁻4, ⁺14, 0

Arrange in order from greatest to least.

**30.** ⁻3, ⁺6, ⁺5

**31.** ⁻6, ⁻3, ⁺4

**32.** ⁻4, ⁺5, ⁺3

**33.** 0, ⁻7, ⁻10

**34.** ⁺8, ⁻8, 0

**35.** ⁺3, ⁺5, ⁺8

**36.** ⁻12, ⁻8, ⁻10

**37.** ⁻15, ⁺6, ⁺8

**38.** ⁺1, ⁻4, ⁺3

Write *always*, *sometimes*, or *never* to make true statements.

**Math Journal** ✓

**39.** A negative integer is _?_ less than a positive integer.

**40.** A negative integer is _?_ greater than 0.

**41.** A negative integer is _?_ less than another negative integer.

**42.** A positive integer is _?_ greater than 0.

**43.** A positive integer is _?_ less than another positive integer.

**PROBLEM SOLVING**

The chart shows the daily average temperature for five days.

**44.** Which day had the coldest average temperature?

**45.** Which day had the warmest average temperature?

**46.** What was the median (middle) temperature?

**47.** Which day was the average temperature between ⁻3°C and ⁺1°C?

| Day | Average Temperature |
|-----------|--------------------|
| Monday | ⁻2°C |
| Tuesday | ⁺5°C |
| Wednesday | ⁻3°C |
| Thursday | ⁺1°C |
| Friday | ⁺2°C |

# 14-10 Adding Integers with Like Signs

An anchor is 2 ft below sea level. It goes down 4 more feet. What is its new depth written as an integer?

2 ft below sea level   ⟶   ⁻2

4 ft down   ⟶   ⁻4

To find the anchor's new depth, add: ⁻2 + ⁻4 = _?_ .

▶ You can use a number line to model the addition of integers.

- Start at 0.
- Move *left* for negative integers.
- Move *right* for positive integers.

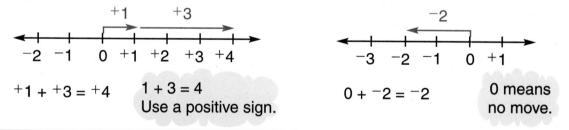

▶ To add integers with *like* signs:

- Add the integers.      2 + 4 = 6
- Use the sign of the addends.    ⁻2 + ⁻4 = ⁻6

> 2 + 4 = 6
> Use a negative sign.

The anchor's depth written as an integer is ⁻6 ft.

**Study these examples.**

⁺1 + ⁺3 = ⁺4     1 + 3 = 4
Use a positive sign.

0 + ⁻2 = ⁻2     0 means no move.

---

Write an addition sentence for each number line.

**1.**

**2.**

**3.**

**4.**

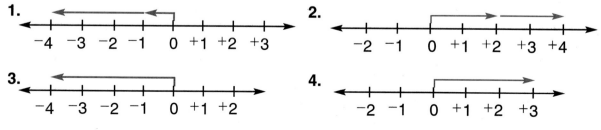

Add. Use a number line to help you.

**5.** $^{+}5 + {}^{+}2$  **6.** $^{+}6 + {}^{+}4$  **7.** $^{+}9 + {}^{+}3$  **8.** $^{+}7 + {}^{+}12$

**9.** $^{-}12 + {}^{-}3$  **10.** $^{-}8 + {}^{-}9$  **11.** $^{-}4 + {}^{-}11$  **12.** $^{-}8 + {}^{-}13$

**13.** $0 + {}^{+}8$  **14.** $^{+}7 + 0$  **15.** $^{-}10 + 0$  **16.** $0 + {}^{-}5$

**17.** Describe a rule for each row of exercises above and give another example.

*Communicate* ✓

Copy and complete each table.

**18.**

| $n$ = integer | $n + 5$ |
|---|---|
| $^{+}3$ | $3 + 5 = 8$ |
| $^{+}6$ | ? |
| $^{+}9$ | ? |
| $0$ | ? |
| $^{+}5$ | ? |

**19.**

| $n$ = integer | $n + {}^{-}4$ |
|---|---|
| $^{-}4$ | $^{-}4 + {}^{-}4 = {}^{-}8$ |
| $^{-}8$ | ? |
| $0$ | ? |
| $^{-}3$ | ? |
| $^{-}7$ | ? |

Find the sum.

**20.** $^{-}5 + ({}^{-}3 + {}^{-}2)$

$^{-}5 + \underline{\ ?\ } = \underline{\ ?\ }$

**21.** $({}^{+}8 + {}^{+}2) + {}^{+}9$

$\underline{\ ?\ } + {}^{+}9 = \underline{\ ?\ }$

**22.** $^{+}3 + ({}^{+}7 + {}^{+}5)$  **23.** $({}^{-}2 + {}^{-}9) + {}^{-}6$  **24.** $({}^{+}4 + {}^{+}1) + {}^{+}13$

**25.** $({}^{-}1 + {}^{-}10) + {}^{-}12$  **26.** $^{-}6 + ({}^{-}3 + {}^{-}3)$  **27.** $({}^{-}5 + 0) + {}^{-}10$

**PROBLEM SOLVING** Write each answer in words and as an integer.

**28.** A geologist worked at a site 3 m above sea level. Later he moved to a site 5 m higher. How far above or below sea level is the new site?

**29.** Team A's score in one game is $^{-}9$. If the team makes another score of $^{-}20$, what is its total score?

**30.** The selling price of stock X fell 8 points one day and 12 points the next day. What was the total change over the two-day period?

**31.** The football team had a gain of 6 yd on one play and a gain of 5 yd on the next play. How many yards were gained on the two plays?

459

# 14-11 Adding Integers with Unlike Signs

Jan lost 8 points in the first round of a game.
He earned 3 points in the second round.
What was his score after the second round?

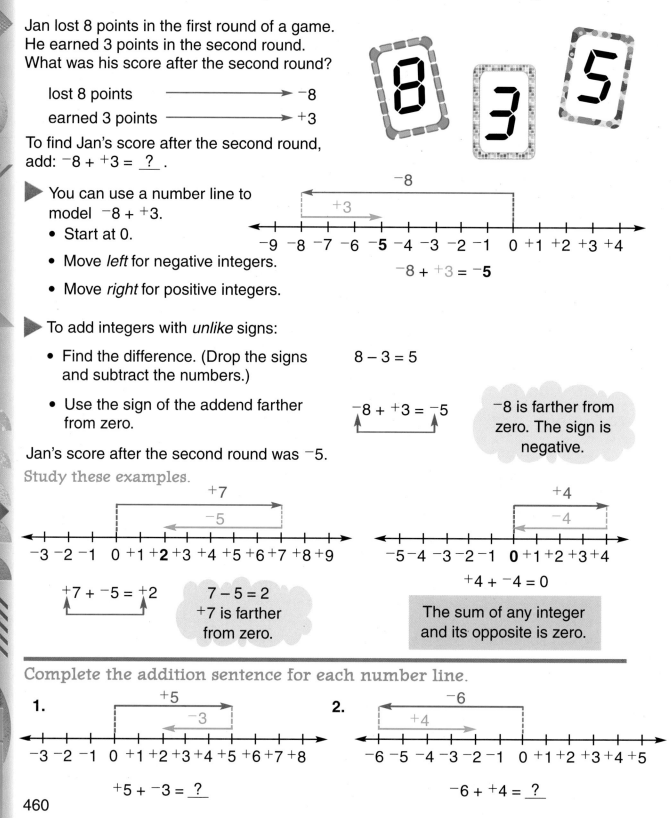

lost 8 points ⟶ ⁻8

earned 3 points ⟶ ⁺3

To find Jan's score after the second round,
add: ⁻8 + ⁺3 = __?__ .

▶ You can use a number line to
model ⁻8 + ⁺3.

- Start at 0.

- Move *left* for negative integers.

- Move *right* for positive integers.

⁻8 + ⁺3 = ⁻5

▶ To add integers with *unlike* signs:

- Find the difference. (Drop the signs
and subtract the numbers.)

  8 − 3 = 5

- Use the sign of the addend farther
from zero.

  ⁻8 + ⁺3 = ⁻5

  ⁻8 is farther from
  zero. The sign is
  negative.

Jan's score after the second round was ⁻5.

Study these examples.

⁺7 + ⁻5 = ⁺2

7 − 5 = 2
⁺7 is farther
from zero.

⁺4 + ⁻4 = 0

The sum of any integer
and its opposite is zero.

Complete the addition sentence for each number line.

**1.**

⁺5 + ⁻3 = __?__

**2.**

⁻6 + ⁺4 = __?__

Write an addition sentence for each number line.

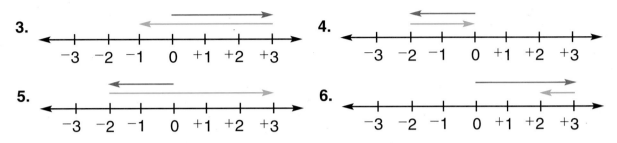

**3.**

−3 −2 −1 0 +1 +2 +3

**4.**

−3 −2 −1 0 +1 +2 +3

**5.**

−3 −2 −1 0 +1 +2 +3

**6.**

−3 −2 −1 0 +1 +2 +3

Find the sum. Use a number line to help you.

**7.** $^+10 + {}^-4$     **8.** $^+9 + {}^-11$     **9.** $^+7 + {}^-1$     **10.** $^+13 + {}^-17$

**11.** $^-9 + {}^+2$     **12.** $^-13 + {}^+15$     **13.** $^-8 + {}^+2$     **14.** $^-7 + {}^+5$

**15.** $^+8 + {}^-8$     **16.** $^-9 + {}^+9$     **17.** $^+11 + {}^-11$     **18.** $^+25 + {}^-25$

**19.** $^-15 + {}^+7$     **20.** $^+21 + {}^-13$     **21.** $^-36 + {}^+25$     **22.** $^+11 + {}^-9$

**PROBLEM SOLVING**  Write each answer in words and as an integer.

**23.** Sally's checking account has a balance of $^-$\$12. If she deposits \$30, what will be her new balance?

**24.** A quarterback gained 16 yd on one play. Then he lost 13 yd on the next play. What was the total gain?

**25.** An anchor hung against the side of a boat 4 ft below sea level. A sailor lowered the anchor 20 ft. What is the total depth of the anchor?

**26.** In March, Ben gained 2 lb. In April, he lost 4 lb. What was Ben's total gain or loss in March and April?

**Connections: Science**

**27.** Electrons have a charge of $^-1$ and protons have a charge of $^+1$. The total charge of an ion is the sum of its electrons and protons. Find the total charge of an ion of:

**a.** 13 protons and 17 electrons

**b.** 9 protons and 4 electrons.

**c.** 8 protons and 8 electrons.

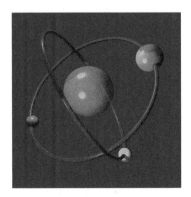

461

# 14-12 Subtracting Integers

While studying about ions, Rachel found a rule for subtracting integers.

**Key**

1 ⚪ = +1

1 ⚫ = −1

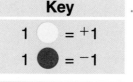

$+1 + {}^-1 = 0$

zero pair: ⚪⚫

---

### Hands-On Understanding

**Materials Needed:** two-color counters, work mat

**Integer Mat**

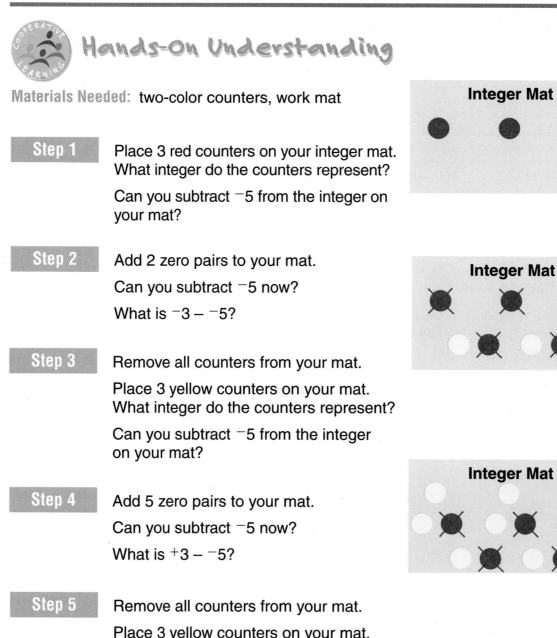

**Step 1** Place 3 red counters on your integer mat. What integer do the counters represent?

Can you subtract ⁻5 from the integer on your mat?

**Step 2** Add 2 zero pairs to your mat.

Can you subtract ⁻5 now?

What is ⁻3 − ⁻5?

**Step 3** Remove all counters from your mat.

Place 3 yellow counters on your mat. What integer do the counters represent?

Can you subtract ⁻5 from the integer on your mat?

**Step 4** Add 5 zero pairs to your mat.

Can you subtract ⁻5 now?

What is ⁺3 − ⁻5?

**Step 5** Remove all counters from your mat.

Place 3 yellow counters on your mat. What integer do the counters represent?

Can you subtract ⁺5 from the integer on your mat?

**Step 6** Add 2 zero pairs to your mat.

Can you subtract $^+5$ now?

What is $^+3 - {}^+5$?

**Integer Mat**

**Step 7** Remove all counters from your mat.

Place 3 red counters on your mat.
What integer do the counters represent?

Can you subtract $^+5$ from the integer
on your mat?

**Step 8** Add 5 zero pairs to your mat.

Can you subtract $^+5$ now?

What is $^-3 - {}^+5$?

**Integer Mat**

## Use counters to subtract.

1. $^-6 - {}^-2$
2. $^+6 - {}^-2$
3. $^+6 - {}^+2$
4. $^-6 - {}^+2$
5. $^-9 - {}^-4$
6. $^+9 - {}^-4$
7. $^+9 - {}^+4$
8. $^-9 - {}^+4$
9. $^-5 - {}^-8$
10. $^+5 - {}^-8$
11. $^+5 - {}^+8$
12. $^-5 - {}^+8$

## Copy and complete the table.

13.

| Subtraction Sentences | Addition Sentences |
|---|---|
| $^-3 - {}^-5 = \underline{?}$ | $^-3 + {}^+5 = \underline{?}$ |
| $^+3 - {}^-5 = \underline{?}$ | $^+3 + {}^+5 = \underline{?}$ |
| $^+3 - {}^+5 = \underline{?}$ | $^+3 + {}^-5 = \underline{?}$ |
| $^-3 - {}^+5 = \underline{?}$ | $^-3 + {}^-5 = \underline{?}$ |

## Communicate

*Discuss*

14. How does adding zero pairs help to model
subtraction of integers?

15. How can you use addition to subtract integers?
Give examples to explain your answer.

463

# 14-13 Multiplying Integers

The stock of the Jones Company drops $3 per share. Ana owns 4 shares. What is the total change in the value of Ana's shares?

To find the total change in value, multiply: $4 \times {}^-3 = $ ___?___

| |
|---|
| An integer with no sign is a positive integer: $4 = {}^+4$, $12 = {}^+12$, and so on. |

$4 \times {}^-3$ means ${}^+4 \times {}^-3$

positive four      negative three

The pattern above shows that ${}^+4 \times {}^-3 = {}^-12$.

The total change in the value of Ana's shares is ${}^-\$12$.

| |
|---|
| $4 \times 3 = 12$ |
| $4 \times 2 = 8$ |
| $4 \times 1 = 4$ |
| $4 \times 0 = 0$ |
| $4 \times {}^-1 = {}^-4$ |
| $4 \times {}^-2 = {}^-8$ |
| $4 \times {}^-3 = {}^-12$ |

▶ To find the product ${}^-4 \times {}^-3$, study the pattern below.

${}^-4 \times {}^-3$

negative four      negative three

So ${}^-4 \times {}^-3 = 12$ or ${}^+12$.

| |
|---|
| $4 \times {}^-3 = {}^-12$ |
| $3 \times {}^-3 = {}^-9$ |
| $2 \times {}^-3 = {}^-6$ |
| $1 \times {}^-3 = {}^-3$ |
| $0 \times {}^-3 = 0$ |
| ${}^-1 \times {}^-3 = 3$ |
| ${}^-2 \times {}^-3 = 6$ |
| ${}^-3 \times {}^-3 = 9$ |
| ${}^-4 \times {}^-3 = 12$ |

▶ The product of two integers:

- is *positive* if they have the *same* sign.
- is *negative* if they have *different* signs.
- is *zero* if one or both is *zero*.

Study these examples.

$${}^+5 \times {}^+6 = {}^+30 \qquad {}^-7 \times {}^+8 = {}^-56 \qquad 0 \times {}^+9 = 0 \qquad 0 \times 0 = 0$$

Use the rules on page 464 to find each product.

1. $^-7 \times \, ^+5$
2. $^+3 \times \, ^-4$
3. $^-2 \times \, ^-5$
4. $^+8 \times \, ^+5$

5. $^+9 \times \, ^-6$
6. $^-5 \times \, ^+5$
7. $^-8 \times \, ^+10$
8. $0 \times \, ^+8$

9. $^-4 \times 0$
10. $^-1 \times \, ^+11$
11. $^+1 \times \, ^-20$
12. $^-7 \times \, ^-7$

13. $(^-8)(^-8)$
14. $^+1 \, (^-1)$
15. $5 \, (^-10)$
16. $(^-12)(^-11)$

Choose the correct answer to complete each statement.
Explain each answer. Let p = positive integer and n = negative integer.

17. $p \times p = \underline{?}$     **a.** positive     **b.** negative     **c.** cannot tell

18. $n \times n = \underline{?}$     **a.** positive     **b.** negative     **c.** cannot tell

19. $p \times n = \underline{?}$     **a.** positive     **b.** negative     **c.** cannot tell

20. $(p \times p) \times p = \underline{?}$     **a.** positive     **b.** negative     **c.** cannot tell

21. $(n \times n) \times n = \underline{?}$     **a.** positive     **b.** negative     **c.** cannot tell

## PROBLEM SOLVING

22. At noon the temperature was 8°C. The temperature dropped 2°C per hour. What was the total change in 6 hours?

23. The Acme Tigers football team loses 8 yards on each of the first 3 plays of the game. Write an integer to express the results.

24. Three times the sum of two positive integers is $^+9$. Find the integers.

25. The difference of two negative integers is $^-4$. Their product is $^+60$. Find the integers.

 **Share Your Thinking**

*Communicate* ✓

26. Discuss with a classmate how to add two negative integers and how to multiply two negative integers. Can you find a number pattern for each that another classmate could use to "discover" the rules? Explain your answer.

## 14-14 Dividing Integers

Pia wants to lose 8 lb in 4 weeks. If she loses the same number of pounds each week, how many pounds will she lose per week?

lose 8 lbs ———→ $^-8$

To find how many pounds Pia loses per week, you can find a missing factor:

$\underline{\ ?\ } \times \ ^+4 = \ ^-8$      $4 = \ ^+4$

$\underline{\ ?\ } = \ ^-2$

or

you can divide, since you are *sharing* a set (8 lb) among equal groups (4 wk).

$^-8 \div \ ^+4 = \ ^-2$

Pia loses 2 lb per week.

Think: What integer times $^+4$ equals $^-8$? $^-2 \times \ ^+4 = \ ^-8$

| Multiplication Sentence | Related Division Sentences |
|---|---|
| $^-2 \times \ ^+4 = \ ^-8$ | $^-8 \div \ ^+4 = \ ^-2$ <br> $^-8 \div \ ^-2 = \ ^+4$ |
| $^+3 \times \ ^-5 = \ ^-15$ | $^-15 \div \ ^-5 = \ ^+3$ <br> $^-15 \div \ ^+3 = \ ^-5$ |
| $^-6 \times \ ^-9 = \ ^+54$ | $^+54 \div \ ^-9 = \ ^-6$ <br> $^+54 \div \ ^-6 = \ ^-9$ |

Division is the inverse of multiplication.

Copy and complete.

1. $^-6 \times \ ^-7 = \ ^+42$

   $^+42 \div \ ^-7 = \underline{\ ?\ }$

   $^+42 \div \ ^-6 = \underline{\ ?\ }$

2. $^-9 \times \ ^+5 = \ ^-45$

   $^-45 \div \ ^+5 = \underline{\ ?\ }$

   $^-45 \div \ ^-9 = \underline{\ ?\ }$

3. $^+8 \times \ ^+3 = \ ^+24$

   $^+24 \div \ ^+3 = \underline{\ ?\ }$

   $^+24 \div \ ^+8 = \underline{\ ?\ }$

Write two related division sentences.

4. $^-5 \times \ ^+6 = \ ^-30$

5. $^+6 \times \ ^-4 = \ ^-24$

6. $^-7 \times \ ^-4 = \ ^+28$

7. $^+9 \times \ ^+8 = \ ^+72$

8. $^-2 \times \ ^-8 = \ ^+16$

9. $^+6 \times \ ^-9 = \ ^-54$

## Rules of Division

Here are rules of division that can help you divide integers quickly and correctly.

- The quotient of integers with *like* signs is positive.

$$^+18 \div {}^+3 = {}^+6 \qquad\qquad ^-20 \div {}^-5 = {}^+4$$

$$^+15 \div {}^+5 = {}^+3 \qquad\qquad ^-54 \div {}^-9 = {}^+6$$

- The quotient of integers with *unlike* signs is negative.

$$^-10 \div {}^+5 = {}^-2 \qquad\qquad ^-30 \div {}^+6 = {}^-5$$

$$^+36 \div {}^-9 = {}^-4 \qquad\qquad ^+42 \div {}^-7 = {}^-6$$

Divide.

**10.** $^+60 \div {}^+5$  **11.** $^+32 \div {}^-8$  **12.** $^-63 \div {}^+9$  **13.** $^-55 \div {}^-11$

**14.** $^+48 \div {}^+12$  **15.** $^+52 \div {}^-4$  **16.** $^-10 \div {}^-10$  **17.** $^-30 \div {}^-6$

**18.** $^+45 \div {}^+9$  **19.** $^+44 \div {}^-11$  **20.** $^-100 \div {}^+5$  **21.** $^-45 \div {}^-45$

## PROBLEM SOLVING

**22.** The quotient is $^+1$. The divisor is $^+9$. What is the dividend?

**23.** The dividend is $^-48$. The quotient is $^+8$. What is the divisor?

**24.** The dividend is $^+40$. The divisor is $^-5$. What is the quotient?

**25.** The divisor is $^+12$. The quotient is 0. What is the dividend?

**26.** The temperature drops 25°F in 5 hours. What is the average change per hour, written as an integer?

**27.** Tony withdraws $180 from his account in 3 weeks. What was the average withdrawal per week, written as an integer?

### Challenge

Tell whether each integer in the pattern is divided/multiplied by a *positive integer* or a *negative integer*. Then complete the pattern.

**Discuss** ✓

**28.** $^-80, {}^+40, {}^-20, \underline{\ ?\ }, \underline{\ ?\ }$

**29.** $^-243, {}^-81, {}^-27, \underline{\ ?\ }, \underline{\ ?\ }$

**30.** $^+3, {}^-12, {}^+48, \underline{\ ?\ }, \underline{\ ?\ }$

**31.** $^+1, {}^+5, {}^+25, \underline{\ ?\ }, \underline{\ ?\ }$

# 14-15 Functions and Coordinate Graphs

You can use a rule or equation to make a function table and use ordered pairs to locate points on a coordinate grid.

$(^-1,^-2)$

Graph the function $y = x + 1$ on a coordinate grid using integer values from $^-2$ to $^+2$. Then use the graph to find the value of $y$ when $x = ^+4$.

$(^+2,^+3)$

▶ **To graph a function on a coordinate grid:**

- Make a function table.
  - Substitute values for $x$ in the rule or equation.
  - Find the corresponding $y$-values.
  - Write an ordered pair for each $x$-and $y$-value.
- Graph each ordered pair.
- Connect the points.

| x | x + 1 | y | (x, y) |
|---|---|---|---|
| $^-2$ | $^-2 + 1 = ^-1$ | $^-1$ | $(^-2,^-1)$ |
| $^-1$ | $^-1 + 1 = 0$ | $0$ | $(^-1,0)$ |
| $0$ | $0 + 1 = ^+1$ | $^+1$ | $(0,^+1)$ |
| $^+1$ | $^+1 + 1 = ^+2$ | $^+2$ | $(^+1,^+2)$ |
| $^+2$ | $^+2 + 1 = ^+3$ | $^+3$ | $(^+2,^+3)$ |

Remember: Start at the origin and move $x$ units to the *right* or *left*. Then move $y$ units *up* or *down*.

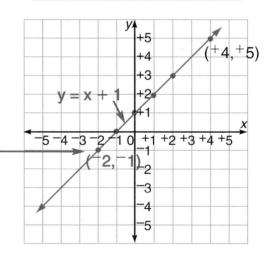

Start at 0. Move *left* 2 units. Move *down* 1 unit.

When $x = ^+4$, $y = ^+5$.

Copy and complete each function table.
Then graph on a coordinate grid.

**1.**

| x | y = x | y | (x, y) |
|---|---|---|---|
| ⁻1 | y = ⁻1 | ⁻1 | (⁻1, ⁻1) |
| 0 | ? | ? | ? |
| ⁺1 | ? | ? | ? |
| ⁺2 | ? | ? | ? |

**2.**

| x | y = x + 2 | y | (x, y) |
|---|---|---|---|
| 0 | ? | ? | ? |
| ⁺1 | ? | ? | ? |
| ⁺2 | ? | ? | ? |
| ⁺3 | ? | ? | ? |

Use the given graph of y = x + ⁻1.

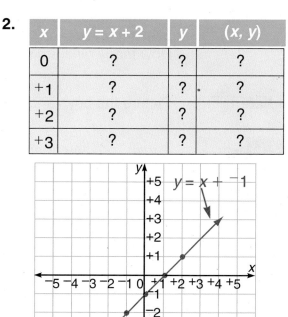

**3.** When x = 0, what is the value of y?

**4.** When x = ⁻1, what is the value of y?

**5.** When x = ⁻3, what is the value of y?

**6.** For what value of x is y = ⁺1?

**7.** For what value of x is y = ⁻3?

**8.** For what value of x is y = 0?

Graph each function on a coordinate grid using integer values from
⁻2 to ⁺2 for x. Then use the graph to find the value of y when x = ⁺3.

**9.** y = x + 3      **10.** y = x + ⁻2      **11.** y = x + ⁻3      **12.** y = ⁻x

## PROBLEM SOLVING

Function tables and coordinate graphs are used in problem solving .

Given a constant wind speed
of 7 miles per hour, a windchill
table shows that the windchill
temperature (y) in °F is equal to
the actual temperature (x) in °F
reduced by 5°F: y = x + ⁻5.

| Actual Temperature x | Windchill Temperature y = x + ⁻5 | (x, y) |
|---|---|---|
| ⁻5 | y = ⁻5 + ⁻5 = ⁻10 | (⁻5, ⁻10) |
| 0 | ? | ? |
| ⁺5 | ? | ? |
| ⁺10 | ? | ? |

**13.** Copy and complete the table.
Then graph on a coordinate
grid. Find x when y = ⁻15
from the graph.

469

# 14-16 Problem-Solving Strategy: Write an Equation

**Problem:** In music class there are 19 boys. This is 5 less than twice the number of girls. How many girls are in the music class?

**1 IMAGINE** Picture the boys and girls in music class.

**2 NAME**

*Facts:* number of boys—19
number of boys—5 less than twice the number of girls

*Question:* How many girls are in the music class?

**3 THINK** First, to find the number of girls, write a word equation, then substitute:

Let $n$ represent the number of girls.

| number of boys | is | 5 less than twice the number of girls |
|---|---|---|
| 19 | = | $2n - 5$ |

Then solve the equation by using the Guess and Test strategy.

**4 COMPUTE**

Solve for $n$:     $19 = 2n - 5$     Use the order of operations.

Try $n = 10$.     $19 \stackrel{?}{=} 2 \cdot 10 - 5$
$19 \stackrel{?}{=} 20 - 5 \longrightarrow 19 \neq 15$

Try $n = 11$.     $19 \stackrel{?}{=} 2 \cdot 11 - 5$
$19 \stackrel{?}{=} 22 - 5 \longrightarrow 19 \neq 17$

Try $n = 12$.     $19 \stackrel{?}{=} 2 \cdot 12 - 5$
$19 \stackrel{?}{=} 24 - 5 \longrightarrow 19 = 19$

So $n = 12$.
There are 12 girls in music class.

**5 CHECK** Begin with 12, multiply by 2 and subtract 5.
$12 \times 2 = 24$ and $24 - 5 = 19$. The answer checks.

*Write an equation to solve each problem.*

**1.** Tresse practiced 30 min longer than Lyle. Together they practiced 1 h 50 min. How long did each one practice?

| | |
|---|---|
| **IMAGINE** | Picture the students practicing. |
| **NAME** | *Facts:* Tresse practiced 30 min longer than Lyle. |
| | Total practice time—1 h 50 min |
| | *Question:* How long did each one practice? |
| **THINK** | Write and solve an equation using Guess and Test. Let *t* represent the time Lyle practiced and *t* + 30 represent the time Tresse practiced. $t + t + 30 = 110$ ⟵ 1 h 50 min |

**COMPUTE** ⟶ **CHECK**

**2.** There are 35 students in chorus. Nine students sing alto, 8 sing tenor, 4 sing bass, and the rest sing soprano. How many sing soprano?

**3.** Josh spent $8.50 for a music lesson. Then he spent $\frac{1}{5}$ of that amount for sheet music. How much did he spend altogether?

**4.** Ms. Murphy teaches 18 students music. Three more than half of them take piano lessons. How many piano students does Ms. Murphy teach?

**5.** There are 18 fifth graders in the band. This is 8 more than one fourth of the students in the band. How many students are in the band?

**6.** Helene has taken flute lessons $1\frac{1}{2}$ years longer than Doug. Lynn has taken flute lessons 1 year less than Doug. If Lynn has taken flute lessons for 2 years, how long has Helene been taking flute lessons?

 **Make Up Your Own**

*Discuss* ✓

**7.** Write an equation. Then write a problem that you can solve using it. Share your work with a classmate.

471

# Moving On: Algebra

## 14-17 Problem-Solving Applications

Solve each problem and explain the method you used.

Imagine

Name

Think

Compute

Check

1. Math-o-Matic is a mathematics video game. Players try to solve equations and puzzles. The Math-o-Matic screen shows two expressions: $5(4 + 4)$ and $150 \div (2 + 1)$. Which expression has the greater value?

2. The Math-o-Matic screen shows this sentence: $5 + 4 \times 3 \underline{\ ?\ } 5 \times 4 + 3$. Should the player input $<$, $=$, or $>$ to make a true sentence?

3. Math-o-Matic asks players to find the missing operation symbol to make the expression $80 \div (10 \underline{\ ?\ } 2)$ equal 10. Which key should the player hit?

4. The Math-o-Matic function machine printed this input and output material. Find its rule.

| $m$ | 12 | 10 | 6 | 3 |
|-----|----|----|----|----|
| ?   | 6  | 5  | 3  | 1.5 |

5. What is the value of $b$ in this Math-o-Matic equation: $17 \times b = 50 + 1$?

6. The variables $c$ and $d$ have the same value in all these equations. Find the values of $c$ and $d$.
$$c + d = 21 \qquad c - d = 1 \qquad c \times d = 110$$

7. Rolland's final Math-o-Matic score is twice Ben's final score, which is 2750. What is Rolland's score?

8. Melanie's score is one third of Loni's score, which is 3327. What is Melanie's score?

9. Tina input these expressions to equal $\frac{1}{2}$. Which one or ones are correct?
$$25\% \text{ of } 8 \qquad 3\frac{1}{8} - \frac{5}{8} \qquad 1.5 \div 3 \qquad \frac{5}{7} \times \frac{7}{10}$$

10. Math-o-Matic shows this series of equations.
$$5e = 7.5 \longrightarrow e + f = 2 \longrightarrow f - g = 0.2$$
Solve the equation to find the value of $g$.

Choose a strategy from the list or use another strategy you know to solve each problem. You may combine strategies.

USE THESE STRATEGIES
Write an Equation
Guess and Test .
Extra Information
Multi-Step Problem
MoreThan One Solution
Use a Model/Diagram
Use a Formula

**11.** The value of $x$ in this magic square is $\frac{3}{4}$ of 12. What is its value?

| 18 | $s$ | $t$ |
|----|-----|-----|
| $q$ | $p$ | $r$ |
| 24 | $x$ | 12 |

**12.** The sum of each horizontal, vertical, and diagonal row in the magic square is the same. What is the sum of each row?

**13.** Write and solve equations to find the value of $p, q, r, s,$ and $t$ in the magic square.

**14.** The game machine prints a 2-digit number. The sum of the digits is 15 and the difference between them is 1. What are the possible numbers?

**15.** Ashlee figures out that 35% of the 60 questions in the Math-o-Matic game involve solving equations. How many of them do not involve solving equations?

**16.** Adam plays 2 rounds of Math-o-Matic. His first score is 24 less than his second score. His total for both rounds is 264. What is his average score for each round?

**17.** The volume of a rectangular prism is 120 in.$^3$ The length, width, and height are whole numbers and each is 1 in. longer than the other. The length is the longest edge. What is the length?

**18.** Glen plays 5 rounds of Math-o-Matic. He answers $\frac{3}{4}$ of the questions correctly in each round. He gets 8 points for each correct answer and finishes with a total of 600 points. How many questions does he miss?

**19.** Pattie moved the entire figure formed by joining the coordinates (5,3), (9,3), (9,6) left 3 and down 2. Name its new coordinates and find its area.

**20.** The perimeter of an isosceles triangle is 18 in. The congruent sides are odd numbers between 4 and 10. What are the lengths of the three sides?

# Chapter Review and Practice

**Evaluate each expression.**                    *(See pp. 440–443.)*

**1.** $a - 6\frac{1}{4}$, when $a = 10$

**2.** $13\frac{1}{8} + c$, when $c = 15\frac{1}{2}$

**3.** $12m$, when $m = \frac{1}{6}$

**4.** $y \div 12$, when $y = 3\frac{1}{2}$

**Copy and complete each table.**                *(See pp. 444–445.)*

**5.**

| $m$ | 2 | 3 | 4 | 5 | 6 |
|-----|---|---|---|---|---|
| $m + 4$ | ? | ? | ? | ? | ? |

**6.**

| $b$ | 9 | 12 | 15 | 18 | 21 |
|-----|---|----|----|----|----|
| $\dfrac{b}{3}$ | ? | ? | ? | ? | ? |

**Solve each equation.**                    *(See pp. 446–451, 454–457.)*

**7.** $x + 2 = 7$

**8.** $5b = 10$

**9.** $\frac{1}{3}s = 18$

**10.** $\frac{2}{3} + p = \frac{2}{3}$

**11.** $\frac{2}{3}d = \frac{2}{3}$

**12.** $\frac{2}{3} \cdot \frac{3}{4} = \frac{3}{4} \cdot c$

**Write the opposite.**

**13.** $^-7$          **14.** $^+5$          **15.** 0

**Compare. Write $<$ or $>$.**

**16.** $^-2 \underline{\ ?\ } {}^+2$          **17.** $0 \underline{\ ?\ } {}^-2$

**Compute.**                    *(See pp. 458–467.)*

**18.** $^+5 + {}^+11$

**19.** $^-12 + {}^+4$

**20.** $^-3 + {}^-5$

**21.** $^-6 - {}^+4$

**22.** $^-2 - {}^-7$

**23.** $^+9 - {}^+10$

**24.** $^-1 \times {}^-18$

**25.** $^+6 \times 0$

**26.** $^+4 \times {}^-12$

**27.** $^-63 \div {}^+7$

**28.** $^-81 \div {}^+9$

**29.** $^+48 \div {}^+3$

**Use the graph on the right.**                    *(See pp. 452–453, 468–473.)*

**30.** Name the point for:

    **a.** $(^-4, {}^-4)$          **b.** $(0, 0)$          **c.** $(^+3, {}^+3)$

**31.** When $x = {}^-1$, what is the value of $y$?

## PROBLEM SOLVING

**32.** There are 46 people on the bus. Five more than half of them will transfer to other buses. How many will transfer?

*(See Still More Practice, p. 488.)*

## RATIONAL NUMBERS

Stock A-B-C fell $8\frac{1}{2}$ points one day and gained $12\frac{1}{4}$ points the next day.

You can write these numbers as positive and negative numbers.

*fell* $8\frac{1}{2}$ points $\longrightarrow$ $-8\frac{1}{2}$

*gained* $12\frac{1}{4}$ points $\longrightarrow$ $+12\frac{1}{4}$

$-8\frac{1}{2}$ and $+12\frac{1}{4}$ are **rational numbers**.

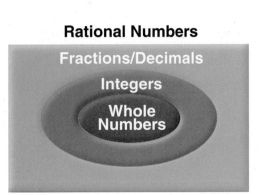

**Rational Numbers**

The diagram above shows that whole numbers, integers, and fractions are rational numbers. Some decimals are also rational numbers.

Like integers, every rational number has an opposite and all rational numbers can be shown on a number line.

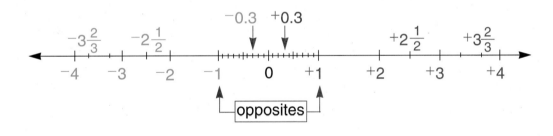

### Write a rational number for each expression.

**1.** a deposit of $20.50

**2.** 2.5 km underwater

**3.** 3 floors up

**4.** a loss of $5\frac{1}{2}$ pounds

**5.** 6.2 m above sea level

**6.** $8.50 profit

### Write the opposite of each rational number.

**7.** $+1.1$

**8.** $-\frac{5}{3}$

**9.** $-\frac{1}{9}$

**10.** $0$

**11.** $+1\frac{1}{7}$

**12.** $-5\frac{2}{5}$

### Draw a number line and locate each rational number.

**13.** $-0.5$

**14.** $+\frac{1}{3}$

**15.** $+1\frac{1}{8}$

**16.** $-2$

**17.** $-2\frac{1}{4}$

**18.** $-4\frac{2}{5}$

## Performance Assessment

**Make and use a coordinate grid.**

1. Draw a rectangle that has one vertex in each quadrant.

2. Name each vertex and give its coordinates.

3. Find the perimeter and area of your rectangle.

4. Name the coordinates of each point where the rectangle crosses:
   **a.** the $x$-axis             **b.** the $y$-axis

**Write an addition and a subtraction expression for each number.**

5. $\frac{2}{3}$        6. 0.21

**Write a multiplication and a division expression for each number.**

7. $1\frac{4}{5}$        8. 0.2

**Evaluate each expression.**

9. $19 + x$, when $x = 32$      10. $5c$, when $c = 6$      11. $50 \div q$, when $q = \frac{1}{2}$

**Solve each equation.**

12. $5 + x = 18$      13. $\frac{1}{4}m = 6$      14. $\frac{4}{7} + \frac{2}{3} = \frac{2}{3} + b$

**Write the rule for the table.**

15.

| $a$ | ? |
|-----|---|
| 20 | 2 |
| 30 | 3 |
| 40 | 4 |
| 50 | 5 |

**Copy and complete the table.**

16.

| $x$ | $y = x + 3$ | $y$ | $(x, y)$ |
|-----|-------------|-----|----------|
| $^+1$ | ? | ? | ? |
| 0 | ? | ? | ? |
| $^-1$ | ? | ? | ? |
| $^-2$ | ? | ? | ? |

**Write as an integer.**

17. a gain of $3      18. 5 floors down

**Order from greatest to least.**

19. $^+7, ^-7, 0$      20. $^+2, ^-6, ^-5$

**Compute.**

21. $^-5 + ^-13$      22. $^+17 + ^-4$      23. $^+6 - ^-8$      24. $^+21 - ^+13$

25. $^+7 \times ^-9$      26. $^-8 \times ^-3$      27. $^+84 \div ^+4$      28. $^-54 \div ^+6$

## Practice 1-1

In the number 9,513,607,482, write the digit in each place. Then give its value.

**1a.** thousands    **b.** tens    **c.** millions

   **d.** ten millions    **e.** billions

Write the number in standard form.

**2.** six billion, twelve million, ninety-eight

**3.** 9,000,000 + 70,000 + 6000 + 70 + 3

**4.** seventy-six and fourteen thousandths

Compare. Write <, =, or >.

**5a.** 326.49 _?_ 326.94    **b.** 0.2 _?_ 0.20

**6a.** 247,913 _?_ 247,193    **b.** 7.05 _?_ 7.5

Round each number to the place of the underlined digit.

**7a.** 7,2̲80,961    **b.** $967.3̲5    **c.** 6.14̲3

Round each number to the greatest place.

  **8a.** 3,498,276    **b.** 459.604    **c.** 0.89

Write in order from least to greatest.

  **9.** 721,056; 702,156; 720,156; 72,156

PROBLEM SOLVING

**10.** Give the value of each 6 in 6326.061.

**11.** Write a number that can be rounded to 0.76 using the digits 5, 7, 9.

**12.** A pecan weighs 31.06 g. A walnut weighs 27.631 g. An almond weighs 30.9 g. Which nut weighs the most? the least?

**13.** Order the following numbers from greatest to least: 739.7, 793.7, 730.9.

**14.** Give the word name for 36.147.

**15.** How are the numbers 96.37 and 963.7 alike? different? Which is the greater number?

## Practice 1-2

Find the missing number.

**1a.** 7 + 6 = ☐ + 7    **b.** 9 = ☐ + 9

**2a.** (4 + 5) + 8 = 4 + (☐ + 8)    **b.** ☐ − 5 = 0

Add or subtract.

**3a.**   34,729   **b.**   48,924   **c.**  $180.77
   + 29,886     + 9,789     +  99.65

**4a.**   6000   **b.**   9103   **c.**  $447.03
   − 2534     − 894     − 195.80

**5a.**   125,704   **b.**   756,183   **c.**  $375.89
   306,199     19,975     46.50
  + 511,111   + 103,078   +  97.28

Estimate. Use front-end estimation.

**6a.**  $74.20   **b.**   2841   **c.**  $946.21
  + 63.81     − 1607     − 370.88

**7a.** 3627 + 9720 + 2156 + 829

  **b.** $947.27 + $635.12 + $47.38

Estimate. Use rounding.

  **8a.** 4732 + 649 + 7893    **b.** 3749 − 2314

Align. Then add or subtract.

  **9a.** 4307 + 75,857 + 212    **b.** 8006 − 3179

PROBLEM SOLVING

**10.** Kyle bought a fishing rod for $18.75, a reel for $27.50, lures for $9.25, and bait for $3.88. How much did he spend?

**11.** A toll machine counted 37,894 cars and 9198 trucks crossing a bridge. How many more cars crossed the bridge?

**12.** Find the difference of $703.07 and $116.98.

**13.** The sum is 97,000. One addend is 42,809. What is the other addend?

**14.** Claire saw this Roman numeral on the court house: MDCCCLXXIX. Write the number in standard form.

## Practice 2-1

Find the missing factor.

**1a.** $7 \times \underline{\ ?\ } = 28$   **b.** $\underline{\ ?\ } \times 4 = 36$

**2a.** $8 \times \underline{\ ?\ } = 56$   **b.** $\underline{\ ?\ } \times 6 = 48$

Name the property of multiplication used.

**3a.** $8 \times 1 = 8$   **b.** $2 \times 6 = 6 \times 2$

**4a.** $5 \times 0 = 0$   **b.** $(3 \times 2) \times 5 = 3 \times (2 \times 5)$

**5a.** $1 \times 6 = 6$   **b.** $3 \times 9 = 9 \times 3$

Find the products.

**6a.** $8 \times 4$   **b.** $3 \times 9$   **c.** $6 \times 5$
$8 \times 40$   $3 \times 90$   $6 \times 50$
$8 \times 400$   $3 \times 900$   $6 \times 500$

Estimate. Then multiply.

**7a.**  10,074   **b.**  9827   **c.**  $14.07
$\times\ \ \ \ \ \ 6$   $\times\ \ \ \ \ 31$   $\times\ \ \ \ \ \ 88$

Multiply.

**8a.**  204   **b.**  375   **c.**  $50.36
$\times\ \ \ 93$   $\times\ \ \ 46$   $\times\ \ \ \ \ \ \ 70$

### PROBLEM SOLVING

**9.** Find the product if the factors are 3807 and 49.

**10.** Each of the 6 parking levels holds 109 cars. What is the total capacity of the parking garage?

**11.** Sharon bought 7 paperback books. Each cost $3.95. How much did she spend?

**12.** About 480 people visit the science museum each day. Estimate how many people visit in a month.

**13.** A jet travels 525 mi an hour. How far can the jet travel in 13 hours?

**14.** A factory produces 1360 boxes in an hour. How many boxes does it make in 12 hours?

## Practice 2-2

Multiply.

**1a.** $6 \times 42,003$   **b.** $37 \times 7018$

**2a.** $473 \times 3219$   **b.** $78 \times \$40.98$

**3a.** $945 \times \$30.88$   **b.** $500 \times 7873$

Estimate. Then multiply.

**4a.**  $11.82   **b.**  $34.03   **c.**  $90.91
$\times\ \ \ \ \ 647$   $\times\ \ \ \ \ 608$   $\times\ \ \ \ \ 356$

**5a.**  7583   **b.**  6108   **c.**  3315
$\times\ \ \ 209$   $\times\ \ \ 978$   $\times\ \ \ 462$

**6a.**  8848   **b.**  2056   **c.**  7902
$\times\ \ \ 729$   $\times\ \ \ 943$   $\times\ \ \ 574$

Find the product.

**7a.**  349   **b.**  3946   **c.**  $34.77
$\times\ 800$   $\times\ 700$   $\times\ \ \ 300$

### PROBLEM SOLVING

**8.** A mason needs 376 bricks to make a fireplace. How many bricks does she need to make 25 fireplaces?

**9.** At a sale, Leslie sold 2000 stickers for $0.25 each. How much money did she collect?

**10.** A ticket agent sold 458 tickets at $16.75 each. How much money did she collect?

**11.** The factors are 3905 and 748. Find the product.

**12.** Marty's heart beats 72 times in one minute. At this rate, how many times will Marty's heart beat in an hour?

**13.** What is the total cost of 394 hats that cost $7.49 each?

**14.** Write a two-digit number and a four-digit number that have a product of 810,000.

## Practice 3-1

Write four related facts using the given numbers.

**1a.** 7, 9, 63    **b.** 4, 9, 36    **c.** 3, 8, 24

Find the quotients.

**2a.**     56 ÷ 7    **b.**    72 ÷ 8
        560 ÷ 7            720 ÷ 80
       5600 ÷ 7           7200 ÷ 800
      56,000 ÷ 7         72,000 ÷ 8000

Estimate the quotient.

**3a.** 2435 ÷ 6    **b.** 8251 ÷ 9    **c.** 5516 ÷ 7

**4a.** 8230 ÷ 19    **b.** 4986 ÷ 23    **c.** 8937 ÷ 34

**5a.** 57,178 ÷ 29        **b.** 78,359 ÷ 42

Divide and check.

**6a.** 7)4963    **b.** 6)7958    **c.** 8)95,104

**7a.** 3)217,916    **b.** 5)372,135    **c.** 4)257,689

**8a.** 6)$10.20    **b.** 9)$79.38    **c.** 3)$156.09

PROBLEM SOLVING

**9.** Ron has saved 1425 pennies. If he divides them equally into 5 piles, how many pennies will go into each pile?

**10.** A store made $9876 in 3 weeks. Find the average amount of money the store made each week.

**11.** One hundred nineteen books are packed in 7 boxes. If the same number of books are packed in each box, how many books are in each box?

**12.** A gift costs $38.00. If 5 friends share the cost equally, how much will each person pay?

**13.** How many nickels are in $17.25?

## Practice 3-2

Divide and check.

**1a.** 40)160    **b.** 50)2500    **c.** 30)90,000

**2a.** 17)399    **b.** 36)780    **c.** 25)906

**3a.** 51)3488    **b.** 82)9486    **c.** 46)7700

**4a.** 62)$45.88        **b.** 13)$44.33

**5a.** 78)69,408        **b.** 46)$175.72

**6a.** 31)624,516        **b.** 16)963,008

Write whether each number is divisible by 2, 3, 4, 5, 6, 9, and/or 10.

**7a.** 1800    **b.** 32,508    **c.** 602,535

Compute. Use the order of operations.

**8a.** 52 + 6 × 7 ÷ 3    **b.** 12 − 8 ÷ 4 + (7 − 3) × 5

**9a.** 8 × 3 − 21 ÷ 7    **b.** (3 × 9) − 8 + (48 ÷ 6)

PROBLEM SOLVING

**10.** Ms. Cooper has 182 markers. If she has 14 students in her art club, what is the greatest number of markers each student can have?

**11.** Fifty-two ticket agents sold 16,640 tickets. If each agent sold the same number of tickets, how many tickets did each sell?

**12.** Elena has 1372 stamps. She has 96 pages in her stamp album. How many stamps can go on each page? How many stamps will be left over?

**13.** Jed consumed 2680 calories yesterday. If he ate an equal number of calories in 3 meals, estimate the number of calories per meal.

**14.** Estimate to compare the quotient of 9158 divided by 38 with the quotient of 10,148 divided by 43.

**15.** How many quarters are in $70.75?

## Practice 4-1

Write whether each is a prime or composite number.

**1a.** 59      **b.** 121      **c.** 309

Find the missing term.

**2a.** $\frac{2}{5} = \frac{?}{10}$      **b.** $\frac{6}{7} = \frac{30}{?}$

**3a.** $\frac{10}{13} = \frac{30}{?} = \frac{?}{65}$      **b.** $\frac{3}{4} = \frac{?}{24} = \frac{27}{?}$

Find the greatest common factor (GCF) for each set of numbers.

**4a.** 6 and 12      **b.** 8, 12, and 32

Write each fraction in lowest terms.

**5a.** $\frac{15}{27}$      **b.** $\frac{24}{36}$      **c.** $\frac{35}{49}$

**6a.** $\frac{18}{48}$      **b.** $\frac{20}{28}$      **c.** $\frac{49}{63}$

Find all the factors of:

**7a.** 40      **b.** 308      **c.** 246

Find the least common denominator (LCD) of each set of fractions.

**8a.** $\frac{3}{5}, \frac{2}{3}$      **b.** $\frac{1}{6}, \frac{3}{4}$, and $\frac{5}{8}$

### PROBLEM SOLVING

9. Use a factor tree to find the prime factorization of 28.

10. Mario has seen 5 of the 8 films at the multiplex. What fractional part of the films has he not yet seen?

11. Liz painted $\frac{3}{12}$ of her design blue and $\frac{2}{8}$ of it red. Did she paint the same amount in each color? Explain your answer.

12. Write $\frac{4}{5}$ as an equivalent fraction with a denominator of 20.

13. Seven tenths is equivalent to how many fortieths?

14. Which fraction is closer to $\frac{1}{2}$: $\frac{5}{6}$, $\frac{6}{13}$, or $\frac{3}{9}$?

15. What number is a common factor of every set of numbers? Why?

## Practice 4-2

Round to the nearest whole number.

**1a.** $3\frac{7}{8}$      **b.** $4\frac{1}{5}$      **c.** $9\frac{3}{7}$

Write each as a whole number or mixed number in simplest form.

**2a.** $\frac{13}{3}$      **b.** $\frac{35}{8}$      **c.** $\frac{49}{7}$

**3a.** $\frac{80}{11}$      **b.** $\frac{29}{2}$      **c.** $\frac{63}{8}$

Compare. Write $<$, $=$, or $>$.

**4a.** $\frac{5}{8}$ __?__ $\frac{1}{8}$      **b.** $3\frac{2}{5}$ __?__ $3\frac{4}{5}$

**5a.** $\frac{3}{5}$ __?__ $\frac{3}{7}$      **b.** $2\frac{1}{2}$ __?__ $2\frac{3}{6}$

Order from least to greatest.

**6a.** $\frac{1}{2}, \frac{3}{12}, \frac{1}{3}$      **b.** $\frac{2}{3}, \frac{7}{8}, \frac{1}{6}$

### PROBLEM SOLVING

7. Peter cut a loaf of bread into 6 equal parts. He ate 2 of these parts. Write a fraction for the parts he did not eat.

8. Thad has read $\frac{5}{8}$ of the book. Mia has read $\frac{3}{4}$ of the same book. Who has read less?

9. Jenna has sanded $3\frac{1}{3}$ boards. Hal has sanded $3\frac{1}{2}$ boards. Who has done more sanding?

10. Rico picked 16 lb of peaches and shared them equally with 6 friends. Write a mixed number to show how many pounds of peaches each person received.

11. A film lasted $1\frac{7}{8}$ hours. About how many hours long was the film?

12. How many half-dollar coins are in three and a half dollars?

## Practice 5-1

Add. Write each sum in simplest form.

**1a.** $\frac{7}{15} + \frac{8}{15}$  **b.** $\frac{7}{8} + \frac{5}{8}$  **c.** $\frac{8}{9} + \frac{5}{9}$

**2a.** $\frac{1}{4} + \frac{1}{3}$  **b.** $\frac{3}{5} + \frac{3}{10}$  **c.** $\frac{1}{6} + \frac{1}{2}$

**3a.** $5\frac{1}{2} + 3\frac{1}{4}$  **b.** $2\frac{1}{6} + 3\frac{1}{2}$

**4a.** $\frac{1}{2} + \frac{1}{4} + \frac{1}{3}$  **b.** $3 + \frac{1}{5} + 1\frac{7}{10}$

Subtract. Write each difference in simplest form.

**5a.** $\frac{5}{8} - \frac{3}{8}$  **b.** $\frac{11}{6} - \frac{5}{6}$  **c.** $\frac{19}{10} - \frac{7}{10}$

**6a.** $\frac{7}{8} - \frac{3}{4}$  **b.** $\frac{7}{10} - \frac{2}{5}$  **c.** $\frac{11}{12} - \frac{3}{4}$

**7a.** $3\frac{3}{4} - 1\frac{1}{2}$  **b.** $4\frac{6}{7} - 2$

**8a.** $5\frac{1}{2} - 1\frac{1}{5}$  **b.** $8\frac{7}{9} - 5\frac{1}{3}$

Add or subtract. Write each answer in simplest form.

**9a.** $\frac{7}{8} + \frac{1}{2} + \frac{3}{4}$  **b.** $10\frac{3}{4} - 4\frac{1}{3}$

PROBLEM SOLVING

**10.** Steve weighs $67\frac{1}{4}$ lb. Mark weighs $\frac{3}{4}$ lb more. Find Mark's weight.

**11.** Rachel sang for $1\frac{1}{3}$ h and danced for $\frac{3}{4}$ h. How much longer did she sing?

**12.** On three hikes, Andrew walked $6\frac{1}{8}$ mi, $7\frac{1}{4}$ mi, and $12\frac{1}{2}$ mi. How far did Andrew hike altogether?

**13.** The sum of two fractions is $\frac{11}{16}$. One fraction is $\frac{3}{8}$. What is the other?

**14.** Liza has $\frac{4}{5}$ yd of ribbon. If she cuts off $\frac{3}{10}$ yd, how much ribbon does she have left?

**15.** Jacob needs 8 pounds of apples. If he has already picked $3\frac{5}{8}$ lb, how many more pounds of apples must he pick?

## Practice 5-2

Add. Write each sum in simplest form.

**1a.** $\frac{9}{12} + \frac{1}{5}$  **b.** $\frac{7}{20} + \frac{3}{8}$  **c.** $\frac{1}{7} + \frac{3}{4}$

**2a.** $\frac{3}{4} + \frac{5}{6}$  **b.** $\frac{2}{3} + \frac{6}{7}$  **c.** $\frac{4}{9} + \frac{3}{7}$

**3a.** $1\frac{5}{9} + 1\frac{3}{4}$  **b.** $10\frac{1}{3} + 4\frac{7}{8}$

**4a.** $6\frac{1}{8} + 8\frac{5}{6}$  **b.** $9\frac{1}{4} + 3\frac{2}{3} + 2\frac{2}{5}$

Subtract. Write each difference in simplest form.

**5a.** $\frac{4}{5} - \frac{2}{3}$  **b.** $\frac{8}{9} - \frac{3}{5}$  **c.** $\frac{5}{6} - \frac{2}{7}$

**6a.** $\frac{11}{12} - \frac{5}{8}$  **b.** $\frac{13}{15} - \frac{1}{6}$  **c.** $\frac{4}{7} - \frac{1}{5}$

**7a.** $3\frac{3}{5} - 1\frac{1}{4}$  **b.** $5\frac{7}{8} - 1\frac{2}{3}$

**8a.** $10 - 3\frac{2}{3}$  **b.** $9\frac{1}{4} - 5\frac{4}{5}$

**9a.** $3 - 1\frac{9}{10}$  **b.** $8\frac{2}{3} - 7\frac{9}{10}$

Estimate. Use front-end estimation.

**10a.** $6\frac{5}{6} + 4\frac{1}{2}$  **b.** $8\frac{1}{6} - 3\frac{7}{8}$

**11.** $13\frac{4}{5} + 9\frac{1}{6} + 7\frac{9}{10}$

PROBLEM SOLVING

**12.** Jeanne is $10\frac{1}{2}$ years old. Her brother Jake is $6\frac{3}{4}$ years old. How much older is Jeanne?

**13.** Maria rode her bike $2\frac{1}{3}$ mi to the store and then another $1\frac{4}{5}$ mi to the library. How far did she ride in all?

**14.** Ellen ordered 6 pizzas for a party. Guests ate $4\frac{7}{8}$ pizzas. How much pizza was left over?

**15.** The theater is showing a double feature. One movie lasts $1\frac{7}{8}$ h. The second movie lasts $2\frac{1}{4}$ h. Estimate the total length of the double feature.

## Practice 6-1

Rename each as an improper fraction.

**1a.** $3\frac{1}{4}$     **b.** $7\frac{2}{5}$     **c.** $6\frac{9}{10}$

Write the reciprocal of each number.

**2a.** $5$     **b.** $3\frac{1}{2}$     **c.** $2\frac{1}{4}$

Draw a diagram to show each product.
Then write a multiplication sentence.

**3a.** $\frac{1}{2} \times \frac{3}{4}$    **b.** $\frac{1}{3} \times \frac{3}{5}$    **c.** $\frac{2}{5} \times \frac{5}{6}$

Multiply.

**4a.** $\frac{3}{4} \times \frac{7}{10}$    **b.** $\frac{5}{8} \times \frac{3}{4}$    **c.** $\frac{1}{8} \times \frac{5}{9}$

**5a.** $4 \times \frac{5}{6}$    **b.** $3 \times \frac{2}{3}$    **c.** $\frac{4}{5} \times 9$

**6a.** $\frac{1}{3} \times 3$    **b.** $\frac{6}{7} \times \frac{9}{8}$    **c.** $8 \times \frac{3}{5}$

Use fraction strips or circles to model each
quotient. Then write a division sentence.

**7.** $4 \div \frac{1}{5}$    **b.** $3 \div \frac{3}{4}$    **c.** $\frac{3}{4} \div \frac{1}{8}$

Divide.

**8a.** $\frac{3}{4} \div 4$    **b.** $\frac{5}{8} \div 10$    **c.** $\frac{3}{4} \div \frac{1}{2}$

**9a.** $\frac{4}{9} \div \frac{3}{5}$    **b.** $\frac{7}{15} \div \frac{3}{5}$    **c.** $\frac{3}{4} \div \frac{5}{8}$

**10a.** $6 \div \frac{2}{3}$    **b.** $5 \div \frac{10}{13}$    **c.** $9 \div \frac{3}{7}$

PROBLEM SOLVING

**11.** In a class of 28 students, $\frac{1}{7}$ wear glasses.
How many students wear glasses?

**12.** Evan swam $\frac{7}{8}$ mi. He broke up the swim into
$\frac{1}{12}$ -mi laps. How many laps did he swim?

**13.** James grew $\frac{2}{3}$ in. each month for the last
five months. How much has he grown?

**14.** Six friends share $\frac{3}{4}$ lb of chocolates. How
much chocolate does each get?

## Practice 6-2

Rename each as an improper fraction.

**1a.** $1\frac{7}{10}$     **b.** $8\frac{11}{12}$     **c.** $9\frac{3}{7}$

Write the reciprocal of each number.

**2a.** $2$     **b.** $\frac{14}{9}$     **c.** $3\frac{8}{11}$

Multiply.

**3a.** $\frac{3}{5} \times 5\frac{1}{3}$      **b.** $\frac{8}{9} \times 4\frac{1}{2}$

**4a.** $7 \times 3\frac{1}{4}$      **b.** $8\frac{2}{3} \times 5$

**5a.** $5\frac{2}{3} \times 4\frac{1}{9}$      **b.** $2\frac{1}{2} \times 6\frac{5}{6}$

**6a.** $9 \times 3\frac{4}{5}$      **b.** $6\frac{1}{3} \times 3\frac{1}{6}$

Divide.

**7a.** $3\frac{1}{3} \div 10$      **b.** $5\frac{2}{5} \div 9$

**8a.** $2\frac{1}{4} \div 3$      **b.** $5 \div 3\frac{3}{4}$

**9a.** $4\frac{1}{5} \div 2\frac{1}{3}$      **b.** $5\frac{5}{6} \div 1\frac{2}{3}$

Estimate.

**10a.** $12\frac{2}{9} \times 3\frac{1}{5}$      **b.** $5\frac{3}{4} \times 6\frac{1}{2}$

**11a.** $2\frac{1}{4} \times 2\frac{7}{8}$      **b.** $8\frac{1}{5} \times 6\frac{2}{3}$

PROBLEM SOLVING

**12.** Eli has $16\frac{1}{2}$ lb of nuts. How many $\frac{11}{12}$ -lb
bags can he fill?

**13.** Katy packed $10\frac{1}{2}$ gal of ice cream into
$1\frac{3}{4}$ -gal cartons. How many cartons
did she fill?

**14.** Lisa ran $2\frac{1}{2}$ times farther than Dana. If Dana
ran $\frac{7}{8}$ mi, how far did Lisa run?

**15.** Karen lives 3 miles from school. Her teacher
lives $3\frac{3}{4}$ times that distance. About how far
from school does the teacher live?

## Practice 7-1

PROBLEM SOLVING

Use the spinner to find the probability of each event.

**1a.** $P$ (even)  **b.** $P$ (<10)

**2a.** $P$ (5 or 10)  **b.** $P$ (8)

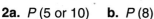

Use the circle graph to solve problems 3–4.

**3.** How many art projects are on display?

**4.** What fraction of the projects is:
   **a.** drawings?
   **b.** clay?
   **c.** paintings

**Art Projects**

Clay, Drawings, Paper Mâché, Puppets, Paintings — 3, 6, 3, 2, 10

Draw a tree diagram and list all possible outcomes.

**5.** Spin a spinner with 3 equal sections marked *A, B, C,* and pick a marble without looking from a bag containing 2 red marbles and 2 green marbles.

**6.** Make a tree diagram to find the probability of rolling a 5 on a cube numbered 1–6 *and* tossing a penny to land on tails.

The table gives class sizes at Nora's school. Use it to solve problems 7–8.

| Class Size | | | | | |
|---|---|---|---|---|---|
| Class | 5A | 5B | 5C | 5D | 5E |
| Number of Students | 32 | 29 | 34 | 32 | 33 |

**7.** Find the range, mean, median, and mode of the class sizes.

**8.** Suppose each class gets one new student. Which would *not* change: range, mean, median, mode? Explain your answer.

**9.** Tom scored 90, 95, 92, and 94 on four tests. After the fifth test the mode of his scores was 92. What did he score on the fifth test?

## Practice 7-2

PROBLEM SOLVING

Copy and complete the frequency table.

| Trees Seen on Hike | | |
|---|---|---|
| **Tree** | **Tally** | **Total** |
| **1.** Hickory | ЖЖ ЖЖ ЖЖ /// | ? |
| **2.** Aspen | ЖЖ ЖЖ // | ? |
| **3.** Maple | ? | 21 |
| **4.** Birch | ? | 29 |

The table shows Andre's pulse rate during a long bike ride. Use it to solve problems 5–8.

| Andre's Pulse Rate | | | | | |
|---|---|---|---|---|---|
| Time | 2:00 | 2:15 | 2:30 | 2:45 | 3:00 |
| Pulse | 72 | 108 | 120 | 96 | 88 |

**5.** Find the range, mean, median, and mode for Andre's pulse rate during his ride.

**6.** Make a line graph to show Andre's pulse rate.

**7.** When was Andre's pulse the fastest?

**8.** What was Andre's pulse at 2:45?

**9.** Make a table to show the following data: A scientist has collected 75 ladybugs, 55 mealworms, 38 centipedes, 40 spiders, and 24 crickets.

**10.** Diane's test scores for the first grading period are; 81, 82, 76, 95, 88, 83, 85, 84, 83, and 93. Draw a line plot for Diane's test scores. Then find the range and mode.

**11.** A bag contains 4 red marbles, 2 green marbles, 6 blue marbles, 3 black marbles, and 1 yellow marble. What is the probability of picking a green *or* a black marble? *not* a blue marble?

**12.** What is the probability of picking 1 blue marble from a bag of 15 green marbles?

**13.** What is the probabilty of picking 1 red marble from a bag of 20 red marbles?

Which type of graph would you use to show:

**14a.** increases or decreases in sales from 1 week to the next?

   **b.** how the sales for each week compare with sales for other weeks?

   **c.** what part of the sales for the month was made during each of the weeks.

## Practice 8-1

Classify each angle. Name its vertex and sides.

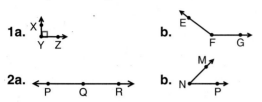

**1a.** X↑ Y Z

**b.** E F G

**2a.** P Q R

**b.** N M P

Are the lines perpendicular? Write *Yes* or *No*.
Use a protractor to check your answers.

**3a.**        **b.**        **c.**

Name each polygon.

**4a.**         **b.**

Classify each quadrilateral.

**5a.**        **b.**

**6a.**        **b.**

### PROBLEM SOLVING

**7.** Draw an isosceles triangle that has a
right angle.

**8.** How would you classify a triangle whose
sides measures 8 m, 8 m, and 8 m?

**9.** Draw a triangle with exactly two congruent
angles. What type of triangle is it?

**10.** A triangle has an obtuse angle and two sides
that are congruent. Is each of the congruent
sides longer or shorter than the third side?

**11.** Explain why triangle *MNO* and triangle *RLP*
are *not* similar.

**12.** Draw two congruent rectangles. How
do you know they are congruent?

**13.** Draw a right triangle. Then draw another
right triangle that is similar to it.

---

## Practice 8-2

Find the perimeter of each polygon.

**1a.** 7 m / 10 m / 3 m / 6 m / 4 m

**b.** 14 ft / 32 ft

Estimate the circumference of each circle.

**2a.** 9 yd

**b.** 45 in.

Is the dotted line a line of symmetry?

**3a.**        **b.**

**4a.**        **b.**
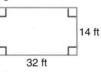

Write *slide*, *flip*, or *turn* to indicate how
each figure was moved.

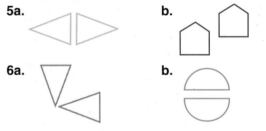

**5a.**        **b.**

**6a.**        **b.**

### PROBLEM SOLVING

**7.** Find the perimeter of a square that measures
35 cm on a side.

**8.** Find the perimeter of a regular hexagon
with a side of 9 m.

**9.** Sue is knitting a baby blanket that is a
rectangle 100 cm by 140 cm. How much
ribbon will she need to trim the edge?

**10.** Estimate the circumference of a circular
clock whose radius is 8 inches.

## Practice 9-1

Write the letter of the best estimate.

**1.** A bed might be 76 _?_ long.
   **a.** ft       **b.** yd       **c.** in.

**2.** A brick might weigh 3 _?_ .
   **a.** lb       **b.** oz       **c.** T

**3.** A coffee pot might hold 2 _?_ .
   **a.** gal      **b.** pt       **c.** qt

**4.** The temperature during a snow
   storm might be _?_ .
   **a.** 20°F    **b.** 40°F    **c.** 60°F

Compare. Write <, =, or >.

**5a.** 6 lb _?_ 86 oz      **b.** 250 min _?_ 4 h

**6a.** 4 gal _?_ 20 qt      **b.** 5 yd _?_ 180 in.

### PROBLEM SOLVING

**7.** Lois bought a bag of ice cubes to keep the punch cold. Would the bag of ice weigh 10 oz or 10 lb?

**9.** Rob estimated the distance he had to walk from the school to his house as 1.2 yd. Would this be a reasonable estimate? Why or why not?

**8.** Moira needs 2 pt of honey for a recipe. She has 3 c of honey. Does she have enough? Explain.

**9.** Ben knitted a scarf that was 70 in. long. Was it more or less than 6 ft long? How much more or less?

**10.** The thermometer says 32°C. Should Sally wear a parka or shorts?

**11.** One moving van holds 1800 lb. Another van holds 1 T. Which holds more?

## Practice 9-2

Copy and complete the chart.

| | Time Zone | Time | | | |
|---|---|---|---|---|---|
| **1.** | Pacific | 4:10 A.M. | ? | ? | ? |
| **2.** | Mountain | ? | 11:00 P.M. | ? | ? |
| **3.** | Central | ? | ? | 1:15 A.M. | ? |
| **4.** | Eastern | ? | ? | ? | 2:00 P.M. |

Add or subtract.

**5a.**   3 d 17 h      **b.**   4 ft 9 in.
    + 2 d 15 h         + 3 ft 7 in.

**6a.**   5 qt 1 c       **b.**   7 T 380 lb
   − 3 qt 3 c         − 3 T 900 lb

**7a.**   2 wk 6 d     **b.**   3 y
    + 7 wk 5 d       − 1 y 7 mo

**8a.**   9 yd 27 in.    **b.**   10 lb 5 oz
    + 3 yd 30 in.      − 5 lb 6 oz

**9a.**   3 gal 3 qt     **b.**   6 y
    + 2 gal 1 qt     − 3 y 280 d

### PROBLEM SOLVING

**10.** Amy has two scarves. One is 5 ft long and the other is 7 ft long. How many yards are there in the combined length of both scarves?

**11.** Amos cut 2 yd 2 ft from a board that was 4 yd long. How long is the remaining piece of board?

**12.** How much more than a gallon is 7 quarts?

**13.** A punch recipe calls for 1 pt grape juice, 1 qt pineapple juice, 1 gal lemonade, and 3 c orange juice. Find the total quantity of punch this recipe makes.

**14.** Three railroad cars measure 19 ft 8 in., 21 ft 3 in., and 20 ft 10 in. Find their total length.

**15.** The carnival began at 11:15 A.M. and ended at 10:45 P.M. How long did it last?

**16.** Karla bought 1lb of cheese. If the cheese cost $1.75 for 8 oz, how much did Karla pay?

## Practice 10-1

Write the place of the underlined digit.
Then write its value.

**1a.** 4<u>9</u>.6    **b.** 0.34<u>8</u>    **c.** 12.6<u>7</u>2

Write each decimal in expanded form.

**2a.** 367.04    **b.** 70.163    **c.** 6.45

Estimate by both rounding and front-end estimation. Between what two numbers will the exact sum or difference be?

**3a.**    0.77     **b.**    3.54     **c.**    0.923
     + 0.586          9.078          − 0.68
                    + 5.166

Estimate. Then add or subtract.

**4a.**    0.473     **b.**    36.3     **c.**    17.004
     + 0.96          + 43.5          + 12.059

**5a.**    0.75     **b.**    1.6     **c.**    17.439
     − 0.2          − 0.74          − 8.8

**6a.** 94.637 + 17.08 + 24.3    **b.** 12 − 7.84

### PROBLEM SOLVING

7. Write the decimal that has seven thousandths, nine tenths, and six ones.

8. Marc rode his bike 4.35 km from home to the park. Then he rode along the park and back home again, a distance of 16.9 km. About how far did he ride?

9. What is 74.16 increased by 9.056?

10. Snow accumulation in March was 1.26 in., 3.75 in., and 2.049 in. Find the total snowfall in March.

11. A board is 36.37 cm long. If Richard cuts off 9.5 cm from it, how much of the board is left?

12. Janis spent $7.99 on invitations, $3.79 on balloons, and $4.75 on streamers for a party. How much change did she get back from a $20 bill?

13. Eleni measured two books. One was 22 mm thick. The other was 18.25 mm thick. How much thicker was the first book?

## Practice 11-1

Find the missing number.

**1a.** <u> ? </u> × 3.7 = 370    **b.** 1000 × <u> ? </u> = 324

**2a.** 42.6 ÷ <u> ? </u> = 4.26    **b.** <u> ? </u> ÷ 1000 = 0.007

Multiply.

**3a.** 7 × 0.65    **b.** 2.7 × 0.8    **c.** 0.16 × 0.9

**4a.** 3.2 × 0.7    **b.** 0.63 × 0.3    **c.** 7 × 0.32

**5a.** 0.6 × 3.74    **b.** 4.3 × 6.92    **c.** 0.08 × 11.5

Divide and check.

**6a.** 0.374 ÷ 2    **b.** 0.3 ÷ 6    **c.** 1.6 ÷ 8

**7a.** 0.64 ÷ 8    **b.** 5.39 ÷ 5    **c.** 1.308 ÷ 6

**8a.** 2.4 ÷ 2    **b.** 0.92 ÷ 4    **c.** 0.744 ÷ 6

### PROBLEM SOLVING

9. Estimate the product of 2.287 and 6.9. Is the actual product greater or less than the estimated product?

10. Is the estimated product of 13.608 and 0.62 greater or less than the exact product?

11. Estimate the quotient of 47.32 and 6 using compatible numbers.

12. The school year has 180 days. If 0.05 of them are missed due to bad weather, how many days are missed?

13. Burritos are $2.79 each. How much do 100 burritos cost?

14. Evan spent $74.33 for 3 video games. Estimate the cost of each game.

15. Liam picked 64.3 pounds of fruit. Three tenths of the fruit were pears. How many pounds of pears did Liam pick?

## Practice 12-1

Write the letter of the best estimate.

**1.** A tree might be __?__ tall.
   **a.** 4m     **b.** 4 cm     **c.** 4 km

**2.** A thumbtack might have a mass of __?__ .
   **a.** 3 g     **b.** 3 mg     **c.** 3 kg

**3.** A medicine dropper might hold __?__ .
   **a.** 5 mL     **b.** 50 L     **c.** 50 mL

**4.** A refrigerator might be __?__ wide.
   **a.** 1 m     **b.** 1 cm     **c.** 1 dm

**5.** A bear might have a mass of __?__ .
   **a.** 4 kg     **b.** 400 g     **c.** 400 kg

Compare. Write $<$, $=$, or $>$.

**6a.** 5 dm __?__ 0.5 m     **b.** 2 kg __?__ 2100 g

**7a.** 870 mL __?__ 8.7 L     **b.** 3.1 km __?__ 310 cm

PROBLEM SOLVING

**8.** A snake measures 89.4 cm. Is this more or less than 1 meter?

**9.** Which holds more: a pitcher whose capacity is 1.5 L or 150 mL?

**10.** Alena bought two bags of nuts. Each weighs 600 g. Will the nuts fit into a box that holds 1 kg of nuts? Explain.

**11.** Sean runs a 1500-m race. Does he finish the race if he runs 1km 500 m? Why or why not?

**12.** A recipe suggests serving 250 g of meat for each person. How many kilograms of meat should Lena buy if she is serving 6 people for dinner?

**13.** David is using a glass that holds 250 mL to fill a 4.5 L fishbowl. How many full glassses will he need to fill the fishbowl?

## Practice 12-2

Estimate the area of each figure.

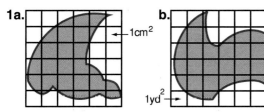

**1a.**     ←1cm$^2$     **b.**     1yd$^2$→

Find the area of each figure.

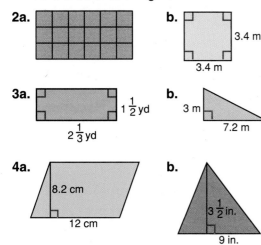

**2a.**     **b.** 3.4 m, 3.4 m

**3a.** $1\frac{1}{2}$ yd, $2\frac{1}{3}$ yd     **b.** 3 m, 7.2 m

**4a.** 8.2 cm, 12 cm     **b.** $3\frac{1}{2}$ in., 9 in.

Find the volume of each figure.

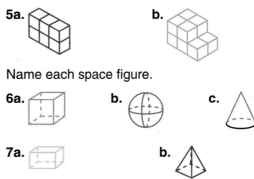

**5a.**     **b.**

Name each space figure.

**6a.**     **b.**     **c.**

**7a.**     **b.**

PROBLEM SOLVING

**8.** Mr. Ruiz bought a carpet that cost $15 a square yard. How much did he pay to cover a floor 5 yd by 4 yd?

**9.** Find the volume of a box that is 3 dm wide, 5 dm deep, and 6 dm high.

**10.** How many cubic centimeters will 7.9 grams of water fill?

**11.** An aquarium measures 3 ft long, 2.5 ft wide, and 3.5 ft high. Is the volume of the aquarium more or less than an aquarium with a volume of 1 yd$^3$? Explain your answer.

## Practice 13-1

Write the ratio of the number of:

☆ ☆ ☆ ☆
○ ○ ○
▽ ▽

**1a.** circles to stars  **b.** triangles to stars

**2a.** circles to triangles  **b.** triangles to circles

Which are proportions? Write = or ≠.

**3a.** $\dfrac{5}{6}$ _?_ $\dfrac{11}{12}$  **b.** $\dfrac{13}{4}$ _?_ $\dfrac{39}{12}$

Find the missing term in each proportion.

**4a.** $\dfrac{4}{5} = \dfrac{n}{20}$  **b.** $\dfrac{7}{8} = \dfrac{49}{n}$  **c.** $\dfrac{5}{n} = \dfrac{25}{40}$

Write as a percent.

**5a.** $\dfrac{39}{100}$  **b.** $\dfrac{78}{100}$  **c.** $\dfrac{9}{100}$

**6a.** 0.46  **b.** 0.7  **c.** 0.05

Write as a fraction in simplest form.

**7a.** 60%  **b.** 85%  **c.** 5%

Write as a decimal.

**8a.** 35%  **b.** 6%  **c.** 10%

PROBLEM SOLVING

9. On a map, 1 cm represents 12 km. What does 5 cm represent?

10. A basement playroom is 10.5 m long. Using the scale of 2 cm = 3 m, what is the length of the playroom in a scale drawing?

11. Seventy-five percent of registered voters cast ballots in the election. If there were 4000 registered voters, how many voted?

12. Tapes that usually cost $8 each are on sale for 30% off. What is the price for a tape on sale?

13. Hank paid $3.50 for 2 hamburgers. How much will he pay for 8 hamburgers?

14. What percent of the letters in *CALIFORNIA* are vowels?

Chapter 14

## Practice 14-1

Label each expression as *numerical* or *algebraic*.

**1a.** $13 - 7$  **b.** $15d$  **c.** $3a + 7$

Evaluate each expression.

**2a.** $9 + c$, when $c = 5$

**b.** $1\frac{1}{2} + b$, when $b = 3\frac{3}{5}$

**c.** $30.5 - n$, when $n = 1.6$

**3a.** $y \div 4$, when $y = 88$

**b.** $\frac{1}{2}x$, when $x = 24$

**c.** $30 \div r$, when $r = \frac{1}{3}$

Solve each equation.

**4a.** $6 + p = 17$  **b.** $\frac{1}{3}z = 8$

**5a.** $h + 3.6 = 10$  **b.** $150 = 50d$

**6a.** $\frac{3}{5}k = \frac{3}{5}$  **b.** $18 = a + 8\frac{1}{2}$

**7a.** $\frac{1}{7} \cdot \frac{1}{3} = \frac{1}{3} \cdot m$  **b.** $\frac{5}{6} + c = \frac{5}{6}$

Copy and complete the table.

8.

| $y$ | 0 | 1 | 3 | 5 | 7 |
|-----|---|---|---|---|---|
| $7y$ | ? | ? | ? | ? | ? |

Name the point for each set of coordinates.

**9a.** (2, 1)

**b.** (0, ⁻2)

**c.** (⁻1, 0)

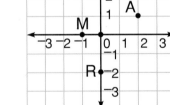

Write an addition and a subtraction expression for each number.

**10a.** 9  **b.** $\dfrac{3}{5}$  **c.** 0.35

Write a multiplication and a division expression for each number.

**11a.** 6  **b.** $\dfrac{5}{6}$  **c.** 0.12

## TEST 1

**1.** 31 + 73 + 69   **2.** 75 − 27 + 40 − 75

**3.** $\frac{1}{4}$ of 1 gal = __?__ qt   **4.** $48.95 − $22.70

**5.** 3002 − 1369   **6.** From $3 take $.08.

**7.** How much less than $20 is $17.95?

Compare. Use <, =, or >.

**8.** 3 million 6 __?__ 1 billion 2

**9.** 60,000 + 7000 __?__ 60,000 + 900

**10.** 3 and 4 hundredths __?__ 3 and 40 thousandths

**11.** John spent $.79 and had $.15 left. How much did he have to begin with?

**12.** A town's population increased by 275. This brought the population to 12,240. What was the population before the increase?

**13.** How many times greater is the digit in the tens place than the same digit in the tenths place?

**14.** What is the total number of days in September, October, and November?

**15.** A four-digit number is odd. It is divisible by 5. The first digit is 4 less than the last digit. The second digit is 4 times the first digit. The third digit is zero. What is the number?

## TEST 2

**1a.** 1428 ÷ 7   **b.** 8 × 292

**2a.** XCVII + LIII   **b.** 90 is __?__ more than 26.

**3.** From $9 take $0.15.

**4.** 78 increased by 46 is __?__ .

Complete the pattern.

**5.** 3.01, 3.0, 2.99, __?__ , __?__

**6.** 4 × (6 × 2) = __?__

**7.** 9 × (3 × 4) = __?__

**8.** $\frac{1}{10}$ is to 0.1 as $\frac{1}{1000}$ is to __?__ .

**9.** CXL is to 140 as MDI is to __?__ .

**10.** How much greater than 6 hundredths is 6 tenths?

**11.** At $0.39 a pt, what is the cost of 24 qt of milk?

**12.** Express in increasing order: 1.5, 0.3, $\frac{4}{10}$, 5 tenths.

**13.** What numbers have a product of 54 and a difference of 3?

**14.** About how much is the cost of 185 headbands at $2.79 each?

**15.** Tom spent $.75 for a ball and $1.75 for a card. What was his change from $3?

## TEST 3

**1a.** 306 × 24   **b.** $7.08 × 35

**2.** How much greater is 30 × 8000 than 3 × 800?

**3.** $9.03 + $0.85 + $0.04

**4.** $5 + $2.35 + $0.08

**5.** $83.16 ÷ 27   **6.** 8)2416

**7.** 6000 × 12 − 100

**8.** 2 + 6 × 3 − 10 ÷ 5

**9.** 12 ÷ 3 + 4 × 2

**10.** 2000 × 9 − 100

**11.** What numbers have a quotient of 9 and a sum of 70?

**12.** A school musical was attended by 250 adults and 120 children. If adults paid $1.50 and children paid $.75 for each admission, how much money was taken in?

**13.** Find the cost of 7 gal of milk at $.74 a quart.

**14.** A car was driven 575 mi last week. This was 275 mi more than it had been driven the previous week. How many miles was the car driven the previous week?

**15.** If the multiplicand is 724 and the multiplier is 608, what is the product?

489

## Challenge

Compare. Use $<$, $=$, or $>$.

**1a.** $\frac{5}{8}$ _?_ $\frac{1}{2}$   **b.** $\frac{9}{10}$ _?_ $1$   **c.** $0$ _?_ $\frac{2}{11}$

**2a.** $\frac{3}{8}$ _?_ $\frac{4}{16}$   **b.** $2\frac{1}{5}$ _?_ $2\frac{2}{10}$   **c.** $\frac{31}{8}$ _?_ $4$

Complete.

**3.** 12 is to $2 \times 2 \times 3$ as 30 is to _?_ .

**4.** $1\frac{1}{8}, 1\frac{3}{8}, 1\frac{5}{8}, 1\frac{7}{8},$ _?_

**5.** Find: **a.** GCF of 12 and 18   **b.** LCD of $\frac{2}{9}$ and $\frac{2}{3}$

**6.** Name the composite numbers between 1 and 20.

**7.** Find the sum of three eighths and one eighth.

**8.** Rename eight sevenths as a mixed number.

**9.** Rename seven eighths as sixteenths.

**10.** $\$7.08 \times 35 - \$100$

**11.** If a dozen pens cost $5.76, find the cost of a single pen.

**12.** Mr. DeMasi's sales for 3 months amounted to $2448. Find his average amount of sales for one month.

**13.** Jason worked $\frac{4}{5}$ of an hour on his homework. Juan worked $\frac{7}{10}$ of an hour on his. Who worked longer?

**14.** Six out of 24 fifth graders are on the football team. What part of the students are not football players?

**15.** How much does Mrs. Lawlor save in buying one 32-oz container of yogurt at $2.29 instead of four 8-oz containers at $.69 each?

**1.** Find the fourth term: 16,000, 4000, 1000, _?_

**2.** $\frac{3}{10} + \frac{9}{10}$   **3.** $6000 - 38$

**4.** $\$0.62 \times 86$   **5.** $2350 \div 47$

**6.** Find the tenth fraction. $\frac{2}{3}, \frac{4}{6}, \frac{6}{9}, \ldots$

Find the missing digits.

**7.**
$$\begin{array}{r} 5\,6\,\square\,7 \\ +\quad 7\,9\,\square \\ \hline 6\,\square\,9\,8 \end{array}$$

**8.**
$$\begin{array}{r} \$5\,\square.7\,\square \\ -\quad 4.2\,6 \\ \hline \$\square\,2.\square\,7 \end{array}$$

**9.** $2\frac{5}{6} + 4\frac{4}{6}$   **10.** $6\frac{1}{3} + 4\frac{3}{4}$

**11.** Larry had $0.95. He had more quarters than dimes. How many dimes did he have?

**12.** How many even three-digit numbers can Bea make using the digits 3, 4, and 5 without repeating any digit?

**13.** The temperature at 6:00 A.M. was 28°F. It rose 2° every hour until 3:00 P.M. What was the temperature at 3:00 P.M.?

**14.** Alan, Bill, and Chad have papers of three different weights: 1.2 g, 0.9 g, 1.05 g. Alan's weighs less than Bill's and Chad's weighs the most. Find the weight of each boy's paper.

**15.** How many different ways can 2 red, 2 blue, and 2 green beads be arranged on a string so no two beads of the same color are side by side?

Compare. Use $<$, $=$, or $>$.

**1.** $\frac{2}{3} \times \frac{1}{4}$ _?_ $\frac{3}{4} \times \frac{1}{2}$

**2.** $\frac{1}{6} \times \frac{1}{2}$ _?_ $\frac{3}{12} \times \frac{1}{6}$

Solve for $n$.

**3.** $\frac{3}{4}$ of $48 = n$   **4.** $9\frac{3}{7} = \frac{n}{7}$

**5.** $3\frac{3}{5} \times 1\frac{1}{9} = n$   **6.** $\frac{1}{5} \div \frac{1}{3} = n$

**7.** $3\frac{1}{2} \div \frac{1}{6} = n$

**8.** From $\frac{2}{5} + \frac{1}{4}$ take $\frac{3}{20}$.

**9.** 1000 cents $=$ _?_ dollars

**10.** Rename the number 8 as twelfths.

**11.** Nine students have a brother; seven have a sister; twelve have neither a brother nor a sister. If there are 25 in the class, how many students have both a brother and a sister?

**12.** A car travels at 55 mph. About how far will it go in $3\frac{3}{4}$ hours?

**13.** Kim used $\frac{1}{2}$ c of flour. She gave 2 c of flour to Fay and then had $3\frac{3}{4}$ c left. How many cups of flour did Kim start with?

**14.** Joan had $60. Each day, starting Monday, she spent $\frac{1}{2}$ of what she had the day before. How much money did she have left on the fourth day?

**15.** One third of what number equals eight?

**1a.** 1003 − 999 **b.** Divide 27,234 by 9.

**2a.** 479 + 963 **b.** $5.80 × 80

**3a.** $\frac{1}{5} + \frac{1}{20} + \frac{1}{10}$ **b.** $5\frac{5}{6} - 2\frac{3}{4}$

**4a.** $8\frac{5}{6} + 10\frac{1}{2}$ **b.** $\frac{8}{12} = \frac{n}{3}$

**5a.** 643 + 872 + 948 **b.** 204 × 700

**6.** How much less than $6\frac{3}{4}$ is $4\frac{2}{3}$?

**7.** How many eighths are there in $\frac{3}{4}$?

**8.** Divide 1296 by 18. **9.** From $9 take $6.35.

**10.** MCMLXV is to 1965 as MMI is to _?_ .

**11.** Justin missed $\frac{1}{8}$ of his spelling words. If 24 was the perfect score, how many did he spell correctly?

**12.** Sue tosses a coin and spins a dial marked 1, 2, 3, and 4. What is the probability of getting heads and an even number? (*Hint:* Use a tree diagram.)

**13.** In each of 5 rounds Shalika scored 16, 20, 13, 24, and 28 points. Find the range, median, and mode of Shalika's scores.

**14.** Find the cost of 5 meters of wire at $0.42 a meter.

**15.** At $1.40 a dozen, how many oranges can be bought for $0.70?

**1.** Draw 2 lines that are perpendicular. How many angles did you form?

**2.** Find the missing angle of △*ABC:*
 **a.** 45°, 45°, _?_ **b.** 65°, 70°, _?_

**3.** Given the radius, find the diameter.
 **a.** 12 cm **b.** $4\frac{3}{8}$ in. **c.** 1.07 m

**4.** How many diagonal lines of symmetry are in:
 **a.** a regular pentagon? **b.** a regular hexagon?

**5.** 840 + 30 + 78 **6.** $\frac{14}{15} ÷ \frac{14}{15}$

**7.** $2\frac{7}{10} ÷ \frac{18}{25}$ **8.** From 6 take $\frac{3}{8}$.

**9.** How much less than $6\frac{3}{4}$ is $4\frac{2}{3}$?

**10.** Write 2028 in Roman numerals.

**11.** Find the perimeter of a rectangular rug 2 yd long and 4 ft wide.

**12.** Tony has 3 pairs of slacks: brown, blue, and black. He also has 4 shirts: white, yellow, pink, and green. How many different outfits can he wear?

**13.** About how many times greater than the diameter of a circle is the circumference?

**14.** A jet averages 675 mph. In how many hours will it fly 22,950 miles?

**15.** Estimate the circumference of a table with a diameter of 5 ft.

**1a.** 136 in. = _?_ ft _?_ in.
 **b.** 2 yd 2 ft = _?_ in.

**2a.** 7 pt 1 c = _?_ c
 **b.** 77 fl oz = _?_ c _?_ fl oz

Compare. Use <, =, or >.
 **3a.** 20 lb 6 oz _?_ 236 oz
 **b.** 2 T 650 lb _?_ 5000 lb

**4a.** 56 d = _?_ wk **b.** 10 min = _?_ s

**5.** 3 lb 12 oz + 1 lb 7 oz

**6.** 5 yd 2 ft + 3 yd 2 ft

**7.** 6 h 20 min − 2 h 35 min

**8.** 12 + 3 × 5 ÷ 5 **9.** (7 × 8) − 14

**10a.** 4.9 + 6.5 **b.** 9.2 − 6.4

**11.** A stained glass ornament in the shape of a regular pentagon has a perimeter of 40 in. What is the length of one side?

**12.** At $.40 a quart, what is the cost of 3 gallons of syrup?

**13.** At $2.28 a yard, find the cost of 2 feet of terry cloth fabric.

**14.** The original temperature was 26°F. The first two hours the temperature dropped 4°F. The next two hours it increased $1\frac{1}{2}$°F. What was the final temperature?

**15.** A waitress earned $265.80 last week in salary and tips. If her tips amounted to $102.20, what was her salary?

## TEST 10

**1a.** $8 \div 2\frac{2}{5}$  **b.** $6\frac{1}{4} \times 240$

**2.** 288 eggs $=$ _?_ dozen

Compare. Use $<$, $=$, or $>$.

**3.** $9.5 - 4.062$ _?_ $7.85 - 2.104$

**4.** $6.004 + 2.003 + 0.864$ _?_ $7.062 + 1.809$

**5.** $4.398 + 6.07$ _?_ $16.09 - 3.42$

**6.** $0.6 + 0.132 + 0.25$

**7.** $14.4 + 21.89$  **8.** $0.731 - 0.209$

**9.** $3.6 \times 0.45$  **10.** $2.864 \div 4$

**11.** Dennis bought 7.5 m of felt to make a banner. If he had 4.6 m left, how much did he use?

**12.** What decimal is one hundredth more than 0.4?

**13.** Ray rode his bicycle 9.6 mi Monday, 4.8 mi Tuesday, and 6.6 mi Wednesday. What is the average distance he traveled each day?

**14.** A bakery's sales for a six-day week are: $525, $720, $625, $475, $588, and $640. It bakes about 35 doz donuts each day. Find its average sales for that week.

**15.** Miguel jogs 6.3 km each day. He swims 12 laps daily. How many kilometers does he jog in four days?

## TEST 11

Divide by 10, then multiply by 100.

**1a.** 0.06  **b.** 2.5  **c.** 0.9

Complete.

**2a.** 3 km $=$ _?_ m  **b.** 180 cm $=$ _?_ dm

**3a.** 62 m $=$ _?_ cm  **b.** 500 mL $=$ _?_ dL

**4a.** 2 dg $=$ _?_ mg  **b.** 2000 L $=$ _?_ kL

**5a.** 4000 g $=$ _?_ kg  **b.** 3 g $=$ _?_ dg

**6a.** 15 cg $=$ _?_ mg  **b.** 12 cm $=$ _?_ mm

**7.** A square pyramid is to 5 vertices as a triangular prism is to _?_ .

**8.** 60% is to $\frac{3}{5}$ as 75% is to _?_ .

Solve for $n$.

**9.** $\frac{8}{36} = \frac{n}{9}$  **10.** $\frac{9}{10} = \frac{81}{n}$

**11.** Find the volume of a cereal box that measures 14 in. by 9 in. by 3 in.

**12.** Seventy-two cakes were sold at a school fair. Each cake costs $3.75. How much was raised?

**13.** If Jan saves $24.60 a month, how much money will she save in one year?

**14.** If Ramón walks 1 km in 12.5 min, how many meters does he walk in 1 minute?

**15.** If the diameter of a circular picture frame is 40 cm, what is the approximate circumference in meters?

## TEST 12

**1.** Which is greater: $7.3 \times 2.04$ or $7.03 \times 2.4$?

Evaluate $a + 3\frac{1}{2}$ when $a$ is:

**2a.** $6\frac{1}{4}$  **b.** $12\frac{1}{2}$  **c.** $26\frac{1}{3}$

Evaluate $b - 0.06$ when $b$ is:

**3a.** 3.7  **b.** 11.03  **c.** 20.192

Evaluate $24x$ when $x$ is:

**4a.** 32  **b.** 16  **c.** 103

Evaluate $\frac{y}{6}$ when $y$ is:

**5a.** $4\frac{1}{3}$  **b.** $12\frac{1}{2}$  **c.** $16\frac{1}{4}$

**6.** Solve for $z$: $3 + 6 + 2 - 7 + z = 12$

**7.** Solve for $a$: $(4 \times 4) + (a \times 5) = 26$

**8.** Solve for $g$: $\frac{1}{g} \times 0.012 = 0.004$

**9.** What percent of a dollar is 2 quarters, 3 dimes, 1 nickel, and 4 pennies?

**10.** Which is greater: $3 \times 10^2$ or $3 \times 20$?

**11.** The sum of a number and twice the number is 21. Find the number.

**12.** On a map, the library is $1\frac{1}{4}$ cm west and $2\frac{1}{2}$ cm north of Ron's house. The scale is 1 cm $=$ 2 km. What is the actual distance from the library to Ron's house? (*Hint:* Use a grid.)

**13.** Find the area of a triangle whose height is $1\frac{1}{2}$ ft and base is 8 in.

**14.** On a grid, connect points (1, 1), (5, 1), (1, 4) to form a polygon. Find its area.

**15.** The base of a triangle is 3 more than its height. Its area is 54 sq units. Find its base and height.

## SET 1

**1.** 9 + 2    6 + 6    7 + 8    4 + 7
16 − 8    13 − 5    14 − 6    12 − 3

**2.** 7 + 0    0 + 12    8 + 5    6 + 9
0 + 7    12 + 0    5 + 8    9 + 6

**3.** 11 − 11    8 − 0    28 − 28    10 − 0
17 − 0    18 − 9    16 − 5    13 − 13

**4.** Estimate: 18 + 21 + 11    38 + 42
807 + 48    281 + 398    97 + 9

**5.** Round to the nearest dollar.
$7.26    $19.84    $148.80    $4.79

**6.** What is seven increased by two?

**7.** Eleven is two greater than what number?

**8.** If Ellen weighs 82 lb, how much must she gain to weigh 91 lb?

**9.** What is 14 increased by 7?

**10.** Estimate the sum of $9.95 + $7.45.

**11.** Dan weighs 40 kg and Terry weighs 12 kg less. What is Terry's weight?

**12.** How many more inches than 2 ft is 30 in.?

**13.** Nine equals 12 minus what number?

**14.** Zero added to 9 equals how much?

**15.** Jon has $0.31, and Ben has $0.49 more than Jon. How much does Ben have?

## SET 2

**1.** 3 + 4 + 7 + 6        9 + 1 + 2 + 6
8 + 4 + 2 + 4        5 + 5 + 4 + 6

**2.** Estimate: 53 − 38    67 − 16    41 − 27
39 − 11    22 − 9

**3.** Add 2 to: 99, 79, 12, 22, 42, 82, 102, 39, 59, 62, 92, 109

**4.** Take 2 from: 91, 71, 61, 21, 11, 111, 51, 41, 101, 31, 81

**5.** Estimate: $16.20 + $23.85
$8.07 + $24.49    $39.75 + $11.66
$42.18 + $28.06

**6.** What is five less than twenty-one?

**7.** How many ten thousands are in 1,352,896?

**8.** 469,210 = 400,000 + _?_ + 9000
+ _?_ + _?_

**9.** What is the value of 8 in 862,004?

**10.** If 245 students are enrolled and 15 are absent, how many are present?

**11.** 16 + 14 + 25 = _?_

**12.** Write the numeral: twenty thousand, five hundred two

**13.** What is 31 decreased by 8?

**14.** What is 33 increased by 8?

**15.** If Jeff has $0.87, how much does he need to make $1.00?

## SET 3

**1.** Give the value of the underlined digit.
8<u>2</u>35    <u>6</u>719    35<u>4</u>2    1<u>1</u>31

**2.** Add 1000 to: 40, 140, 240, 340, 440, 540, 640

**3.** Compare. Use < or >: 3781 _?_ 3187
13,482 _?_ 13,284    7532 _?_ 7352

**4.** Order least to greatest: 87, 81, 89; 136, 361, 316; 2460, 2640, 2046

**5.** Read: 0.7    0.68    0.003    0.1    0.259
0.99    0.06

**6.** Write the numeral: six thousand, two

**7.** In the number 60,543, what is the value of 5?

**8.** What decimal is one tenth more than 7.1?

**9.** A fish weighs 16.07 lb. Another weighs 16.7 lb. Which weighs more?

**10.** Complete the pattern.
0.01, 0.03, 0.05, _?_, _?_, _?_

**11.** Jen has $12.75 in quarters and $6.15 in nickels. About how much money does she have?

**12.** Which is less: 456,017 or 465,007?

**13.** In the number 26,908, the 2 means
2 × _?_ .

**14.** Write 3,628,405,012 in expanded form.

**15.** Write the Roman numeral for 1946.

 **SET 4**

1. Give the value of the underlined digit.
   1.<u>6</u>6, <u>7</u>.394, 35.9<u>8</u>, <u>4</u>0.136, 1<u>2</u>.41

2. Compare. Use <, =, or >: 8.89 _?_ 8.8
   2.3 _?_ 0.3   6.60 _?_ 6.6   2 _?_ 1.8

3. Round to the nearest ten thousand.
   12,365   38,114   75,489   31,777
   57,261   44,119   67,123   25,986

4. Give the standard numeral: XXXV
   CXLIII   DCCVII   MCMXCIII   MDL

5. Add 8 to: 3, 33, 93, 43, 13, 83, 53,
   63, 73, 23

6. What is the value of 4 in 3456?

7. Write the numeral: seven million,
   six hundred thousand, forty-three

8. Write LXXV as a standard numeral.

9. Write the numeral: six and four tenths

10. Which is greater: 3 tenths or 3 hundredths?

11. Write each as a decimal:
    $1\frac{7}{10}$, $3\frac{1}{10}$, $2\frac{3}{10}$

12. Eighteen is how many less than 2 dozen?

13. What is 49 increased by 3?

14. What is 22 decreased by 3?

15. If 18 cards were left in the box after Cindy
    used 4, how many were there at first?

**SET 5**

1. Multiply by 6, then add 3: 2, 0, 4, 1,
   3, 6, 5, 7, 8

2. 8 × _?_ = 72   9 × _?_ = 36
   _?_ × 7 = 28   _?_ × 6 = 66
   7 × _?_ = 49   9 × _?_ = 81

3. 3 × 2 × 4   6 × 2 × 2   3 × 0 × 8
   4 × 1 × 7   5 × 2 × 7   2 × 3 × 10

4. Multiply by 40: 3, 5, 9, 6, 4, 8, 0, 7, 1, 2

5. Multiply by 1000: 6, 12, 24, 32, 8, 16, 44,
   58, 63, 15

6. Estimate the product: 38 × 24

7. At $0.20 each, find the cost of 6 rulers.

8. One of the factors of 18 is 9. What is
   the other factor?

9. One cassette costs $9.95. Estimate the cost
   of 5.

10. Estimate the product: 425 × 29

11. How many days are in 9 weeks?

12. Estimate the cost of 6 games, if one game
    costs $8.98.

13. There are 60 books on each of 4 shelves.
    How many books are there in all?

14. The Tran family traveled 105 mi each day of
    vacation. If they traveled for 3 days, how
    many miles did they travel?

15. How much greater is 4 × 6 than 3 × 7?

**SET 6**

1. Round to the nearest hundred: 623, 755,
   288, 143, 892, 324, 509

2. Give the first 10 multiples of: 3, 2, 6, 4, 5,
   7, 9, 8, 1

3. Multiply by 5, then add 4: 80, 90, 40, 70,
   60, 30, 20, 50

4. Take 8 from: 11, 41, 91, 61, 21, 81, 71,
   51, 31, 101

5. Add 4 to: 57, 97, 67, 14, 84, 24, 74, 44,
   37, 77, 54, 17

6. The sum of Tanya's and Paul's ages is 18
   years. If Tanya is six, how old is Paul?

7. Estimate the cost of 6 boxes of cards, if
   one box costs $4.99.

8. At $8 an hour, how much will a worker earn in
   4 hours?

9. If one factor of 32 is 4, what is the other factor?

10. What is the standard numeral for CXX?

11. What is 27 increased by 6?

12. Write the numeral: three million, four hundred
    fifty thousand, ninety

13. Ann has 7 dimes, 5 nickels, and 13 pennies.
    How much money does Ann have?

14. At 50 mph, how far can a train travel in
    8 hours?

15. Glenn bought 9 pencils at $0.30 each. What
    was his change from $3.00?

**SET 7**

1. Give 4 related facts.
   2, 7, 14   7, 5, 35   6, 8, 48   4, 9, 36
2. Divide by 2: 4, 40, 400, 4000, 40,000, 400,000
3. $7 \div 1$   $0 \div 84$   $34 \div 34$
   $62 \div 1$   $0 \div 17$   $20 \div 20$
4. $72 \div \underline{\;?\;} = 9$   $64 \div 8 = \underline{\;?\;}$
   $27 \div \underline{\;?\;} = 9$   $42 \div 6 = \underline{\;?\;}$
   $25 \div \underline{\;?\;} = 5$   $20 \div 2 = \underline{\;?\;}$
5. Divide by 4: 8, 4, 24, 40, 36, 12, 32, 28
6. The quotient is 7. The dividend is 56. What is the divisor?
7. If the quotient is 6 and the dividend is 6, what is the divisor?
8. If the cost of 9 folding chairs is $54, what is the cost per chair?

9. If two balls cost $2.80, what is the cost of one ball?
10. When 27 is divided by 8, what is the remainder?
11. If 3 workers each earned $15.20 in one hour, what was their total earnings?
12. At $9 each, how many blankets can be purchased for $108?
13. Estimate the quotient: $39,798 \div 8$
14. Which is cheaper: $0.30 each or $3.50 a dozen?
15. At $45 per day, how many days must Adam work to earn $450?

**SET 8**

1. Divide by 8: 71, 68, 72, 73, 75, 76
2. Which are divisible by 3?
   15, 22, 39, 45, 32, 61, 53, 57, 72, 87, 92
3. Divide by 5: 6, 7, 11, 12, 16, 34, 42, 27
4. $45 \div 15$   $24 \div 12$   $42 \div 14$
   $39 \div 13$   $64 \div 16$   $33 \div 11$
5. Take 6 from: 15, 35, 95, 85, 45, 25, 55, 75, 65
6. How many dozen in 120 eggs?
7. A 2-lb box of nuts costs $18.60. What is the cost per pound?
8. Divide 639 by 3.
9. Compute: $6 \times 2 \div 4 + 7 = \underline{\;?\;}$
10. At $1.20 each, how many pairs of socks can be bought for $8.40?

11. There are 24,000 seats in the stadium with 24 seats in each row. How many rows are there?
12. Ted picked 7 baskets of 20 apples each and 5 baskets of 20 peaches each. How much fruit did Ted pick?
13. Frank paid $5.40 for 9 bottles of spring water. What is the cost per bottle?
14. An airplane traveled 30,600 mi in 30 days. How many miles did it travel each day?
15. Jan paid $1.80 for 9 bran muffins. How much does one muffin cost?

**SET 9**

1. Add 3 to: 99, 29, 33, 53, 69, 83, 103
2. Take 5 from: 12, 32, 72, 92, 82, 62, 22
3. Is the fraction closer to 0 or 1?
   $\frac{1}{3}, \frac{5}{6}, \frac{2}{8}, \frac{4}{10}, \frac{6}{7}, \frac{2}{5}, \frac{1}{4}, \frac{2}{3}$
4. Which fractions are in lowest terms?
   $\frac{1}{2}, \frac{2}{3}, \frac{3}{6}, \frac{5}{10}, \frac{4}{7}, \frac{6}{15}, \frac{3}{5}, \frac{7}{8}$
5. Which have a GCF of 2?
   8 and 10   21 and 24   16 and 30
6. How much is 48 decreased by 6?
7. Which fraction has a different denominator:
   $\frac{3}{7}, \frac{2}{7}$, or $\frac{3}{5}$?

8. What is the numerator in $\frac{5}{8}$?
9. How many fifths are in one whole?
10. What fractional part of an hour is 10 minutes?
11. Express $\frac{30}{54}$ in simplest form.
12. $\frac{3}{4} = \frac{?}{8} = \frac{9}{?} = \frac{?}{16} = \frac{18}{?} = \frac{?}{36}$
13. Sean read for $\frac{2}{3}$h and Sara for $\frac{3}{4}$h. Who read longer?
14. Rename $\frac{31}{6}$ as a mixed number.
15. Which is the greatest: $\frac{1}{4}, \frac{1}{6}, \frac{1}{8}, \frac{1}{10}$, or $\frac{1}{2}$?

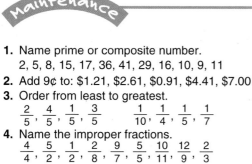 

1. Name prime or composite number.
   2, 5, 8, 15, 17, 36, 41, 29, 16, 10, 9, 11
2. Add 9¢ to: $1.21, $2.61, $0.91, $4.41, $7.00
3. Order from least to greatest.
   $\frac{2}{5}, \frac{4}{5}, \frac{1}{5}, \frac{3}{5}$     $\frac{1}{10}, \frac{1}{4}, \frac{1}{5}, \frac{1}{7}$
4. Name the improper fractions.
   $\frac{4}{4}, \frac{5}{2}, \frac{1}{2}, \frac{2}{8}, \frac{9}{7}, \frac{5}{5}, \frac{10}{11}, \frac{12}{9}, \frac{2}{3}$
5. Express as a mixed number.
   $\frac{11}{7}, \frac{13}{8}, \frac{21}{2}, \frac{15}{6}, \frac{8}{3}, \frac{25}{4}, \frac{31}{5}, \frac{19}{9}$
6. Which is the smallest: $\frac{1}{4}, \frac{1}{2}$, or $\frac{1}{8}$?
7. How much more than $\frac{1}{14}$ is $\frac{2}{7}$?

8. Rename $4\frac{5}{6}$ as an improper fraction.
9. Express $\frac{30}{48}$ in simplest form.
10. Name the prime numbers between 1 and 20.
11. What fractional part of 1 year is 4 months?
12. Which of these fractions is not in simplest form: $\frac{5}{8}, \frac{5}{9}, \frac{6}{9}$, or $\frac{8}{11}$?
13. Express $\frac{74}{9}$ as a mixed number.
14. $\frac{4}{5} = \frac{?}{20} = \frac{?}{35} = \frac{24}{?} = \frac{36}{?} = \frac{?}{50}$
15. Find the LCD of $\frac{1}{9}$ and $\frac{1}{12}$.

---

### SET 11

1. Express as an improper fraction.
   $3\frac{1}{6}, 2\frac{1}{6}, 4\frac{5}{6}, 8\frac{4}{6}, 9\frac{1}{6}, 5\frac{1}{6}, 7\frac{5}{6}, 10\frac{1}{6}$
2. Express as a mixed number.
   $\frac{31}{6}, \frac{19}{6}, \frac{11}{6}, \frac{25}{6}, \frac{13}{6}, \frac{27}{6}, \frac{61}{6}, \frac{17}{6}, \frac{29}{6}$
3. $\frac{2}{6} + \frac{1}{6}$    $\frac{3}{7} + \frac{2}{7}$    $\frac{8}{18} + \frac{5}{18}$    $\frac{3}{8} + \frac{4}{8}$
4. Add $\frac{1}{2}$ to: $\frac{1}{2}, 2, 3\frac{1}{4}, 1\frac{1}{2}, 4, 5\frac{1}{4}$
5. $1\frac{3}{8} + 2\frac{4}{8}$    $3\frac{7}{12} + 4\frac{2}{12}$    $5\frac{6}{11} + 1\frac{3}{11}$
6. What is the sum of $1\frac{3}{8}$ and $\frac{5}{8}$?
7. Joe worked for $3\frac{1}{2}$ h. Sam worked for $4\frac{1}{4}$ h. How much time did they work altogether?

8. Add: $\frac{1}{6} + \frac{1}{3} + \frac{1}{2}$    9. $9\frac{5}{4} = \frac{?}{4}$
10. Len weighed $78\frac{1}{2}$ lb and then gained $1\frac{1}{2}$ lb. How much does he weigh now?
11. How much larger than $\frac{1}{4}$ of a circle is $\frac{3}{4}$ of the same circle?
12. How much less than 3 is $2\frac{1}{3}$?
13. From $1\frac{1}{5}$ subtract $\frac{1}{10}$.
14. Hikers are $7\frac{1}{2}$ m from camp. After walking $3\frac{1}{4}$ m back, how far do they have to go?
15. From 2 take $\frac{4}{9}$.

---

### SET 12

1. $\frac{5}{9} - \frac{3}{9}$    $\frac{17}{21} - \frac{9}{21}$    $\frac{18}{19} - \frac{10}{19}$    $\frac{13}{15} - \frac{6}{15}$
2. Take $\frac{1}{10}$ from: $\frac{9}{10}, \frac{2}{10}, \frac{6}{10}, \frac{3}{10}, \frac{4}{10}, \frac{7}{10}$
3. $6\frac{4}{10} - 2\frac{2}{10}$    $5\frac{3}{4} - 3\frac{2}{4}$    $4\frac{4}{5} - 1\frac{1}{5}$
4. Express in lowest terms.
   $\frac{7}{14}, \frac{7}{21}, \frac{7}{28}, \frac{7}{63}, \frac{7}{35}, \frac{7}{56}, \frac{7}{49}, \frac{7}{42}$
5. Express as an improper fraction.
   $2\frac{1}{9}, 3\frac{4}{9}, 5\frac{2}{9}, 8\frac{1}{9}, 9\frac{4}{9}, 7\frac{1}{9}, 4\frac{4}{9}, 6\frac{5}{9}$
6. $7\frac{3}{5} = 6\frac{?}{5}$
7. What is $\frac{1}{10}$ less than $\frac{1}{5}$?
8. How much greater than $\frac{3}{4}$ is 2?

9. If Lee weighed $90\frac{1}{2}$ lb and lost $2\frac{1}{2}$ lb, how much does he weigh?
10. How many yards of cloth are there in two remnants, one of which contains $\frac{5}{8}$ yd and the other $\frac{3}{8}$ yd?
11. Estimate the cost of 8 mugs at $2.89 each.
12. Take $\frac{8}{9}$ from 6.
13. Fay had 4 yd of tape. She used $3\frac{7}{8}$ yd. How many yards does she have left?
14. Ralph studied $1\frac{1}{2}$ h on Monday and $2\frac{1}{2}$ h on Tuesday. How many hours did Ralph study?
15. Add: $\frac{2}{5} + \frac{1}{10} + \frac{3}{5}$

**1.** $\dfrac{1}{2} \times \dfrac{3}{4}$   $\dfrac{2}{3} \times \dfrac{1}{3}$   $\dfrac{3}{4} \times \dfrac{1}{8}$   $\dfrac{2}{3} \times \dfrac{4}{5}$

**2.** $6 \times \dfrac{1}{2}$   $12 \times \dfrac{1}{3}$   $\dfrac{2}{3} \times 30$   $\dfrac{2}{5} \times 20$

**3.** Find the GCF:   3 and 9   2 and 6
   7 and 21   8 and 16   3 and 16

**4.** Express as an improper fraction.
   $6\dfrac{1}{8}, 5\dfrac{3}{8}, 3\dfrac{5}{8}, 4\dfrac{3}{8}, 8 = \dfrac{?}{10}$   $5 = \dfrac{?}{8}$

**5.** Round to the nearest whole number.
   $4\dfrac{1}{3}, 10\dfrac{6}{11}, 8\dfrac{6}{7}, 11\dfrac{1}{5}$

**6.** How much greater than $\dfrac{1}{6}$ of 24 is $\dfrac{1}{6}$ of 36?

**7.** Express $9\dfrac{3}{7}$ as an improper fraction.

**8.** Jim lives $\dfrac{3}{4}$ km from the zoo. Tom lives 4 times that distance from the zoo. How far from the zoo does Tom live?

**9.** How many minutes are in $\dfrac{1}{3}$ of an hour?

**10.** Find $\dfrac{5}{8}$ of 40.

**11.** At the rate of 40 mph, how far will a car travel in $\dfrac{3}{4}$ of an hour?

**12.** Tim needs to study 3 h. He has studied $\dfrac{2}{3}$ of that time. How much more of that time does he have to study?

**13.** How many $\dfrac{3}{4}$- c portions can be made from $3\dfrac{3}{4}$ c of pudding?

**14.** Jill worked $\dfrac{1}{2}$ h. Sue worked $\dfrac{2}{3}$ h. How long did they both work?

**15.** Divide 6 lb into $\dfrac{1}{2}$-lb packages.

---

## SET 14

**1.** $9 \div \dfrac{1}{4}$   $3 \div \dfrac{1}{4}$   $4 \div \dfrac{1}{2}$   $6 \div \dfrac{1}{3}$   $2 \div \dfrac{1}{8}$

**2.** Give the reciprocal: $6, \dfrac{1}{7}, \dfrac{2}{3}, 8, 3\dfrac{1}{4}$

**3.** $\dfrac{2}{5} \div \dfrac{1}{5}$   $\dfrac{7}{9} \div \dfrac{1}{9}$   $\dfrac{5}{6} \div \dfrac{1}{6}$   $\dfrac{3}{8} \div \dfrac{1}{8}$

**4.** $\dfrac{1}{6} \div 6$   $\dfrac{7}{10} \div 7$   $\dfrac{1}{3} \div 6$   $\dfrac{2}{3} \div 2$   $\dfrac{6}{7} \div 6$

**5.** $\dfrac{3}{4} = \dfrac{?}{16} = \dfrac{?}{24} = \dfrac{?}{36} = \dfrac{?}{8} = \dfrac{?}{12} = \dfrac{?}{32}$

**6.** Is the reciprocal of 8: $\dfrac{1}{8}$ or 8?

**7.** Which is greater: $\dfrac{4}{5} \div \dfrac{1}{2}$ or $\dfrac{4}{5} \times \dfrac{1}{2}$?

**8.** Dividing a number by $1\dfrac{1}{4}$ is the same as multiplying it by $\underline{\ ?\ }$.

**9.** Is $\dfrac{33}{4}$ greater or less than 8?

**10.** Divide 8 yd into $\dfrac{1}{4}$- yd pieces.

**11.** Three people divided $\dfrac{1}{2}$ of a pizza. How much did each person receive?

**12.** How many sixths are there in $\dfrac{1}{3}$?

**13.** Express $\dfrac{46}{9}$ as a mixed numeral.

**14.** Write XCVIII as a standard numeral.

**15.** How many minutes are there in $\dfrac{3}{4}$ hour?

---

## SET 15

**1.** Add 7 to: 9, 19, 39, 79, 89, 49, 59, 29

**2.** Multiply by 9 and add 2: 8, 3, 10, 2, 0, 9, 5

**3.** Express as a mixed number.
   $\dfrac{37}{6}, \dfrac{35}{6}, \dfrac{43}{6}, \dfrac{49}{6}, \dfrac{19}{6}, \dfrac{29}{6}, \dfrac{25}{6}, \dfrac{55}{6}, \dfrac{61}{6}$

**4.** $2 = 1\dfrac{?}{8}$   $2 = 1\dfrac{?}{6}$   $2 = 1\dfrac{?}{4}$   $3 = 2\dfrac{?}{9}$

**5.** $2 - \dfrac{3}{8}$   $2 - \dfrac{5}{6}$   $2 - \dfrac{3}{4}$   $2 - \dfrac{1}{5}$

**6.** On three days, Meg worked 8 h, 6 h, and 10 h. What was the average number of hours worked?

**7.** How many books are there?

□□ □□ □□ □□ □□

Key: □□ = 25 books

A bank contains 5 quarters, 3 dimes, and 2 nickels. Pick a coin at random. Find the probability.

**8.** P(quarters)   P(nickels)

**9.** P(dimes)   P(quarters or dimes)

**10.** P(pennies)   P(coins)
   Danielle's math test scores were: 82, 86, 86, 90, 93, 95.

**11.** Find the median of the scores.

**12.** Find the range of the scores.

**13.** Find the mode of the scores.

**14.** Ned is 63 in. tall. Nell is $\dfrac{2}{3}$ as tall. How tall is Nell?

**15.** What is 32 decreased by 5?

## SET 16

1. Give the range.
   6, 11, 8, 15   17, 5, 9, 20   10, 12, 18, 9
2. Give the median: 86, 74, 81, 87
   92, 87, 96   72, 80, 76, 84
3. $\frac{1}{8} = \frac{?}{16} = \frac{?}{32} = \frac{?}{40} = \frac{?}{72} = \frac{?}{48} = \frac{?}{64}$
4. Express in simplest form.
   $\frac{8}{16}, \frac{8}{32}, \frac{8}{64}, \frac{8}{72}, \frac{8}{56}, \frac{8}{24}$
5. Divide by 8, then subtract 2: 48, 56, 64, 32,
   72, 80, 16, 40, 24
6. In a 6-h school day there are 8 equal time
   periods, including 6 subjects, a study period,
   and lunch. What part of an hour is
   there for lunch?
7. Write XCIII as a standard numeral.

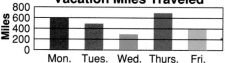

8. When were more than 500 mi traveled?
9. How many more miles were traveled on
   Thursday than on Tuesday?
10. How many miles were traveled on the last
    three days?
11. A square playpen measures 4 ft on each
    side. Find the perimeter.
12. If a bird flies 10 mph, how far can it fly in
    30 minutes?
13. At $0.30 each, what will 7 rolls cost?
14. What is 85 decreased by 6?
15. How many days are in 9 weeks?

## SET 17

1. Name each symbol: $\overleftrightarrow{AB}$, $\overrightarrow{DE}$, $\angle XYZ$,
   $\overline{KL}$, $\angle T$, $\overleftrightarrow{RS}$, $\overrightarrow{TU}$, $\overleftrightarrow{JC}$,
2. Name parallel, perpendicular, or neither.
3. Acute, right, or obtuse angle? 27°, 174°, 90°, 45°,
   12°, 115°, 5°, 162°
4. Identify:
5. Congruent? Yes or No.
6. A _?_ is used to measure angles.

7. An _?_ triangle has at least 2 congruent sides.
8. What is the perimeter of a room 20 ft long
   and 12 ft wide?
9. A triangular field measures 30 yd by 42 yd
   by 60 yd. What is the perimeter?
10. Find the perimeter of a picture frame
    12 in. long and 10 in. wide.
11. A _?_ is a rectangle with 4 congruent sides.
12. The diameter of a circular clock is 15 in. What
    is the radius?
13. Estimate the circumference of a
    merry-go-round whose diameter is 50 ft.
14. Draw two congruent figures.
15. A circular table has a radius of 3 ft. Estimate
    the circumference.

## SET 18

1. Which are divisible by both 2 and 5?
   25, 10, 8, 20, 12, 35, 40, 30, 18, 24
2. Choose fractions close to 1.
   $\frac{1}{9}, \frac{7}{8}, \frac{3}{16}, \frac{23}{24}, \frac{18}{20}, \frac{2}{9}, \frac{3}{11}, \frac{16}{17}$
3. Express in feet: 60 in., 36 in., 84 in.,
   48 in., 96 in., 72 in., 108 in.
4. Compare. Use <, =, or >.
   20 c _?_ 5 qt   6 c _?_ 50 oz
   3 qt _?_ 8 pt   16 oz _?_ 1 lb
5. 2 T = _?_ lb   80 oz = _?_ lb
   2000 lb = _?_ T   2 lb 3 oz = _?_ oz
6. What speed must a boat maintain in order
   to go 54 miles in 6 hours?
7. How many yards equal 21 ft?

8. What is the best estimate of weight for an
   elephant: 200 lb, 6000 oz, or 2 T?
9. How many feet are in 1 yd 2 ft?
10. At the rate of 500 mph, how far does a jet
    travel in 30 minutes?
11. $\frac{1}{4}$ lb = _?_ oz
12. Which is more and by how much:
    2 ft or 26 in.?
13. $6\frac{2}{7} = \frac{?}{7}$
14. Al weighed 7 lb 9 oz at birth. At 1 year he
    weighed 21 lb 13 oz. How much weight
    did he gain?
15. Express $\frac{36}{72}$ in simplest form.

498

## SET 19

1. Is the temperature hot or cold?
   5°C, 80°C, 25°F, 250°F, 32°C, 100°C
2. Give the number of minutes in: 4 h, 2 h, 120 s, 300 s, 5 h, 420 s, 1 h
3. Add 15 minutes to: 9:45, 12:15, 7:30, 2:00, 10:05, 3:25, 11:30
4. Give the number of days in: 3 wk, 8 wk, 2 wk, 10 wk, 5 wk, 7 wk, 4 wk
5. Express in quarts: 16 pt, 36 pt, 20 pt, 8 pt, 28 pt, 12 pt, 32 pt
6. What fractional part of a day is one hour?
7. How many minutes are there in 9 hours?
8. 1 century = $\underline{\ ?\ }$ years
9. It is 3 hours earlier in California. When it is 1 P.M. in New York, what time is it in California?
10. The temperature at 12 noon was 24°F. By 8 P.M. it had dropped 30°. What was the temperature at 8 P.M.?
11. Water freezes at $\underline{\ ?\ }$ °C and $\underline{\ ?\ }$ °F.
12. Could you swim in water heated to 100°C?
13. If the game began at 8:00 P.M. and ended $2\frac{1}{4}$ h later, what time did the game end?
14. If a jet leaves Oregon at 9 A.M. and travels 5 hours to Florida, what time will the plane land in Florida?
15. How many pints are contained in a 9-quart jug?

## SET 20

1. Give the value of the underlined digit.
   0.1$\underline{3}$5, 0.4$\underline{8}$, 0.$\underline{7}$, 0.25$\underline{9}$, 0.$\underline{6}$10, 0.$\underline{1}$, 0.7$\underline{3}$,
2. Add 6 to: 9, 19, 39, 79, 99, 69, 29, 49
3. Express in simplest form: $\frac{32}{40}, \frac{24}{72}, \frac{16}{32}, \frac{40}{48}$
4. Compare. Use <, =, or >.
   0.62 $\underline{\ ?\ }$ 0.26    0.9 $\underline{\ ?\ }$ 0.90
   1.345 $\underline{\ ?\ }$ 1.435    0.519 $\underline{\ ?\ }$ 0.159
5. Order from least to greatest.
   0.2, 0.02, 0.21    0.36, 0.63, 0.33
   5.111, 5.101, 5.110    0.429, 0.492, 0.9
6. What decimal is one thousandth more than 0.05?
7. Which is greater: 36.08 or 36.80?
8. Complete the pattern.
   1.3, 1.6, 1.9, $\underline{\ ?\ }$, $\underline{\ ?\ }$, 2.8, $\underline{\ ?\ }$
9. Round 45.629 to the nearest hundredth.
10. Round $50.51 to the nearest dollar.
11. Kay bought 0.25 lb of ham, 0.5 lb of cheese, and 0.3 lb of bologna. Is the total more or less than a pound and by how much?
12. Compare. Use <, =, or >.
    0.203 + 0.650 $\underline{\ ?\ }$ 0.808
13. Place the decimal point in the answer.
    1.734 + 2.15 = 3884
14. Ned bought stickers for $1.80 and a stamper for $2.05. Find the total cost.
15. Find the difference between 0.01 and 0.001.

## SET 21

1. 0.3 + 0.4    0.24 + 0.6    0.54 + 0.05
   0.8 + 0.08    0.2 + 0.13
2. 0.9 − 0.2    0.38 − 0.07    0.07 − 0.03
   0.66 − 0.3    0.74 − 0.03
3. Express as a whole or mixed number.
   $\frac{56}{8}, \frac{57}{8}, \frac{59}{8}, \frac{61}{8}, \frac{63}{8}, \frac{64}{8}, \frac{65}{8}, \frac{67}{8}$
4. Express as an improper fraction.
   $8\frac{1}{9}, 4\frac{5}{9}, 6\frac{2}{9}, 3\frac{7}{9}, 9\frac{4}{9}, 7\frac{5}{9}, 5\frac{4}{9}, 2\frac{2}{9}$
5. Divide by 9: 64, 57, 29, 22, 83, 73, 69, 50, 14, 19, 30, 85, 47
6. Chad saved $42.75. He bought a computer game for $38.75. How much money does he have left?
7. What fractional part of a foot is 6 inches?
8. One day Pat earned $42. The day before she earned $13 less. How much did she earn the day before?
9. At a speed of 7 mph, how long will it take a boat to travel 154 miles?
10. If the length of a rug is 21 ft, what is its length in yards?
11. At $2.03 each, find the cost of 7 pens.
12. Take 2.4 from 6.7. Then add 1.2 to the difference.
13. Donna's times on her runs were 0.25 h, 1.4 h, and 0.75 h. What was her total time?
14. At $1.20 each pair, how many pairs of socks can be bought for $8.40?
15. The class collected 22.75 lb of newspapers on Mon. and 14.15 lb on Tues. About how many pounds did they collect?

 **SET 22**

1. Multiply by 10: 0.6, 0.05, 1.02, 36.3, 0.009, 2.103, 0.013
2. Multiply by 100: 0.2, 0.43, 0.6, 4.01, 6.005, 24.3, 71.8, 0.09
3. Multiply by 1000: 0.1, 0.04, 2.3, 0.003, 49.7, 52.34, 0.016
4. Estimate: $0.62 \times 0.29$  $3.1 \times 4.6$ $0.08 \times 1.4$  $50.3 \times 2.2$  $19.7 \times 0.94$
5. Estimate: $2.431 \div 6$  $561.9 \div 7$ $36.22 \div 9$  $4.49 \div 15$  $605.14 \div 8$
6. Don makes $4.85 an hour. How much will he make in 10 hours? in 100 hours?
7. A spool of ribbon has 9.2 yd. How many yards are there on 1000 spools?

8. If 9 identical items cost $724.62 about how much does one item cost?
9. If 12.5 kg of popping corn is put equally into 100 bags, how much corn will there be in each bag?
10. Estimate the cost of 4 shirts at $49.75 each.
11. Each glass holds 8.3 oz of milk. How much milk is in 5 glasses?
12. Which is less: $0.06 \times 0.4$ or $0.60 \times 0.4$?
13. Jill had 1.5 lb of cheese. She used 0.75 lb in lasagna. How much did she have left?
14. At $2.80 a pound, what will 0.5 lb of tea cost?
15. How much less than $2 \times 8$ is $7 \times 2$?

**SET 23**

1. Multiply each by 2: 0.3, 0.02, 0.4, 0.08, 0.5, 0.07, 0.6, 0.2
2. Divide by 10: 2.6, 0.8, 3.5, 7.34, 0.03, 15.9, 24.7
3. Divide by 100: 13.7, 51.1, 0.9, 0.6, 422.9, 27.5, 43.8, 0.7
4. Divide by 1000: 5000, 4500, 300, 380, 60, 65, 5
5. $0.18 \div 9$  $0.08 \div 2$  $0.32 \div 4$ $3.12 \div 3$  $4.016 \div 8$
6. Estimate the cost of 5 blank cassette tapes at $2.99 a tape.
7. How many kilometers are in 1000 m?

8. Eight notebooks all the same price cost $7.20. How much does 1 notebook cost?
9. How many centimeters are in 3 m?
10. If there are 100 cm in 1 meter, how many meters are there in 500 cm?
11. Which is more and by how much: 2 kg or 1800 g?
12. Which is more and by how much: 220 cm or 2 m?
13. How many grams are there in 1 kg?
14. What decimal part of a meter is 1 cm?
15. Find the area of a rectangular rug 8 ft by 4 ft.

**SET 24**

1. Complete: $2 L = \underline{\ ?\ } dL$  $60 dL = \underline{\ ?\ } L$ $12 dL = \underline{\ ?\ } mL$  $3000 L = \underline{\ ?\ } kL$
2. Complete: $1 m = \underline{\ ?\ } cm$  $1 L = \underline{\ ?\ } cL$ $1000 g = \underline{\ ?\ } kg$  $1 dm = \underline{\ ?\ } cm$
3. Take 7 from: 16, 56, 86, 26, 96, 76
4. $\frac{4}{9} = \frac{?}{63} = \frac{?}{45} = \frac{?}{81} = \frac{?}{72} = \frac{?}{36} = \frac{?}{27}$
5. Divide by 7, then subtract 3: 21, 42, 63, 28, 56, 70, 35, 49
6. How many square feet of plastic are needed to cover the bottom of a square playpen that measures 5 ft on each side?
7. What space figure has 6 faces, 12 edges, and 8 vertices?

8. A triangle has a base of 9 ft and an altitude of 6 ft. What is its area?
9. A stack of newspapers measures 20 in. long, 10 in. wide, and 30 in. high. Find the volume.
10. Write MCCXL in standard form.
11. Estimate the area of a tile floor that measures 9.7 ft by 13.2 ft.
12. Which is more and by how much: 3 L or 2800 mL?
13. If 25 raisins weigh about 25 g, how many milligrams is that?
14. If a ship sails 270 km in 9 h, what is its average speed per hour?
15. What is the cost of 2 basketballs at $18 each?

**1.** Complete: $\frac{1}{3} = \frac{?}{9}$   $\frac{1}{2} = \frac{?}{20}$

$\frac{2}{3} = \frac{?}{18}$   $\frac{3}{5} = \frac{?}{25}$   $\frac{3}{4} = \frac{?}{24}$

**2.** Read the ratio: 17:24, 8:12, 36:5, 1:18, 2:27

**3.** Express as = or ≠: $\frac{6}{8} \underline{\ ?\ } \frac{3}{4}$

$\frac{10}{20} \underline{\ ?\ } \frac{2}{3}$   $\frac{45}{30} \underline{\ ?\ } \frac{3}{2}$   $\frac{2}{1} \underline{\ ?\ } \frac{6}{10}$

**4.** Express as a percent.
$\frac{45}{100}, \frac{16}{100}, \frac{7}{100}, \frac{92}{100}, \frac{71}{100}, \frac{10}{100}, \frac{14}{100}$

**5.** Express as a fraction.
63%, 85%, 5%, 28%, 1%, 98%, 11%

**6.** What percent of a dollar is $.25?

**7.** On a map, City A is $3\frac{1}{2}$ in. from City B.
The scale is 1 in. = 20 mi. What is the actual distance from City A to City B?

**8.** 16 is $\frac{4}{5} \times 20$ as 4 is $\underline{\ ?\ }$ of 20.

**9.** What percent of a dollar is 1 penny?

**10.** Of 60 animals in the pet shop, 20 percent are dogs. How many are dogs?

**11.** Sneakers are on sale at 60% off the original price of $80. How much is the discount?

**12.** Of 100 children, 32% wear sneakers. How many children is that?

**13.** If a map scale reads 1 cm = 5 km, how many centimeters would represent a distance of 60 km?

**14.** 7 is to 1 as 35 is to $\underline{\ ?\ }$.

**15.** 39 in. = 3 ft $\underline{\ ?\ }$ in.

---

**1.** Express as a percent.
0.4, 0.73, 0.05, 0.91, 0.1, 0.88, 0.56

**2.** Express as a decimal.
38%, 4%, 10%, 52%, 44%, 30%, 60%

**3.** Express as a whole or a mixed number.
$\frac{63}{9}, \frac{64}{9}, \frac{67}{9}, \frac{70}{9}, \frac{72}{9}, \frac{73}{9}, \frac{76}{9}$

**4.** Express in lowest terms: $\frac{36}{45}, \frac{27}{36}, \frac{45}{54}, \frac{54}{63}, \frac{63}{72}, \frac{72}{81}$

**5.** Double, then add 0.1 to: 0.9, 0.04, 0.13, 0.20, 0.25, 0.7, 0.31

**6.** If 6 h are spent sleeping, what percent of the day is that?

**7.** $\frac{5}{8} = \frac{?}{16} = \frac{?}{32} = \frac{?}{40} = \frac{?}{56} = \frac{?}{48} = \frac{?}{64}$

**8.** Of 100 people surveyed, 42 voted yes, 34 voted no, and the rest were undecided. What percentage was undecided?

**9.** If 1 in. represents 60 ft, how many feet will 7 in. represent?

**10.** Jean and Joe used 4 yard of ribbon to make 5 bows. If the bows were all the same size, what part of a yard was used for each?

**11.** At a speed of 48 mph, how far will a car travel in 15 minutes?

**12.** Which of the following fractions is not in simplest form: $\frac{7}{17}, \frac{9}{35}, \frac{14}{35}$, or $\frac{15}{34}$?

**13.** How much less than 3 dozen is 28?

**14.** How many pints are in 10 quarts?

**15.** If one notepad costs $0.90, find the cost of 4 notepads.

---

**1.** Evaluate $a + 7$ when $a$ is: 9, 15, 3, 7, 11, 21, 32, 40, 54

**2.** Evaluate $b - 10$ when $b$ is: 56, 72, 84, 96, 25, 11, 38, 47

**3.** Evaluate $6x$ when $x$ is: 9, 7, 10, 12, 13, 15, 11, 8, 6

**4.** Evaluate $\frac{y}{5}$ when $y$ is: 35, 45, 50, 60, 75, 90, 25, 15

**5.** $3 + 4 + 5$      $2 \times 3 \div 3 \times 4$
$4 \times 1 + 6 - 7$      $5 \times 2 - 8 + 4$

**6.** Express Joan's age 4 years from now. Let $z$ = Joan's age now.

**7.** $\frac{4}{7} = \frac{?}{14} = \frac{?}{28} = \frac{?}{35} = \frac{?}{21} = \frac{?}{70} = \frac{?}{63}$

**8.** Solve for $n$: $n + 3 + 5 = 18$

**9.** Solve for $c$: $5c = 40$

**10.** Express as an equation: The cost of 1 lb of peaches is 3 times the cost of 1 lb of apples. Let $g$ = cost of apples.

**11.** Solve for $m$: $\frac{1}{m} \times 72 = 9$

**12.** Solve for $p$: $\frac{4}{5} = p + \frac{1}{5}$

**13.** Express as an equation:
Ted's height is 3 in. less than Bob's. Let $d$ = Bob's height.

**14.** Write XCII in standard form.

**15.** Ben's total of 4 scores was 30. He remembers three scores: 9, 7, 8. What score did he forget?

**acute angle**   An angle that measures less than 90°. (p. 270)

**acute triangle**   A triangle with three acute angles. (p. 276)

**addend**   Any one of a set of numbers to be added.

**algebraic expression**   A mathematical expression that contains one or more variables. (p. 440)

**A.M.**   Abbreviation that indicates time from midnight to noon.

**angle**   A figure formed by two rays that have a common endpoint.

**area**   The number of square units needed to cover a flat surface.

**associative (grouping) property**   Changing the grouping of the addends (or factors) does not change the sum (or product). (pp. 44, 68)

**average**   The quotient obtained by dividing a sum by the number of addends. Also called *mean.*

**axis**   The horizontal or vertical number line of a graph.

**balance**   The tool used to measure mass.

**bar graph**   A graph that uses bars to show data. The bars may be of different lengths. (p. 252)

**base**   One of the equal factors in a product; a selected side or face of a geometric figure. (pp. 93, 396)

**BASIC**   An acronym for Beginner's All-purpose Symbolic Instruction Code; a computer language used to process information.

**benchmark**   An object of known measure used to estimate the measure of other objects.

**cancellation**   The dividing of any numerator and denominator of a set of fractions by their greatest common factor before multiplying. (p. 204)

**capacity**   The amount, usually of liquid, a container can hold.

**Celsius (°C) scale**   The temperature scale in which 0°C is the freezing point of water and 100°C is the boiling point of water.

**chord**   A line segment with both endpoints on a circle. (p. 284)

**circle**   A set of points in a plane, all of which are the same distance from a given point called the *center.*

**circle graph**   A graph that uses the area of a circle to show the division of a total amount of data. (p. 252)

**circumference**   The distance around a circle.

**closed curve**   A path that begins and ends at the same point and may not intersect itself.

**clustering**   To find addends that are nearly alike in order to estimate their sum. (p. 75)

**common denominator**   A number that is a multiple of the denominators of two or more fractions.

**common factor**   A number that is a factor of two or more numbers.

**common multiple**   A number that is a multiple of two or more numbers.

**commutative (order) property**   Changing the order of the addends (or factors) does not change the sum (or product). (pp. 44, 68)

**compass**   An instrument used to draw circles.

**compatible numbers**   Numbers that are easy to compute with mentally. (p. 112)

**composite number**   A whole number greater than 1 that has more than two factors. (p. 136)

**cone**   A space, or solid, figure with one circular base, one vertex, and a curved surface. (p. 400)

**congruent figures**   Figures that have the same size and shape. (p. 274)

**coordinates**   An ordered pair of numbers used to locate a point on a grid. (p. 452)

**cross products**   The products obtained by multiplying the numerator of one fraction by the denominator of a second fraction and the denominator of the first fraction by the numerator of the second fraction. (p. 418)

**cube**   A space, or solid, figure with six congruent square faces. (p. 398)

**cubic measure**   A measure of volume.

**customary system**   The measurement system that uses inch, foot, yard, and mile; fluid ounce, cup, pint, quart, and gallon; ounce, pound, and ton. (See *Table of Measures*, p. 515.)

**cylinder**   A space, or solid, figure with two parallel, congruent circular bases and a curved surface. (p. 398)

**data**   Facts or information.

**database**   A group of facts and figures that are related and can be arranged in different ways.

**decagon**   A polygon with ten sides. (p. 273)

**decimal**   A number with a decimal point separating the ones from the tenths place.

**decimal point**   A point used to separate ones and tenths in decimals.

**degree (°)**   A unit used to measure angles.

**degree Celsius (°C)**   A unit used to measure temperature.

**degree Fahrenheit (°F)**   A unit used to measure temperature.

**denominator**   The number below the bar in a fraction.

**diagonal**   A line segment, other than a side, that joins two vertices of a polygon. (p. 279)

**diameter**   A line segment that passes through the center of a circle and has both endpoints on the circle. (p. 284)

**difference**   The answer in subtraction.

**digit**   Any one of the numerals 0, 1, 2, 3, 4, 5, 6, 7, 8, or 9.

**discount**   A reduction in the regular, or list, price of an item.

**distributive property**   Multiplying a number by a sum is the same as multiplying the number by each addend of the sum and then adding the products. (p. 69)

**dividend**   The number to be divided. (p. 96)

**divisible**   A number is divisible by another number if the remainder is 0 when the number is divided by the other number.

**divisor**   The number by which the dividend is divided. (p. 96)

**double bar graph**   A graph that uses pairs of bars to compare two sets of data. (p. 263)

**double line graph**   A graph that uses pairs of line segments to compare two sets of data. (p. 263)

**edge**   The line segment where two faces of a space figure meet.

**elapsed time**   The amount of time that passes between the start and end of a given period. (p. 311)

**electronic spreadsheet**   A computer-based software program that arranges data and formulas in a column-and-row format. (p. 406)

**END**   A BASIC command that tells the computer it has reached the end of a program.

**endpoint**   The point at the end of a line segment or ray.

**equation**   A number sentence that shows equality of two mathematical expressions. (p. 446)

**equilateral triangle**   A triangle with three congruent sides and three congruent angles. (p. 276)

**equivalent fractions**   Different fractions that name the same amount. (p. 9)

**estimate**   An approximate answer; to find an answer that is close to the exact answer.

**evaluate**   To find the number that an algebraic expression names.

**even number**   A whole number that is divisible by 2.

**event**   A set of one or more outcomes of a probability experiment.

**expanded form**   The written form of a number that shows the place value of each of its digits. (p. 34)

**exponent**   A number that tells how many times another number is to be used as a factor. (p. 93)

**face**   A flat surface of a space figure.

**factor**   One of two or more numbers that are multiplied to form a product.

**factor tree**   A diagram used to find the prime factors of a number. (p. 137)

**Fahrenheit (°F) scale**   The temperature scale in which 32°F is the freezing point of water and 212°F is the boiling point of water.

**flip**   To turn a figure to its reverse side. (p. 290)

**formula**   A rule that is expressed by using symbols. (p. 394)

**FOR...NEXT loop**   BASIC statements that tell the computer to repeat program lines a given number of times.

**fraction**   A number that names a part of a whole, a region, or a set.

**frequency table**   A chart that shows how often each item appears in a set of data. (p. 246)

**front-end estimation**   A way of estimating by using the front, or greatest, digits to find an approximate answer.

**GOTO**   A statement in BASIC that tells the computer to branch to a specific line.

**graph**   A pictorial representation of data.

**greatest common factor (GCF)**   The greatest number that is a factor of two or more numbers. (p. 138)

**grid**   A network of perpendicular lines used to locate points.

**half-turn symmetry**   The symmetry that occurs when a figure is turned halfway (180°) around its center point and the figure that results looks exactly the same. (p. 288)

**height**   The perpendicular distance between the bases of a geometric figure. In a triangle, the perpendicular distance from the opposite vertex to the line containing the base. (p. 396)

**hexagon**   A polygon with six sides. (p. 13)

**hexagonal prism**   A prism with two parallel hexagonal bases. (p. 398)

**hexagonal pyramid**   A pyramid with a hexagonal base. (p. 398)

**identity property**   Adding 0 to a number or multiplying a number by 1 does not change the number's value. (pp. 44, 68)

**improper fraction**   A fraction with its numerator equal to or greater than its denominator. (p. 150)

**inequality**   A number sentence that uses an inequality symbol: $<$, $>$, or $\neq$.

**INPUT**   A BASIC command that tells the computer to wait for a response from the user.

**integers**   The whole numbers and their opposites. (p. 456)

**intersecting lines**   Lines that meet or cross. (p. 12)

**interval**   The number of units between spaces on a graph.

**inverse operations**   Mathematical operations that *undo* each other, such as addition and subtraction or multiplication and division.

**isosceles triangle**   A triangle with two congruent sides. (p. 276)

**least common denominator (LCD)**   The least common multiple of the denominators of two or more fractions. (p. 147)

**least common multiple (LCM)**   The least number, other than 0, that is a common multiple of two or more numbers. (p. 146)

**line**   A set of points in order extending indefinitely in opposite directions.

**linear measure**   A measure of length.

**line graph**   A graph that uses points on a grid connected by line segments to show data. (p. 252)

**line of symmetry**   A line that divides a figure into two congruent parts. (p. 288)

**line plot**   A graph that uses Xs to show information and to compare quantities. (p. 250)

**line segment**   A part of a line that has two endpoints.

**lowest terms**   A fraction is in lowest terms when its numerator and denominator have no common factor other than 1. (p. 142)

**mass**   The measure of the amount of matter an object contains.

**mathematical expression**   A symbol or a combination of symbols that represents a number.

**mean**   The average of a set of numbers. (p. 248)

**median**   The middle number of a set of numbers arranged in order. If there is an even number of numbers, the median is the average of the two middle numbers. (p. 248)

**metric system**   The measurement system based on the meter, gram, and liter. (See *Table of Measures*, p. 515.)

**minuend**   A number from which another number is subtracted. (p. 45)

**mixed number**   A number that is made up of a whole number and a fraction.

**mode**   The number that appears most frequently in a set of numbers. (p. 248)

**multiple**   A number that is the product of a given number and any whole number. (p. 146)

**multiplicand**   A number that is multiplied by another number.

**multiplier**   A number that multiplies another number.

**net**   A flat pattern that folds into a space figure. (p. 399)

**number sentence**   An equation or an inequality.

**numeral**   A symbol for a number.

**numerator**   The number above the bar in a fraction.

**numerical expression**   A mathematical expression that contains only numbers. (p. 440)

**obtuse angle**   An angle with a measure greater than 90° and less than 180°. (p. 270)

**obtuse triangle**   A triangle with one obtuse angle. (p. 276)

**octagon**   A polygon with eight sides. (p. 273)

**odd number**   A whole number that is not a multiple of 2.

**ordered pair**   A pair of numbers that is used to locate a point on a coordinate grid. (p. 452)

**order of operations**   The order in which operations must be performed when more than one operation is involved.

**outcome**   The result of a probability experiment.

**palindrome**   A number or word that reads the same forward or backward.

**parallel lines**   Lines in a plane that never intersect. (p. 12)

**parallelogram**   A quadrilateral with two pairs of parallel sides. (p. 278)

**pentagon**   A polygon with five sides. (p. 13)

**pentagonal prism**   A prism with two parallel pentagonal bases. (p. 398)

**pentagonal pyramid**   A pyramid with a pentagonal base. (p. 398)

**percent**   The ratio or comparison of a number to 100. (p. 422)

**perimeter**   The distance around a figure.

**period**   A set of three digits set off by a comma in a whole number.

**perpendicular lines**   Lines that intersect to form right angles. (p. 270)

**pi ($\pi$)**   The ratio of the circumference of a circle to its diameter. An approximate value of $\pi$ is 3.14, or $\frac{22}{7}$. (p. 299)

**pictograph**   A graph that uses pictures or symbols to show data. (p. 252)

**place value**   The value of a digit depending on its position, or place, in a number.

**plane**   A flat surface that extends indefinitely and has no thickness.

**plane figure**   A two-dimensional figure that has straight or curved sides.

**P.M.** Abbreviation that indicates time from noon to midnight.

**point** An exact location, or position, usually represented by a dot.

**polygon** A closed plane figure made up of line segments that meet at vertices but do not cross. (pp. 13, 273)

**prime factorization** Expressing a composite number as the product of prime numbers. (p. 137)

**prime number** A whole number greater than 1 that has only two factors, itself and 1. (p. 136)

**PRINT** A BASIC statement that tells the computer what to display on the screen.

**prism** A space figure with two faces called *bases* bounded by polygons that are parallel and congruent. (p. 398)

**probability** A branch of mathematics that analyzes the chance that a given outcome will occur. The probability of an event is expressed as the ratio of a given outcome to the total number of outcomes possible.

**product** The answer in multiplication.

**proportion** A number sentence that shows that two ratios are equal. (p. 418)

**protractor** An instrument used to measure angles. (p. 268)

**pyramid** A space figure whose base is a polygon and whose faces are triangles with a common vertex. (p. 398)

**quadrilateral** A polygon with four sides. (p. 13)

**quotient** The answer in division.

**radius** (plural *radii*) A line segment from the center of a circle to a point on the circle. (p. 284)

**range** The difference between the greatest and least numbers in a set of numbers. (p. 248)

**ratio** A comparison of two numbers or quantities by division. (p. 416)

**ray** A part of a line that has one endpoint and goes on indefinitely in one direction.

**READ** A statement in BASIC that takes entries one-by-one from a DATA statement and assigns each entry to a variable.

**reciprocals** Two numbers whose product is 1. (p. 214)

**rectangle** A parallelogram with four right angles. (p. 278)

**rectangular prism** A prism with six rectangular faces. (p. 400)

**rectangular pyramid** A pyramid with a rectangular base. (p. 398)

**regular polygon** A polygon with all sides and all angles congruent.

**regular price** The original, marked, or list price of an item before a discount has been given.

**REM** A statement in BASIC that describes what a program or section of a program will do; a remark.

**remainder** The number left over when a division computatuion is completed.

**repeating decimal** A decimal with digits that from some point on repeat indefinitely. (p. 377)

**rhombus** A parallelogram with all sides congruent. (p. 278)

**right angle** An angle that measures 90°. (p. 270)

**right triangle** A triangle with one right angle. (p. 276)

**Roman numerals** Symbols for numbers used by the Romans. (p. 54)

**rounding** To approximate a number by replacing it with a number expressed in tens, hundreds, thousands, and so on.

**RUN** A BASIC command that tells the computer to process a program.

**sale price** The sale price is the difference between the list price and the discount.

**scale** The ratio of a pictured measure to the actual measure; the tool used to measure weight.

**scale drawing** A drawing of something accurate but different in size.

**scalene triangle** A triangle with no congruent sides. (p. 276)

**set** A collection or group of numbers or objects.

**side** A line segment that forms part of a polygon; a ray that forms one part of an angle.

**similar figures** Figures that have the same shape. They may or may not be the same size. (p. 274)

**simple closed curve** A path that begins and ends at the same point and does not intersect itself.

**simplest form** The form of a fraction when the numerator and denominator have no common factor other than 1. (p. 142)

**slide** To obtain another figure by moving every point of a figure the same distance and in the same direction. (p. 290)

**space (or solid) figure** A three-dimensional figure that has volume.

**sphere** A curved space figure in which all the points are the same distance from a point called the *center*. (p. 398)

**spreadsheet** A computer program that arranges data and formulas in a grid of cells.

**square** A rectangle with all sides congruent.

**square measure** A measure of area.

**square pyramid** A pyramid with a square base. (p. 398)

**standard form** The usual way of writing a number using digits.

**statistics** The study of the collection, interpretation, and display of data.

**straight angle** An angle that measures 180°. (p. 270)

**subtrahend** A number that is subtracted from another number. (p. 45)

**sum** The answer in addition.

**surface area** The sum of the areas of all the faces of a space figure.

**symmetrical figure** A plane figure that can be folded on a line so that the two halves are congruent. (p. 288)

**terminating decimal** A decimal in which digits do not show a repeating pattern. A terminating decimal results when the division of the numerator of a fraction by the denominator leaves a 0 remainder. (p. 377)

**tessellation** The pattern formed by fitting plane figures together without overlapping or leaving gaps. (p. 321)

**transformation** A flip, slide, or turn that changes the location of a figure on a plane without changing its size or shape. (p. 290)

**trapezoid** A quadrilateral with only one pair of parallel sides. (p. 278)

**tree diagram** A diagram that shows all possible outcomes of an event or events. (p. 240)

**triangle** A polygon with three sides. (p. 13)

**triangular prism** A prism with two parallel triangular bases. (p. 398)

**triangular pyramid** A pyramid with a triangular base. (p. 398)

**turn** To obtain a new figure by rotating a figure into a different position. (p. 290)

**unit fraction** A fraction with a numerator of 1. (p. 193)

**unit price** The cost of one item.

**variable** A symbol, usually a letter, used to represent a number. (p. 129)

**Venn diagram** A drawing that shows relationships among sets of numbers or objects. (p. 63)

**vertex** (plural *vertices*) The common endpoint of two rays in an angle, of two line segments in a polygon, or of three or more edges in a space figure.

**volume** The number of cubic units needed to fill a space figure.

**weight** The heaviness of an object.

**whole number** Any of the numbers 0, 1, 2, 3, . . . .

**zero property** Multiplying a number by 0 always results in a product of 0. (p. 68)

514

## Mathematical Symbols

| | | | | | |
|---|---|---|---|---|---|
| $=$ | is equal to | $\cdot$ | decimal point | $\overleftrightarrow{AB}$ | line $AB$ |
| $\neq$ | is not equal to | $\circ$ | degree | $\overline{AB}$ | line segment $AB$ |
| $<$ | is less than | $+$ | plus | | |
| $>$ | is greater than | $-$ | minus | $\overrightarrow{AB}$ | ray $AB$ |
| $\approx$ | is approximately | $\times$ | times | $\angle ABC$ | angle $ABC$ |
| | equal to | $\div$ | divided by | $ABC$ | plane $ABC$ |
| $\ldots$ | continues without end | $P(E)$ | probability of an event | $\sim$ | is similar to |
| $\%$ | percent | $cm^2$ | square centimeter | $\cong$ | is congruent to |
| $2:3$ | two to three (ratio) | $in.^3$ | cubic inch | $\parallel$ | is parallel to |
| $\$$ | dollars | | | $\perp$ | is perpendicular to |
| ¢ | cents | | | $(3, 4)$ | ordered pair |

## Table of Measures

### Time

| | | | |
|---|---|---|---|
| 60 seconds (s) | = 1 minute (min) | 12 months (mo) | = 1 year (y) |
| 60 minutes | = 1 hour (h) | 52 weeks | = 1 year |
| 24 hours | = 1 day (d) | 365 days | = 1 year |
| 7 days | = 1 week (wk) | 100 years | = 1 century (cent.) |

### Metric Units

**Length**

| | |
|---|---|
| 10 millimeters (mm) | = 1 centimeter (cm) |
| 100 centimeters | = 1 meter (m) |
| 10 centimeters | = 1 decimeter (dm) |
| 10 decimeters | = 1 meter |
| 1000 meters | = 1 kilometer (km) |

**Capacity**

| | |
|---|---|
| 10 milliliters (mL) | = 1 centiliter (cL) |
| 100 centiliters | = 1 liter (L) |
| 10 centiliters | = 1 deciliter (dL) |
| 10 deciliters | = 1 liter |
| 1000 liters | = 1 kiloliter (kL) |

**Mass**

| | | | |
|---|---|---|---|
| 10 milligrams (mg) | = 1 centigram (cg) | 10 decigrams | = 1 gram |
| 100 centigrams | = 1 gram (g) | 1000 grams | = 1 kilogram (kg) |
| 10 centigrams | = 1 decigram (dg) | 1000 kilograms | = 1 metric ton (t) |

### Customary Units

**Length**

| | |
|---|---|
| 12 inches (in.) | = 1 foot (ft) |
| 3 feet | = 1 yard (yd) |
| 36 inches | = 1 yard |
| 5280 feet | = 1 mile (mi) |
| 1760 yards | = 1 mile |

**Capacity**

| | |
|---|---|
| 8 fluid ounces (fl oz) | = 1 cup (c) |
| 2 cups | = 1 pint (pt) |
| 2 pints | = 1 quart (qt) |
| 4 quarts | = 1 gallon (gal) |

**Weight**

| | |
|---|---|
| 16 ounces (oz) | = 1 pound (lb) |
| 2000 pounds | = 1 ton (T) |

### Perimeter

Rectangle: $P = (2 \times \ell) + (2 \times w)$

Square: $P = 4 \times s$

### Circumference of Circle

$C \approx 3 \times d \approx 3 \times 2 \times r$

### Area

Rectangle: $A = \ell \times w$

Square: $A = s \times s = s^2$

Parallelogram: $A = b \times h$

Triangle: $A = \frac{1}{2} \times b \times h$

### Volume

Rectangular Prism: $V = \ell \times w \times h$

Cube: $V = e \times e \times e = e^3$

## Other Formulas

**Distance** $=$ Rate $\times$ Time: $d = r \times t$

**Discount** $=$ List Price $\times$ Rate of Discount: $D = LP \times R$ of $D$

**Sale Price** $=$ Regular Price $-$ Discount: $SP = RP - D$

**Sales Tax** $=$ Marked Price $\times$ Rate of Sales Tax: $T = MP \times R$ of $T$

## Percent Table

| | | |
|---|---|---|
| $10\% = \frac{1}{10}$ | $70\% = \frac{7}{10}$ | $4\% = \frac{1}{25}$ |
| $20\% = \frac{1}{5}$ | $80\% = \frac{4}{5}$ | $5\% = \frac{1}{20}$ |
| $30\% = \frac{3}{10}$ | $90\% = \frac{9}{10}$ | $25\% = \frac{1}{4}$ |
| $40\% = \frac{2}{5}$ | $1\% = \frac{1}{100}$ | $50\% = \frac{1}{2}$ |
| $60\% = \frac{3}{5}$ | $2\% = \frac{1}{50}$ | $75\% = \frac{3}{4}$ |